Springer Geology

The book series Springer Geology comprises a broad portfolio of scientific books, aiming at researchers, students, and everyone interested in geology. The series includes peer-reviewed monographs, edited volumes, textbooks, and conference proceedings. It covers the entire research area of geology including, but not limited to, economic geology, mineral resources, historical geology, quantitative geology, structural geology, geomorphology, paleontology, and sedimentology.

More information about this series at http://www.springer.com/series/10172

Mihir Deb · Sanjib Chandra Sarkar

Minerals and Allied Natural Resources and their Sustainable Development

Principles, Perspectives with Emphasis on the Indian Scenario

Springer

Mihir Deb
Department of Geology
University of Delhi
Delhi
India

Sanjib Chandra Sarkar
Department of Geological Sciences
Jadavpur University
Kolkata, West Bengal
India

ISSN 2197-9545 ISSN 2197-9553 (electronic)
Springer Geology
ISBN 978-981-10-4563-9 ISBN 978-981-10-4564-6 (eBook)
DOI 10.1007/978-981-10-4564-6

Library of Congress Control Number: 2017937703

Printed on acid-free paper

This Springer imprint is published by Springer Nature
The registered company is Springer Nature Singapore Pte Ltd.
The registered company address is: 152 Beach Road, #21-01/04 Gateway East, Singapore 189721, Singapore

Dedicated to

The memory of Professor Sivasunder Deb (1905–1979),
a pioneer teacher and researcher of Ore Geology in India,
and a guiding light for our lifelong interest in the subject

Preface

Human civilization since its inception has been tied to mineral extraction in some form or the other and mineral wealth has been considered by many as the starting point of economic growth. Thus, from the early days of Natural Science to the complex science and technology that foster human societies today, the discipline of Earth Science has played a key role in human development. During the eighteenth and nineteenth centuries, the fledgling science of Geology provided the basic inputs for the exploitation of coal and other raw materials which brought about the Industrial Revolution and subsequently, rapid economic growth in Europe and North America. During the twentieth century the increasingly robust Geoscience fuelled the "Hydrocarbon Age" and facilitated the so-called second industrial revolution based on microelectronics and information technology. This period saw unprecedented energy consumption and massive advancement in transportation and communication, which has shrunk our globe in relative terms. It also guided the exponential growth in the recovery of various metals and other mineral commodities for the development of infrastructure for our increasingly prosperous societies. Changes in some fields in the last century have been more substantial than those that occurred during the entire preceding millennium. On the flip side, however, along with the obvious benefits there were many negative side effects. The pattern of growth in so-called developed, actually "overdeveloped" countries requires excessive, often wasteful consumption of natural resources entailing damage to the world's ecosystem and environment. It is also easily overlooked that these georesources which are essential for poverty alleviation and in achieving a better quality of life in the developing world, as well as for all modern technologies that we are so dependent upon, have to be searched, found and finally extracted from different levels in the earth's crust.

From the initial stage of exploration to mining, processing and finally marketing, the exploitation of georesources is fraught with several critical issues. *First*, the georesources are distributed unevenly globally and also regionally, within a large country like India. To keep up the continuous supply of these resources according to the growing needs of the society, it is not only essential to know ***where*** they occur and their geological characteristics, but also ***why***, ***how***, and ***when*** did they form. This understanding will form the basis for most of the future discoveries that will ensure the long-term availability of these essential commodities. *Second,* to satisfy the ever-increasing demand of raw material in different industries, as well as to augment the foreign exchange reserve of the country, large-scale mining and increase in necessary imports are usually preferred. Besides, with exponential growth of mineral production in the last century, most near-surface deposits have either been exhausted or are nearing exhaustion. Consequently, exploration and exploitation at greater depths are fast becoming essential though they are rather expensive. For some commodities, improved technologies allow lower grade, high tonnage, and near-surface deposits to be exploited, but obviously at the cost of higher energy consumption. All these factors make the large producers (often multinational corporates) the most suitable players in the georesource extraction industry. Consequently, sustainability is an important issue and can be achieved by such players only by embracing ideas of corporate and social responsibility and by internalizing the costs of mineral production. *Third,* metal deposits and thick forest cover have a close association in many parts of the world, as in the five mineral-rich states of Chhattisgarh, Odisha, Andhra Pradesh, Jharkhand and Madhya Pradesh in east-central India. This region hosts extensive deposits of iron, manganese, chromium, and aluminum – all important metals that drive the consumer industries. But these forests also provide habitat to wildlife and forest dwellers, mostly indigenous people. In many places, the hills with forests containing the deposits within also act as the catchment areas for important rivers. Though a developing country like India needs its mineral resources, including energy resources, for its industrial growth, and water and fertile soil for food security of its growing population, the complex relationship that emerges between mining, forests, wildlife, change in life and livelihood, and displacement of tribals, along with rampant environmental degradation of air, water, and soil in mining areas produce an extremely complex socioeconomic challenge. Lack of inclusive growth of poverty-trapped people in mining areas over protracted periods and large land acquisition for mining and its infrastructure have fuelled discontent and even anger in large parts of peninsular India turning a socioeconomic problem into a political one through the ultra-left Maoist (Naxalite) movement. Mineral industry, however, can play a vital role in turning the situation around provided there is good and fair governance and sustainable mineral resource development.

This book attempts to look at the georesources of India in a holistic way. Georesources are first defined as the nonrenewable materials in the earth's crust which are essential for human civilization and human sustenance. Nonrenewability is here estimated against average human life-span. Thus, water and soil that we have long overlooked, along with the common minerals and industrial rocks, as well as energy-producing (both organic and inorganic) commodities come under the ambit of this treatment in the first two parts of the book, comprising eight chapters. The content is organized in a format that will help the reader to extract information on any aspect readily. For the uninitiated reader, an introductory chapter will look into some basic definitions as well as nature and characteristics of mineral deposits followed by a chapter on the different crustal processes that produce the various ore deposits in the endogenous and exogenous environments. The third part of the book containing three chapters will reflect on the issues of sustainability and their relevance to the mineral industry of the country. Sustainability in this context is not only the durability of a deposit but also its relation with the ambient society. The various socioeconomic as well as regulatory and environmental issues that face the Indian mining industry today are highlighted and analyzed in the backdrop of different contemporary case studies. Finally, a roadmap of sustainable mining framework, which may not be free of the authors' logical bias and prejudices, is put forward. The environmental issues related to the exploitation of various energy resources are also discussed at length.

We have included Indian illustrations as examples in fair number in the interest of Indian students. However, we have not overlooked classical examples from around the world to aid our discussions.

The authors' efforts will be amply rewarded if the scientific fraternity finds the book useful.

Delhi, India
Kolkata, India
February 2017

Mihir Deb
Sanjib Chandra Sarkar

Acknowledgements

The authors are obliged to their students and colleagues for encouraging them – in fact, even prodding them, to write a book on Resource Geology that will critically review the nature of the geological resource material, their genesis and their sustainable development keeping in view the trans-generational needs. The treatment was required to be informative, critical and at the same time to be appreciated by the student beginners. With due humility, we submit that we have tried our best to do the same.

We record our appreciation for our spouses who ungrudgingly lent us their support at the expense of family commitments to see that we reach our goal in good time. We also invoke the blessings of our late parents in our endevour.

We are thankful to our friends and co-workers who have shared with us their data, field experience, and critical opinion and to our students, who raised pertinent questions in class and in the field, which we have tried to address in this book. We are sorry that we are unable to mention each one of them by name. Special thanks are due to Prof. D.M. Banerjee, Dr. Anupendu Gupta, Dr. Anupam Chattopadhyay, Dr. Somnath Bandyopadhyay, Dr. Tamal Pal, and Ms. Indrani Mukherjee for their individual inputs to the book. However, we alone are responsible for any commission and omission that may still be there.

We are obliged to the Geological Survey of India, Atomic Minerals Directorate, National Geophysical Research Institute, Hindustan Copper Ltd., Hindustan Zinc Ltd., Metal Mining India Ltd., Centre for Interdisciplinary Research and Education, Kolkata and several other firms and agencies in the mining industry for their help, cooperation and encouragement over the years in our research activities, some results of which appear in this book. We also appreciate our erstwhile employers (universities) for giving us required facilities and support over the years.

We are highly thankful to Shri Rajesh Mishra, a former student at Delhi University, who very ably and painstakingly drafted a major part of the figures presented in the book.

Last but not least, we are thankful to Springer for undertaking the publication of this book.

About the Book

Nonrenewable natural resources – metallic and nonmetallic minerals, industrial rocks and energy resources (both organic and inorganic) – have been treated in a holistic manner in this book, including two important resources (soil and water), not commonly covered in most books on this topic. For the uninitiated reader, an introductory chapter looks into some basic definitions as well as nature and characteristics of mineral deposits followed by a chapter on the different crustal processes that produce the various ore deposits in the endogenous and exogenous environments. The strength of the book lies in its critical treatment of the genetic processes of the mineral deposits, their classification and the geodynamic context of metallogeny, and coverage of sustainable development of mineral deposits with special reference to various socioeconomic as well as regulatory and environmental issues that face the Indian mining industry today. The text is punctuated with examples of Indian deposits, balanced with classical deposits around the world, to cater to the interests of Indian students and the international readership. This is a book for advanced undergraduate and postgraduate students of Geology, Environmental Sciences and Natural Resource Management.

Contents

About the Authors

Mihir Deb's research interests for close to five decades have centered on nonrenewable natural resources, their environment of formation and sustainability of extraction. He has published over 100 scientific articles in peer-reviewed national and international journals and has edited three scientific volumes (published by Narosa Publishing House). With an undergraduate degree from Presidency College, Calcutta and Master's and Ph.D. in Ore Geology from Jadavpur University, he joined as faculty in the University of Delhi in 1970 and then pursued postdoctoral research in the City University of New York. He was an Alexander von Humboldt fellow at the University of Göttingen, Germany and at the University of Leicester, U.K. He served as the Head, Department of Geology (2000–2003) and as Director of the School of Environmental Studies (2004–2009) at Delhi University where he initiated an innovative postgraduate course in Environmental Studies with a holistic approach. He has guided a dozen students for Ph.D. in these two disciplines and conducted several basic and applied scientific projects and international collaborative research. His geological and environmental field work have made him travel extensively in North and South Americas, Europe, Mongolia, Vietnam, Indonesia, and parts of Australia and Africa. He also served as Director of the University Science Instrumentation Centre during 2000–2006. He superannuated from Delhi University in 2010 and continued his research and teaching as INSA senior scientist till August, 2014. He joined the newly established Nalanda University at Rajgir, Bihar in September, 2014 and served as a Visiting Professor and the first Dean of the School of Ecology and Environment Studies for a full academic year and initiated an interdisciplinary postgraduate course.

Professor Deb worked in different capacities in several national bodies. These included: Member, State level Environmental Expert Appraisal Committee, NCT of Delhi; Member, Geoscience Advisory Committee of the Ministry of Mines, Govt. of India; Member, Apex Exploration Research Advisory Committee, Department of Atomic Energy, Govt. of India; Member, Working Group on Mineral exploration and Development, Planning Commission (2006); Member, Academic Council of TERI University; Member, Council, Geological, Mining, Metallurgical Society of India. In the International sphere, Mihir Deb served as the Indian representative to

Communities and Small scale Mining (CASM), a World Bank initiative; Asian representative in IUGS-GEM (Geoscience for Environmental Management, an ICSU commission) and Member, UNECE Ad Hoc Group of Experts on Harmonization of Fossil Energy and Mineral Resources Terminology. Mihir Deb was awarded N.N. Chatterjee medal of the Asiatic Society, Calcutta in 1976, the National Mineral Award (Govt. of India) in 1996 and a gold medal of the Indian Geophysical Union in 2013. He is a Fellow of the Indian National Science Academy since 2002 and a Fellow of the Society of Economic Geologists, U.S.A., since 2004. He is also a Life Fellow of the Geological Society of India and the Geological, Mining and Metallurgical Society of India.

Sanjib Chandra Sarkar has taught and conducted research in the Department of Geological Sciences, Jadavpur University, Calcutta, India, for more than four decades. He has published a large number (+60) of research papers in national and international journals, particularly in his area of interest, Precambrian geology and Metallogeny. He has authored two books (One published by Jadavpur University Press, 1984 and the other, co-authored by A. Gupta, by Cambridge University Press, 2012) and an international conference Proceedings Volume (published by Oxford-IBH and Balkema, 1992). His present interest centers on India's sustainable development in the context of its natural resources. Sanjib Sarkar was an undergraduate student in Presidency College and did his Master's degree from Jadavpur University, Calcutta. Then he proceeded to Moscow State University (former USSR) for his Ph.D. degree in Geology & Mineralogy under the supervision of the eminent Academician, V.I. Smirnov. He also participated in a course in Experimental Sulfide Mineralogy at Lehigh University, Pennsylvania, U.S.A. He was the Head of the Department at Jadavpur University during 1981–1984 and was also the Coordinator of the UGC Center of Advanced Study in Economic Geology, Structural Geology and Petrology from 1984 to 1997. He guided 10 Ph.D. candidates during his tenure.

He was the regional councillor for Asia in the International Association on Genesis of Ore deposits (1984–89). He also co-chaired the Working Group on Ore and Metamorphism (1992–99) and served as a Leader of an International Geological Correlation Programme (IGCP-247) on Precambrian Tectonics and Metallogeny (1987–92). Professor Sarkar who also served as a member of the International Commission on Tectonics and Metallization has visited many mineral deposits around the world. He is a Fellow of the Society of Economic Geologist (SEG), USA; Fellow, West Bengal Academy of Sciences and Technology, Life Fellow of the Geological Society of India and the Geological, Mining, Metallurgical Society of India. Professor Sarkar was awarded Prof. N.N. Chatterjee medal for important contribution to Economic Geology by the Asiatic Society, Calcutta in 1969. He was also awarded the Geological Survey of India Sesquicentennial Commemorative Gold medal of the Asiatic Society in 2012 for "significant contribution in the field of Earth Science". He was a DST National Lecturer during 1997–98.

Abbreviations/Acronyms

AD	Anno Domini (Year of our Lord in Latin)
AMD	Atomic Minerals Directorate
Amu	Atomic Mass Unit
BC	Before Christian Era
BCM	Billion Cubic Meter
BP	Before Present
CBM	Coal Bed Methane
CLM	Continental Lithospheric Mantle
CSR	Corporate Social Responsibility
CWC	Central Water Commission
EEZ	Exclusive Economic Zone
EGMB	Eastern Ghat Mobile Belt
EIA	Environmental Impact Assessment
EPA	Environmental Protection Agency
FICCI	Federation of Indian Chamber of Commerce and Industries
FPIC	Free and Prior Informed Consent
GDP	Gross Domestic Product
GOI	Government of India
GSI	Geological Survey of India
GST	Goods and Services Tax
Gt	Giga (billion) metric ton
g/t	gm per ton
IBM	Indian Bureau of Mines
ILR	Inter-linking of Rivers
IOCG	Iron Oxide-Copper-Gold
LIP	Large Igneous Province (Oceanic Plateau)
M & M sector	Mines and Minerals Sector
MOEF & CC	Ministry of Environment and Forests and Climate Change

Mscmd	Million standard cubic meter per day
Mt	Million metric ton
Mtoe	Million tons oil equivalent
MP	Madhya Pradesh
MPa	MegaPascal = 10^6 Pascals
Mtpa	Million ton per annum
MVT	Mississippi Valley Type
Pa (Pascal; SI unit of pressure)	1 Neutron/meter^2 (n/m^2) = 1 kg/cm^2 (kg/cm^2) = lb/in.^{2}
PGE	Platinum Group Elements
PIL	Public Interest Litigation
Ppm	Parts per million
PSU	Public Sector Undertaking
R & D	Research and Development
REE	Rare Earth Elements
SCLM	Sub-Continental Lithospheric Mantle
SEDEX	Sedimentary Exhalative
SSC	Sediment-Hosted Stratiform Copper
TARWR	Total Actual Renewable Water Resource
TCF	Trillion Cubic Feet
TCM	Trillion Cubic Meter
TOI	Times of India
TTG	Tonalite-Trondhjemite-Granodiorite
UCIL	Uranium Corporation of India Ltd.
UP	Uttar Pradesh
UNFC	United Nations Framework Classification
UNCTAD	United Nations Conference on Trade and Development
USD	United States Dollar
VMS	Volcanogenic Massive Sulfides
WRI	Water Resource Institute

Part I
Mineral Resources

Chapter 1
Introduction

Abstract Minerals, both metallic and nonmetallic, and building stones have played key role in shaping human civilization since ancient times. Demand for and production of these earth resources increased manifold during the Industrial Revolution and both have seen an exponential growth during the last century, keeping pace with growing national and global economies. Ever-widening base of mineral use has made minerals an integral part of our everyday life. Mineral deposits are basically geochemical anomalies of one or more metals in a section of the earth's crust produced by igneous, metamorphic, and sedimentary processes, either individually or in conjunction. They become economic deposits or ore deposits if they can be worked at a profit, given the economic, sociopolitical, and technological situation at a place. Also crucial in this respect are the grades and tonnage of the deposit, and the cut-off grade of the metal in question. The process of ore formation is best understood by assessing the nature and geological characteristics of an ore deposit, among other features. The deposit may be dispersed or confined, epigenetic or syngenetic, and stratiform or stratabound. Each of such features, as well as the morphology of the ore body, will throw light on the endogenous or exogenous ore-forming environment prevailing in the lithosphere during the ore-forming time. Any large body of information about a population in Natural Sciences needs to be grouped or classified in terms of similarities and contrast. In a population of ore deposits, no two deposits are exactly identical. When we group them according to their characteristics, we have a classification scheme. But when we specify which characteristics are essential to belong to a group, we have the basis for a mineral deposit model. Many classification schemes, mostly genetic, have been proposed since the beginning of the last century, culminating in the UNFC scheme of 2009. A classification scheme that integrates host rocks and tectonic setting are most useful. Thus, tectono-lithological classification schemes are now more popular since genetic classifications are often fraught with controversies. However, magmatic, metamorphic, and sedimentary ore deposits still provide the best overall classification. Based on the current understanding of mineral-forming processes and products, a new scheme of classification has been proposed in this book. The mineral deposit models, on the other hand, are based on both regional and local attributes of deposits. They are divided into two end members: (1) the empirical or

M. Deb and S.C. Sarkar, *Minerals and Allied Natural Resources and their Sustainable Development*, Springer Geology,
DOI 10.1007/978-981-10-4564-6_1

occurrence model, based on observational data; and (2) the genetic or conceptual model, based on theoretical concepts. The models are useful in regional exploration programs. However, some models are still not mature (e.g., Iron oxide-copper-gold deposits) while others are fairly mature, based on robust databases (e.g., placer gold, sediment-hosted Pb–Zn deposits).

Keywords Mineral use in history · Economic mineral deposits · Ore deposits · Ore and gangue minerals · Grade and tonnage · Cut-off grade · Reserves and resource · Stratabound and stratiform · Dispersed and confined · Syngenetic and epigenetic · Deposit classification · Mineral deposit modeling · Empirical models · Conceptual models

1.1 Mineral Use and the Human Civilization

Mineral(-rock) use by human beings dates back to a very distant past, i.e., Paleolithic Age (beginning ~400,000 years) when flint or pieces of other minerals and rocks of comparable properties (vein quartz, quartzite, chert, obsidian, basalt, etc.) were first honed to hunting tools, and then on a later date in making fire by rubbing. This continued for quite some time, until people also learnt the use of burnt clay, making potteries, tiles, and even arty figurines. Burnt clay figurines, believed to be 30,000–20,000 BCE in age have been reported from Moravia, south central Europe. Gemstones and noble metals like gold fascinated people in ancient civilizations of Egypt, Babylonia, Assyria, and India, even several thousand years ahead of the Christian era. Egyptians are known to have sunk shafts around 2000 years BCE to collect emerald from the Egyptian coast of Red Sea. Gold use continued alongside Copper Age (3900–2100 BCE), Bronze Age (2100–1200 BCE) and Iron Age (1200–50 BCE) when the utility of several metals were already recognized. Most of the radio-carbon dates of charcoal, wood, bamboo, used in zinc-lead mines of Zawar, Rajpura-Dariba, and Rampura-Agucha in Rajasthan, India dated somewhat pre-/post-beginning of the Christian era. One timber even dated 3040 $\pm$ 150 years BP suggesting that an advanced stage of zinc smelting was achieved in the Indian subcontinent in the third millennium BCE (Gandhi 2000). The importance of minerals and their exploitation found place in the learned works of the Pre-Christian era, the two important among which are the great Indian scholar Kautilya's *Arthasastra* (fourth century BCE) in Sanskrit and the *Book of Stones* in Greek by the Greek scholar Theophrastus (fourth-third century BCE).

There was indeed some Science in those magnum opuses, but studies in ore geology in the real scientific sense appeared several centuries later. Georgius Agricola (1494–1556 CE) was one of the pioneers amongst early earth scientists. He is credited with proposing a fairly reasonable theory of ore genesis, if we keep in view the overall state of development of Science at that time. Earth Science lagged behind most of the other Physical Sciences in that its conclusions were generally deductive than inductive. Therefore controversies raged. But the situation started turning in the twentieth century as relevant experiments were conducted and results of experiments from other related subjects were assimilated.

Mineral industry went on without a landmark change until about the Industrial Revolution in Europe during the eighteenth–nineteenth centuries. Augmentation of the mineral industry as a result of industrial revolution led some interested persons to look for mineral deposits not known earlier. Successful search needed a priori knowledge of where and how such deposits could form. The development was much more spectacular in the nineteenth and early twentieth century when the Western society strived hard to make itself an essentially urban, industrialized, and technocratic one. They went out searching for mineral deposits. Their colonies were the favored pastures in this respect. The two great world wars during the last century accelerated exploration for mineral resources. But the post-second World War period has possibly been outstanding in exploration and exploitation of mineral deposits as demand and production sky-rocketed. As a spin off, there has been considerable advancement in Economic or Ore Geology in particular and Earth Sciences in general.

A wide variety of minerals are used by industry to manufacture consumer goods that are used in everyday life. We generally take for granted the importance of such minerals and their presence in the goods we buy. In many cases, they are not listed as ingredients in the packaging. For example, hardly any user of high-end smartphones know that the device they are holding can contain precious metals like gold (0.034 g), silver (0.34 g), copper (15 g), platinum (0.001 g), and palladium (0.015 g) as well as some rare earth elements. A composite diagram of some of the commonly used mineral raw materials and their end products is shown in Fig. 1.1

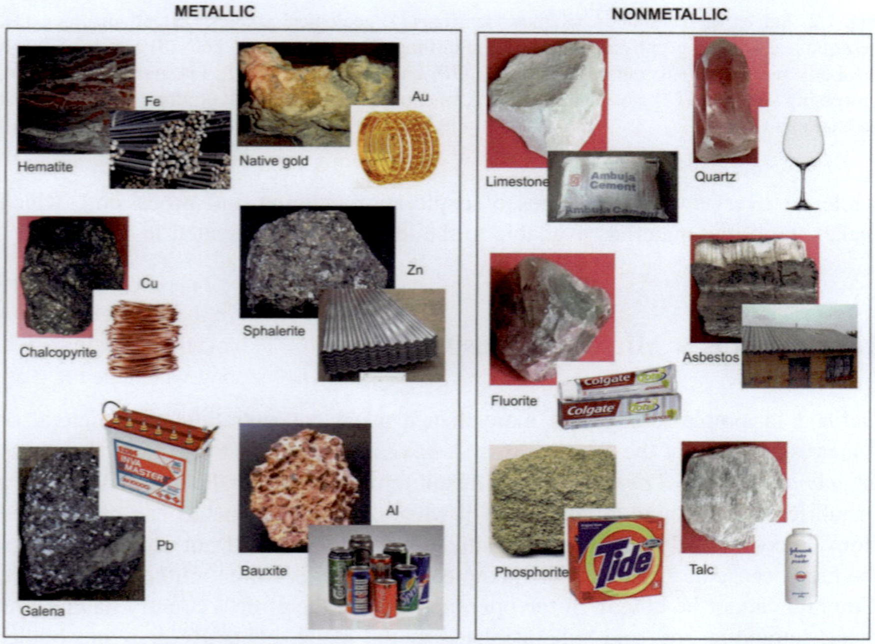

Fig. 1.1 Some minerals and rocks and their products in everyday life. We are so used to these products that most of us are not bothered about their source and value

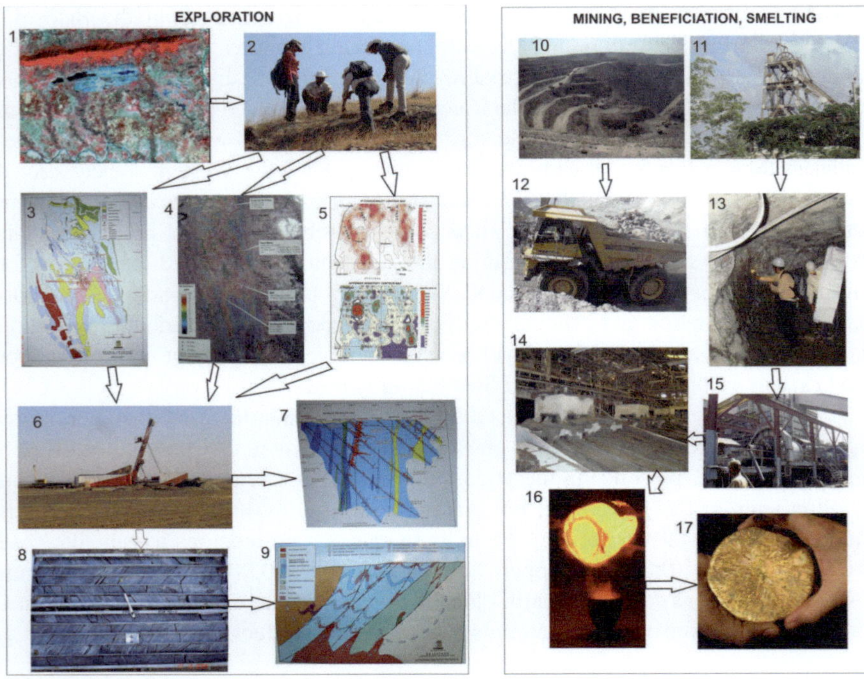

Fig. 1.2 Stages of exploration [Remote sensing (*1*), geological studies (*2*) and mapping (*3*), geochemical (*4*) and geophysical (*5*) surveys, drilling (*6*) and bore hole core (*8*) sections (*7*) and ore body modeling (*9*)], mining [open cast (*10*) and underground (*11*, *13*), transporting (*12*)], processing [washing (*14*), concentrating (*15*)], smelting (*16*) and the final product, gold bullion, as an example (*17*)

while an overview of the stages of exploration, mining and processing, which makes these raw materials available to the industries, is presented in Fig. 1.2.

1.2 What is a Mineral Deposit?

Defined in simple terms, it is more than a usual concentration of a mineral or minerals in a part of the Earth's crust. A *mineral deposit*, or to be more particular, an *economic mineral deposit* may be metalliferous or nonmetalliferous. We call it a metalliferous economic mineral deposit when one or more metals can be obtained from it economically and in special situations strategically, if not economically, in the literal sense of the word. An example of the latter is a uranium-bearing deposit. Uranium cannot be bought in the open market and therefore a country having low quality uranium ores and still interested in the metal, generally does not bother about the economic viability of the deposit. The exploitable metal is generally

Table 1.1 Average concentration of some useful elements in earth's crust and their concentration required for the minimum ore grade (data from multiple sources)

(a) Element (alphabetically arranged)	(b) Crustal average (ppm)	(c) Concentration in lowest ore grade (times b)
Alumnium (Al)	82.3×10^3	5
Antimony (Sb)	0.2	10×10^6
Beryllium (Be)	2.8	200
Chromium (Cr)	100.0	45×10^2
Cadmium (Cd)	0.2	10×10^2
Cobalt (Co)	25.0	40
Copper (Cu)	55.0	70
Fluorine (F)	625.0	5×10^3
Gold (Au)	0.004	250
Iron (Fe)	56.0×10^3	7
Lead (Pb)	13.0	15×10^2
Lithium (Li)	20.0	250
Manganese (Mn)	950.0	300
Mercury (Hg)	0.08	25×10^3
Molybdenum (Mo)	1.5	600
Nickel (Ni)	75.0	14
Niobium (Nb)	20.0	35
Phosphorus (P)	10.5×10^2	200
Platinum (Pt)	0.01	200
Selenium (Se)	0.05	4×10^3
Silver (Ag)	0.07	200
Tin (Sn)	2.0	10×10^2
Tungsten (W)	1.5	10×10^2
Tantalum (Ta)	2.0	50
Thorium (Th)	8.0	5×10^3
Titanium (Ti)	57×10^2	15
Uranium (U)	2.0	500
Vanadium (V)	135.0	20
Zinc (Zn)	70.0	430

accumulated many times its crustal average (Table 1.1) in an ore by igneous, metamorphic, or sedimentary processes in the lithosphere (crust and upper mantle) and is thus a geochemical anomaly. How many times concentration (cf. Fig. 1.3) will qualify it to be a mineral deposit depends on how costly the metal is. Concentrations of elements may have remained at a place without being considered a mineral deposit. But with time it may be recognized as an *ore deposit* based on its location and changes due to changing societal needs, technology, usage, change in economy, politics, taxation, infrastructure, etc. Thus, the economic aspect of a

Fig. 1.3 Anomalous metal concentration defines an ore deposit. Note the rising values in the wall rock compared to the country rocks

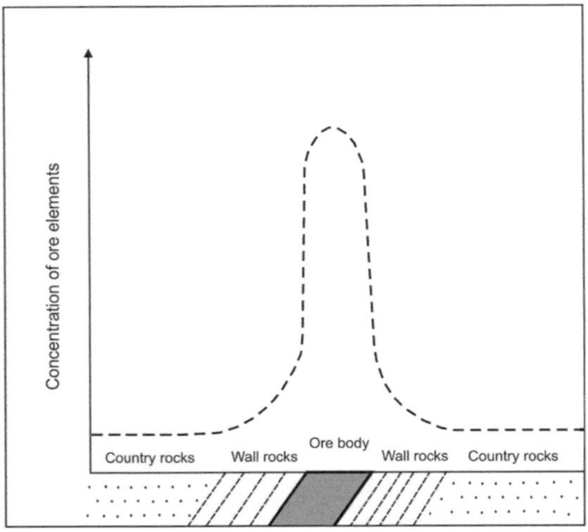

mineral deposit has a time connotation as well. What was not an economic mineral deposit yesterday could be an economic/strategic deposit today because of advancement of technology and increased market demand. And what is not economic today may turn out to be economic tomorrow because of the same reasons. A responsible society (Government) generally has this perspective, while estimating mineral resources for the future. The terms mineral deposit or ore deposit are used interchangeably in geological literature unless the metallic nature of its economic content is emphasized, when the latter term is preferred.

A mineral from which one or more metals are extracted economically or strategically is called an *ore mineral*, which is generally a compound of the metal(s) with one or more anions like sulfur, oxygen, arsenic, antimony, tellurium, etc.[1] Noble metals in ores occur generally in the native state (whence the attribute "noble") or as alloys, however. Another word often used in this context is *gangue*, which is commonly one or more of rock-forming minerals associated with the ore mineral and is generally unwanted, from the economic point of view. An ore mineral and a gangue mineral, or several of them together, make an aggregate which is called *ore*. An ore is a piece of valuable rock according to some people. The above definitions of an ore are not subscribed to by all authors today. According to some usage, the word ore is used only when there is a connotation of profit. Some others use the word more comprehensively to include wide ranging commodities (e.g., Evans 2001) from bauxites to kyanite or sillimanite which are clearly industrial rocks and minerals. Robb (2005), for example, even included fossil fuels under the ambit of ore. A common confusion in the above context arises in the two terms, grade and ore. The two do not equate at all times. For example,

[1]For a list of common minerals of economic interest, their chemical composition and use as source of metals/nonmetals, see Appendix A.

economic Au grade today is >1–2 g/t, whereas in pre-1990s it was >10 g/t. This is the concept of *cut-off grade* – a grade below which it is not possible to mine an ore deposit profitably at a particular time. In the 1960s, the cut-off grade of Cu deposits varied generally between 1.5 and 2.0 wt% Cu. But when the porphyry copper deposits were discovered in the 1970s, the cut-off grade of Cu fell to as low as 0.5 wt% Cu. Another important relationship is between grade and tonnage of ore deposits – high grade deposits generally have low tonnage while low grade deposits have high tonnage, and are preferred these days for open cast mining.

The nonmetalliferous economic minerals inter alia include gemstones and semi-precious stones, industrial minerals and rocks including refractory minerals (kyanite, sillimanite, etc.), rocks and minerals for the cement industry (Ca–Mg carbonates, etc.), ceramic industry (kaolin, quartz, feldspar), chemical industry (sulfur, pyrite, barite, fluorspar, various salts), fertilizer industry (phosphorites), insulation and electrical industry (mica, asbestos), pigments and filler material (clays), and abrasives (garnet and corundum) and of course the fossil fuels (coal, natural gas and petroleum) (cf. Fig. 1.4). Although much more R&D efforts are directed toward the metallic minerals relative to industrial minerals, the value of the two commodities extracted vary with the level of development in the countries concerned. In a less developed country, the value of metallic minerals extracted is commonly higher than those of industrial minerals, the two values are about the same in newly developed countries, but in industrialized developed countries, the value of industrial minerals far exceed those of metallic minerals (cf. Evans 2001).

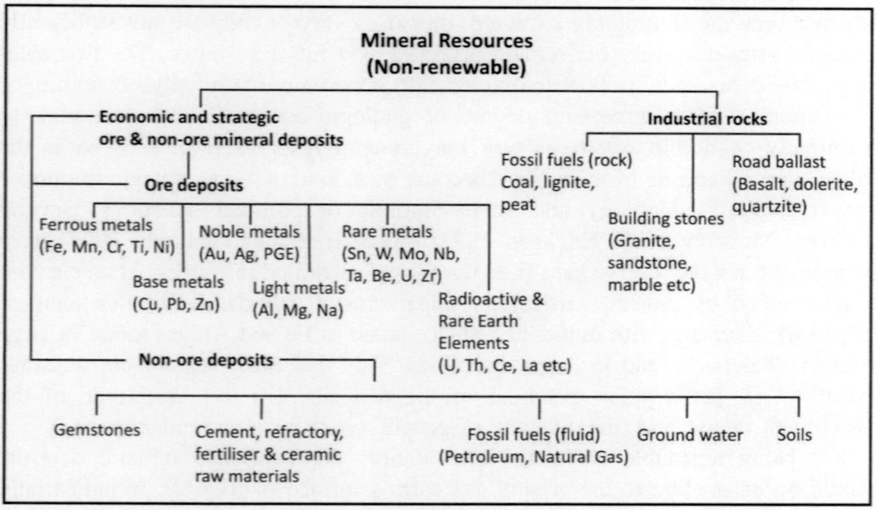

Fig. 1.4 Basic classification of nonrenewable mineral resources

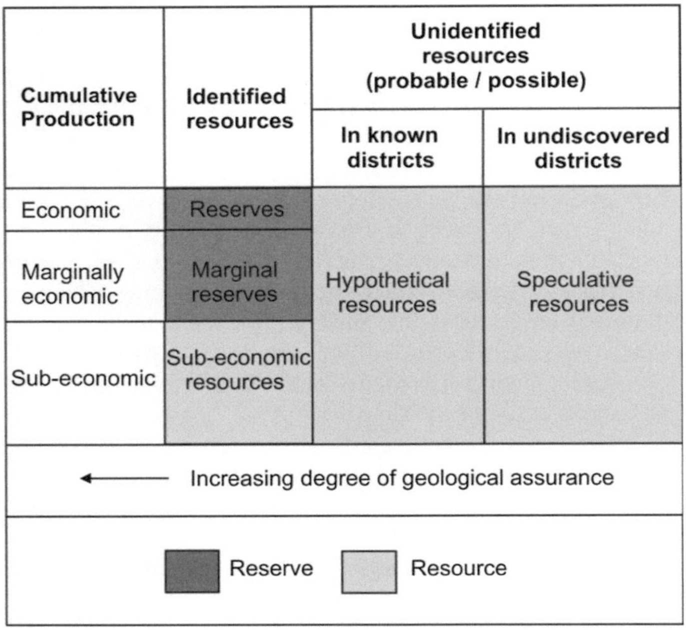

Fig. 1.5 Framework of distinction between reserves and resources (after USGS 1980)

Mineral commodities are categorized into two major groups depending on whether they are thoroughly explored after discovery or they are not sufficiently explored after discovery but remain potential as a future resource. The first category, called *reserves*, is further divided into economic, marginally economic, or subeconomic. With increasing degree of geologic assurance, the economic or marginally economic *resources* turn into mineral *reserves* (Fig. 1.5), while the subeconomic deposits of today may become economic or even strategic tomorrow with changing technology and socioeconomic or political situations. Several workers (Mckelvy 1960; Erickson 1973; Mookherjee and Panigrahi 1994) have pointed out that the reserve base (i.e., reserves + identified resources) of an element is determined by crustal, particularly upper crustal abundances of that element (Fig. 1.6). Elements with higher abundance, such as Fe and Al, are found in large number of deposits and in larger quantities. This first order relationship is influenced by (i) preservation potential of the deposits, (ii) the magnitude of the enrichment factor, and (iii) diversity of genetic types for a particular element.

For being amenable to human exploitation, the economic mineral deposits should occur as close to the present day earth's surface as possible. In industrially developed countries (or in their earlier colonies) near surface mineral deposits have been mostly explored, or are in advanced stages of exploitation, if not already exhausted. Therefore, the major concern of the mineral industry today is to explore the mineral deposits at relatively deeper levels, particularly in those countries, an

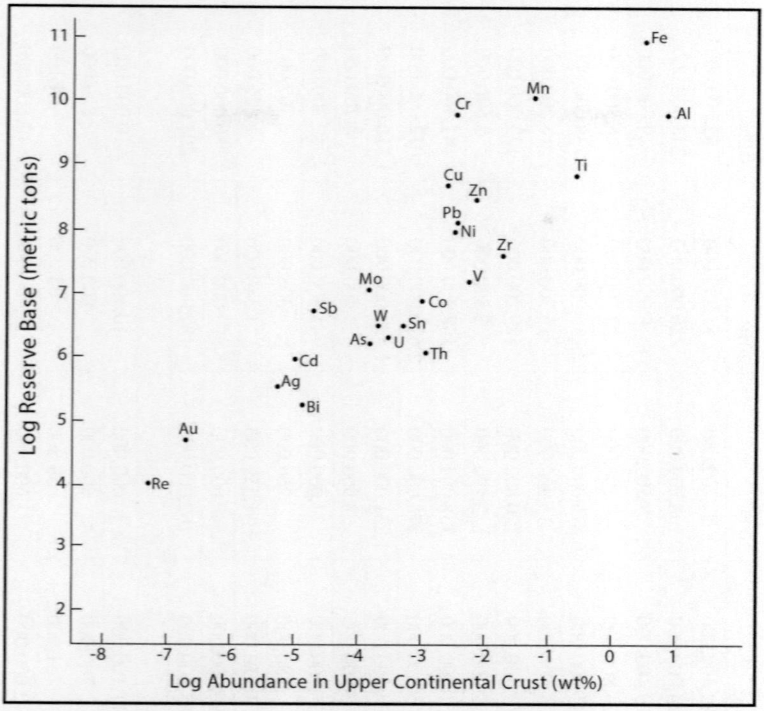

Fig. 1.6 Log–log plot of upper crustal abundances of 24 metals against their "reserve base" (explained in text) (after Mookherjee and Panigrahi 1994)

exercise that would be technologically sophisticated and consequently costlier. Today the deepest mines in the world are all gold mines located in South Africa: the Tau Tona (3.9 km), Savuka (3.7 km), and East Rand (3.5 km). Champion Reef mine of Kolar gold field, India, was being mined at 3200 m depth when mining ceased due to adverse working conditions, such as high overhead cost and technical hazards, like high temperature, rock bursts, etc. But we still need minerals if we want to sustain our industrial civilization. This statement holds good even in the context of the fact that we have already found some replacement for metals and such substitution is likely to increase in future.

The study of mineral deposits covered under *Economic Geology* has a strong interface with several other allied disciplines, such as, *Mineralogy, Mining Geology*, *Beneficiation,* and *Mineral Processing* and *Metallurgy*. But the moot question today is: can we extract natural resources and conserve the environment at the same time? Sustainability, environmental protection and reclamation are therefore important factors in the exploitation programs of mineral deposits today. We generally are not accustomed to see soil and water included in a book of Economic geology, although we do not have objection to calling them natural

Table 1.2 World production of a few important metals, nonmetals and fuels during 2007–2011 (after British Geological Survey 2013) (Unless stated, all figures are in metric ton)

S. No.	Commodity	2007	2008	2009	2010	2011
1	Alumina (tons)	77,600,000	82,800,000	77,200,000	85,200,000	82,100,000
2	Bauxite	213,000,000	214,000,000	198,000,000	226,000,000	248,000,000
3	Baryte	7,900,000	9,400,000	7,500,000	9,000,000	9,000,000
4	Chromite	25,200,000	24,800,000	20,100,000	27,500,000	26,300,000
5	Coal (total)	6,587,000,000	6,811,000,000	6,876,000,000	7,235,000,000	7,739,000,000
6	Copper ore (metal content)	15,500,000	15,600,000	15,900,000	16,200,000	16,200,000
7	Diamond (carats)	169,600,000	164,900,000	120,800,000	129,200,000	123,600,000
8	Gold (kg)	2,350,000	2,290,000	2,490,000	2,590,000	2,600,000
9	Gypsum	157,200,000	151,200,000	138,000,000	138,300,000	141,300,000
10	Iron ore	2,052,000,000	2,205,000,000	2,275,000,000	2,620,000,000	3,012,000,000
11	Crude steel	1,348,000,000	1,334,000,000	1,231,000,000	1,429,000,000	1,516,000,000
12	Lead	3,700,000	3,800,000	3,900,000	4,400,000	4,700,000
13	Manganese	35,900,000	38,100,000	34,000,000	41,900,000	47,300,000
14	Mica	340,000	380,000	260,000	308,000	307,000
15	Nickel	1,585,000	1,556,000	1,359,000	1,540,000	1,826,000
16	Petroleum crude (tons)	3,904,000,000	3,798,000,000	3,840,000,000	3,923,000,000	3,980,000,000
17	Phosphate rock	159,000,000	165,000,000	162,000,000	182,000,000	203,000,000
18	*Titanium*					
	Ilmenite	11,300,000	11,100,000	10,000,000	10,400,000	10,900,000
	Rutile	600,000	700,000	600,000	800,000	800,000
19	Uranium (ton) (U_3O_8 equiv.)	48,800	51,900	59,800	63,000	64,100
20	Zinc	11,200,000	12,000,000	11,600,000	12,400,000	12,800,000

resources. People tend to forget that soil is a breakdown product of rocks and thereby is a legitimate issue for geologists. There would be no plants, no food grown without soil. There would be no life on earth without water, not to speak of civilization. Thus, since they satisfy the definition of mineral resource in a broad sense, soil, and ground water are also grouped under nonrenewable[2] mineral resources in this book (cf. Craig et al. 2011).

Any text book on *Mineral Resources* in the context of the foregoing discussion should, therefore, contain the essentials of the scientific aspects of the subject and at the same time cover the principles of sustainable development of these resources, keeping in mind the limits to their exploitation, their replaceability and recyclability. The authors here cannot but recall the great saying of a great man of the twentieth century, M.K. Gandhi that "*the world possibly has enough for meeting everybody's need, but not everyone's greed.*"

1.3 Minerals in National and Global Economy

Minerals have a huge impact on national and global economy. The value of world mining of metals and industrial minerals was worth ~910 billion USD in 2005 (UNCTAD data), out of which the metals accounted for 27%, i.e., ~250 billion USD. World production of a few important metals, nonmetals and fuel over a period of 5 years (2007–2011) is presented in Table 1.2. The production figures for these same commodities in India for the same period are shown in Table 1.3. A conservative estimate of the value of minerals and fuels produced in 2011 is depicted in Fig. 1.7.

A good indication of the role of mineral production in the economic activity of a country can be obtained by comparing the value of mineral production and gross domestic product (GDP). A perusal of the World Bank data on major metals (Fig. 1.8) shows that developed countries like Canada (1%), USA (0.2%), and Japan (0%) have very little contribution of mineral production to their GDP whereas for some developing countries with good mineral wealth, a large fraction of GDP is contributed by their mineral production, such as, in Papua New Guinea (32.5%), Zambia (25.6%), DR Congo (22%), Chile (18.6%), and Mongolia (17.2%). The only developed country with substantial contribution to GDP through mineral production is Australia (7.2%). Countries like Brazil, China, and India contribute between 2.8 and 2.6% to their GDP through mineral production. In a classical sense, economic activity depends on domestic mineral resource availability. While this is true to some extent for mineral-rich countries like South Africa, it is not true for countries with strong economy like Japan and Korea with hardly any to limited mineral resource availability at the moment. Raw material export is important in a nation's early stages of development when mineral

[2]As we shall see later, ground water is not entirely nonrenewable.

Table 1.3 Indian production of a few important metals, nonmetals and fuels during 2007–2011 (after British Geological Survey 2013) (Unless stated, all figures are in metric ton)

S. No.	Commodity	2007	2008	2009	2010	2011
1	Alumina	3,320,169	3,620,252	3,432,716	3,576,917	3,970,000
2	Bauxite	22,624,960	15,460,202	14,124,093	12,640,785	12,992,000
3	Baryte	1,076,290	1,686,000	2,152,552	2,333,805	1,716,042
4	Chromite	4,872,847	4,073,479	3,425,580	4,262,207	3,735,000
5	Coal (Bit)	457,082,000	492,757,000	532,042,000	533,223,000	546,884,000
	Coal (lig)	33,980,000	32,421,000	34,071,000	37,651,000	80,686,000
6	Copper ore (metal content)	32,726	30,528	28,491	31,696	30,340
7	Diamond (carats)	586	536	16,891	19,774	18,489
8	Gold (kg)	2969	2438	2084	2239	2174
9	Gypsum	3,400,050	3,876,671	3,370,322	4,346,700	3,323,000
10	Iron ore	213,246,000	212,960,000	218,553,000	207,998,000	168,534,000
11	Crude steel	53,48,000	57,791,000	63,527,000	68,321,000	72,200,000
12	Lead	77,717	82,053	82,629	83,342	91,493
13	Manganese	2,696,980	2,789,000	2,491,950	2,881,080	2,387,000
14	Mica	1242	1462	1061	1293	1736
15	Petroleum crude (tons)	34,118,000	33,506,000	33,691,000	37,705,000	37,887,000
16	Phosphate rock	1,849,188	1,803,954	1,605,489	2,152,215	2,297,000
17	Kyanite Sillimanite	5,102 40,537	4,620 33,702	5,496 33,687	5,569 47,671	4064 58,579
18	Titanium					
	Ilmenite Rutile	675,967 20,518	676,003 19,098	767,355 18,573	663,217 26,593	700,000 27,000
19	Uranium (ton) (U_3O_8 equiv.)	270	271	290	400	400
20	Zinc	551,268	647,537	677,824	726,526	713,416

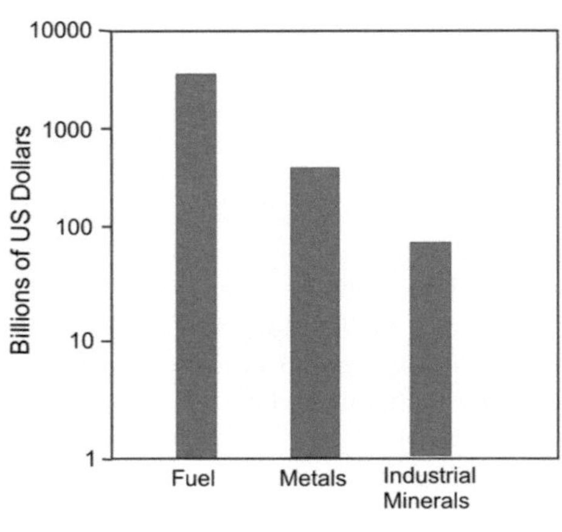

Fig. 1.7 A rough estimate of the value of fuel (petroleum and crude oil), metals and industrial minerals produced in the world in 2011 (*Source* USGS mineral summary)

resources are discovered. With the money earned from such exports, or otherwise, over the years, the country develops its industrial infrastructure and starts producing processed minerals, or separated metals which when exported fetch more money. The key issue in mineral trade for developed countries is import reliance. To be assured of steady supply of important raw materials, countries like Japan and Korea invest in mineral-rich countries for the development of mineral resources from exploration to mining to processing. Other countries like USA, however, conserve their mineral resources and apparently find it cheaper to import from developing countries.

India is well endowed with deposits of many minerals. India is ranked fourth among the mineral producer countries, behind China, United States, and Russia, in terms of volume of production. It ranked eighth on the basis of value of mineral production. The country produces as many as 87 minerals, including 4 fuel minerals, 10 metallic minerals, 47 nonmetallic minerals, 3 atomic minerals, and 23 minor minerals (including building and other materials). The mineral development and mining sector is a significant contributor to India's GDP growth, augmenting basic industries like steel, power, cement, etc. However, the contribution of this sector to GDP has been on the decline in the recent decades. The sector contributed 3.4% to India's GDP in 1992–1993, which declined to 3% in 1999–2000 and further to 2.3% in 2009–2010. It has further contracted to 2% in 2012–2013 (FICCI 2013) because of various socioeconomic and political reasons. This declining trend is clearly manifested in the production figures in Table 1.3 wherein important ores of iron, manganese, chromite, copper, gold, and alumnium (bauxite), along with nonmetals like barite, kyanite and gypsum show a trend of reduced production from 2007 to 2011. However, coal and petroleum, along with metals like lead and zinc, uranium, and titanium show an increase in production in this time period as do nonmetals, phosphate rock, sillimanite, and diamond. Interestingly, although productions of iron ore and bauxite have reduced, the productions of steel and alumina have increased in this period, suggesting enhanced import of the respective raw materials concerned.

1.4 Nature and Geological Characteristics of Ore Deposits

There are two main characteristics of ore deposits that are of prime concern to a geologist: its form and geologic setting. Both these features vary widely from deposit to deposit but some basic common trend can be seen in groups of deposits which help to provide a basis for classification. While describing the form of a deposit, one needs to first see if it is *dispersed* or *confined* and if latter, whether it is *discordant* or *concordant* with its host rock. The dispersed deposits have a low concentration in a relatively large volume of rock and are unlikely to have a clear-cut boundary, such as the interlaced veinlets producing a stockwork (cf. Fig. 2.18a, b), whereas the confined ores commonly show higher concentration within a structure or strata. The latter is represented often by a vein that is a fissure

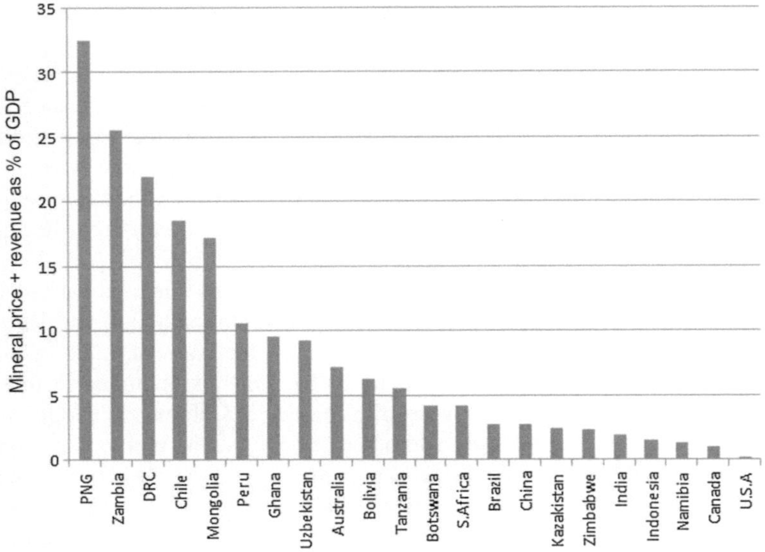

Fig. 1.8 Mineral Rent (% of GDP) for different countries based on calculations with 2011 data pertaining to Sn, Au, Pb, Zn, Fe, Cu, Ni, Ag, Al and P. Mineral Rent is the difference between the value of production of a stock of minerals at world prices and their total cost of production (after Worldbank.org)

in the rocks filled with mineral matter. They may be simple or complex, producing a pinch and swell vein, a saddle reef, a chambered vein, a finned vein or a ladder vein. The discordant deposits are commonly younger than their host rocks and cut across the strata. The concordant deposits on the other hand are subparallel to the layering or the prevailing planar surface in the host rock and may or may not be younger than the host rock. This leads to two terms in very common usage in ore geology: *epigenetic*, a deposit formed after the formation of the host rock(s) and is therefore commonly discordant but may also be concordant if the ore has been emplaced parallel to the stratification of the host rock. The other word is *syngenetic*, used when a deposit has formed at about the same time as the host rock and is, therefore, generally concordant with it. Epigenetic deposits may form through the action of hydrothermal fluids which are juvenile in nature, that is, related to a magmatic or metamorphic source and ascended from deep within the crust and referred to as *hypogene*. But epigenetic deposits may also form from hydrothermal (hot water) solutions which are *supergene* that is derived from descending meteoric water in the crust or from dewatering of connate water in an over pressured sedimentary pile. A set of ore deposits with varying form and characteristics is shown in Fig. 1.9. It may be a useful exercise to see how each one of them measure up to the definitions of the terms mentioned above.

Two other terms of common use in the description of ore bodies are stratiform and stratabound (Stanton 1972; Canavan 1973). A *stratiform* orebody has the form

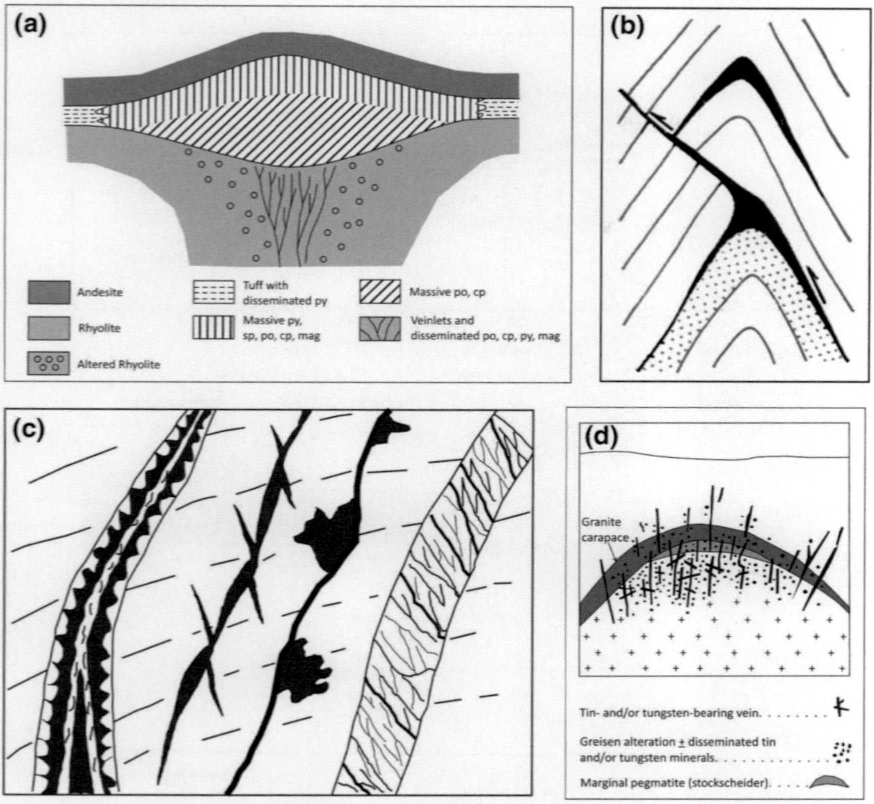

Fig. 1.9 Sketches showing the nature of mineral deposits: **a** concordant versus discordant, lensoid, and stockwork ores; **b** syngenetic versus epigenetic; also concordant saddle reef versus discordant ores; **c** different vein morphologies (from *left* to *right*): vein with comb structure, finned pinch and swell vein, chambered vein, ladder vein; **d** dispersed versus confined, disseminated and vein ores in gresenized granite

of a stratum and occurs within and is conformable with the enclosing sedimentary strata. It may be formed by syngenetic, diagenetic, or epigenetic mineralization within sediments of all kinds or within sills, conformable lava flows or conformable veins. It may be metamorphosed and retain its identity as stratiform provided its original characteristics are recognizable. A *stratabound* orebody, on the other hand, occurs within a rock layer or group of layers, whether composed of sedimentary or igneous rock, but excluding layers formed by magmatic differentiation. The ore may have any form of distribution: disseminated in, concordant with or transgressive to the layer or group of layers. Thus, as Stanton (1972) has emphasized, all stratiform orebodies are stratabound but not all stratabound orebodies are stratiform. Examples of stratabound and stratiform ores are shown in Fig. 1.10a–c.

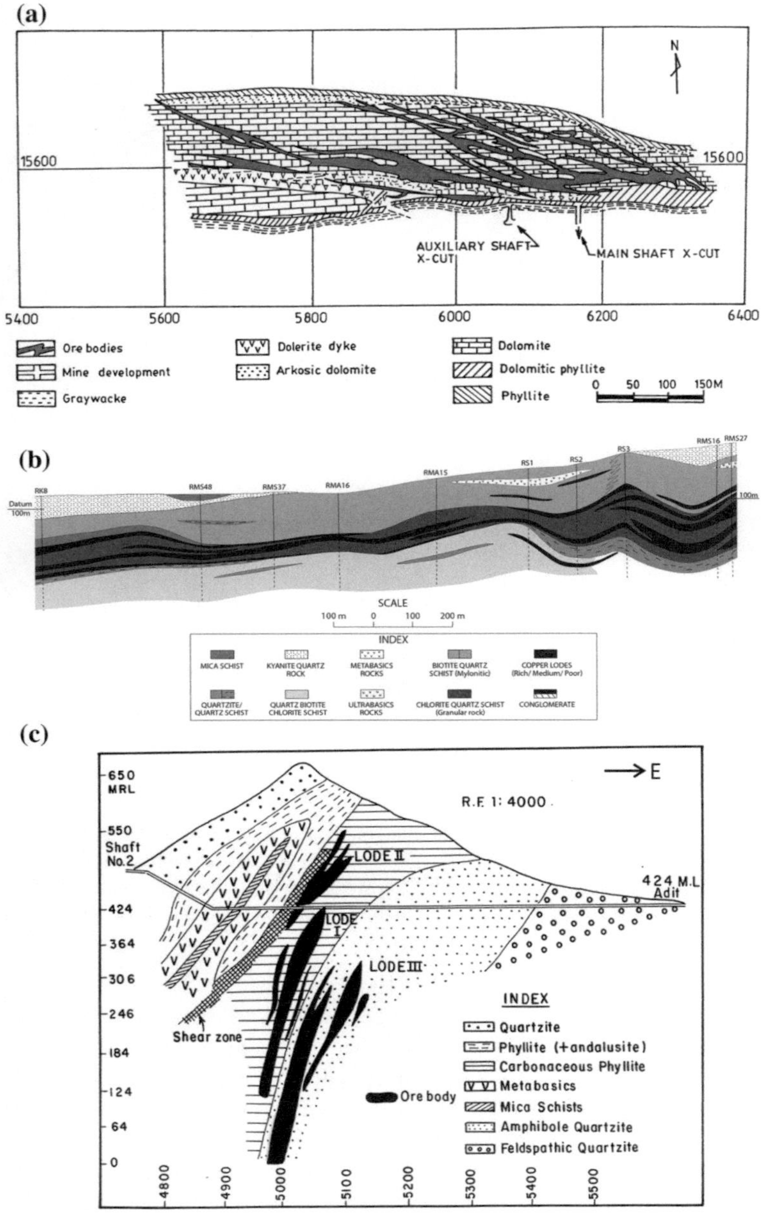

Fig. 1.10 Examples of stratabound ore deposit: **a** Mochia Magra mine (Geological plan at 240 m level) (after HZL); and stratiform ore deposits: **b** Rakha copper deposit, Singhbhum (after Deb 1971; Sarkar 1984), and **c** Kolihan copper deposit, Khetri copper belt, Rajasthan (after HCL)

1.5 Classification and Models of Mineral Deposits

Classification of mineral deposits is almost as old as the study of mineral deposits itself. To classify means to assign or to arrange into an order (Webster Dictionary). In other words, it is to divide a number of items in a population into groups (and subgroups), based on their similarity and contrast. Numerous ore deposits occur in the crust. So, a means of organizing them is clearly required. However, no two ore deposits are identical. Similarities notwithstanding, each ore deposit has some uniqueness like a person's finger print. This uniqueness arises out of fundamental differences in earth processes and environments of mineral deposit formation and local, site-specific geologic variations. These two factors produce an extremely large variation in the characteristics of mineral deposits. Thus, we need to separate the essential from incidental attributes and classification aids in this process and highlights what is important. If we group the deposits according to their characteristics, a *classification* scheme is produced. If we specify which characteristics are essential to belong to a group, we have the basis of a *model* (Barton 1993).

Historically, the *neptunist* and *plutonist* concepts were developed in the late 1700s by two almost contemporary geologists, Abraham Werner in Freiburg, Germany, and James Hutton in Scotland, respectively. While the former considered all ore deposits (perception apparently derived mainly from the study of Kupferschiefer deposits in central Europe) to have formed in an *exogenous* environment from sedimentation out of sea water, the latter held (from observations in the Cornwall tin deposits of UK) that all ore deposits are *endogenous* and products of igneous processes. Though both the apparently dogmatic ideas cannot be accepted as such, if generalized, however, they still make the two ends of the spectrum of ideas on ore genesis we deal with today (cf. Fig. 1.11).

By now, there are many classification schemes available for ore deposits, like in most other branches in Natural Sciences. Most of these classifications are based on genetic concepts prevalent at that time or believed by the authors. The major advantage of a genetic model is that it increases the genetic concepts of mineral deposits of the reader. Consequently, while working on a mineral deposit, he is deductive rather that inductive. But Science, needless to say, prefers an inductive conclusion to a deductive one. In fact that is the basic approach of Science. Some classification schemes are more frequently cited in text books and used by exploration geologists. Those are briefly discussed below.

Lindgren (1907) was the first in the early part of twentieth century to bring out a detailed genetic classification of ore deposits which he himself modified later (Lindgren 1933) through his own experience (Appendix B). Graton (1933) suggested two modifications to Lindgren's classification by adding a group *telethermal* above epithermal and another group *leptothermal* in between *epithermal* and *mesothermal*. The renowned German geologist Schneiderhohn suggested a classification of mineral deposits in 1932 and later modified it in 1962 (Appendix C). Around the same time in USA, Bateman (1950) proposed a classification of mineral

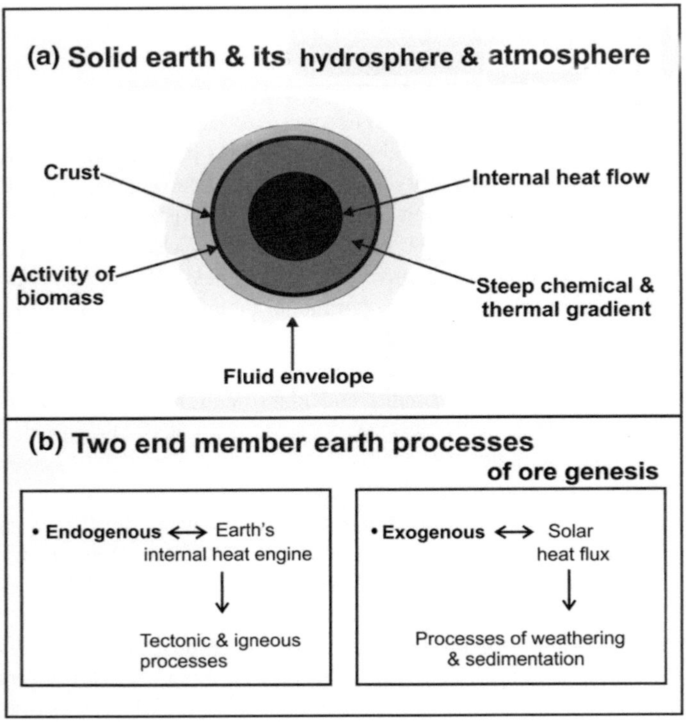

Fig. 1.11 The environment of exogenous and endogenous deposits. The former is characterized by the effect of solar heat flux while the latter is driven by earth's internal heat engine

deposits which was few decades later modified and elaborated upon (Jensen and Bateman 1981) (Appendix D). These three classification schemes are not completely different from one another, as is obvious to a discerning eye. However, from the point of a modern approach, these classifications are not sufficiently adequate. A classification scheme proposed by the present authors can be seen in Table 2.1. The "early magmatic" and "late magmatic" are abandoned, as liquation involving separation of a sulfidic melt and a silicate melt is more a function of sulfide saturation than fall of temperature (vide Chap. 2). Hydrothermal deposits may form from a large number of sources. This should be remembered while subdividing them into hypothermal, mesothermal and epithermal deposits. SEDEX, SSC, and MVT (see List of abbreviations) are new inclusions here because of their frequent use today. Bio-geochemical degradation of plants and phyto-planktons have been included in this classification as we have included fossil fuels, such as coal, oil, and natural gas. Iron oxide-copper-gold (IOCG) is a relatively new grouping of mineral deposits, the origin of which is still shrouded in controversy. So also are oxidic deposits of Fe (±Ti). A more recent tectono-lithological classification scheme is now widely accepted and is based on the association of deposits with their host rocks, and also with the broad tectonic environment (Cox and Singer 1986). This widely accepted and used classification scheme (Table 1.4), which remains a bench mark on the topic, however, may raise some doubt in the mind of the reader because while the lithologic divisions are clear, the tectonic aspect is only implicit. It is used in the scheme in the broadest possible sense and the geotectonic environments identified are the resultant of tectonic activity and do not indicate the tectonic settings (vide Table 8.1). Thus, it uses sub-aerial versus marine volcanic environment, clastic versus carbonate sedimentary environment and the presence or absence of particular type of intrusions to classify deposit types (cf. Poulsen et al. 2000).

Throughout the last century, as stated above, many different systems were developed for resource classification, reflecting the different physical characteristics of the resources as well as the geographic and socioeconomic diversity of the

Table 1.4 Lithotectonic classification of mineral deposits (after Cox and Singer 1986)

GEOTECTONIC ENVIRONMENT				
Igneous	Intrusive	Mafic-Ultramafic		
		Alkaline and basic		
		Felsic	Phanerocrystalline	Calcareous wall rocks (1)
			Porphyroaphanitic	Co-eval volc. w.rocks (2)
			Ign/sed wall rocks	(3)
	Extrusive	Mafic		
		Felsic-Mafic	Sub-aerial volcanic wall rocks (4) – (8)	
			Calcareous w.r.	(9)
			Marine	(10)
Sedimentary	Clastic		Conglomerate	(11)
	Carbonate			
	Chemical			
Regional Metamorphic	Metavolcanic/ Metasedimentary			(12)(13)
	Metapelite/ metaarenite			
Surficial	Residual			(14)
	Depositional			(15)

Deposit models (examples from gold deposits, after Poulsen et al. 2000)

(*1*) Polymetallic replacement; (*2*) porphyry Cu–Au; (*3*) Au–Ag–Te veins; (*4*) hot spring Au–Ag); (*5*) creede epithermal vein; (*6*) Comstock epithermal vein; (*7*) Sado epithermal vein; (*8*) Epithermal Qtz–alunite gold; (*9*) Sediment–hosted gold; (*10*) Kuroko massive sulfide; (*11*) Quartz–pebble–congl. Au–U; (*12*) Low sulfide Au–Qtz veins; (*13*) Homestake gold; (14) laterite-saprolite Au; (15) Placer Au-PGE producing areas (ECE Energy series 2010). However, no effort was made to harmonize the terminologies in the different classification systems. More recently, with globalization of commodity trading and financial markets firmly established, a harmonized framework classification system was considered to be of great value. Thus, the UNFC-2009 classification was developed for fossil energy and mineral reserves and resources located on or below the Earth's surface for applications to energy and mineral studies, resource management functions, corporate business processes, etc. UNFC-2009 is a generic principle-based system in which resources are classified on the basis of three fundamental criteria: economic and social viability (E), field project status and feasibility (F) and geological knowledge (G), using a numerical coding system. A class of resource is uniquely defined by selecting from each of the three criteria a particular combination of a category or a subcategory. Since the codes are quoted in the same sequence (E, F, G), the letters are dropped and just the numbers retained, e.g., 211 = E2F1G1, the highest category being 111 (Fig. 1.12). For further clarity in global communication, additional generic subclasses are defined as shown in Table 1.5.

A *mineral deposit model* on the other hand is a systematically arranged body of information that describes the essential attributes of a mineral deposit type. It may be presented as a narrative, a list of attributes, a set of equations, a cartoon, a collection of relations between objects or properties of those objects, or any combination of the above (Barton 1993). A *mineral deposit type* on the other hand is a collective term for mineral deposits that (a) share a set of geological attributes, (b) contain a particular mineral commodity/s such that (a) and (b) together distinguish them from other types of mineral deposits.

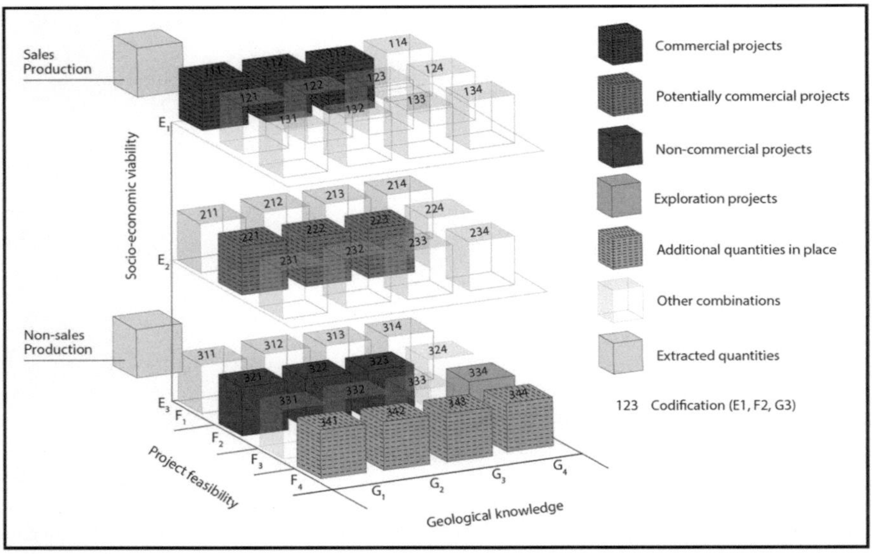

Fig. 1.12 United Nations Framework Classification (after UNFC-2009)

Table 1.5 UNFC scheme of classification

<table>
<tr><td rowspan="15">Total Commodity initially in place</td><td rowspan="2">Extracted</td><td colspan="5">Sales production</td></tr>
<tr><td colspan="5">Non-sales production</td></tr>
<tr><td colspan="2" rowspan="2">Class</td><td rowspan="2">Sub-class</td><td colspan="3">Categories</td></tr>
<tr><td>E</td><td>F</td><td>G</td></tr>
<tr><td rowspan="8">Known deposit</td><td rowspan="3">Commercial projects</td><td>On production</td><td>1</td><td>1.1</td><td>1,2,3</td></tr>
<tr><td>Approved for development</td><td>1</td><td>1.2</td><td>1,2,3</td></tr>
<tr><td>Justified for development</td><td>1</td><td>1.3</td><td>1,2,3</td></tr>
<tr><td rowspan="2">Potentially commercial projects</td><td>Development pending</td><td>2</td><td>2.1</td><td>1,2,3</td></tr>
<tr><td>Development on hold</td><td>2</td><td>2.2</td><td>1,2,3</td></tr>
<tr><td rowspan="2">Non-commercial projects</td><td>Development unclarified</td><td>3.2</td><td>2.2</td><td>1,2,3</td></tr>
<tr><td>Development not viable</td><td>3.3</td><td>2.3</td><td>1,2,3</td></tr>
<tr><td colspan="2">Additional quantities in place</td><td>3.3</td><td>4</td><td>1,2,3</td></tr>
<tr><td rowspan="2">Potential deposit</td><td>Exploration Projects</td><td>[No sub-classes defined]</td><td>3.2</td><td>3</td><td>4</td></tr>
<tr><td colspan="2">Additional quantities in place</td><td>3.3</td><td>4</td><td>4</td></tr>
</table>

The attributes in an ore deposit model are dealt with in two scales: observations in *local scales*, made directly in the field (mineralogy, zoning, alteration haloes, etc.). Also in *Regional geologic scale*: concerning features which may be interpreted through local observations or through global tectonic considerations (e.g., back arc environment). The attributes generally considered are:

Geological Environment	Deposit Description
Regional geology	Ore body shape, size & disposition
Rock types	Mineralogy
Textures	Texture / Structure
Age range	Alterations
Depositional environment	Ore controls
Tectonic setting (s)	Weathering
Associated deposit types	Geochemical signature

Not all the attributes will be equally developed in all the ore deposits in an ore district. Combining each of the attributes from individual deposits, a comprehensive model can be developed for a group of deposits in a district. A good example of this approach is presented by Cunningham et al. (1991) for the volcanic dome-hosted precious metal deposits in Bolivia, South America (Fig. 1.13).

However, there remains a problem of distinguishing critical from incidental attributes. For example, the common presence of secondary dolomites in carbonate-hosted Pb–Zn deposits (MVT-type) (cf. Chap. 2), requires us to consider if (a) this dolomite is essential to the model; (b) what happens if limestone is present in a deposit instead of dolomite; (c) or is permeability a more critical factor? The formation of dolomite either enhances permeability or may reflect preexisting permeability which is exploited by both the dolomite and the ore-forming solutions.

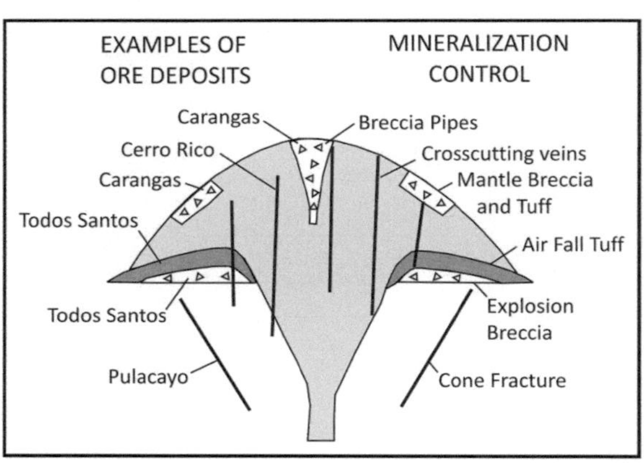

Fig. 1.13 Empirical model of volcanic dome-hosted precious metal deposits in Bolivia (after Cunningham et al. 1991)

Perhaps the dolomite records a particular range of Ca/Mg ratio which in turn reflects the characteristics of the basinal brines that constitute the ore fluid. Thus, in any case, the dolomite is a powerful ore guide and finds a place in the "Final model" (Cox and Singer 1986).

Based on our understanding of the attributes and origins of mineral deposit types, we may be permitted to conclude that some models are still not in a mature stage (e.g., Olympic Dam type IOCG deposits) while others are fairly mature and robustly founded on excellent databases (e.g., placers and marine evaporate deposits). Ore deposit models can be divided into two end members (Thompson 1993): (1) the empirical or occurrence model, based on observational data; and (2) the genetic or conceptual model, based on theoretical concepts. Thus the different types of ore deposit models in vogue are (cf. Cox and Singer 1986):

Empirical (Descriptive) model: where the various attributes are recognized as essential but their relationship is unknown (cf. Fig. 1.13). Direct observations and experience are critical to the application of an empirical model. These models form the basis for successful exploration and must have been in use since modern prospecting began.

Grade-tonnage model is an important subset of empirical data for given economic and technological conditions. About 60 grade-tonnage models are available from data of 3900 well-explored deposits. Grade or tonnage on the horizontal axis and cumulative percentage of deposits in the vertical axis are plotted in some models, whereas in others the vertical axis shows the grade and the horizontal displays the tonnage (Fig. 1.14). These statistics provide a general guide to the size

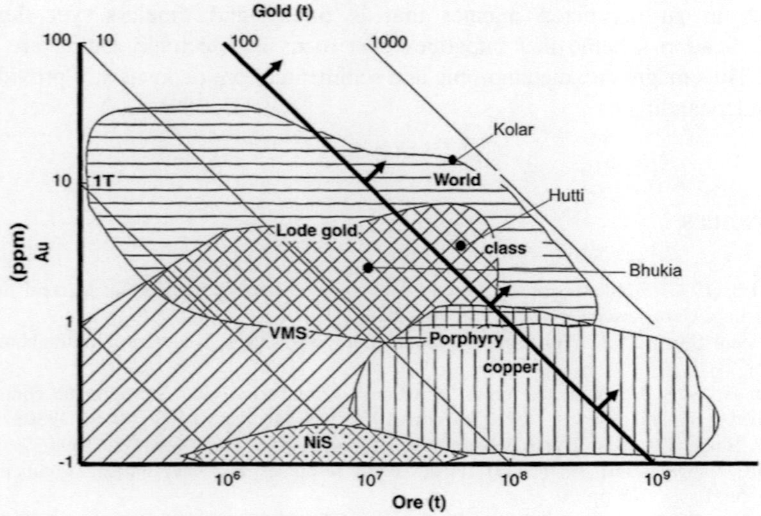

Fig. 1.14 Grade and tonnage model of some gold deposit types. Deposits falling above the bold line are considered world class. Note the position of three Indian deposits: Kolar, Hutti and Bhukia (after Poulsen et al. 2000)

and grade of a potential mineral deposit target that can influence and help to modify exploration strategy.

Theoretical (Genetic) model: where the attributes are interrelated through some genetic concept. Such models attempt to describe the physical and chemical processes that led to the formation of an ore deposit, using experimental or calculated constraints. However, undue emphasis on certain elements within a genetic model and identifying them as the most important criteria, based on the author's experience, become counter-productive at times.

Quantitative process model, e.g., numerical analysis of flow, chemistry, and thermal behavior of fluids around a cooling pluton or sequence and amounts of minerals deposited from evaporating sea water, etc.

Occurrence–probability model: forms the ultimate goal of an exploration geologist and is clearly the most difficult one to develop.

Thompson (1993) used five deposit types to illustrate the effectiveness of empirical and genetic models toward exploration. Features of empirical model led to numerous discoveries of sediment-hosted gold, Archean lode gold and porphyry copper deposits (cf. Chap. 2 for details of these deposit types). However, when the empirical database is limited, such as in Platinum group deposits, confidence in such models is reduced and their genetic models are more suited for use in exploration. In reality, the origin of different mineralization remains controversial in varying degrees and hence, genetic models are less useful for exploration, except for a limited number of deposit types, such as the VMS deposits, and to a large extent, porphyry Cu(–Mo) deposits, etc.

In summary, mineral deposit models summarize features that are common or shared by group of mineralized systems, whereas classification means grouping deposits in an organized manner that is timely and timeless, yet flexible. A classification scheme that integrates host rocks and tectonic setting are most useful. Thus magmatic, metamorphic and sedimentary ore deposits still provide the best understanding.

References

Barton PB (1993) Problems and opportunities for mineral deposit models. In: Mineral Deposit modeling. Geol Assoc Can Spec Pap 40:7–14

British Geol Surv (2013) World mineral production, 2007–2011. Keyworth, Nottingham, UK, 76 pp

Canavan F (1973) Notes on the terms 'stratiform', 'stratabound' and 'stratigraphic control' as applied to mineral deposits. J Geol Soc Aust 19(4):543–546. doi:10.1080/00167617308728822

Cox DP, Singer DA (eds) (1986) Mineral deposit models. U.S. Geol Surv Bull 1693:379 pp

Craig JR, Vaughan DJ, Skinner BJ (2011) Earth resources and environment. Prentice Hall, Pearson, Essex 528 pp

Cunningham CG, Mcnamee J, Vásquez JP, Ericksen GE (1991) A model of volcanic dome-hosted precious metal deposits in Bolivia. Econ Geol 86:415–421

Deb M (1971) Aspects of ore mineralization in the central section of the Singhbhum copper belt, Bihar. Unpublished Ph.D. thesis, Jadavpur University, Calcutta, 187 pp
ECE Energy series (2010) United Nations framework classification for fossil energy and mineral reserves and resources 2009. 39:20
Erickson RL (1973) Crustal abundance of elements, and mineral reserves and resources. U.S. Geol Surv Prof Pap 820:21–25
Evans AM (2001) An introduction to economic geology and its environmental impact. Blackwell Science, London, UK 364 pp
FICCI (2013) Development of Indian mining industry: the way forward. Non-fuel minerals. FICCI Mines and Metals division, 119 pp
Gandhi SM (2000) Ancient mining and metallurgy in Rajasthan. In: Deb M (ed) Crustal evolution and metallogeny in the northwestern Indian shield. Narosa Publ House, New Delhi, pp 29–48
Graton LC (1933) The depth zones in ore deposition. Econ Geol 28:513–555
Harris DC (1989) Mineralogy and geochemistry of the Hemlo gold deposits, Ontario. Geol Surv Can, Econ Geol Rep 38:88
Jensen ML, Bateman AM (1981) Economic mineral deposits, 3rd edn. Wiley, New York 593 pp
Lindgren W (1933) Mineral deposits. McGraw-Hill Book Co., New York 930 pp
Mckelvy VE (1960) Relation of reserves of the elements to their crustal abundance. Am J Sci 258A:234–241
Mookherjee A, Panigrahi MK (1994) Reserve base in relation to crustal abundance of metals. J Geochem. Expl 51:1–9
Poulsen KH, Robert F, Dubé B (2000) Geological classification of Canadian gold deposits. Geol Surv Can Bull 540:106 pp
Robb L (2005) Introduction to ore-forming processes. Blackwell Publishing, Oxford 373 pp
Sarkar SC (1984) Geology and ore mineralization of the Singhbhum copper-uranium belt. Jadavpur University Press, Eastern India 263 pp
Schneiderhohn H (1962) Erzlagerstätten—Kurzvorlesungen zur Einführung und Wiederholung. Fischer, Jena 371 pp
Stanton RL (1972) Ore petrology. McGraw-Hill, New York 713 pp
Thompson JFH (1993) Application of deposit models to exploration. In: Mineral Deposit modeling. Geol Assoc Can Spec Pap 40:51–67
U.S. Geological Survey (1980) Principles of a resource/reserve classification for minerals. U.S. Geol Surv Bull 1450-A, 5 pp
Vaughan DJ, Craig JR (1978) Mineral chemistry of metal sulfides. Cambridge University Press, Cambridge 493 pp

Chapter 2
How Do Mineral Deposits Form and Transform? A Systematic Approach

Abstract Formation and transformation of mineral deposits are interactions of geospheres, one including the atmosphere, hydrosphere, biosphere, lithosphere, and asthenosphere and the other involving the mantle and the core of the earth. Complex chemical and thermal interactions between these two geospheres have led to distribution and concentration of elements and even, later modifications, producing the mineral or ore deposits of today. The essential processes involve magmatism, hydrothermal, and sedimentary processes with a strong impact of tectonism and in places, of weathering and erosion. The genetic processes vary in details. The principal ones are outlined below with the principal products in parentheses: (1) Essentially magmatic processes (Ni, Cu, PGE Cr, Fe–Ti); (2) Pegmatitic processes (rare metals, ceramic, and radioactive elements); (3) Essentially magmatic hydrothermal processes (Sn, W, U, Cu, Mo, REE); (4) Essentially amagmatic hydrothermal processes (Cu, Pb–Zn, Au, U); (5) Sedimentary (-diagenetic) processes (Fe, Mn, U, Sn, Ti, monazite, phosphorite, carbonate rocks, rock salt gypsum); (6) Lateritic and non-lateritic residual processes (Fe, Mn, Al, Ni, and clays); (7) Supergene oxidation and enrichment (Cu, Ag, Au, U); (8) Biogeochemical degradation of biomass (peat-lignite-coal, natural gas, and oil).

Keywords Magmatic process of ore genesis · Pegmatitic process of mineral formation · Hydrothermal process of ore formation · SEDEX deposits · VMS deposits · MVT deposits · Metamorphism and ore genesis · Sedimentary-diagenetic processes of ore formation · Placerization · Lateritic process of ore formation · Supergene enrichment of ores · Biogeochemical degradation of biomass

2.1 Introduction

The earth's total stock of ore metals is found in two geospheres or physicochemical systems (Brimhall 1991): the outermost geosphere is a thin reactive shell near the surface of the earth which includes the atmosphere, biosphere, hydrosphere, lithosphere, and the asthenosphere on which the oceanic and continental plates

M. Deb and S.C. Sarkar, *Minerals and Allied Natural Resources and their Sustainable Development*, Springer Geology,
DOI 10.1007/978-981-10-4564-6_2

move. This complex geosphere is underlain by another, comprising the mantle and the core (cf. Fig. 1.11). It is believed that all metals are primordial, that is, they have neither been created nor destroyed since the beginning of the earth, except for a few metals like lead, which is formed in part by the radioactive decay of uranium and thorium. Complex thermal and chemical interactions between the two geospheres have redistributed the elements from time to time through Earth history. During this migration and redistribution, mineral concentrations have taken place in rocks of all ages by primary igneous and metamorphic processes in deeper crust or upper mantle or by secondary processes resulting from fluid movement and weathering on or close to the surface. Understanding the genesis of ores through their characterization and obtaining the precise information of magmatic, hydrothermal, tectonic, and sedimentary events that produce the ore-forming crustal fluids and metal sources that form these economic mineral deposits is a basic approach in ore genetic studies. Thus many ore deposits are produced at depth in the endogenous environment, characterized by the earth's internal heat and its dissipation. They are eventually exhumed or brought close to the surface either by erosion of the cover or by tectonic uplift or both. A large number of economically important mineral deposits also form in the exogenous environment where weathering and sedimentation are major geological processes in water-dominant systems that are driven by the solar heat flux and where biological mediation is common. Between these two end members, there are also processes which in various combinations contribute to the formation of ore deposits or their transformation. The ore-forming process may be initiated in the endogenous realm but produce the deposit at or near the surface, or the other way round, where meteoric $\pm$ connate waters penetrate the rocks, undergo gravity-driven flow through the strata, or get convected upwards from depth, leach metals during passage, and deposit them in specific locales in the crust, commonly oceanic crust, thereby having the signatures of both the environments. Presently active ore-forming processes in marine environment produce deposits of this type and provide a clearer perception of the geological setting and processes responsible for the generation of the ancient analogs.

We list below (Table 2.1) the mineralizing processes of eight major types. This list includes only those mineral deposit types which are economically important for the particular metal and account for its major reserve. For an exhaustive list of various mineral deposit types of the different metals, the reader may refer to the recent publication of Dill (2010). It must have been noted that we have divided the mineralogenetic processes into several types, eight to be precise, in Table 2.1. But why have we put the attribute "essentially" in some of the above types? Let us try to explain. As will be obvious from the discussion that follows, endogenous mafic/ultramafic magmas may be important contributors to the formation of Ni–Cu, Cr, PGE ores for attaining critical composition for ore genesis. But such potentially ore-bearing magmas generally require magma contamination or country rock assimilation for reaching the critical composition i.e., supersaturation for chromite deposition or Ni–Cu–Fe sulfide melt separation. Thus, the process responsible for ore generation is mainly magmatic, though not totally. Again intrusion-related magmatic fluids are usually the principal contributors to the formation of hydrothermal ore deposits. But in some cases, aqueous fluid contribution from the country rocks might have been substantial. The attribute, "essentially amagmatic" is given to SEDEX, MVT and VHMS deposits also. Fluids from which they were deposited were essentially hydrothermal following the basic definition of hydrothermality. Only there is little or no magmatic contribution. Some other deposits, such as orogenic Au, unconformity type U deposits, and some rare base metal deposits have been included in this type. Sedimentary(-diagenetic) processes have several subtypes, as shown under (5) in Table 2.1. Their characters are straight forward. Residual processes may give rise to both lateritic and non-lateritic (Al-rich clay) deposits. Biogeochemical degradation produces fossil fuels from micro- and macro-plant remains under suitable geological environments. Salient aspects of these processes have been discussed in this chapter, except the ones at 5 (c) and (d) and (8), which are discussed in the chapters on nonmetallic minerals (Chap. 4) and Energy Resources (Chap. 6) respectively.

The formation of most of the earth's mineral resources requires the presence of a fluid phase to extract the ore elements from their hosts by dissolution, to promote transport (mechanical and/or chemical) and ultimately to deposit them in suitable locales. Therefore, the knowledge and understanding of the fluid phase are critical to the development of any model of ore genesis. These fluids, which may vary widely in terms of composition, temperature, physical state, and flow characteristics, represent a state of matter in which the molecules are able to flow past one another without any limit or without any fracture or dislocation. Their state as gas, vapor, or liquid exhibits a progressively closer association of molecules and is characterized by different mechanical and thermodynamic properties. At 374 °C and 225 kg/cm^3, that is, at critical temperature and pressure, the most common solvent, water, changes to a supercritical fluid which has both the gaseous property of being highly mobile and the liquid property of dissolving various components (Dill 2010). The different kinds of fluids, in this context, normally reside in, are generated within or enter into the crust from below (hypogene) or above (supergene). They may be mantle-derived, crustally generated or meteoric in origin. Thus, their flow system is either internal energy-driven (endogenous) or solar-energy-driven (exogenous), that is, hydrologic (cf. Fig. 1.11). Both the flow systems are local to regional in extent and under particular circumstances, are capable of horizontal fluid movement on continental scale (Norton 1977; Mookherjee 2000). Their movement is driven by various kinds of forces, such as, heat, tectonic deformation, gravity, buoyancy, capillary action, osmotic pressure, and/or surface tension along thermal/pressure/compositional/permeability/chemical potential gradients. At shallow depths, the fluid movement takes place through fractures, dissolution cavities, and interconnected pore spaces. At greater depths, on the other hand, fluid movement takes place in tectonically active regions within interstratified rocks of differing competencies, which confines the movement within brittle layers acting as "metamorphic/tectonic" aquifers.

Several types of geological situations bring about the flow of crustal fluids either in a near-hydrostatic fluid pressure regime or in an overpressured fluid regime (Cox 2005). Hydrothermal systems develop in both magmatically active and amagmatic

Table 2.1 Major processes and products in mineral deposit formation (Authors' unpublished work)

Ore genetic process*	Products
1. Essentially magmatic processes	Ni–Cu, PGE, Cr, Fe–Ti, Fe–V deposits
2. Pegmatitic process	Rare metals, ceramic, and radioactive minerals,
3. Essentially magmatic hydrothermal processes	
a. Mineralization associated with quartz-rich leucogranite	Sn, W, U, Mo
b. Mineralization associated with porphyry systems	Cu, Mo, Au
c. Skarn- and greisen-related mineralization	Fe, W, Au, Cu, Pb–Zn, Mo, Sn
d. IOCG type mineralization	Cu, U, Au, REE
4. Essentially amagmatic hydrothermal process	
a. Volcanic-hosted massive sulfides	Cu, Zn–Pb (including present marine metallogenesis)
b. Sedimentary exhalative (SEDEX) deposits	Pb–Zn, Cu
c. Mississippi valley type (MVT) ores	Pb–Zn
d. Sediment-hosted stratiform ores	Cu (Zambian Cu belt; Kupferschiefer)
e. Metamorphogenic ore formation	Au (orogenic), U (Unconformity type), rare base metals (e.g., Mt. Isa, Australia)
f. Sandstone-hosted U-V	Colorado plateau type
5. Sedimentary(-diagenetic) processes	
a. Placerization	Au, PGE, monazite (Th, REE), Sn, Ti (rutile, ilmenite), zircon, precious stones (diamond, ruby)
b. Sedimentary-diagenetic	Fe, Mn (including present marine metallogenesis), U
c. Evaporative deposition	Rock salt (NaCl), gypsum, K-salts
d. Diagenetic modification of organic remains, carbonates	Phosphate deposits, magnesite, dolomite rocks
6. Lateritic and non-lateritic residual processes	Fe, Mn, Al, Ni, and clays
7. Supergene oxidation and enrichment	Cu, Ag, U, Au
8. Biogeochemical degradation of biomass	Peat–lignite–coal, natural gas, and oil.

*Here, "ore genetic processes" may be read as "Economic mineral deposit formation processes"

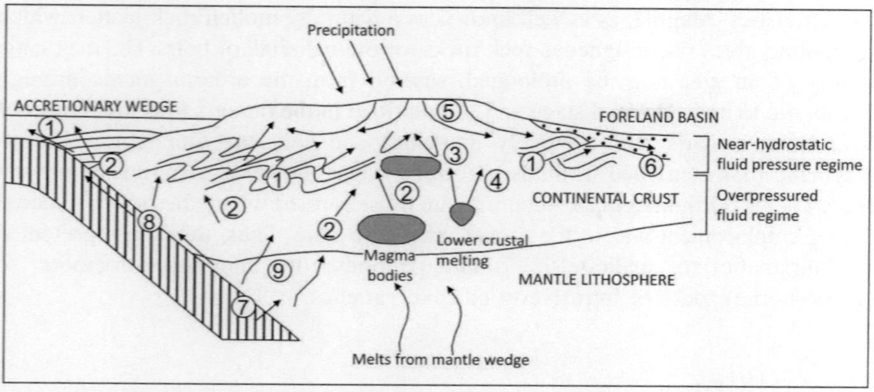

Fig. 2.1 Sketch of the distribution of various fluid sources and pathways in a convergent regime. Refer to text for pathway numbers (after Cox 2005)

environments, particularly in parts of accretionary and collisional orogens. Various fluid sources and pathways (Fig. 2.1) in this context are: (1) structurally controlled fluid flow; (2) metamorphic devolatilization; (3) thermally driven convection; (4) fluid exsolution from magma bodies; (5) topographically driven (gravity) flow of meteoric fluids; (6) basinal flow; (7) devolatilization of subducting slab; (8) fluid escape along slab interface and into slab hanging wall and (9) devolatilization of hydrated mantle wedge.

Having formed by such right combination of processes and often being modified later on near the surface of the earth, the ore deposits remain transitory in the geological sense and reflect the dynamic processes within and outside the earth. These include deep weathering over a protracted period of time eroding the ore deposit or subduction of the plate on which the ore deposit lies. They also serve as important geochemical sensors providing useful record and history of transport paths and forces operative in the crust.

2.2 Essentially Magmatic Processes

It is a common geological knowledge that different igneous rocks host different associations of ore deposits and particular metal associations are found in specific igneous rocks, e.g., Cr, V, Ni, PGE, Cu, Zn, and Au (both siderophile and chalcophile) are associated with basic-ultrabasic igneous rocks in which they show the maximum crustal abundance. Similarly, elements like Be, Li, Sn, W, U, and Th, as well as F (all lithophile elements) are associated with acidic igneous rocks. This implies that there is clearly a strong first-order correlation between magma composition and metal enrichment, and a particular metal or metal association has a relationship to the environment of magma generation and its chemical characteristics. Magma, as is well known, is a naturally molten rock matter, which on cooling gives rise to igneous rock/rocks, or ore material, or both. The magmatic history of an area may be prolonged, starting from the orthomagmatic through pegmatitic to hydrothermal stage and as described in the famous Reaction Series of N.L. Bowen, starts with relatively low-silica and ends up with more siliceous magmatic rocks enriched in fluids. All three stages, however, may not be present everywhere. "Orthomagmatic" deposits are those formed where the source material during emplacement was in the proper magmatic state. Thus, they are represented by ultramafic to mafic-felsic (dunite-peridotite to anorthosite-troctolite to gabbro-norite) rocks of intrusive to effusive varieties.

2.2.1 Magmatic Ore Deposits

The geologic processes mentioned above produce "orthomagmatic" ore deposits of various transition metals (Ti, V, Cr, Fe, Ni, Cu ± PGE) in different combinations. The three main groups associated with basic to ultrabasic magmatic rocks are:

(i) Cr ± PGE (Pt, Pd, Ir, Os, Re, Rh), Fe–V
(ii) Ni, Cu, PGE
(iii) Fe–Ti.

The first group may or may not have either PGE or Fe–V ores or both in the sequence and is represented by stratiform chromite ores in large layered intrusions, such as, Stillwater Complex, Montana, USA; Kemi, Finland; Fiskenaesset, Greenland; and or funnel-shaped intrusions, such as, Bushveld in South Africa; Great Dyke, Zimbabwe; Muscox and Bird River Sill in Canada; Dore Lake, Australia and Bacuri Complex, Amapa state, Brazil (Cawthorn 1996). The Bushveld Complex and the Great Dyke of Zimbabwe account for more than 90% of world resource of chromium. The rocks of these intrusions include dunite, peridotite, pyroxenite and anorthosite, and less commonly, gabbroic rocks. Podiform chromite deposits in ophiolite complexes, often dismembered, occurring mainly in Kazakhstan, Phillipines, Turkey, Cyprus, Greece, and Albania, belong to the first group as well. Also referred to as the "Alpine type", their irregular shape and limited reserve is not generally conducive to mining.

In the large intrusives, the chromite-rich layers, a few mm to several meters thick, are laterally persistent over long distances and alternate repetitively with silicate layers. They show regular changes in mineral compositions (e.g., En-variation in pyroxenes and An-variation in plagioclase) and cryptic layering. In the enormous Bushveld Complex near the city of Pretoria in the Republic of South Africa, 29 chromitite seams (Fig. 2.2a) occur in three segments (Fig. 2.2b), eastern, western, and northern. The igneous stratigraphy sits over the Pretoria Group basement separated by a Marginal zone, and followed upward by the Lower zone, the Critical zone hosting the chromitite layers and the Merensky Reef rich in PGE with some Ni–Cu, the Main zone showing little differentiation and the upper zone with the vanadiferous magnetite layers (Fig. 2.2c). The largest orebodies are the LG3 and LG4 chromitite (a rock composed essentially of chromite) layers in western Bushveld. The Merensky Reef is a thin (0.3–0.6 m) sheet of pegmatitic pyroxenite layer in an igneous zone of the complex, traceable over 200 km. Thin chromitite bands rich in PGE mark the top and bottom of the Reef (Fig. 2.2d). The Great Dyke of Zimbabwe is 532 km-long and 5–9.5 km wide and hosts as many as eleven persistent chromitite seams. As discussed below, the chromite accumulations in such intrusives have taken place through fractional crystallization, gravitative settling, flow differentiation, and filter pressing.

In deposits of the second group, Ni:Cu ratio may vary drastically and PGE may also be totally absent. The sulfidic nickel ores with PGE may be concentrated in the basal zone of mafic to ultramafic intrusions, such as, in Sudbury deposits, Ontario, Canada (Fig. 2.3a), in the Stillwater Complex, Montana, USA or Nkomati mine, Mpumalanga, RSA or they may be hosted by rift-related volcanic rocks, such as, in the Duluth complex, USA; in Norilsk-Talnakh deposits in Russia; at Jinchuan, China; at Muscox, Nunavut, and Crystal Lake, Canada. The Cu–Ni sulfide ores are common in the early cycles of some greenstone belts where they may occur in three different subtypes: the komatiitic subtype (e.g., Kambalda deposits in Yilgarn craton in Western Australia; Langmuir, Ontario, Canada; Selebi-Pikwe, Botswana), the dunitic subtype (Mt. Keith, Western Australia) or the picritic subtype (Pechenga, Kola Peninsula, Russia).

This second group of deposits commonly forms massive, network or disseminated ores (cf. Fig. 2.4) with rather simple mineralogy of pyrrhotite, pentlandite, and chalcopyrite. They are mostly of Archean to Proterozoic age found in unfractionated Mg-rich basic igneous rocks in old continental setting. In all these deposits, separation of a sulfur-rich liquid containing Fe–Ni–Cu by liquation from a parental magma and its sinking to the bottom of the magma chamber produced the orebodies (Fig. 2.3b).

As pointed out by Cawthorn et al. (2005), some fundamental differences exist between the base metal-rich (Ni–Cu-rich) and PGE deposits. The former are commonly localized and are discontinuous near the base of intrusives. Important are the three criteria pointed out by Naldrett (1989) for the magmatic base metal ores. These are: (1) an olivine-rich magma; (2) proximity to a major fault; (3) presence of sulfide/sulfate-bearing country rocks. PGE deposits hardly meet these criteria. Rather they occur in rocks that have pyroxenes with a Mg number of 0.8 (suggesting that the liquid contained only about 6% MgO) when PGE were concentrated. Their great lateral continuity means freedom from structural control. They also occur in the middle of intrusions, where the role of country rock assimilation is probably not important. However, as PGE occur in very small quantity (in ppb) in the mafic-ultramafic magmas, large volumes of the latter are needed for the development of these ores. The role of this volume factor is obvious from the formulation of Campbell and Naldrett (1979) which controls the grade (tenor) of a sulfide liquid:

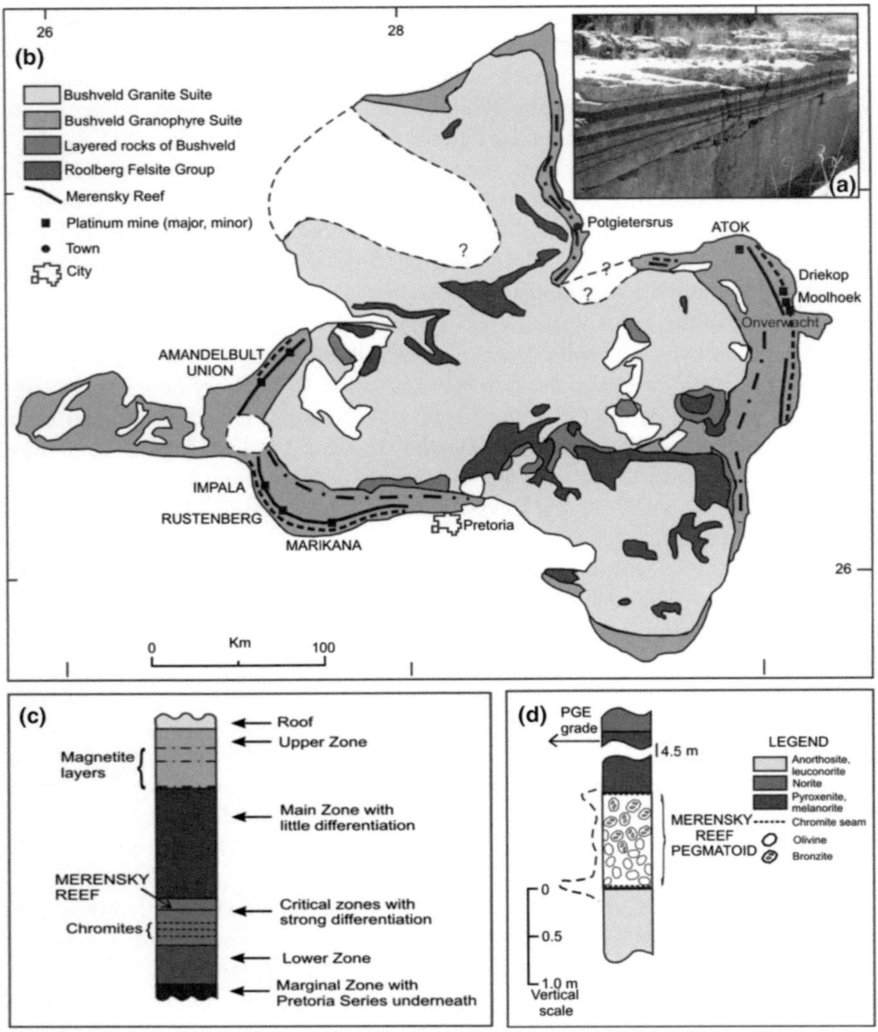

Fig. 2.2 **a** Chromitite seams of Bushveld complex (after pinterest.com); **b** Geological map of Bushveld Complex showing the disposition of chromite, PGE (Merensky Reef) and Fe–Ti–V orebodies in the different sectors; symbols for different mineralization same as in (**c**) (after Campbell et al. 1983); **c** generalized stratigraphic log showing the different zones and their mineralization; **d** an enlarged view of the PGE-enriched Merensky Reef (after Naldrett 1989)

$$C_{\mathrm{Sul}} = [C_o D\,(R+1)]/(R+D),$$

where C_{Sul} is the concentration of the element in the sulfide fraction; C_{o} is the original trace element concentration in the host magma; D is the sulfide-silicate

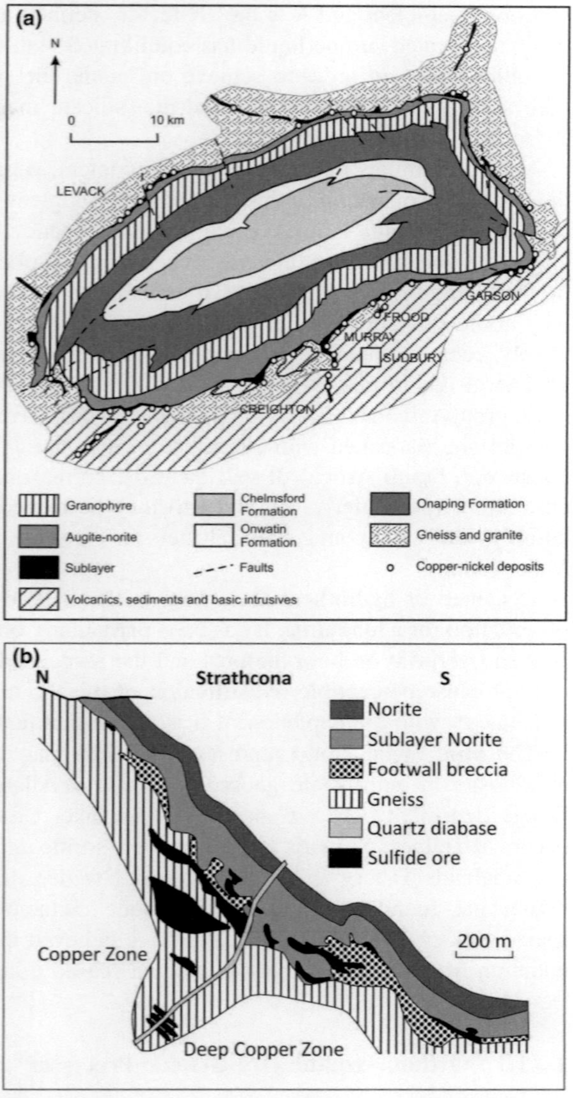

Fig. 2.3 **a** Geological map of the Sudbury complex, Ontario, Canada; **b** Cross section of the Strathcona mine showing the concentration of the sulfide ore at the bottom of the Ni-eruptive (after Barnes and Lightfoot 2005)

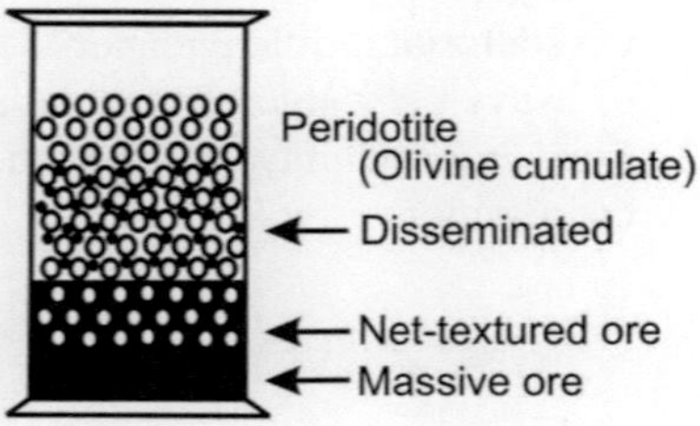

Fig. 2.4 The billiard ball model of Naldrett (1989) showing the generation of different ore textures by liquid immiscibility

partition coefficient and *R* is the "R factor" defined as the mass of silicate magma that a segregated sulfide liquid has equilibrated with (mass ratio of silicate magma to sulfide melt). In order to achieve ore grade, the sulfide liquid must equilibrate with significant quantities of metal-rich silicate magma, that is, have very high R-factors ($\sim$10,000 or higher).

Most chromitites have elevated PGE content, suggesting a correlation between the two. An explanation of this relationship is contentious. High partition coefficient of PGE in chromite is unexpected. Recent tendency has been to explain the association by mechanical adherence rather than as solid solution. Hiemstra's (1985) preferred suggestion was that PGE crystallized as very small grains, so small that they could not gravitate down and thus redeposited on chromite grains and ultimately became constituents of the chromitite layers. Mungall (2002) perceived change in fO_2 in magma around the chromite grains to initiate formation of platinum group minerals. It may be pointed out that there are many PGE-bearing reefs that are not associated with chromite, such as the Great Dyke, J-M Reef, Platreef (Bushveld, South Africa). It will therefore be reasonable to conclude that PGE in chromitites (particularly, Ru and Rh) may be related both to the sulfide accumulation in the matrix and metal clusters in chromite, as suggested by Barnes and Maier (2002).

Evidence of hydrothermal activity in the PGE ore zone has been a common observation for a long time. It has been particularly observed in the Merensky Reef. It is an overprint on both the ores and the associated silicates. But interestingly it did not cause perceptible redistribution of metals. Instead modification of the ore mineralogy with the depletion of S is a common feature (Cawthorn et al. 2005).

The third metal group represents the orthomagmatic ilmenite deposits within anorthosite or anorthosite gabbro, such as, at Allard lake, Lac Tio, Lac du Pin Rouge deposit, Quebec, Canada; Sanford lake, USA. The largest ilmenite body occurs at Tellnes, Norway, at the base of a noritic anorthosite body. Other deposits are Smalands Taberg and Ulvö deposits, Sweden. Interestingly, these anorthosite bodies are found to occur in high-grade metamorphic terrains with gneisses, granulites, schists and amphibolites. It is believed that residual melts after differentiation from anorthosite–norite magma caused a late enrichment in Fe and Ti.

2.2.1.1 Orthomagmatic Ore Genetic Processes

These are broadly categorized into two types:

1. Crystallization-differentiation, leading to disseminations or gravitative settling of heavy crystallites into cumulates.
2. Fluid Immiscibility or liquation, leading to segregation of melts of contrasted composition.

Crystallization-Differentiation

In a liquid multicomponent chemical system, differential crystallization may take place within the system due to the variation of physicochemical parameters, solubility limit being exceeded at a given T and P for some phases, leading to aggregation of the crystals into separate cumulates, commonly in the form of layers and lenses (Fig. 2.5). This differentiation is dominated by gravity, but can as well take place, or rather be augmented by filter pressing or tectonic squeezing across layering, diffusion, or even fluid flow, when a part of the material is still in the fluid state.

Monomineralic layers of chromite (chromitite) or magnetite (magnetitite) are often found in the layered mafic intrusions, represented commonly by mafic and felsic (gabbro-anorthosite) plutonic rocks, sometimes modified to ferrogabbro and ferrodiorite (cf. Robb 2005). They contain a fairly large portion of Cr and Fe–Ti–V ores in the mineralized system. The development of these layers suggest as if these were formed in the interludes when the silicate crystallization was switched off by a natural mechanism.

Opinions, however, vary over a wide range regarding the origin of the chromitite layers. Some major views include:

1. Crystal sinking and sorting
2. Injection of chromite-porphyritic magma
3. Liquid immiscibility
4. Variation in oxygen fugacity (fO_2) and total pressure (P_T)
5. Crustal contamination
6. Magma mixing.

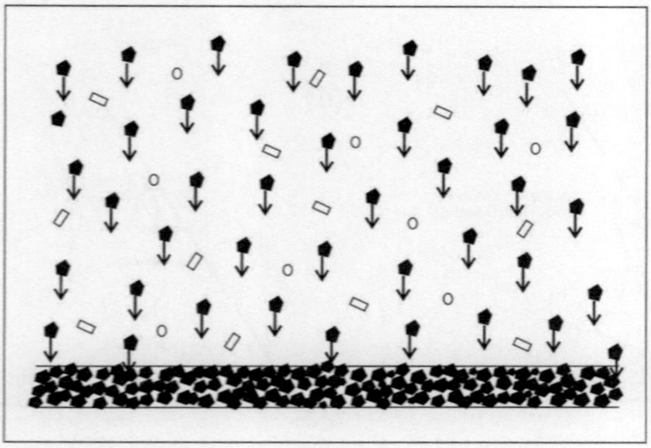

Fig. 2.5 A sketch of gravity separation of chromite cumulates (*black*) producing a chromitite seam. Silicates (olivine, circle, and pyroxene, plagioclase, rectangle) are also present in the system

Irvine (1977) provided a logical explanation for this feature in chromite deposits, based on the processes of magma mixing and magma contamination. His models are portrayed in the ternary diagrams a–d in Fig. 2.6 in which the end members are quartz, olivine and chromite (Fig. 2.6a). Irvine's contention has been experimentally confirmed by Murck and Campbell (1986) and accepted by most. In the normal crystallization process of a magma whose composition is represented by A in Fig. 2.6b, the only mineral to appear on the liquidus will be olivine. Settling down of olivine will produce the ultramafic rock dunite. The changed magma composition will now move toward the cotectic line, meeting it at B. Here, a small proportion of chromite ($\sim 1\%$) will start crystallizing along with much olivine. The magma composition will now evolve along the cotectic toward C, where olivine will be replaced by orthopyroxene. From C the magma composition will move toward D where plagioclase will be a new entrant to the system. This system of crystallization of the basic magma will not lead to formation of chromite layers and would instead form chromite as an accessory phase.

Development of a chromite seam is an extraordinary situation which is explained by two somewhat different models:

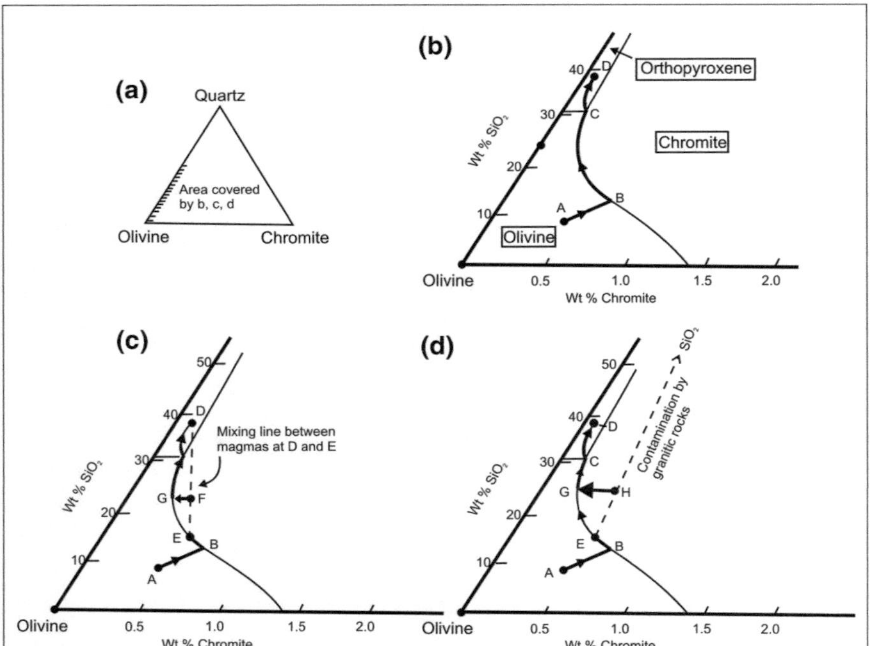

Fig. 2.6 **a** A portion of the ternary system Quartz–Olivine–Chromite; **b** The path of crystallization of a mafic magma; **c** effects of magma mixing and **d** magma contamination (after Irvine 1977)

1. Magma mixing
2. Magma contamination (by country rocks).

Magma mixing is introduction of a neomagma (not as primitive as the starting one) to the system. Let us assume (Fig. 2.6c) that a neomagma of composition E is introduced at D. Depending on the proportion of D and E, the composition of the mixture will lie on any point on DE. In this diagram, it is at F. In the interval F-G chromite only will crystallize and ultimately be affected by gravitative settling and forming a seam. On reaching G, the system will evolve along the cotectic and accessory chromite only will form along with much olivine.

Contamination of magma by siliceous country rocks is another possible mechanism of producing a monomineralic cumulate layer of chromite (in the interval H-G) (Fig. 2.6d). Here magma at the assumed point E on the cotectic becomes contaminated by the assimilation of crustal materials on the way. The contaminated magma will have a bulk composition anywhere on the join between E and the SiO_2 apex of the ternary diagram and accordingly located on it. This composition would transiently though lie on the chromite field and would produce a monomineralic cumulate of chromite (between H and G).

Cawthorn et al. (2005) also discuss these models and conclude that the "magma mixing" model is more acceptible. A mixture at M1 (Fig. 2.7) of primitive chromite-saturated magma P with its fractional derivative D is supersaturated with chromite and would crystallize chromite till it reaches M2. This mixing model (after Murck and Campbell 1986) provides a mechanism to cause crystallization of chromite in the absence of olivine and pyroxene.

Many of the debates and controversies that surround the genesis of layered chromite deposits concern the podiform chromite deposits also. Additional problem here is the genesis of podiform or nodular structure. Suggestion of Matveev and Ballhaus (2002) that involve equilibrium between an olivine–chromite-saturated basaltic melt and a H_2O-rich fluid at high temperature and pressure, explains the problem better. Postdepositional petrochemical changes in chromitite (in terms of texture and composition) are more common than generally believed.

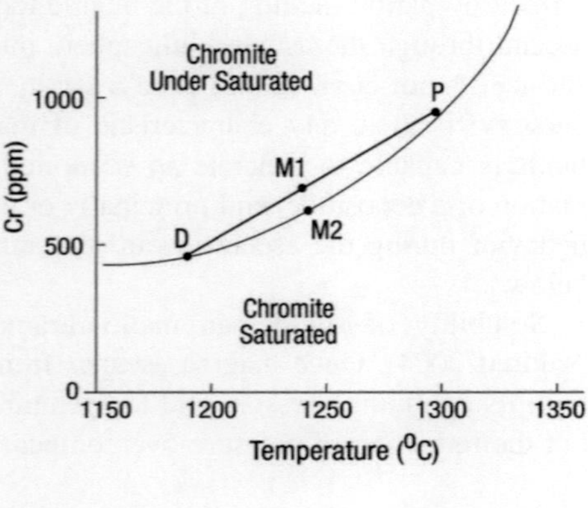

Fig. 2.7 A magma mixing model in the context of chromitite layer formation in the absence of olivine and pyroxene. See text for more details (after Murck and Campbell 1986)

V–(–Ti) bearing magnetite (-hematite) deposits in gabbro-anorthosite rocks, found at different places of the world also formed by crystallization-differentiation process.

Another example of crystallization-differentiation is the development of magnetite or hemo-ilmenite (depending on the magma composition) in association with anorthosite ± gabbro or ferrodiorite massifs. Early crystallization of plagioclase feldspar enriches the residual magma with Fe and Ti ultimately precipitating titaniferous magnetite or hemo-ilmenite. The precipitates accumulate on the floor of the magma chamber or are pressed out to a neighborhood site as slurry, should the situation so compel.

Liquid Immiscibility

Petrogenetic processes involving magma display a common phenomenon in which a homogeneous magma is split into two, initially by chemical disintegration and then by physical segregation. Both the neomagmas could be silicate–silicate, silicate–sulfide or even silicate–oxide. We are generally more concerned with the silicate-sulfide immiscibilities in magmatic ore genetic processes. The evidence of silicate–sulfide immiscibility was first noted in petrography of ores or ore-gangue masses, followed by observations in quenched volcanic rock material (Skinner and Peck 1969). Maclean (1969) confirmed the phenomenon with his experiment in the simple system SiO_2–FeO–FeS. From a homogeneous melt, consisting of silicates and sulfur, a sulfide melt will appear as soon as the magma attains sulfur saturation. Sulfide solubility in the composite material decreases with increasing O_2 content.

$$2\text{FeO}\,(\text{melt}) + \text{S}_2 \leftrightarrow 2\text{FeS}\,(\text{melt}) + \text{O}_2,$$

(Naldrett 1989).

Sulfide solubility increases with the increase of temperature and f_{S2} and decreases with pressure, a_{SiO2} and a_{Na2O} + a_{K2O}.

Mafic-ultramafic magmas genetically associated with Ni–Cu–(PGE) ores, formed by partial melting of the mantle rocks and leaving behind the solid residue, ascend through the astheno-lithosphere into the crust (Fig. 2.8) or erupt on to the surface. Arndt et al. (2005) pose a pertinent question on behalf of students of ore geology: is there any characteristic of mantle-derived magma that makes it particularly capable to generate an economic Ni–Cu–(PGE) deposit or does the formation of a deposit depend principally or entirely on events that control the magma behavior during the ascent toward the surface? The question is briefly addressed below.

Solubility of sulfides in mafic-ultramafic magmas is low (1000–3000 ppm, Naldrett 2004). Once magma escapes from its mantle source and moves upward, both its confining pressure and temperature decrease. Cooling decreases solubility but the reduction of pressure overcompensates it (Mavrogenes and O'Neill 1999).

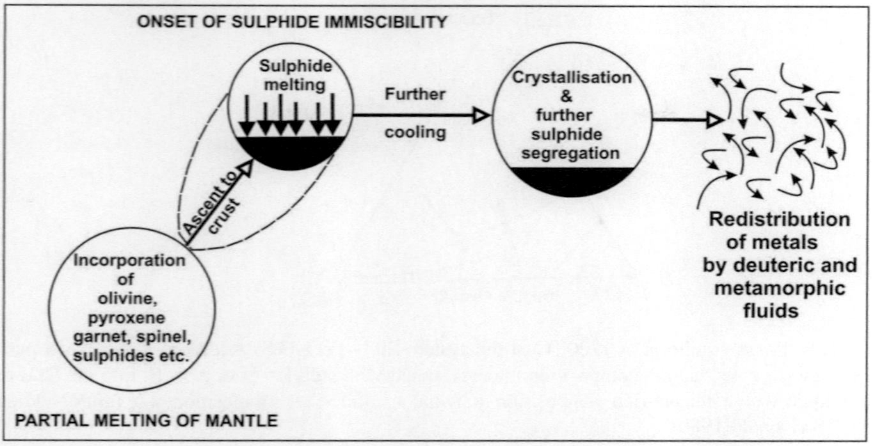

Fig. 2.8 Generation of a mafic magma by partial melting of a portion of the mantle and its emplacement in the crust followed by ore deposition by liquid immiscibility, segregation, and limited redistribution (after Naldrett 1989)

Ultimately, the mafic-ultramafic magma produced becomes more sulfide undersaturated as it rises upward (Lesher and Groves 1986). Normally, this situation will not lead to sulfide accumulation into an ore deposit, unless the magma is close to sulfide saturation and it quickly enters the lithosphere and begins to cool. Obviously, it is an uncommon process and sulfide ore [Ni–Cu–(PGE)] deposits form under specific circumstances.

Volume of the ore is disproportionately large compared to its normal solubility in the magmas of the intrusions/host flows in nature. The most realistic process is the assimilation of sulfur (sulfide/sulfate) by the upcoming mantle-derived magma from the wall rocks (sulfur addition) which makes it sulfide saturated. This is supported by field observations as well as stable isotope studies. Inflow of large volume of magma of comparable composition (magma mixing) is also expected to cause sulfide saturation. Assimilation of upper crustal rocks, incorporating SiO_2, Na_2O, K_2O (magma contamination) will lead to the production of sulfide-rich melt (cf. Fig. 2.9). The initial precipitate is a monosulfide solid solution (Fe, Ni, Cu)S, which scavenged Ni and Cu from the magma, and which on sub-solidus breakdown produces the mineral assemblages (pentlandite + pyrrhotite + chalcopyrite) we find in nature.

Magmas most likely to form sulfide deposits are tholeiitic picrites, because the degree of melting that leads to their formation is sufficient to remove all of the sulfides at the source and yet make it not too undersaturated in sulfides when they reach the crust (Mungall 1999). Sulfide saturation will thus not need much S-assimilation from the wall rocks.

Komatiitic magmas are less qualified to form magmatic Ni–Cu(–PGE) deposits because they would be very much undersaturated in sulfides at crustal levels.

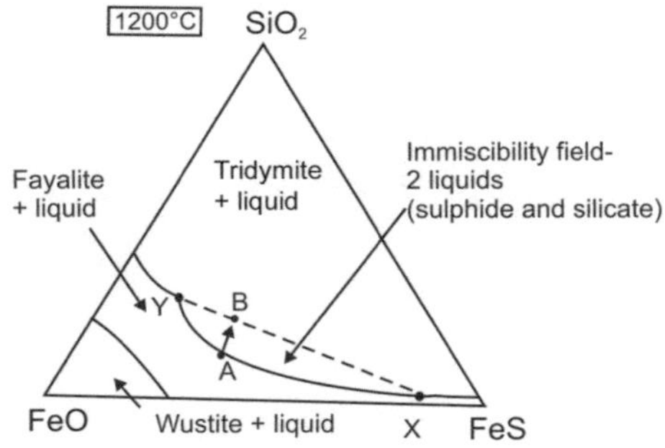

Fig. 2.9 Phase equilibria at 1200 °C in the system SiO_2–FeO–FeS. Addition of silica will push the homogeneous magma composition undersaturated with sulfides from A to B, into the field of two liquid with a silicate-rich composition at Y and a sulfide-rich composition at X (after Naldrett and MacDonald 1980)

However, komatiitic magmas are more capable of melting wall rocks, including sulfur, at appropriate situations (Lesher 1989). This explains the origin of a number of komatiite–hosted Ni–Cu(–PGE) deposits around the world, including Kambalda in Western Australia.

The Sudbury Ni–Cu deposits in Ontario, Canada (Fig. 2.3a, b) are genetically unique in that they are believed to have little or no mantle contribution. They were derived from remelted crustal rocks, remelting triggered by a meteorite impact. The compressed lopolithic intrusion/structure now consists of norite to granophyres, with hugely brecciated footwall rocks injected by sulfidic material. There is a zonal mineral distribution of pyrrhotite + pentlandite + chalcopyrite at proximal location and low-temperature chalcopyrite + bornite at distal. The latter contains highest PGE grades (Barnes and Lightfoot 2005; Cawthorn et al. 2005). This is due to hydrothermal overprinting of the magmatic ores.

Liquid immiscibility is a phenomenon that in ore geology is commonly attributed to sulfidic ores as outlined above. However, Philpotts (1967) reportedly produced two immiscible melts experimentally: one on cooling produced magnetite and apatite (2:1) and the other, a rock of dioritic composition. Nashlund et al. (2002) reported to have obtained such magnetite ores from El Laco volcano, northern Chile. Sillitoe and Burrows (2002), on the other hand, believe these to have formed hydrothermally.

2.2.2 *Pegmatitic Deposits*

Pegmatite commonly is a coarse grained (≥ 3 cm) igneous rock, with extreme grain sizes of 10 cms or more and a heterogenous texture, occurring in a variety of geological settings. Conventionally, it is a granite analog in composition and hence

Fig. 2.10 Hand specimen photograph of a common mineral assemblage in pegmatite from Bhunas mine, Bhilwara district, Rajasthan (Photo: M. Deb)

consists essentially of quartz and feldspar, with varying amounts of muscovite ± biotite micas and a number of other accessory minerals (Fig. 2.10). In rare cases they may be amphibolitic, gabbroid, calc-silicate, or even sulfidic in composition. The latter being very rare, hereafter, by the word "pegmatite," we will mean granitic pegmatite only.

The pegmatites show concentration of elements, such as, Li, F, B, Cs, Ta, Nb, etc., which are not found in abundance in the crust. Some pegmatites are thus hosts to the following economic and strategic minerals (both metallic and industrial minerals) (see Appendix A for mineral compositions):

Alkali feldspars	Tantalite
Muscovite and biotite micas	Columbite
Quartz	Cassiterite
Beryl	Wolframite
Pollucite	Scheelite
Bertrandite	Uraninite
Corundum	Zircon
Gemstones	Monazite
Spodumene	Allanite
Lepidolite	Petalite
Amblygonite	Ilmenite

2.2.2.1 General Characteristics of Pegmatites

Granitic pegmatites are commonly associated with granites and granite gneisses, but are rarely hosted by the parent granite. If they do, they form swarms and network of fracture-filling dykes, the fractures having been produced by cooling or post-consolidation stresses. Granitic pegmatites, which generally cluster near the K_2O end of a $CaO–Na_2O–K_2O$ compositional triangle, are commonly hosted by schists and gneisses (Fig. 2.11). Being commonly syn- or late-orogenic in emplacement they may be foliation-parallel, folded, faulted as well as torn-apart. In size they may be measureable in terms of meters to kilometers, the Greenbushes pegmatites of Western Australia providing an example of the latter (Partington et al. 1995) (Fig. 2.12). Wall rock metasomatism is absent to weak around most pegmatites (Cerný et al. 2005).

Mineral zoning, as mentioned above, is common in pegmatites, but not invariably present. The ones with zoning are termed complex pegmatites while the others without are called simple pegmatites. The number and types of minerals seen in the border or wall zones gradually decrease inward, referred to as "core" or "core zones". The wall zone typically consists of quartz–plagioclase–microcline–muscovite–biotite–garnet–tourmaline(–beryl-apatite). This may be followed by a zone of microcline, ending up with coarse grained quartz at the core. Rare metal

Fig. 2.11 A view of the Bhunas pegmatite mine within Bhilwara gneisses and schists, Rajasthan (Photo: M. Deb)

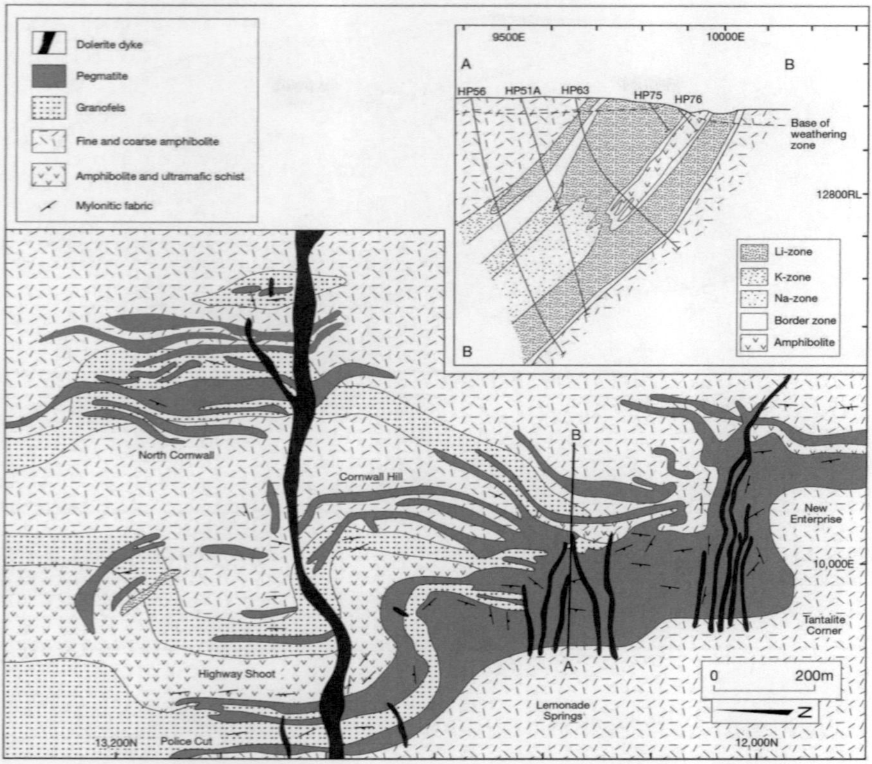

Fig. 2.12 Geological map and cross section of Greenbushes pegmatite showing zonation (after Partington et al. 1995)

concentration may take place at or off the core (Fig. 2.13). Zoning of pegmatites on regional scales, in terms of their elemental composition and mineralogy, may be present around almost barren/barren granite pluton (Fig. 2.14).

2.2.2.2 Classification of Pegmatites

Because of varieties of granitic pegmatites in nature it has been felt necessary to classify or group them according to certain distinctive criteria. Cerný and Ercit (2005) and Cerný et al. (2005) provided recent updates in this respect (Table 2.2). Notably, rare element granitic pegmatites, host to a large proportion of rare metals and rare earth elements (RM and REE), are subdivided into two principal families, LCT and NYF (Table 2.3). Of these the LCT pegmatites are most important in terms of diversity, tonnage, and relationship to the associated rocks.

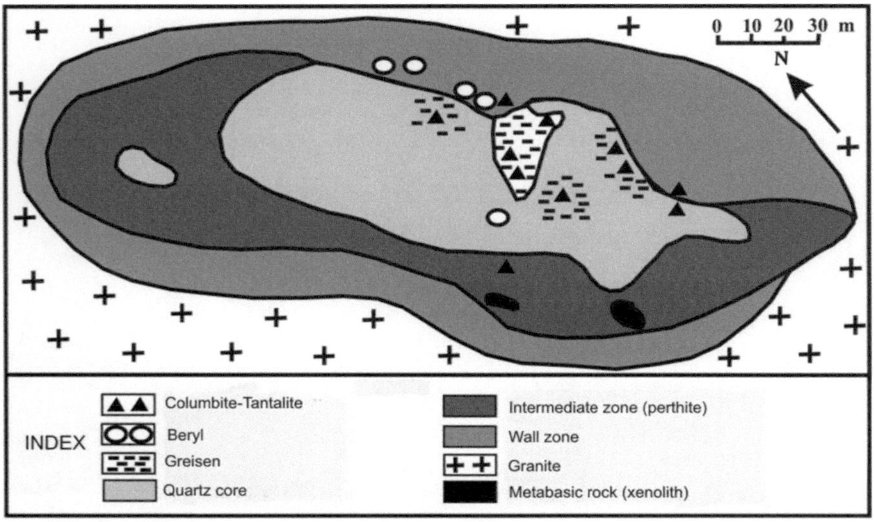

Fig. 2.13 Geological map of the Metapal zoned pegmatite, Bastar-Malkangiri pegmatite belt, Chhattisgarh (after Ramesh Babu 1999)

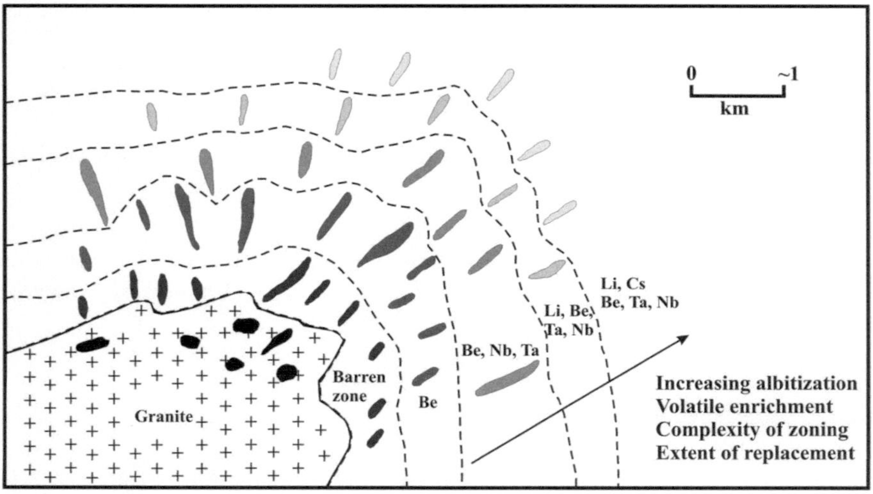

Fig. 2.14 Schematic representation of the regional zonation of pegmatites around a granite intrusion (after Trueman and Cerny' 1982)

2.2.2.3 Origin of Pegmatites and Pegmatitic Deposits

That granitic pegmatites form from plutonic granitic intrusions is proved, inter alia, by a number of observations, including (i) granite–pegmatite suites, intruded into

Table 2.2 Major classes of granitic pegmatites (after Cerny' et al. 2005)

Class	Typical minor elements	Metamorphism of host rocks	Relationship to granites	Structural features wrt host rock foliation
Abyssal	U, Th, Zr, Nb, Ti, Y, REE, Mo. Rarely Be, B. Poor to moderate mineralization	Upper amphibolite to granulite facies (~400–900 MPa/~700–800 °C)	None (?) (segregation of anatectic leucosomes?)	Conformable to mobilized cross-cutting veins
Quartzo-feldspathic	Metallic minerals absent. Micas and ceramic minerals determine value	High pressure Barrovian amphibolite facies (ky-sill) (500–800 MPa/520–650 °C)	None (anatectic bodies) to marginal and exterior	Quasi-conformable to cross-cutting
Muscovite-rare element	Li, Be, Y, REE, Ti, U, Th, Nb > Ta. Rarely Li, Be. Poor mineralization	Moderate to high-P, amphibolite facies 300–700 MPa/520–650 °C)	Interior to exterior poorly defined	Quasi-conformable to cross-cutting
Rare element	Li, Rb, Cs, Be, Ga, Sn, Hf, Nb–Ta, B, P, F, or Be, Y, REE, U, Th, Nb > Ta, F. Poor to abundant mineralization. Gemstones, industrial minerals (andalusite-sillimanite)	Low-P Abukuma-type amphibolites to upper greenschist facies (andalusite-sillimanite) (200–400 MPa/500–650 °C)	Exterior, some interior to marginal	Quasi-conformable to cross-cutting
Mariolitic	Li, Be, B, F, Ta > Nb or Be, Y, REE, Ti, U, Th, Zn, Nb > Ta, F. Poor mineralization. Gemstock	Shallow to subvolcanic (~100–200 MPa)	Interior to marginal	Interior pods, and cross-cutting dikes

Table 2.3 Subdivisions in the rare element class of granite pegmatites (after Cerný et al. 2005)

Family	Geochemical signature	Bulk granite composition	Pegmatite types
LCT	Li, Rb, Cs, Be, Sn, Ga, Ta > Nb (BPF)	Peraluminous, S, I or mixed S + I types	Beryl, complex albite spodumene, albite, elbaite
NYF	Nb > Ta, Ti, Y, Sc, REE, Zr, U, Th, F	Peraluminous to sub-aluminous and metaaluminous; A and I types	Rare earth

N.B. There is another subclass or family in this class called "Mixed Family," which, as the name suggests, have characteristics intermediate between LCT and NYF. As pointed out by Cerny et al. (2005), all the granite phases are weakly to strongly peraluminous and show prominent enrichment in LCT elements that increase in the late and more pegmatitic facies

low-P greenschist to amphibolites facies country rocks; (ii) highly evolved peripheral pegmatite dikes, locally physically linked to the interior or marginal parts of granitic plutons; (iii) rare element pegmatites form aureoles of marginal and exterior dikes, surrounding the granite intrusions; (iv) continuity of geochemical signatures in numerous cases of granite–pegmatite suites. No less important is experimental evidence that the liquidus and solidus of pegmatite forming melts decrease with the increasing contents of Li, Rb, Cs, F, B, and H_2O (Cerný 1991). There is also a decrease in density with depolymerization of the melt, which helps the melt to be more mobile and rise. There is also a decrease in the number of nucleation sites of crystals and an increase in diffusion which helps large crystals to grow.

It was the belief of many geologists until recently that pegmatites crystallize at temperatures of the water-rich granite solidus near 650–700 °C, that is, are products of an intermediate stage of evolution from orthomagmatic to hydrothermal (or pneumatolytic) stage of magmatic evolution. However, it does not agree with the P-T estimates derived from mineral assemblages, isotopic estimates, or fluid inclusion studies. In the latter studies, the crystallization temperatures vary in the range of 450–250 °C, which is well below the solidus temperatures of hydrous granite melts. The most plausible explanation of this phenomenon is the presence of fluxing components, H_2O, B, F, and P in the pegmatitic magma wherein, as a result, the melting and crystallization temperatures are reduced (London 1997). The fluxing component needed is small in quantity. For example, the Tanco pegmatite in Manitoba, Canada, which may be the most fractionated igneous body on earth, contains <2 wt% of (Be_2O_3 + P_2O_5 + F) (Cerný 1991; London 1995). The upper thermal stability of Li-aluminosilicates in the presence of quartz, however, put the upper crystallization temperature of Li-pegmatites at ~700 °C. There was a common belief until recently that the pegmatites cooled slowly. Recent experimental studies suggest that it possibly happened otherwise. The large grain sizes are in fact the effect of fluxes. Crystallization at conditions far below the liquidus (i.e., crystal-melt equilibrium) generates a sequence of crystallization that matches the natural zoning patterns exactly (London 2005).

2.3 Hydrothermal Processes of Ore Formation

A. Introduction

Etymologically, the word "hydrothermal", long in use in ore geology, means "hot aqueous". So, hydrothermal mineralization means "mineralization from a hot aqueous solution." But, how hot? Any temperature between a few tens of centigrade to below solidification of granitic magmas, or beginning of low-T granulite metamorphism set in a volcano-sedimentary sequence. Ambient pressure may be anywhere between a few bars to a maximum of a few kilobars (3 kb at ~10 km depth).

Lindgren (1933), Emmons (1936), Jensen and Bateman (1981), Guilbert and Park (1986), Robb, (2005) and others have discussed about hydrothermal ore deposits in fair details and divided them initially into three broad divisions, namely, hypothermal, mesothermal and epithermal. Their characteristic T-P fields, as suggested, are:

Epithermal: 50–250 °C, <0.5 kb (<1500 m depth)

Mesothermal: 250–500 °C, 0.5–1.5 kb (1500–4500 m depth)

Hypothermal: 500–600 °C, >1.5 kb (>4500 m depth)

It may be pointed out that these P-T values are only suggestive rather than prescriptive. More restricted attributes will not be realistic in geology. For example, Hedenquist et al. (2000) from a study of the modern analog of epithermal mineralization at Tampo, New Zealand, suggest a temperature range of 160–270 °C and pressure equivalents of 50–1000 m. The crustal environments of these three major classes of hydrothermal deposits are shown in Figs. 2.15 and 2.16, with reference to Au mineralization. Not that all the above subclasses of hydrothermal mineralization would always be found in one deposit/ore district in such orderly disposition. It will depend on the original P-T of the ore fluids, their migration and cooling, and of course, later erosion and superimposition.

There was a time when most ore deposits, particularly those formed as veins or irregular bodies and accompanied by a granitic intrusive in the neighborhood, were thought to have formed from fluids given out by the cooling granitic magma body. If the intrusive was not detectable, it was still imagined to be there. Today, ore geology, like most other branches of Earth Sciences, has become more precise. Now we know that hydrothermal fluids for ore deposition may be derived from a cooling and depressurizing granitic magma, as well as from the fluids given out by a volcano-sedimentary pile due to an overpressured uplifted region, rock compaction in an orogenic belt, thermal perturbations and metamorphism (Skinner 1979; Robb 2005), or a combination of these conditions. To the latter group are included the meteoric or connate water getting heated by geothermal heat, or the heat radiated by

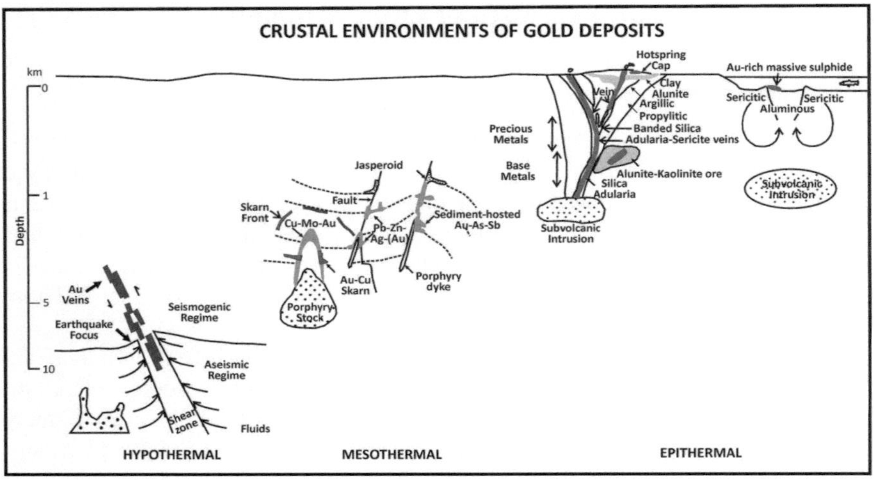

Fig. 2.15 Schematic models of three different crustal environments of gold deposits formed by hydrothermal fluids. In the hypothermal environment, a steep shear zone transects the boundary between seismogenic and aseismogenic crust and controls the fluid movement. For the mesothermal environment, porphyry Cu–Mo–Au, Au skarn and distal "Carlin type" Au–As–Sb mineralizations are shown. In the epithermal environment, the relative positions of subareal hot-spring mineralization and the deeper epithermal veins (cf. Fig. 2.16 for more details) are shown. In the shallow marine exhalative environment, Au-rich volcanogenic massive sulfides are depicted (after Poulsen 1995)

Fig. 2.16 Geological model of hot-spring low sulfidation epithermal gold deposits of McLaughlin mine, California, USA, showing zones of alteration (after Sillitoe 1995)

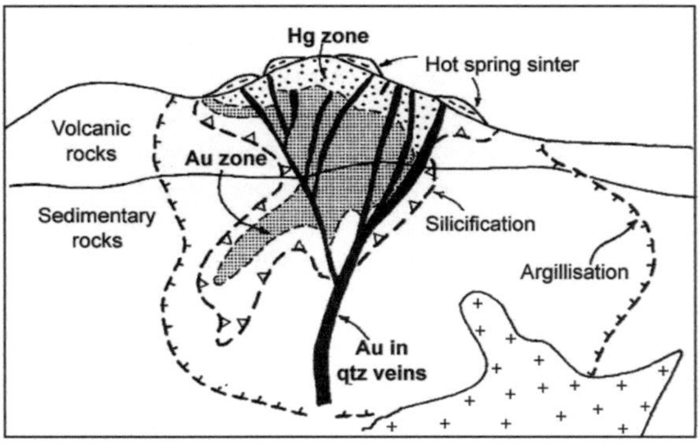

an igneous intrusive body. However, this external contribution need not necessarily be overemphasized (Sillitoe 2010). Hydrothermal mineral deposits can thus be divided into two main categories (Fig. 2.17):

1. Essentially magmatogenous, or magmatic hydrothermal
2. Essentially non-magmatogenous, or amagmatic hydrothermal.

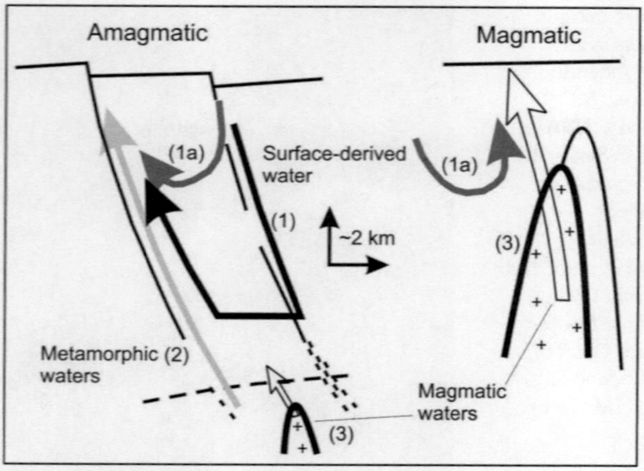

Fig. 2.17 Conceptual models of magmatic and amagmatic fluids in hydrothermal deposits. (*1*) shows surface-derived water (shallow marked *1a*) descending down normal faults toward the brittle-ductile transition zone and returning to the surface to form a deposit; (*2*) shows metamorphic fluids rising along faults to form a deposit; small arrow with (*3*) shows possible subordinate magmatic input to the amagmatic model; (*3*) shows magmatic–hydrothermal ore-forming system related to multiple intrusions (after Seedorf and Barton 2004)

Besides, based on whether the hydrothermal fluid, whatever its source, forms a deposit within the crust at depth or exhales on to the surface, commonly on the sea floor, and forms a mineralized mound, hydrothermal deposits are further categorized into two divisions respectively:

(i) Intracrustal deposits
(ii) Exhalative deposits.

Some aspects of the above outline will be elaborated in the following few pages.

B. **Structures of hydrothermal deposits**

Hydrothermal deposits may assume different forms depending on the dilatant structures in rocks they occupy by fracture-filling and crystallization [such as, in saddle reefs (Fig. 1.9b), fissure veins, ladder veins (Fig. 1.9c), stockwork (Figs. 1.9a and 2.18a, b), breccias-filling (Figs. 1.13 and 2.18c), or by replacement as in shear zones. Vein filling is by growth of new crystals from the walls to the core of the vein. It is open space filling and hence is emplaced at shallow levels, compared to those developed by replacement. Veins may show pinch and swell structures (Fig. 1.9c) and temporally may be emplaced more than once. They may be intersecting or sub-parallel (Fig. 2.19b). The veins may extend for several km along strike (such as the Mother Lode vein system in California) and about a km down the dip. In replacement, new mineral phases grow by reaction. One may be superimposed on the other. In case of regional scale dislocations, second-and third-order structures are generally better locales for ore deposition as these are sites of lower pressure. Ore

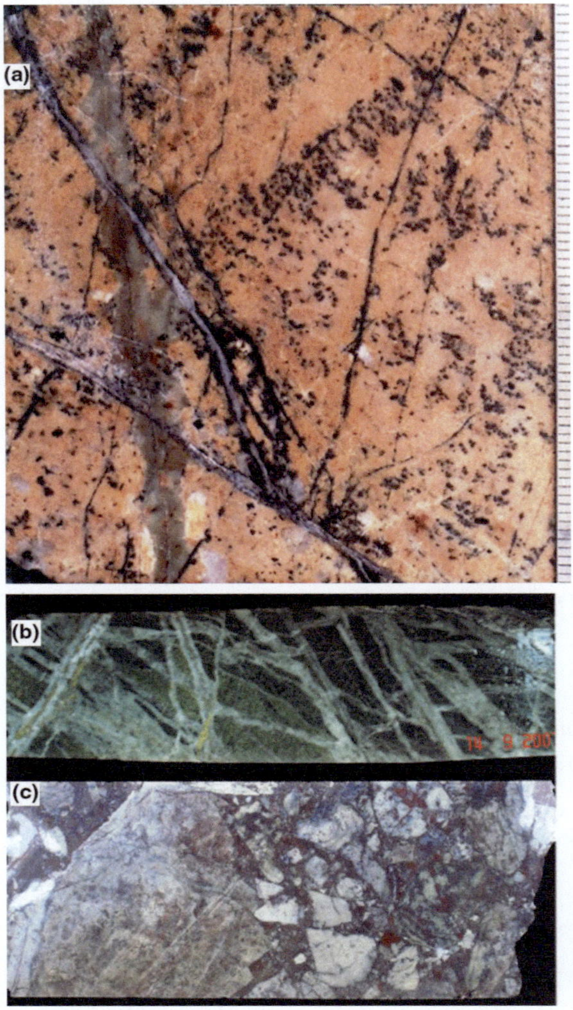

Fig. 2.18 **a** Cross-cutting stockwork veinlets with magnetite ± molybdenite in Keystone intrusion, Mt. Emmons, Colorado, USA (Slab gifted to SC Sarkar by Holly Stein; slab width 10 cm). **b** Silicified stockwork with chalcopyrite mineralization, drill core, KH prospect, Mongolia (Photo: M. Deb). **c** Breccia filling by bornite, drill core, Oyi Tolgoi Cu–Au porphyry deposit, Mongolia (Photo: M. Deb)

controlling structures are either penecontemporaneous with ore mineralization, or are older structures that reopened during mineralization.

Some spectacular examples of Sn–W-bearing greisenized sheeted vein system are seen in parts of Cornwall, SW England (Fig. 2.19a, b, c) and of Cu in northern Chile. The veins are normally controlled by steep to moderately dipping faults that transect granitic intrusions and their wall rocks (Fig. 2.19c). These veins may accompany other intrusion-related deposit types but the larger ones generally occur alone.

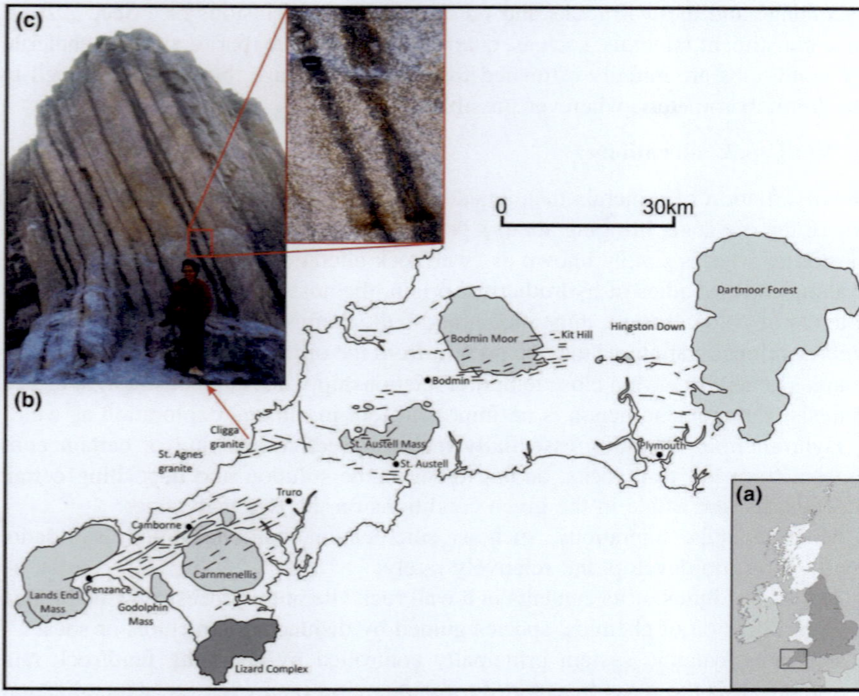

Fig. 2.19 **a** Index map of United Kingdom showing the location of Cornwall. **b** Map of Cornwall granites and their relationship with Sn–W vein system (after Inst Mining and Metall 1985). **c** Outcrop of greisenised granite at Cligga Head, St. Agnes district, western Cornwall, cut by a sheeted vein system comprising greisens-bordered veins between which the granite shows incipient greisenisation. Close up shows tourmaline-rich margins of the veins carrying wolframite-cassiterite mineralization associated with sulfides like stannite, sphalerite, and chalcopyrite (Photo: M. Deb)

C. **Common minerals of hydrothermal origin**

Ore minerals are hardly free from gangue minerals in ores. This is particularly true of hydrothermal ores. A list of common ore and gangue minerals in these ores is given below.

Important ore minerals	Common gangue minerals
Chalcopyrite, pyrite, pyrrhotite, molybdenite, galena, sphalerite, enargite, huebnerite, scheelite, cassiterite, gold, acanthite, uraninite, cinnabar, columbite, tantalite, stibnite	Quartz, calcite, sericite/muscovite, biotite, epidote, chlorite, alkali feldspars, barite, fluorite, chalcedony, carbonate, clay minerals, alunite, tourmaline

A hydrothermal mineralization is identified as such by its P-T conditions of formation within the hydrothermal range, presence of hydrous minerals in the assemblage and the wall rocks and presence of fluid inclusions (see Sect. 2.2e) in some constituent minerals, such as, quartz, calcite, fluorite, barite, pyrite, sphalerite. P-T conditions are initially estimated from the assemblage, but later confirmed by geo-thermobarometers, wherever possible.

D. **Wall rock alterations**

The distribution of minerals mentioned under "gangue" may be spread into either side of the ore zone, bringing about a petrochemical change in the wall rocks. This constitutes what is simply known as "wall rock alterations" (cf. Fig. 2.16). Around or alongside orebodies of hydrothermal origin, the host rocks are variously affected in terms of color, texture, mineralogy, and bulk chemistry due to these alterations. Well-developed variations may be present from the ore zone outwards. Such spatial changes generally have a close temporal relationship with ore deposition. A critical analysis of this phenomenon is an important tool in mineral exploration as well.

Hydrothermal alteration essentially means, selective leaching of certain components from the wall rocks, adding them to the solution and depositing certain minerals that are stable in the given conditions on the two wall zones.

Monomineralic alterations, such as silicification, carbonatization, sulfidation (pyritization) do develop, but relatively rarely.

Transfer of fluids or its contents in a wall rock alteration process may be by fluid flow, or diffusion of chemical species guided by disjunctive structures or shears. It is sort of a dynamic system principally controlled by changing fluid/rock ratio (Reed 1997). If the mass transport during alteration (and even ore mineralization) takes place on a larger scale moving through large distances, the process is commonly known as infiltration. It is generally unidirectional. In virtually all cases of ore deposition, both infiltration and diffusion are operative at the same time. Wall rock alterations may be polychronous, being pre-, syn- and post- ore mineralization.

A good example of hydrothermal alteration can be seen in the Hutti gold mine in Karnataka. Mineralization is confined to "laminated" quartz veins in discrete shears hosted by metabasalts and minor meta-dacites. Wall rock alteration is characterized by narrow but distinct proximal biotite-K-feldspar zones and distal chlorite zones that are symmetrically distributed around the quartz veins of varying widths (Fig. 2.20). Pal and Mishra (2002) record the alteration reactions in the ore zone as follows:

(i) Formation of chlorite and calcite by carbonation-hydration reactions of amphiboles in the distal chlorite zones;
(ii) Sulfidation of Fe-amphiboles to form pyrrhotite.
(iii) Formation of biotite from chlorite.
(iv) Sulfidation of biotite to form pyrite and gold in the proximal zone.

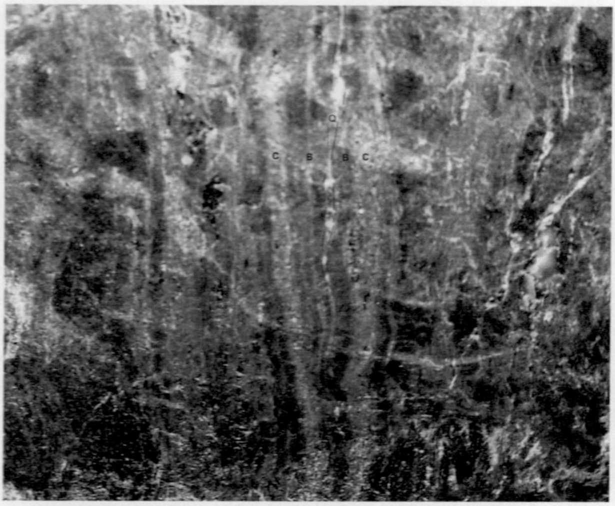

Fig. 2.20 Symmetrical zones of wall rock alteration around a thin quartz vein (*Q*) in Hutti gold mine, Karnataka. Hydrothermally altered meta-basalt host rock shows biotite-rich (*B*) and chlorite-rich (*C*) zones, which display repetition and overlap due to emplacement of successive quartz veins (Photo: M. Deb)

E. Fluid Inclusions

Fluid inclusions (FI), tiny capsules of liquid, gas, or melt, trapped in minerals, provide the most direct clue to the nature of fluids associated with different geological processes, including ore mineralization by hydrothermal processes. Microthermometry is the most widespread technique of FI investigations commonly confined to transparent gangue minerals, such as, quartz, calcite, fluorite, barite, apatite, dolomite, topaz, beryl, etc. associated with the ore. Translucent ore minerals like sphalerite and cassiterite can also be studied. Recent development of Infrared microscopy allows FI studies to be carried out in opaque minerals such as, pyrite, as well. The textural relationships amongst the mineral phases in a paragenetic association[1] must be explained as logically as possible before enough meaningful interpretations can be drawn from FI studies. Most commonly retrieved information from fluid inclusion microthermometry includes temperature, pressure, fluid composition (including salinity), and density of the fluids. Techniques other than microthermometry involve chemical and isotopic analyses of the extracted inclusion fluids. The subject is somewhat elaborated below.

[1]Paragenetic association: a set of genetically related minerals.
Paragenetic sequence in a mineral association connotes sequence of mineral deposition in an association of minerals, where the constituent phases bear evidence of being genetically related.

The inclusions may vary in size from a few microns, commonly about 0.1 mm, to much larger ones and may be trapped during the growth of the mineral (i.e., primary) or during later deformation and recrystallization (i.e., secondary) in post-crystalline fractures. Pseudo-secondary inclusions are quite similar to secondary but fracturing and healing takes place before the growth is complete (Fig. 2.21a). The inclusions may contain different combinations of phases in them, which when identified under the petrographic microscope, provides an important nongenetic classification. By far the most common is a low viscosity liquid and a gas or vapor bubble occupying a volume less than that of the liquid. The inclusions may also contain a liquid + solid, liquid + gas, and liquid + liquid (much rarer). CO_2 is often present in inclusions as both liquid (CO_2-L) and gas (CO_2-G). If the liquid phase has a soluble salt and becomes supersaturated with it, the excess salt may crystallize out as a mineral grain, such as, halite (NaCl) or sylvite (KCl) (Fig. 2.21b). Sometimes, some unrelated pre-existing minerals are trapped in the inclusions which are known as "captive minerals."

It is generally assumed that when the liquid was trapped in the solid mineral, the volume of the liquid filled the total volume of the inclusion cavity. Later, the gas bubble was formed as a result of differential shrinkage of the liquid and the host mineral from the higher temperature of formation to the temperature of observation, i.e., indicating the post-entrapment changes (Fig. 2.22). Therefore, a temperature close to that of trapping (T_t) can be estimated by heating the mineral sample with the inclusion to the point at which the gas bubble disappears completely, this being the temperature of homogenization (T_h). The obtained temperature marks the lower limit for the temperature of formation of the mineral. If after homogenization the inclusion is heated further, a temperature will be obtained when the inclusion cavity bursts or decrepitates. This temperature of decrepitation implies that the host mineral could not have formed at a temperature above the decrepitation condition,

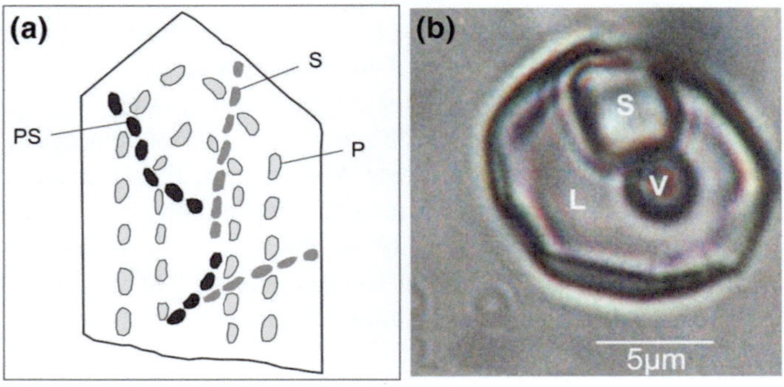

Fig. 2.21 **a** Primary (*P*), secondary (*S*) and pseudo-secondary (*PS*) fluid inclusions (after Shepherd et al. 1985). **b** A primary fluid inclusion in Bhukia gold ore prospect, South Rajasthan. Note the liquid (*L*), vapor (*V*), and daughter crystal (*S*), probably halite, in the inclusion

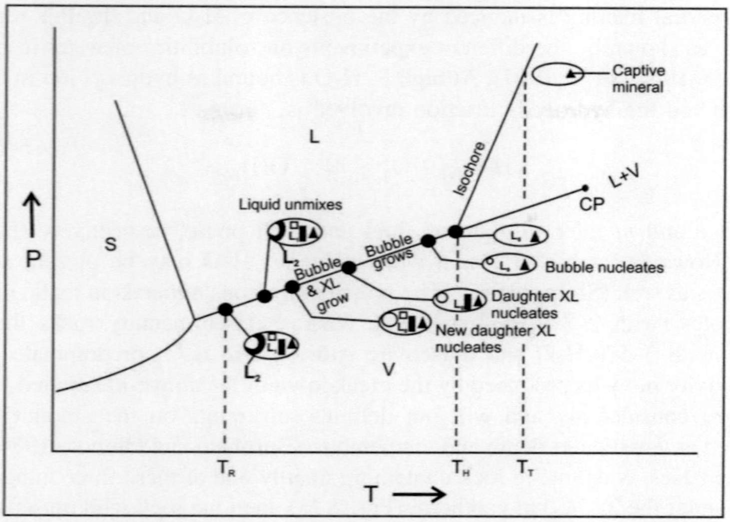

Fig. 2.22 Post-entrapment changes in fluid inclusions T_T = Temperature of trapping; T_H = Temperature of homogenization; T_R = Room temperature. CP = Critical point (after Mookherjee 1999)

i.e., the maximum temperature of formation. Detailed discussion on various aspects of FI studies is available in Roedder (1984) and Shepherd et al. (1985), amongst others.

2.3.1 *Essentially Magmatic Hydrothermal Processes*

The formation of several hydrothermal ore deposits is dependant largely on fluids of magmatic origin and processes that operate during the late stages of felsic magmatism. We discuss such processes and deposits in this present section.

2.3.1.1 Ore-Forming Fluids in the Hydrothermal Processes

The magmatic hydrothermal process leading to ore formation commences with the generation of magmas by partial melting of subducting mafic oceanic crust or anatexis of amphibolites and/or felsic metasedimentary rocks in the lower continental crust. Depending on the source rock composition, mafic or felsic, the partial melting process concentrates different suites of ore elements in the melt produced, as we shall see in somewhat more details shortly. Water plays a crucial role in all stages of the process from putting constraints on melt generation to dissolving in the melts of different compositions at different proportions and finally, separating from the melt with the ore elements to form the hydrothermal ore deposit.

The partial melting is induced by the presence of H_2O and the P-T regime of melting, as shown by the different experiments on solubilities of water in magmas (Burnham 1967, 1979, 1997). At high P, H_2O is bound as hydroxyl ion in hydrous minerals and the hydrolysis reaction involved is:

$$H_2O_{(v)} + O^-_{2(m)} \Leftrightarrow 2(OH)^-_{(m)}$$

Here v and m refer to aqueous fluid and melt phase, respectively (Burnham 1979). However, at higher P and water contents, H_2O may be present as water molecules as well (Stolper 1982). The principal hydrous minerals in mafic rocks are amphiboles (with 2–3% H_2O) while in felsic metasedimentary rocks the micas (biotite with 3–5% H_2O and muscovite with 8–10% H_2O) predominate. Hence, H_2O-activity in melts produced by the breakdown of the above-mentioned minerals will vary considerably and will put definite constraints on the amount of melt produced at a particular depth and temperature (Burnham and Ohmoto 1980). If we consider cases, with source rock containing mainly one of these three minerals and buried under the 25 °C/km geotherm (Fig. 2.23), then the rock with muscovite will start melting at A with the H_2O-content of the neomagma being 7.4 wt% H_2O, the rock with biotite at B with the neomagma having 3.3 wt% H_2O and the rock with amphibole at C with only 2.7 wt% H_2O in the neo-magma. Note that the P is higher and the H_2O content is decreasing from A to C. It is expected that these magmas produced will rise upward in the crust until they intersect the water-saturated granitic or basaltic solidus at A′, B′, or C′ (Fig. 2.23), respectively, when they will get emplaced and solidify. It also follows that the amphibole-sourced melt will be

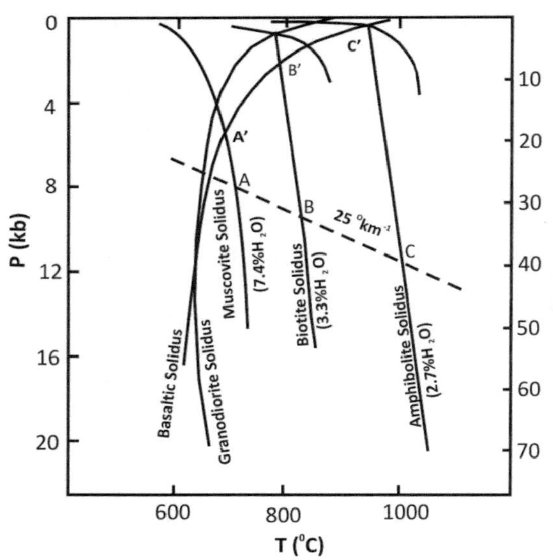

Fig. 2.23 P-T diagram showing the approximate conditions where dehydration melting of amphibole-rich (*A*), biotite (*B*), and muscovite (*C*)-bearing protoliths would occur along a 25 °C/km geothermal gradient and the expected levels of the earth's crust to which melt fractions would rise adiabatically, as a function of the water-saturated solidus of granodiorite and basalt (after Burnham 1997)

emplaced at the shallowest level in the crust. Such magma may incidentally be H_2O-undersaturated, inviting H_2O-vapor from the heated adjacent country rocks.

While tracing the role of water further, experimental studies confirmed that a granite melt will dissolve more water than a basaltic one (Fig. 2.24). The solubility of H_2O in silicate magmas of different compositions is determined mainly by pressure and to a lesser extent by temperature. Thus, a granitic magma at the base of the crust (~30 km and ~10 kb) should be able to dissolve around 15% H_2O. A plausible explanation of the varying solubility of H_2O is found in the differential polymerization of granitic and basaltic magmas. Granitic melts have a viscosity of the order of 10^8–10^{10} poise at 1000 °C and 1 atm. An addition of 6.4 wt% water would, however, lower the viscosity by about 10^5poise (Burnham 1967). Added water will interact with the melt in a way that breaks oxygen bridges and depolymerizes it. Basaltic magma, on the other hand, accommodates fewer $(OH)^-$ group in O^{2-} sites. That is why basic magmas on cooling produce less hydrothermal solutions compared to the granitic magmas.

As the neo-magmas formed at different crustal levels move up to different depths from the surface, they would exsolve the dissolved aqueous fluid due to depressurization. The I-type granitoid magmas are generated in the deeper parts of the lithosphere near subduction zones at higher temperatures and low water contents with contributions from mantle-derived mafic melts. When they move up to the shallow levels, they will exsolve the magmatic fluid by boiling and hydrofracturing and induce vigorous circulation of hydrothermal solutions of different derivations in the endo- and exo-contacts. This environment will encourage the formation of porphyry Cu deposits and epithermal Au–Ag deposits. In contrast to this scenario

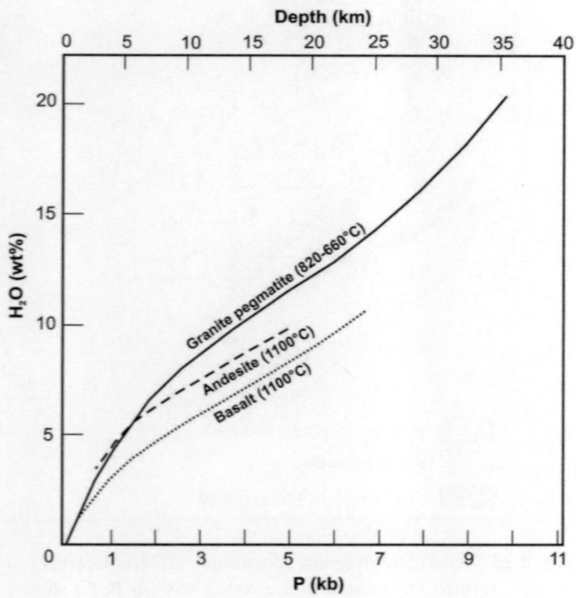

Fig. 2.24 Experimentally determined solubilities of H_2O (in wt%) in silicate melts (of basalt, andesite, and granite/pegmatite) as a function of pressure (after Burnham 1979)

(Fig. 2.25), S-type granites are generated in lower to mid-crustal levels by anatexis of metasedimentary lithologies. They will form at lower P-T conditions but will contain significant amount of dissolved H_2O due to their source material. They will be emplaced not too far from their source at mid-crustal levels (4–5 kb) and are often barren. Only if fractionation occurs substantially, such magmas may produce pegmatitic bodies, or hydrothermal Sn+/W+/U mineralization through their residual melts (Fig. 2.25).

How about the tenure of a hydrothermal system? Mineral calculations show that convective circulation that produce near-surface hydrothermal systems are expected to cool their intrusive heat source in a period of time not exceeding a few tens of thousands of years, large though the intrusion may be. Long-lived ($\sim$10 My) hydrothermal systems are likely to be the result of repeated intrusions (Cathles et al. 1997). Petrochemical alterations and ore mineralization were confined to amagmatic intervals between porphyry intrusions. All porphyry intrusions take place in the temperature range of 400–800 °C. Alteration ages, however record a much younger age corresponding to 250–400 °C, when the system cooled through the blocking temperature of biotite, K-feldspar, and sericite (Sillitoe and Mortensen 2010).

Magmatic hydrothermal ore fluids may be brine or vapor or a combination of both, depending on the specifics of the situation. Fluid inclusion studies indicate that partitioning of base and precious metals between magmatic vapor and

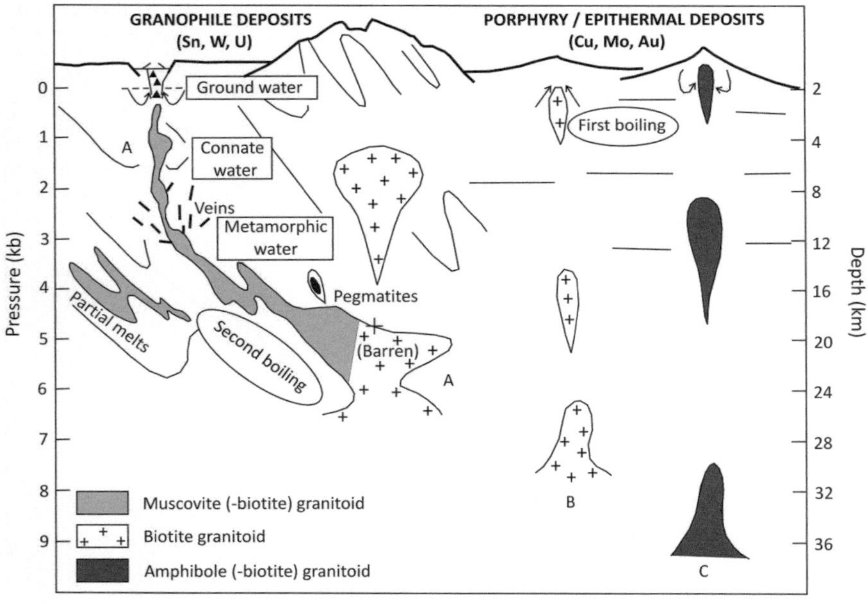

Fig. 2.25 Schematic diagram illustrating the emplacement style and metallogenic character of granites formed under each of the conditions (*A*, *B*, *C*) shown in Fig. 2.23 (after Strong 1988)

hypersaline brine may be substantial and highly element specific, apparently due to contrasting metal complexing properties of the two phases. Cu, and by inference, Au complexing with S partition into vapour and the other heavy metals complexing with chloride partition into the coexisting brine (Heinrich et al. 1992; William-Jones and Heinrich 2005). Cu in the presence of high Cl^- concentration will however behave differently. As defined, the solubility of a mineral is the upper limit of the dissolved metal contained in the mineral that a hydrothermal fluid can transport, assuming thermodynamic equilibrium. However, solubilities of common ore minerals as simple ions have been known to be insignificant for a long time. To attain solubilities necessary for ore deposit formation ore metal ions must form complexes. A *complex* ion is a combination of one or more metal ions (including protons, i.e., H^+) with one or more anionic or neutral species. The latter is called *ligand*. The most important ligands are Cl^-, HS^-, or H_2S and OH^- (Czamanske 1959; Barnes 1979; Wood and Sampson 1998). Several factors, such as fall of temperature, fluid mixing (dilution), and "secondary boiling", reaction with wall rocks play important role in precipitation of ore minerals from hydrothermal fluids. Pressure variation and change of oxidation have conditional effects.

Fall of temperature is generally effective in destabilizing metal ligands, particularly, the chloride complexes. Mixing of two fluids, one being the ore fluid and the other the groundwater may precipitate ores due to two reasons: dilution of the ore fluid will break the complex after a certain value and if the adulterant fluid is relatively cool, which often it is, mixing will reduce the temperature of the mixed fluid and help in ore precipitation. Reduction of fluid pressure along fractures at depth may cause "secondary boiling" and effervescence, giving out acidic components. This will raise the pH of the solution, cool it faster and cause ore mineral precipitation. This is particularly effective in "crack-seal" systems. Oxidation reduces the solubilities of metal sulfide complexes, depositing ore minerals. Examples:

$$PbCl_4^{2-} + H_2S^o = PbS\,(s) + 2H^+ + 4Cl^-$$

$$Zn(HS)_3^- + 4O_{2(aq)} \leftrightarrow ZnS(s) + 3H^+ + 2SO_4^{2-}$$

The first reaction will move from left to right, i.e., precipitating PbS (galena) by the satisfaction of one or more of the following conditions: (i) if there is a loss of temperature (assuming the mineral has prograde solubility), (ii) there is an increase in the activity of H_2S^o, (iii) there is a decrease in the activity of H^+ (increase in pH), (iv) decrease in the activity of Cl^-. In the second, ZnS (sphalerite) precipitates when the complex $Zn(HS)^{3-}$ is oxidized. There are many more plausible reactions in which metal complex breaks down to ore minerals. More discussion on this is beyond the scope of this text.

2.3.1.2 Granitoids and Metal Deposits

Granitic rocks spanning a wide range of compositions in terms of silica contents, low to high-K calc-alkaline or alkaline character and varying from metaluminous to peraluminous compositions, are genetically related to a spectrum of precious, chalcophile and lithophile metal deposits. The metals present are dictated by composition, degree of fractionation, and the redox state of associated intrusions (Fig. 2.26) (Sillitoe 1996). The granites are classified into I-type or S-type which reflects the composition of source regions (Blevin and Chappel 1992) or as magnetite and ilmenite series which point to the redox state of the rocks themselves (Ishihara 1981). All S-type intrusions, peraluminous granites formed by melting of muscovite or (muscovite + biotite)-rich source rocks, essentially belong to the moderately reduced ilmenite series. Sn, W, and U deposits are generally associated with S-type granites. I-type, on the other hand, includes both magnetite (≫)-ilmenite and less commonly ilmenite series granites. This comparatively oxidized calc-alkaline granite magma is related to porphyry copper systems.

A-type granites (Collins et al. 1982) are rather abnormal or atypical in the granite family. Typical A-type granites are the alkali granites of continental rifts. Two different ore associations occur with A-type granitoids: (i) Na-rich granites, containing concentrations of Nb, U, Th, REE, and some Sn and (ii) K-rich granites with profuse hydrothermal silicification and tourmalinization, associated with the mineralization of Sn, W, Pb, Zn, and F. Volcanic equivalents of this granite type include Sn-rich topaz rhyolites in the fields of crustal distension, as in the Tertiary volcanics in Mexico (Pohl 2011). A-type granite-associated mineral deposits are rarely very large.

The metals and associated elements are extracted from the magma by exsolution of an aqueous fluid phase, with the exception of certain rare metals, such as, Ta and Li which are concentrated by direct crystallization from evolved melts (Pollard 1995). Partitioning of metals like Cu into a vapor-rich or hypersaline magmatic fluid from wet magmas with high Cl/H_2O ratios is maximum at shallow crustal depths of 3–4 km (Candela and Piccoli 1995). The metal-rich fluid concentrates in

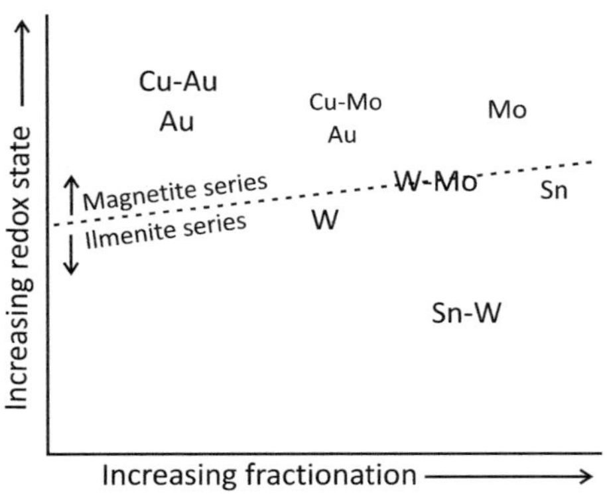

Fig. 2.26 Diagram showing the control of redox state and degree of fractionation of granitic rocks on some selected metal deposits, principal, and subsidiary concentrations differentiated by size of the alphabets (after Blevin and Chappell 1992)

the upper parts of the crystallizing granitic intrusion and is channeled upward through the magma column. It also undergoes admixture with deeply convective groundwater. Metal precipitation takes place in the mixing zone provided by the cylindrical cupola, in the adjacent reactive carbonate horizon and in dilatant fault zones (Fig. 2.27) (Sillitoe 1996). Sulfides (Cu, Mo, Pb, Zn, Ag, Bi, Sb), oxides (Fe, Sn, W), fluorcarbonate (REE) and native metals are thus deposited from the hydrothermal fluid by destabilization of aqueous, most commonly chloride complexes brought about by cooling, reaction with the wall rocks and mixing with circulating groundwater. Sequential precipitation of metals in space (zoning) and in time (paragenetic sequence) takes place due to the physicochemical evolution of the ore fluid and is noted at both deposit and district scales (Evans 1997). In a mixed system Au-mineralization follows Sn-, W-, and Mo-mineralization. Bulk of base metal deposition takes place still later, mostly in an outer zone occupied by metasediments, where reduced sulfur present is likely to help in sulfide precipitation. If Cu is present in the system, it generally precedes deposition of Pb and Zn, as in Cornwall, England. On a district scale Zn–Pb ± Ag occur on the peripheries of Cu, Mo, Sn, W deposits.

There are six broad types of ore deposits generated by intrusion-related magmatic fluids (Sillitoe 1996): the largest of these (up to several thousand million tons) is the porphyry type, while the smallest (<10 Mt) are vein deposits. Intermediate sizes are represented typically by skarns, carbonate-replacements, greisens and breccia-hosted deposits. The spatial interrelationship of all these types is depicted in Fig. 2.28.

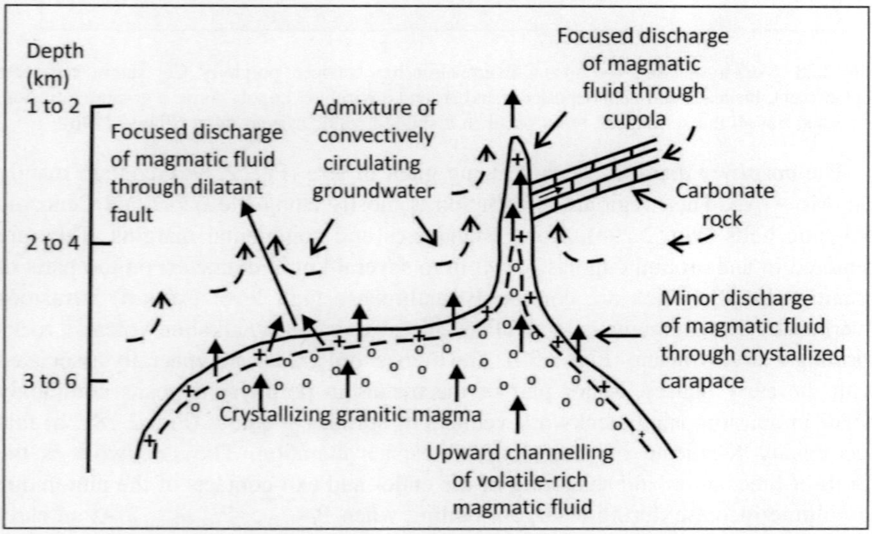

Fig. 2.27 Sketch showing the generation, accumulation and focused release of magmatic fluid along the cylindrical cupola during the crystallization of a granitic intrusion. Note the mixing zones with deeply convecting groundwater, the reactive carbonate horizon and the dilatants fault zone (after Sillitoe 1996)

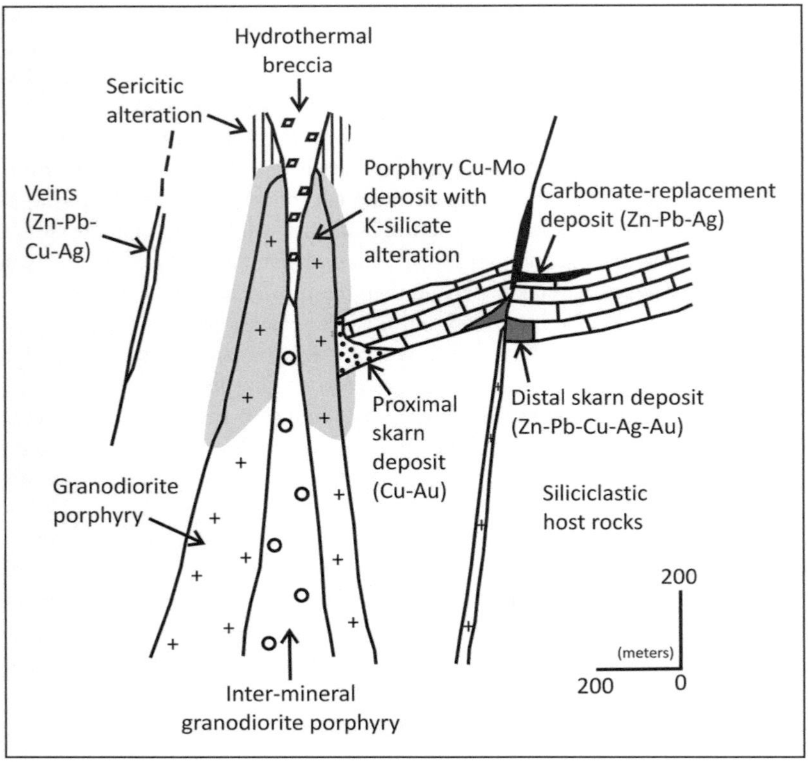

Fig. 2.28 Sketch showing the spatial interrelationship between porphyry Cu, skarn, carbonate replacement, breccias, and vein deposits in and around a porphyry cupola above a granitic intrusion. Note that not all the six deposit types occur in a single granitic system (after Sillitoe 1996)

The ***porphyry deposits***, several being giant in size (Fig. 2.29a), contain mainly Cu, Mo $\pm$ Au. Their regional distribution is mostly along Mesozoic and Cenozoic orogenic belts (Fig. 2.29b), along island arcs and continental margins. They are centered in and around cupolas (<100 m to several km in diameter) on top parts of granitic plutons which are commonly multiphase high level (<6 km) intrusions overlain by co-magmatic, calc-alkaline, or less commonly, alkaline volcanic rocks forming a stratovolcano (Fig. 2.30). The highest ore grades are generally associated with the early phases. Major part of the metals in porphyry deposits commonly occur in anastomosing stockwork veinlets of sulfides $\pm$ quartz (Fig. 2.18a, b) that accompany K-silicate or potassic (K-feldspar) alteration. The stockwork is the result of brecciation and shattering of the endo- and exo-contacts of the pluton due to volume increase during resurgent boiling when $P_{(fluid)} > P_{(confining)}$. At an early high temperature stage, a single phase low- to moderately saline liquid exits the magma and undergoes phase separation during ascent, due to P-dependence of H_2O solubility, to generate an immiscible hypersaline liquid and vapor which produce potassic alteration ($\sim$500 °C) along with the contained low (or high?) sulfidation

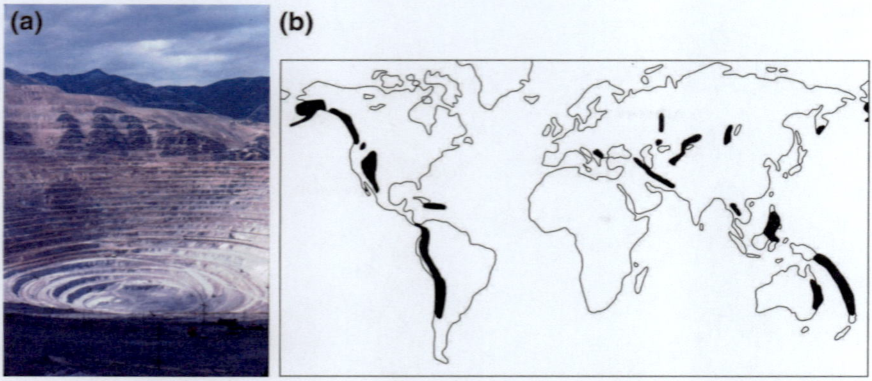

Fig. 2.29 **a** View of the giant porphyry Cu deposit at Bingham, Utah, SW USA (Photo: M. Deb). **b** Global distribution of porphyry deposits worldwide

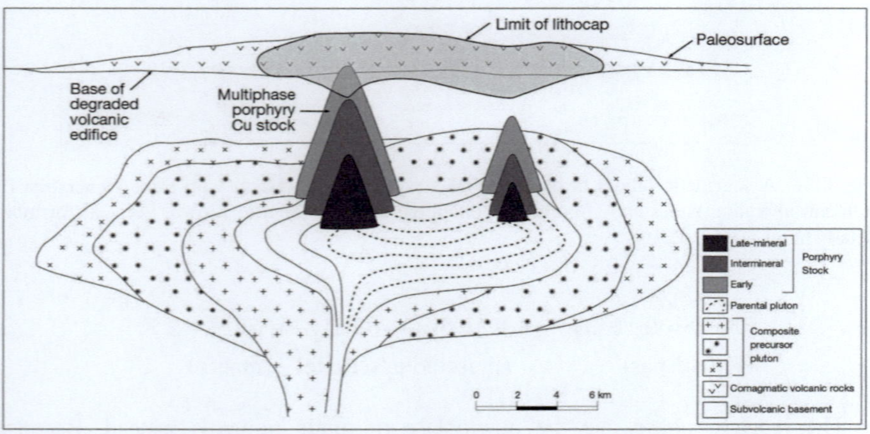

Fig. 2.30 Spatial relationship between porphyry Cu stocks, underlying pluton, overlying co-magmatic volcanic sequence and the lithocap. The precursor pluton is multiphase whereas the parental pluton is a single body undergoing inward consolidation shown by concentric dotted lines (after Sillitoe 2010)

state porphyry Cu $\pm$ Au mineralization (Fig. 2.31) (Sillitoe 2010). The characteristic mineral in the potassic alteration zone is new-formed K- feldspar and/or biotite, brought about by K^+ metasomatism and hydrolysis. Sericite, chlorite, quartz, and iron oxides may be present in small to minor proportions. In a few cases (for example, Yerrington porphyry copper deposit, Nevada, U.S.A.) the position of K may be partly taken up by Na and Ca. Phyllic or sericitic alteration is a very common type of alteration in hydrothermal mineralization, where white potassic mica develops from K-feldspar by reaction with water:

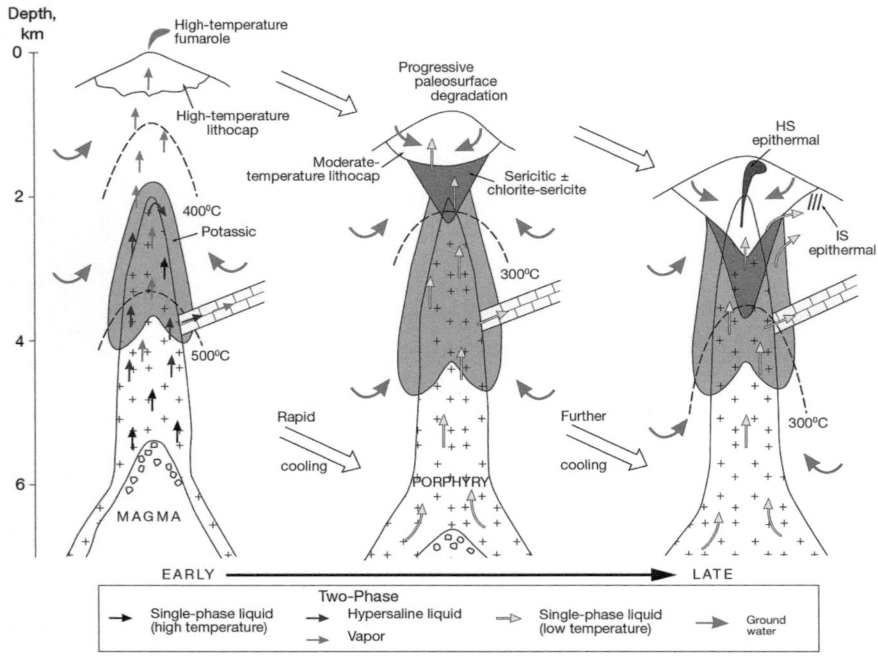

Fig. 2.31 A schematic model of evolution of a porphyry Cu system with time. Generation of different alteration zones and HS and LS epithermal deposits are also shown. See text for more details (after Sillitoe 2010)

$$\begin{array}{lll} 3KAlSi_3O_8 + 2H^+ \leftrightarrow & KAl_3Si_3O_{10}(OH)_2 + 6\ SiO_2 + 2\ K^+ \\ \text{(K-feldspar)} & \text{(muscovite/sericite)} \quad \text{(quartz)} \end{array}$$

The reactant phase may be plagioclase or mafic minerals instead. Because temperature controls are less stringent, this type of alteration is more common than the potassic alteration. Propylitic alteration, characterized by the development of chlorite, epidote, zoisite, albite, and calcite is common as an outer envelope to the porphyry system. Unless set in the geological context, the petrographic character of this zone may be distinguished from greenschist metamorphism only with difficulty. The zone is characterized by low temperature (200–350 °C) minerals and a low fluid/rock ratio. Besides, base leaching related to epizonal precious metal mineralization in porphyry systems often gives rise to intense argillic alteration in the near-surface zone (cf. Fig. 2.16). The involved fluid is acidic and is of low temperature (<250 °C). It may be noted that these alterations are interpreted as a temporal sequence, either from early propylitic to late silicic and argillic, or as spatial progression from a potassic zone outward to a propylitic halo (Henry et al. 1997).

Skarn deposits develop where carbonate rocks are converted to Ca–Mg–Fe–Mn silicate assemblages by reactive fluids in the vicinity of intermediate to granitic intrusions. Their development is controlled by a combination of favorable lithology and structures, and therefore are mostly stratabound and vein and fracture-filled. They may be closely associated with porphyry deposits (Fig. 2.28) or may remain largely unmineralized. Early prograde silicate assemblages of garnet and pyroxenes in calcic systems are deficient in ore metals but the later retrograde stage brings most of the metals in association with hydrous phases such as, actinolite, biotite, muscovite, chlorite, talc, quartz, and carbonates. Skarn deposits can be classified based on the dominant economic metal as Fe, W, Cu, Zn–Pb, Mo, and Sn skarns (Einaudi et al. 1981). Skarn deposit formed at or close to the inner contact zone of an intrusive is called endoskarn, and that formed at the outer contact is called exoskarn. Examples of some well-known skarn deposits are Memé mine, northern Haiti (Cu), Sarbai, Kazakstan (Fe), King Island, Tasmania (W), MacMillan Pass, Canada (W), Pine Creek, California (W), Cantung, NWT, Canada (W), Atamina, Peru (Zn, Cu).

It may be noted that the economic metal concentration in skarns has some general relationship with the composition of the associated intrusive body, oxidation state, tectonic setting, or country rocks: Fe and Au skarn deposits, as a general rule appear to be associated with intrusions of mafic to intermediate compositions; Cu, Pb, Zn, and W skarns to calc-alkaline I-type intrusions and Mo and Sn to S-type, i.e., ilmenite series granitoid. Further, interesting is the case of Bingham Canyon, Utah, USA, which is not only the host to a very large porphyry Cu deposit but also the largest Cu skarn deposit of the world.

Carbonate replacement deposits are formed where magmatic hydrothermal fluids interact with carbonate rocks beyond the skarn front and produce semi-massive to massive sulfides (Fig. 2.28). These deposits, commonly of Sn or Zn–Pb–Ag, have stratabound *manto* or steep chimney configuration and may occur alone or as distal extension to skarn deposits (Einaudi et al. 1981). Several examples of this type are found in Rocky Mountains, Colorado.

Greisens are a secondary mineral assemblage formed by alteration of either the granitic, or the country rocks by the addition of elements, such as, Si, Li, Be, B, F, and H_2O. The resultant mineral assemblage commonly consists of white mica (generally Li-rich), quartz, topaz, tourmaline, and fluorite. The greisenized area may show some kind of zoning. There is often evidence of feldspathization before greisenization, and kaolinization following it, the effect being reflective of varying fluid–rock equilibria with falling temperatures (Strong 1988). Greisen deposits are related to shallow level felsic granites (<6–8 km) to subvolcanic setting and are represented by various metals such as Sn, W (as wolframite) and sometimes Mo, Bi, or Be accompanied by base metals which are paragenetically late in the sequence and occur as disseminated mineralization in pervasively altered cupolas and their wall rocks (Fig. 2.32) (Sillitoe 1996). They commonly show a close relationship to quartz-dominated sheeted or stockwork vein system, where pressure permits fractures to dilate. A continuum in mineralization is noted from pegmatites to veins/greisens in many places.

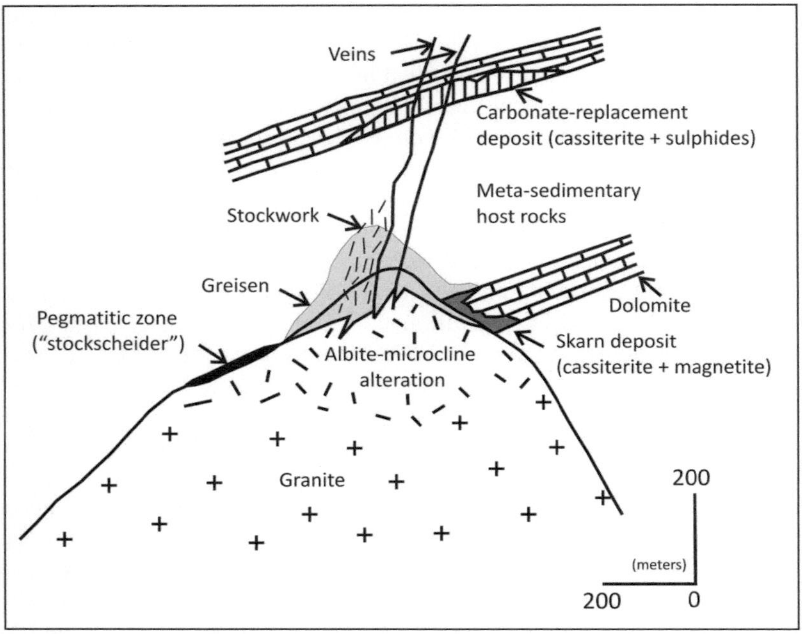

Fig. 2.32 A sketch of a Sn-dominated system centered on a granitic cupola, showing the relationship between pegmatite, greisen, skarn, carbonate replacement, and sheeted veins (after Sillitoe 1996)

The ***Breccia deposits*** comprise vertically extensive pipes that develop either within the granitic intrusions or in overlying wall rocks. Cu-bearing breccias occur within or around porphyry Cu deposits (Fig. 2.28) but the breccias may contain one or more of the other magmatically concentrated metals (Sillitoe 1985).

2.3.1.3 Iron Oxide–Copper–Gold (IOCG) Deposits

This is a recently perceived ore-type, characterized above all by the presence of iron oxide, Cu, and Au, along with U and REE. The nomenclature relates to mineralization in the Olympic Dam deposit, discovered in 1975 (with production starting in 1988), about 560 km north of Adelaide in South Australia. This was the end result of a concept-oriented exploration work based on a Ph.D thesis in Australian National University, Canberra (Haynes, 1972 in Haynes 2006) which conjectured that continental tholeiite basalts, when altered to albite-hematite-phyllosilicate-epidote-carbonate assemblages, become potent copper source rocks dependant on their Fe^{3+}/Fe^{2+} ratios and total Na (as albite) content. The Cu–Au–Ag–U–REE mineralization at Olympic Dam occurs in a hematite-rich granite breccia complex within the Mesoproterozoic (1600–1585 Ma) Roxby Downs Granite with A-type affinities in the Gwaler craton (Solomon and Groves 2000). The complex is overlain by the flat-lying sedimentary sequence of the Stuart shelf geological province. The ore mineralization is structurally controlled and is of hydrothermal origin. Owned totally by BHP-Billiton, the Olympic Dam deposit with its polymetallic ore composition is the world's fourth largest remaining copper deposit, fifth largest gold deposit, and the largest uranium deposit. It also contains significant quantities of silver. The REE minerals won as a by-product of the ore in this deposit are monazite, bastnaesite, fluocerite, and florencite.

Since the discovery of Olympic Dam deposit, exploration has been spearheaded worldwide on deposits that varyingly resemble it, as the Olympic Dam deposit still stands unique. Some deposits having varying resemblance to this type are Salobo, Cristalino in Carajas region of Brazil (2.6 Ga), Ernest Henry in Cloncurry district, Australia (1.7 Ga), Wernecke breccias, Yukon (1.3 Ga), Michelin and Sue-Dianne in Canada, Candelaria (115 Ma) and El Laco deposits in Coastal Cordillera, Chile and some deposits surrounding the Kaapvaal craton in South Africa. A somewhat similar apatite-magnetite deposit in Sweden is known as the Kiruna-type. It is now understood that IOCG represents a diverse variety of ore systems, formed in a variety of tectonic settings, geological environments, and with differing ore components but with a sort of commonality that binds them together. IOCG ore bodies range from around 10 million tons to 4000 million tons or more of contained ore, and have a grade of between 0.2 and 5% copper, with gold contents ranging from 0.1 to 3+ grams per ton. The large size, relatively simple metallurgy and relatively high-grade IOCG deposits can produce extremely profitable mines.

A few years ago Groves et al (2010) made a critical review of the deposits commonly referred to as IOCG in modern geological literature. They believe the earlier definition of IOCG deposits has been a little too liberal. Many of the smaller deposits ascribed to this group and in which Cu and/Au grades are much higher, i.e., typical of skarns, should be left out of the domain of IOCG deposits. The deposits which have been included through this liberal definition of IOCG group are:

1. Fe-oxide Cu–Au deposits (IOCG deposits *sensu stricto*)
2. P-rich iron oxide deposits
3. F- and REE-rich iron oxide deposits
4. Carbonatite-hosted (Cu, F, REE)-rich deposits
5. Cu–Au and Fe-skarns deposits
6. High-grade Fe-oxide-hosted Au($\pm$ Cu)-magnetite deposits

We may exclude subgroup 5 from this list as they constitute well-defined deposit types themselves. Precambrian deposits are better qualified for inclusion in the IOCG group. These authors hold and have recently reiterated (Groves et al. 2016) that IOCG (sensu stricto) deposits are magmatic-hydrothermal in origin. A recent approach in this context (Richards 2016) is to differentiate between intrusion-related IOCG from non-intrusion-related IOCG deposits, because in several deposits the association of a granitic intrusive remains elusive. This also reduces difficulties in developing a robust exploration model for the mineral deposit type.

The complex character of the IOCG deposits may be outlined as follows (Hitzman et al 1992; Papers in Porter 2000, 2002; Williams et al. 2005; Groves et al. 2010, 2016; Sarkar and Gupta 2012):

1. Cu and Au are the principal economic metals in this type of ore. The Cu-grade in >60% of Cu–Au deposits range between 0.5 and 1.5 wt%, averaging about 1 wt%. Chalcopyrite is the principal Cu mineral. Au commonly is <1 ppm. U content variable, not large except in bulk.
2. Iron oxides (magnetite, hematite) present are intimately associated with Fe–Cu sulfides (Fig. 2.33) and are of low-Ti variety. Fe silicates (grunerite, Fe actinolite, and fayalite) may be present instead. Magnetite may be accompanied by a phosphate (apatite).
3. Sulfides are low-S variety (pyrrhotite, chalcopyrite, bornite, chalcocite). Sulfur isotope data (δ^{34} S) are typically, but not invariably, close to 0 ± 5 per mil.
4. Ores are generally enriched (0.5–10 wt%) in LREE.
5. Structurally controlled, commonly with breccias; pervasive alkali (Na ± K ± Ca) metasomatism is commonly present in the ore zone. IOCG mineralization and alkali metasomatism is co-itinerant but not isochronous, metasomatism preceding by 10–20 Ma.
6. IOCG mineralization is not time specific, the major ones however, being Proterozoic in age.
7. Ore zone minerals show high temperature (>250 °C), hypersaline, and CO_2-rich fluid inclusions.

Fig. 2.33 Drill core from Olympic Dam deposit, Australia showing brecciated ore with magnetite (*M*) and chalcopyrite (*C*) in a hematitic (*H*) groundmass (Photo courtesy Indrani Mukherjee)

8. The deposits show clear temporal but not necessarily spatial association with causative intrusions which formed from hybrid mantle-crustal magmas sourced from metasomatized SCLM. Thus, there are felsic intrusions, commonly A-type granites with alkaline to sub-alkaline affinity in the IOCG districts.
9. The ore zone is characterized by low silicification, in contrast to many admittedly hydrothermal ore deposits.
10. The deposits are located in areas that are cratonic or of continental margin environment, characterized by extensional tectonics (Fig. 2.34).

A review of the descriptions of this type of deposits leads to the following definition: IOCG deposits have Cu and Au as the principal economic metals, are structurally controlled, and surrounded by wall rock alteration and/or brecciation, both regional in scale. The ore zone is depleted in SiO_2 content and pyrite, notwithstanding the mineralization is hydrothermal. The Cu mineral is chalcopyrite ± bornite ± chalcocite. The IOCG deposits differ from typical hydrothermal deposits in the magmatism involved. (Partington and Williams 2000). Moreover, there is enough evidence for the involvement of ultrabasic to basic mantle-derived magmas, some with alkaline affinity, participating in the regional magmatic event.

IOCG deposits generally occur inboard of lithospheric boundaries. Kerrich et al. (2005, 2010) explained the intra-continental occurrence of the IOCG system as due to focusing of lithospheric extension and accompanying melting of metasomatized subcontinental lithospheric mantle (SCLM) at necked transition between thick Archean and thinner Proterozoic SCLM that guided mantle underplating. This process produced basic and ultrabasic melts probably of alkaline affinity and enriched in volatiles, Cu and Au. The melt ponds at the crust-lithosphere boundary,

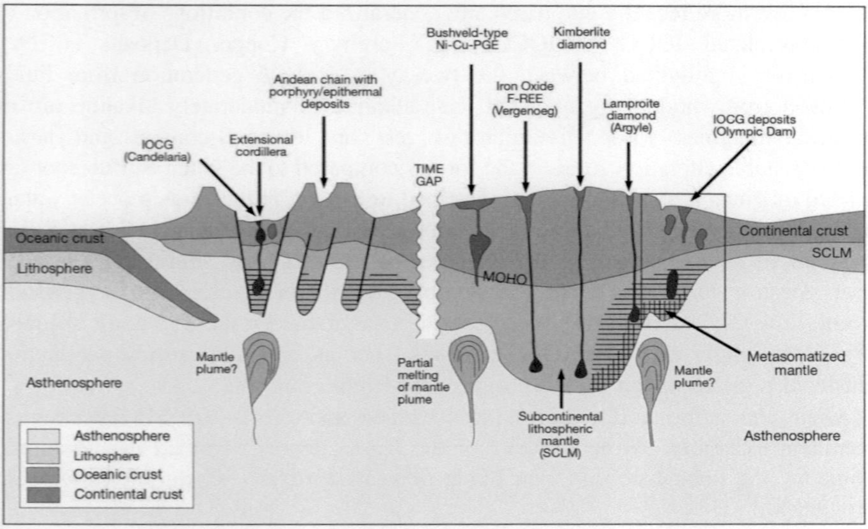

Fig. 2.34 Tectonic and lithospheric setting of IOCG deposits (after Groves et al. 2010)

causes partial melting and production of felsic melts. Deep volatile exsolutions produce giant breccias pipes of silicate rocks replaced by iron oxides followed by Cu, Au and other enriched elements (Hart et al. 2004).

Origin of IOCG deposits has been controversial right from the beginning (Refs. above). Earlier, two models held the sway. One held that the hydrothermal fluids causing wall rock alterations, or producing the mineralization exsolved from cooling granitic intrusives below, though, as indicated earlier, such granitic bodies remain elusive in some cases. The other major model held that the hypersaline fluids owe their origin to (an invisible) evaporate unit in the sequence. The most logical explanation, based on petrologic and isotopic evidence, the LREE and volatile enrichment, the giant size of the host breccia – bodies in some deposits and the intense SiO_2-depletion in the alteration halos put forward by Groves et al. (2010) suggest devolatilization of deep, volatile-rich, mantle-derived magmas as the primary energy- and fluid driving source for the IOCG ore systems, which produced wall-rock alteration and metal deposition at greater depths than normal porphyry Cu–Au producing dioritic to granodioritic magmas. Presence of Ni and Co in economic/subeconomic proportions further bear evidence of the possible involvement of mafic-ultramafic magmatism in the process. Uranium could be leached from the crustal rocks the fluids passed through. On the other hand, evaporative and mixed magmatic and non-magmatic brines are only responsible for pre-ore regional scale alteration and have nothing to do with the mineralization itself. Majority belief at the present moment is that the IOCG systems are produced not only by deep-seated fluid-rich sources but also by fluid circulations controlled by crustal scale dislocations. Further, Groves et al (2010) emphasize that size of the associated alteration is a defining characteristic of the IOCG deposits.

Richards and coworkers (Richards 2013; Richards and Mumin 2013; Richards 2016) have more recently compared and generalized the conditions of formation of intrusion-related IOCG (IR-IOCG) and Porphyry Copper Deposits (PCDs). Similarities highlighted between the two types include generation from fluids exsolved from moderately oxidized, calc-alkaline to moderately alkaline arc or orogenic magmas while dissimilarities rest on lower S-content and larger feldspar-stable alteration zones in the former compared to the latter. Sulfur-poor arc related magmas in the Precambrian, ascribed to S-poor, reduced deep ocean waters in the subduction zones at these times, produced IR-IOCG deposits whereas S-rich magmas in similar setting in the Phanerozoic produced the sulfidic PCDs when deep ocean water sulfate levels rose sharply after the Neoproterozoic oxygenation event. Low fS_2 and High fO_2 conditions were favorable for the formation of low-S (Sn-W) porphyry and some IOCG deposits (such as, in Chile) in post-subduction tensional regimes of the back arc in the mid-Cretaceous (Fig. 2.35).

Giant Precambrian IOCG provinces formed about 100–200 My after super-continent assembly. While majority of the IOCG deposits formed in anorogenic domains, the phosphate-rich iron oxide deposits formed late, during convergent orogenies.

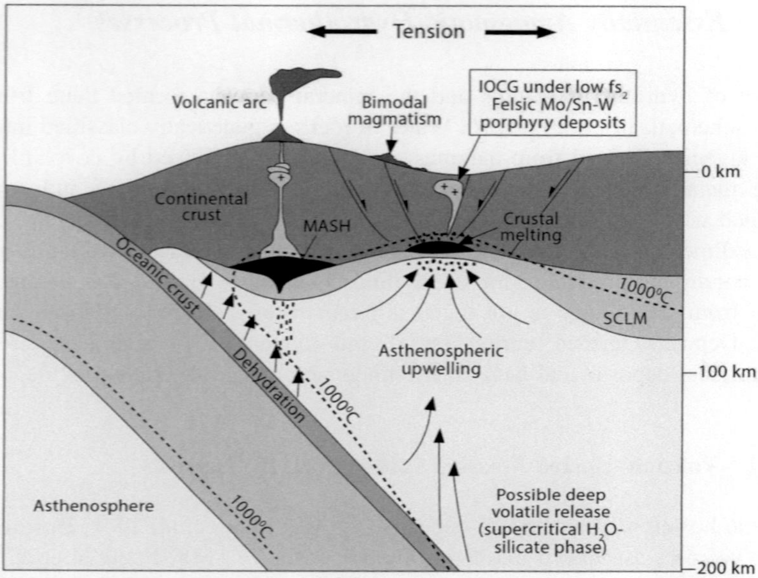

Fig. 2.35 Sketch showing the development of IR-IOCG deposits in transtensional back arc settings in the mid-Cretaceous (after Richards 2016)

Summing up, it may be concluded that there is no straight jacket view about the IOCG deposits, although they share amongst them many things in common. Similarly, there is similarity and difference in their genetic history. It is possible that in future the IOCG deposits will be considered to have formed by interplay of several processes. Some of the persisting unresolved questions about IOCG deposits include: (a) the source of the large amount of Fe in the deposits; (b) role and type of magmatism; (c) presence of a large variety of elements (U, P, REE, Cu, Au, Mo, F, Ba, etc.). Do they indicate multiple source and mixing of reservoirs? (d) Deficiency of sulfides in some deposits; (e) presence of low-Ti–magnetite. Could it be from a non-mafic magma, formed by low-T solutions? High salinity of fluid inclusions in the ore zones is rather suggestive of this.

2.3.1.4 Rare Hydrothermal Nickel Deposits

The Avebury nickel deposit, Tasmania, which contains a resource of 2,60,000 tons of Ni at a grade of 0.9% and associated with an ophiolite sequence is an unconventional deposit of nickel. The ophiolite consists of cumulate peridotite and dunite of Middle Cambrian age. The principal nickel mineral is pentlandite. Modest to strong correlation between Au, Pd, REE, Sn, Mo, W with Ni suggests that the nickel mineralization was caused by the hydrothermal solutions given out by a Devonian granite intrusion at the site (Keays and Jowitt 2013). This should be an eye-opener for other explorers.

2.3.2 Essentially Amagmatic Hydrothermal Processes

Sources of hydrothermal fluids and the mineral deposits formed there from are shown schematically in Fig. 2.36. Water in rocks is genetically classified into four types: juvenile, derived from magmas; metamorphic, produced by devolatilization during metamorphism; meteoric, contributed by the atmosphere; and connate, contained within sediments. Metamorphogenic hydrothermal solutions are formed from sedimentary(-volcanic) rocks undergoing metamorphism, with greenschist facies conditions producing more ore fluids than other grades. The ore mineralization from such fluids is not much different from that produced from juvenile fluids. Deposits formed during burial and diagenesis of sediments are also hydrothermal deposits and have many things common with epithermal deposits.

2.3.2.1 Volcanic-Hosted Massive Sulfide (VHMS) Deposits

Volcanic-hosted massive sulfide deposits (VHMS: Large et al. 2001; Huston et al. 2010), volcanic-associated massive sulfide deposits (VAMS: Franklin et al. 2005) and volcanogenic massive sulfide deposits (VMS: Franklin et al. 2005) mean the same. Only in the last, the connotation is to genesis and in the other two, it is rather descriptive.

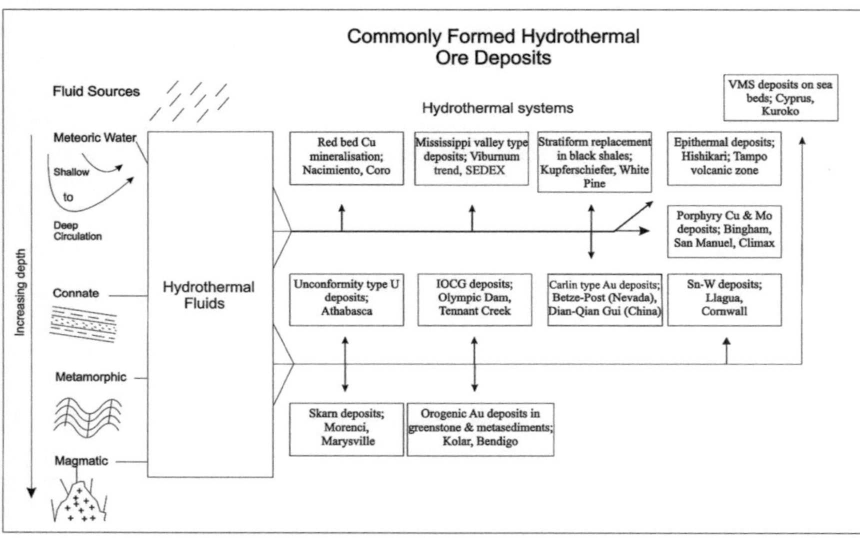

Fig. 2.36 Different fluid sources and different types of hydrothermal ore deposits with examples

These are deposits formed at or near the sea floor in a submarine volcanic setting, either in ocean spreading ridge or in the back arc of an island arc (Fig. 2.37). They are characterized by massive sulfides, that is, $\geq 60\%$ of sulfides of Fe–Cu–Zn–Pb (Fig. 2.38) with additional quantities of Au, Ag, Cd, In, Sn, Ge, etc. The mineralogy is generally simple: sphalerite, galena, chalcopyrite. Fe is present mostly as sulfides (pyrite, pyrrhotite, arsenopyrite) and less often as magnetite. Morphologically, most of these deposits can be divided into two parts: the concordant sulfidic upper part, underlain by a stockwork of ore minerals and silicates (Fig. 2.39). The upper part is generally stratiform and in fair measures, stratified and composed of the massive ores. They generally occur as member(s) of a dominantly volcanic sequence. There is also the presence of thin, but aerially extensive units of ferruginous chemical sediments formed from exhalations in the topmost part of these deposits. This picture is mainly based on the deposits in the Kuroko district of Japan where the role of felsic centere, explosive activity at the vent, brecciation, stringer zone, exhalites and metal zonation are conspicuous. In Canada, based on the Archean deposits of the Noranda camp, an upward stratigraphy has been traced (Lydon 1984) (Fig. 2.40) which involved alteration zone in the footwall with veins and stockwork, and layering, breccias, massive sulfides, and exhalites in the hanging wall. These zones vary in shape and size depending on the host rocks, geologic setting, water depth, etc.

Temporal distribution of VHMS is wide, starting from Archean until the present, interrupted by some breaks or lean periods. These periods are between 2500–2000,

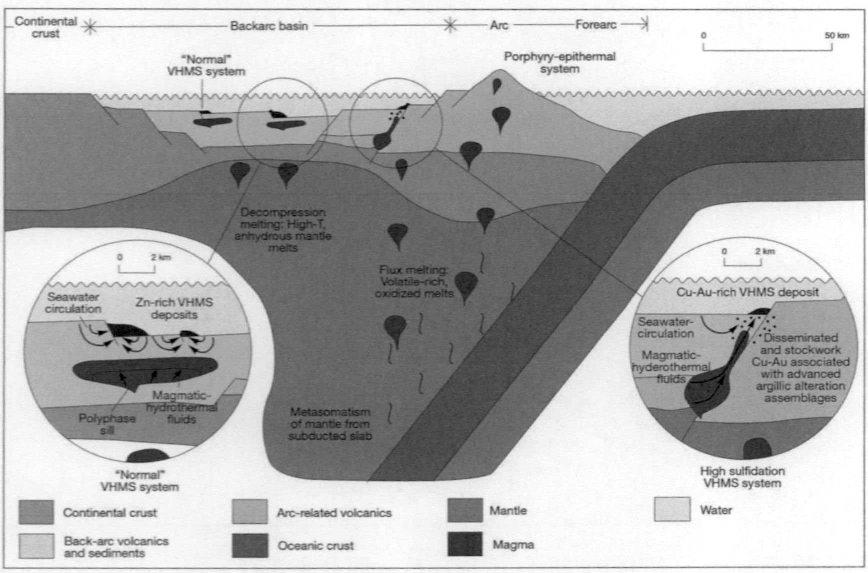

Fig. 2.37 Tectonic settings of volcanic-hosted massive sulfide deposits in convergent margins (after Huston et al. 2010)

Fig. 2.38 Banded massive sphalerite-pyrite ore; drill core from Deri mine, Southern Rajasthan (Photo: M. Deb)

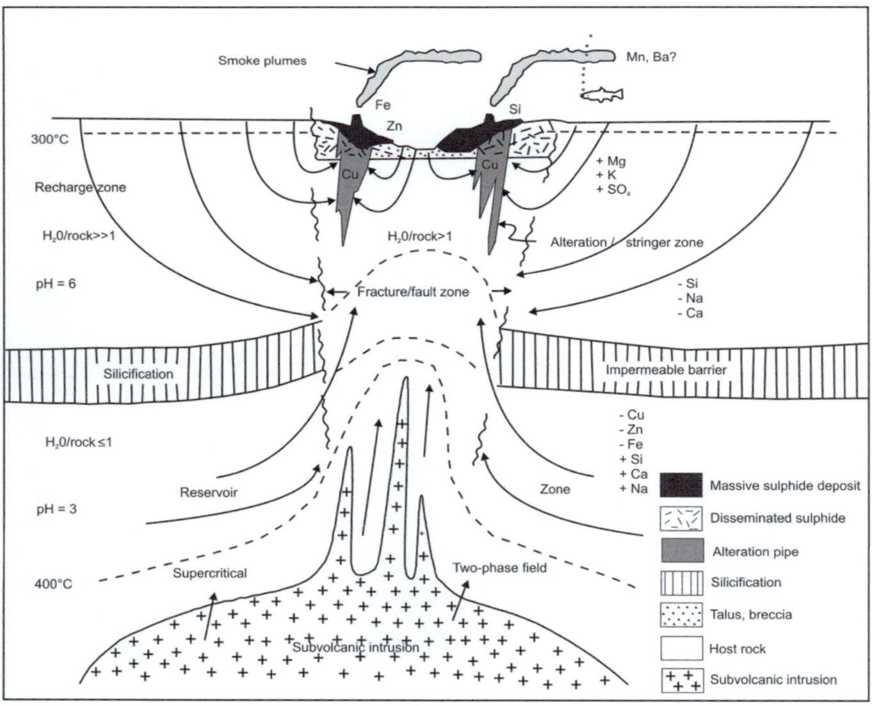

Fig. 2.39 A model of VMS ore formation (after Franklin et al. 2005)

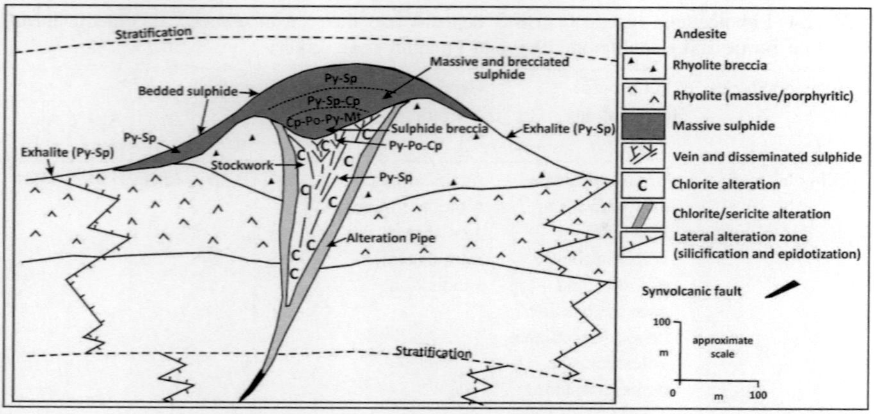

Fig. 2.40 A schematic section across a Noranda type VMS deposit (after Franklin 1995)

1700–1400, and 1000–750 Ma. They are believed to correspond to periods of supercontinent stability or breakup (Huston et al. 2010). Rodinia is however, an exception. There are about 800 deposits of this class worldwide. Some are forming on the ocean floor at present. The average size of a VMS deposit is $\sim$10 Mt, with the modern sea floor deposits being $\sim$0.2 Mt. Any deposit >50 Mt is considered to be a giant deposit of the kind.

VHMS deposits do not occur in any unique geological situation. Rather, they occur in a couple of different settings, represented by varying lithotectonic associations. The following five lithostratigraphic associations of VHMS are recognized (Barrie and Hannington 1999; Franklin et al. 2005):

1. Bimodal mafic
2. Mafic
3. Pelitic-mafic
4. Bimodal-felsic/alkaline
5. Siliciclastic felsic.

The first three types are related to ocean–ocean subduction and represent nascent arc rifting (Type 1) to mature back arc development (Types 2 and 3). In Archean greenstone terrains, environments of Type 1 also include a mantle plume (komatiite) association. Bimodal-felsic and siliciclastic felsic types (Type 4 and 5) formed in ocean-continent margins to continental back arcs. Bimodal alkaline lithofacies have Pb $\geq$ Zn > Cu and its geodynamic setting involves paleo-plume producing sea mount and submarine epicontinental rift. Characters of these types are discussed somewhat elaborately in Table 2.4. As we shall see later, these petrologic associations have genetic significance also.

These lithostratigraphic associations are represented by a number of rock types in different proportions, having formed in varied tectonic settings (Table 2.4). To facilitate discussion of the criteria for classification of VHMS deposit and introduce

Table 2.4 Lithotectonic types of VHMS deposits and their metal contents (geometric mean). (Based on Barrie and Hannington, 1999 and Franklin et al., 2005)

Petrogenetic types	Petrologic associations	Geodynamic settings	Ore metals				
			Cu %	Pb %	Zn %	Au ppm	Ag ppm
1. Bimodal mafic	Basalt-dominated (but with up to 25% felsic volcanic rocks); pillowed and massive basaltic flows; subordinate volcaniclastic rocks; immature wacke; sandstone and argillite; hydrothermal chert in immediate hanging walls in some deposits	Incipient-rifted bimodal volcanic arcs above intra-oceanic subduction	1.24	0.30	2.32	0.81	21.14
2. Mafic	Ophiolite and ophiolite-like assemblages, dominated by pillowed and massive basaltic flows, minor felsic volcanics; minor ultramafic flows; synvolcanic mafic dykes and/or sills up to 50%; sedimentary rocks minor	Mature intra-oceanic back arc, some transform fault related (oceanic mature back arc)	1.82	0.02	0.84	1.40	10.62
3. Pelitic-mafic	Basalt and pelite subequal, or pelite dominant; mafic sills up to 25%; felsic volcanic rocks absent/negligible. Sediments: carbonates, argillite, subordinate siltstone and wacke; mafic ultramafic sills and flows in sedimented	Oceanic mature back arc. Co may be present in the ore metal association	1.23	0.68	1.58	0.75	19.29

(continued)

Table 2.4 (continued)

Petrogenetic types	Petrologic associations	Geodynamic settings	Ore metals				
			Cu %	Pb %	Zn %	Au ppm	Ag ppm
	mid-oceanic ridge and back arc basins						
4. Bimodal-felsic	Felsic volcanic rocks constitute 35–70% of volcanic strata and basalt 20–50%; terriginous sediments ~10%. Some portions may be subaerial	Continental margin arcs and related nascent or primitive epicontinental back arcs, including submarine continental rift settings	1.04	1.14	4.36	1.06	56.35
5. Siliciclastic-felsic	Siliciclastic rocks dominant (~80%); felsic volcaniclastic rocks (with minor flow domes, etc.) constitute about ~25%. Mafic (tholeiitic to alkaline) flows, volcaniclastic rocks and sills in the hanging wall succession (~10%) Fe-, Mn-, Ca-, Ba-rich rocks in the hanging wall	Mature epicontinental back arc	0.62	1.09	2.70	0.59	38.54

the influence of the form and composition of volcanism on the development of VHMS deposits, the five lithostratigraphic types are further subdivided on the basis of their occurrence in successions dominated by one of the three end member lithofacies: flow, volcaniclastic, and sedimentary (Franklin et al. 2005).

Based on metal association (Cu:Pb:Zn) in the VHMS deposits (Fig. 2.41), and considering their type area of occurrence, they are classified into following four types:

(1) Cu (mafic-dominated) (Cyprus type)
(2) Cu–Zn (Archean greenstone terranes) (Noranda type)
(3) Zn–Cu–Pb (Phanerozoic terranes) (Kuroko type)
(4) Pb–Zn–Cu (sediment-dominated settings) (Besshi type)

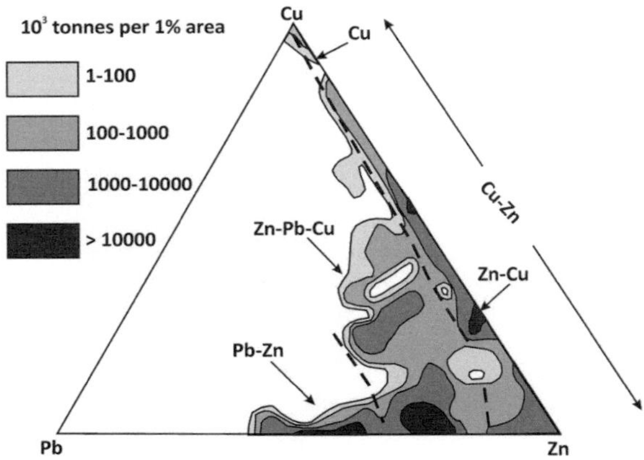

Fig. 2.41 Triangular classification of VMS deposits based on the metal content (after Franklin 1995)

Deposits of felsic-dominated associations have significantly lower 100 Zn/(Zn + Pb) and these associations become important as the age of the deposit decreases.

Origin of VHMS Deposits.

Geological features, results of laboratory experiments and direct observations on ore formation on sections of sea floor (see a later section for more details) convincingly established that these deposits have close relationships in space, time, and genesis with the volcanic rocks with which they are associated. Geological observations further suggest that they are products of volcanically generated hydrothermal systems. Sawkins (1990) argue that the genesis and distribution of VMS deposits should be viewed as integral parts of the spectrum of magmatic-hydrothermal ores related to subduction-related magmatism in arc environments. He suggests that the variations in both metal composition of such ores, as mentioned above and the petrochemistry of coeval intrusions relate in a fundamental way to the siting of their emplacement in arc and back arc systems and the duration of the subduction process.

Sources of Ore Fluids, Metals, and Sulfur

Close association of VHMS with volcanic rocks has been generally taken to suggest that the fluids that produced the deposits were evolved seawater (Franklin et al. 1981), formed by downward penetration of cold oxidized seawater into fractured oceanic crust ± magmatic hydrothermal fluids. It is also generally accepted that these systems have important subvolcanic intrusions that actually work as heat engines causing/enhancing hydrothermal fluid flow (Fig. 2.39). The data from Kuroko deposits in Hokuroku district of Japan including the range of δD values (–10 to –40‰) obtained from fluid inclusions in these ores, however, lend support to a major involvement of magmatic hydrothermal fluids in their genesis (Urabe 1987). The released magmatic fluids merge with modified seawater convection system after breaching through the silicified impermeable barrier which acts as a reservoir cap. Metal enrichment in the magmatic fluid may also supplement the metal content of the seawater convection system. This flow system has two-way movements: the hot lighter fluid moves upward and draws down the cold seawater. Up flow of vent fluids that react with the wall rocks form distinct alteration zones in the footwall with strong depletion of Si, Na, and Ca and enrichment of Mg, K, and SO_4. Thus, a sericite zone forms outwards and Fe/Mg-chlorite inwards, depending on the host rocks being altered. A number of studies of the deeper parts of the semi-conformable regional alteration zones, produced by large-scale fluid flow in VHMS districts, showed that these are leached of metals and at places sulfur, with/without input from below. Host rock contribution is indicated by some data in Table 2.4. The ore-bearing hot fluids ultimately vent out on to the sea floor along dislocation zones (Fig. 2.39), producing the sulfide mound. The stages of evolution of the mound and its zone refining are presented in Fig. 2.42 (Huston and Large 1989). These include: (1) Low T, Zn-rich mound formation at ~200 °C with sericitic alteration of wall rocks by acid fluid. (2) At ~250 °C the mound has grown and sericite zone is broader. (3) Higher T fluids (300–350 °C) move up and get Cu-rich solutions encroaching and formation of chlorite alteration in the pipe

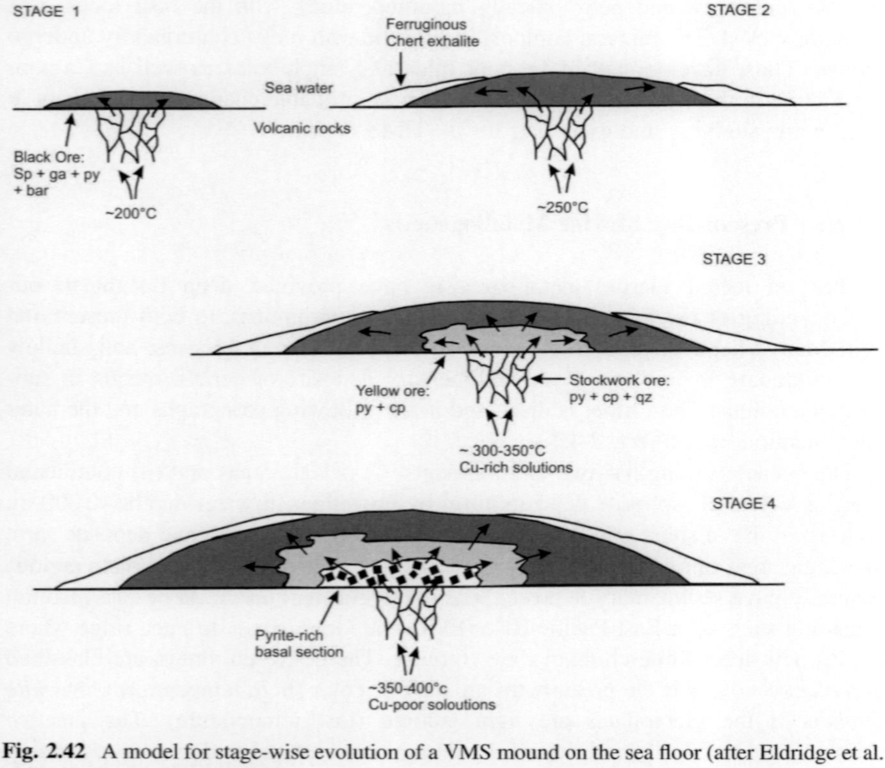

Fig. 2.42 A model for stage-wise evolution of a VMS mound on the sea floor (after Eldridge et al. 1983; Huston and Large 1989)

zone (Fe from fluid reacts with wall rock); (4) Large mound has formed, zone refining occurs; late Fe-rich solutions invade and replace chalcopyrite formed above.

Ore Deposition

Deposition of VHMS is due to mainly two reasons:

1. Mixing between upwelling hot ore-bearing fluids and the cold down-moving water.
2. Cooling of the upflowing hot hydrothermal solution.

Both will destabilize the ore-bearing complexes and precipitate ore.

Sulfur isotope values of VHMS varied with time, in both sulfide and sulfate minerals. Sulfide minerals from Archean and Proterozoic VHMS deposits are characterized by $\delta^{34}S$ values $\simeq 0$ per mil, with little variation between the deposits. The picture changed completely during the Phanerozoic when $\delta\,^{34}S$ became highly variable (Franklin et al. 1981; Huston 1999).

It is obvious from the above discussion that the VHMS deposits are at least penecontemporaneous with host rocks, if not exactly syngenetic in all cases. Most of the geological environments in which they are deposited are, in due course of time, subjected to thermo-tectonic modifications in different degrees. As a result, they are deformed and petrologically modified, along with the host rocks, i.e., metamorphosed. The mineral compositions of the wall rocks consequently undergo change. Thus, development of Fe-poor mica and amphiboles as well as Ca-poor plagioclase in the wall rocks are common. These probable changes may be kept in mind while studying and exploring for a VHMS deposit.

2.3.2.2 Present-Day Marine Metallogenesis

Studies on recent marine metallogenesis have provided deep insight to our understanding of ore-forming processes in such environment, in both present and past. Many chemical processes operate in the oceans, proto-oceans, and shallow seas producing hydrothermal and sedimentary deposits of certain metals in substantial amounts. The former is discussed in the following paragraphs and the latter finds mention under Sect. 2.4.3.

The oceanic setting has two environments: (i) pelagic areas and (ii) continental margins where phosphorite deposits form by upwelling in water depths <1000 m, such as, in the offshore of SW Africa and Peru-Chile. Hydrothermal deposits form in pelagic areas of divergent as well as convergent plate margins. Intra-plate regions generally have sedimentary deposits. The divergent margins could be fast/medium spreading such as in East Pacific Rise (EPR) and Gorda-Juan de Fuca ridge where Fe, Zn, Cu-rich sulfide chimneys are forming. The active chimneys are classified into *black smokers* if the precipitates are dark in color (high temperature), or *white smokers* if the precipitates are light colored (low temperature). The inactive chimneys are physically and chemically unstable and are reduced to rubble, forming part of the basal mounds (Zierenberg et al. 1984). Petroleum reserves may also be generated in thick sedimentary pile of such setting, such as, in Guayamas basin on EPR at 72°N. Slow spreading divergent margins are characterized by Mn and Fe oxide crusts and Fe–Zn−Cu chimneys as in Mid-Atlantic Ridge (MAR), Snake pit at 23°N, TAG field at 26°N, Broken Hill spur at 29°N and Lucky strike at 37°N. The relatively scarce sediment-filled troughs on MOR host large deposits of massive sulfides (Fe, Zn, Pb, rarely with As, Sb) formed by reactions of hydrothermal solutions with the sediments, such as, in Escanaba trough, Gorda ridge; Middle valley, Juan de Fuca ridge (Zierenberg et al. 1993; Goodfellow and Franklin 1993). Hydrothermal processes in convergent margins produce Mn-rich crusts such as in Tonga-Karmadec ridge in SW Pacific and Fe, Cu, Zn, Pb, $\pm$ Au accumulations in extensional regimes of back arc basins related to subduction zones. Both intra-oceanic (e.g., Lau basin, Manas basin in SW Pacific) and intra-continental (e.g., Okinawa trough and Izu-Ogasawara trough in the Japan Sea) settings produce sea floor deposits (Roy 1999).

The Red Sea, which represents an early stage of opening of an ocean (proto-ocean), is a large repository of massive sulfide deposits. In this slow spreading center, seawater penetrates into basic volcanic and Miocene (halite-bearing) evaporites, thereby gaining extreme salinity (9–10x seawater). Dissolved metals as chloride complexes in hot (~200 °C), circulating modified seawater are discharged as solutions into isolated troughs (deeps) rather than on ridge crests (as in MAR). The density-stratified hydrothermal fluid ponds into the deeps at low temperatures (~65 °C) due to slow rate of discharge in a stagnant basin (Fig. 2.43a). The well studied "Atlantis II" deep is 7 km in diameter at a water depth of 2000 m, hosts polymetallic sulfides (Fe, Zn, Cu, Pb, Ag, Au) estimated at 100 Mt (Shanks and Bischoff 1980). Mn in the hydrothermal solution diffuses upward into the oxidizing zone without precipitating in the sulfidic zone, around which it forms a geochemical halo, just like the plume-driven Mn haloes around black smoker fields in EPR and MAR.

Examples of shallow seas are the Baltic and Black seas. The former has oxic sea floor with Fe–Mn nodules and sub-oxic to anoxic deeps with Fe–Mn carbonates and sulfides. The latter has an oxygen stratified water column which hosts Fe sulfides on deep sea floor and Fe–Mn oxides in the shallow margins (Roy 1999).

The ore-forming process in all the oceanic situations mentioned above begins with the seawater penetrating the permeable oceanic crust for several kms downwards, getting heated by the prevailing geothermal gradient and probably by the heat from a subvolcanic magma chamber (Hutchinson et al. 1980). Geophysical studies have confirmed this last conjecture in the superfast-spreading EPR by identifying relatively shallow (1.2–2.4 km below sea floor) axial magma chamber (Detrick et al. 1987). Magmatic contribution to the ore-forming fluid, besides supplying heat, has also been identified based on the high ratios of volatile (^3He, CH_4, H_2S) to non-volatile (Mn, Fe) in the plumes of EPR (Urabe et al. 1995) just like in the eastern Manus back arc basin, W. Pacific. The circulating seawater is reduced by reaction with Fe^{2+}-silicates with its sulfate forming pyrite and magnetite. As it penetrates deeper, reaction between the modified seawater and basalts at

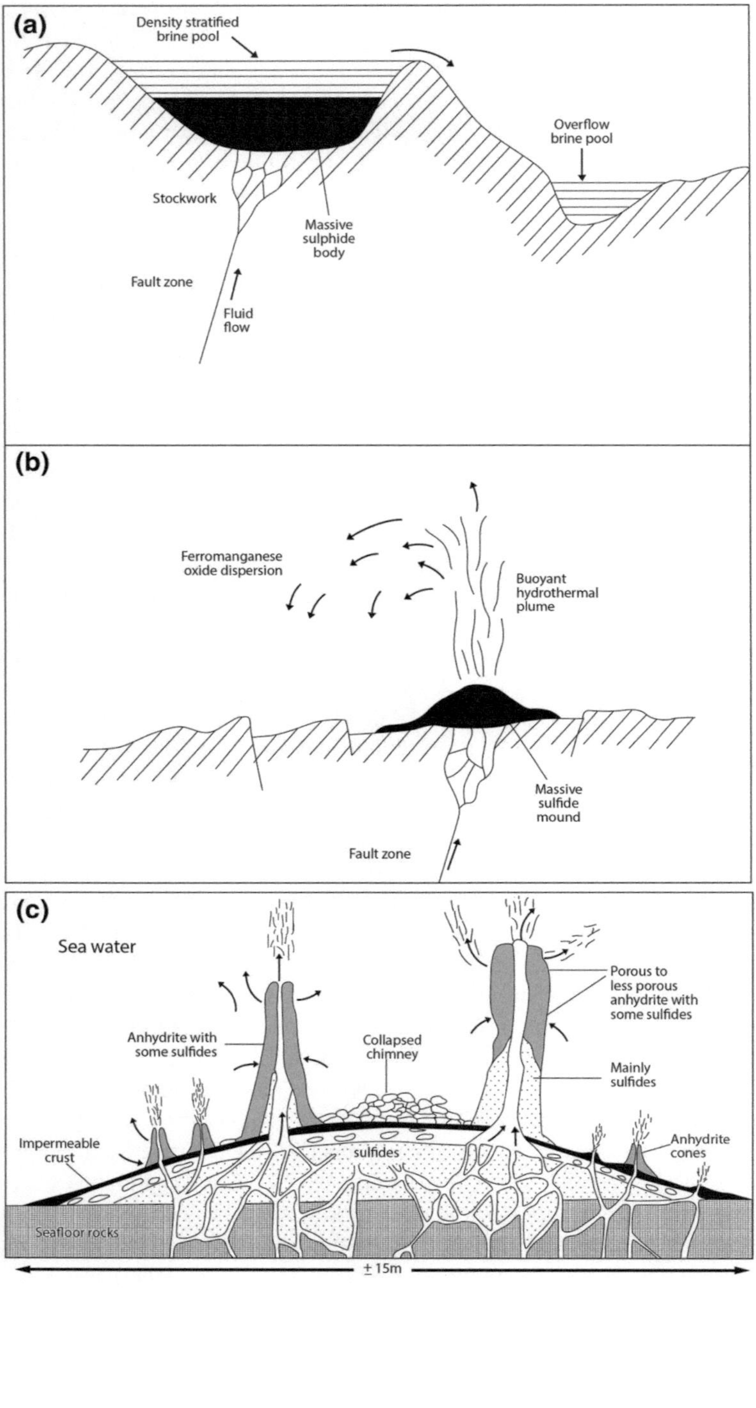
(a)
Density stratified brine pool
Overflow brine pool
Stockwork
Massive sulphide body
Fault zone
Fluid flow
(b)
Ferromanganese oxide dispersion
Buoyant hydrothermal plume
Massive sulfide mound
Fault zone
(c)
Sea water
Porous to less porous anhydrite with some sulfides
Anhydrite with some sulfides
Collapsed chimney
Mainly sulfides
Impermeable crust
sulfides
Anhydrite cones
Seafloor rocks
± 15m

◂**Fig. 2.43** Development of massive sulfide deposits on the sea floor. **a** A density-stratified pond of hydrothermal solution giving rise to a bowl-shaped deposit; **b** Hydrothermal fluid, less dense than seawater, forms a sulfide mound and rises buoyantly to form an exhalative plume that settles on the sea floor to produce ferromanganese oxide crust and other exhalites (after Rona 1988); **c** Formation of chimney and sulfide mound on the sea floor (after Barnes 1988)

temperatures in the range of 350–400 °C lowers the pH further and the water-rock interactions leach metals like Fe, Mn, Cu, Zn, Si, etc. in substantial amounts. At still deeper levels of penetration, H_2O and CO_2 get reduced to H_2 and C_2, respectively in the presence of magnetite. The heated water is buoyant at temperatures around 400 °C and starts convecting upward along fractures and dislocation zones when the reduced carbon and hydrogen combine to produce CH_4 and C_2H_6 in the enriched seawater. In the vent, as pressure is released, it induces boiling, steam generation, phreatic explosions, brecciation, etc. and the hydrothermal fluid starts mixing with shallow circulating SO_4-rich brine. Finally, it exhales on to the sea floor near the ridge crest through black or white smokers (depending on the temperature) at a wide range of temperature (~10–400 °C) and flow rate (1–2 cm to 1–2 m s^{-1}) and mixes with cold, alkaline, and oxygenated ambient seawater. The sulfides precipitate from the turbulent waters to form a mound which holds chimneys, both upright and collapsed, (Fig. 2.43b, c). The picture is very similar to that described for ancient VMS deposits in Fig. 2.39. However, it must be pointed out that information on black smokers at present-day active spreading (as opposed to back arc) ridges cannot be directly applied to the genesis of VMS deposits even though the mechanism of final sulfide deposition in both may be essentially similar (Lydon 1988). This is because subduction-related magmatism is an essential attribute in all major deposits of this type (cf. Sawkins 1990).

2.3.2.3 SEDEX Deposits

SEDEX deposits are defined as sediment-hosted sulfide deposits that form from the discharge of metal-rich hydrothermal fluids on the sea floor, commonly by venting of fluids into reduced sedimentary basins on continental margins or in intra-continental rift settings. They have been referred to by a variety of terms including shale-hosted, stratiform sediment-hosted Zn-Pb and sedimentary exhalative (SEDEX), the last being the most popular general descriptor of this economically important class of deposits. SEDEX deposits are an important resource type and comprise more than 50% of the world's Zn and Pb reserves and more than 25% of the world's Zn and Pb production. The grades and tonnage of Pb and Zn in this class of deposits (Fig. 2.44) are an order of magnitude greater than VHMS deposits. Pb + Zn grades vary between 5 and 20% commonly. Some of the large (+100 Mt) deposits of this type are Broken Hill, H.Y.C., Mt. Isa, and Century in Australia, Gamsberg, and Big Syn in South Africa, Fuenteheridos in Spain, Sullivan (now closed) and Howards Pass in Canada and Red Dog in Alaska. SEDEX deposits are noted first in the early Proterozoic, at about 1800 Ma and are somewhat rare after

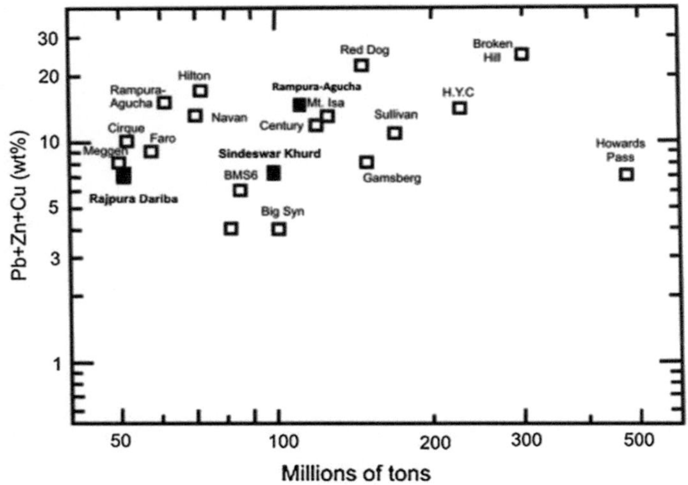

Fig. 2.44 Grade vs tonnage plot for well-known SEDEX deposits worldwide. Note the position of Rampura-Agucha and Rajpura-Dariba deposits

the end of the Mississippian Epoch, except for minor deposits in Jurassic and Cretaceous Periods. They show two major peaks of ore formation – one in the later part of Paleoproterozoic (1800–1600 Ma) and another in the Phanerozoic, between Cambrian and Carboniferous (Fig. 2.45). The major characteristics of SEDEX deposits are treated at length by Goodfellow (2004), Lydon (2004) and Leach et al. (2005), and summarized below.

Most SEDEX deposits are hosted by basinal sediments deposited within intra-continental rifts or fault-bounded grabens of reactivated rifted continental margins. They occur commonly in second or third order basins, spatially associated with syn-sedimentary faulting. The extent of rifting can vary; while some failed to develop an oceanic crust, others terminated after forming an oceanic lithosphere. This gave rise to highly variable hydrothermal reaction zones and composition of the hydrothermal fluids which are commonly saline and show a temperature range between >200–250 °C.

The ambient sedimentary rocks consist of autochthonous carbonaceous chert, shale, siltstone and coarser clastics, sedimentary breccias and carbonate rocks, and allogenic clastic sediments. Highly carbonaceous non-bioturbated laminated sediments, absence of benthic fauna, anomalously high sulfur-carbon ratio and upward increasing positive $\delta^{34}S$ trends in pyrite in host sedimentary rocks in many deposits suggest that they formed during periods when the oceans were stratified with reduced H_2S-rich bottom waters. Many SEDEX deposits are composed of two major hydrothermal facies, a sedimentary hydrothermal facies and a vent facies (Fig. 2.46). However, the vent complexes in some deposits are absent, which is probably due to poor preservation or accumulation of sulfidic products in the vent. The sedimentary hydrothermal facies generally comprise sphalerite, galena, pyrite,

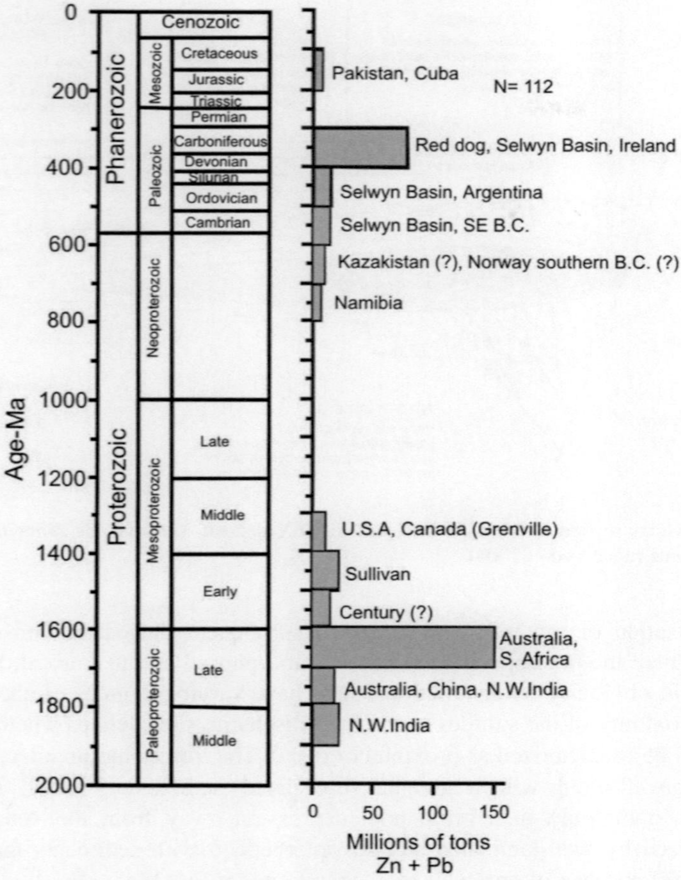

Fig. 2.45 Age distribution of SEDEX deposits worldwide (after Lydon 2004)

barite, and carbonate, and are interbedded with allogenic and autochthonous sediments. The ores are interbedded with Fe sulfides (pyrite, pyrrhotite) and basinal sedimentary rocks which points toward a syn-sedimentary origin. In the vent facies, the underlying sediments are veined, brecciated, infilled, and variably replaced by quartz, carbonate, micaceous minerals, and some sulfides. The hydrothermal alteration generally consists of silicification, carbonatization, sercitization, and chloritization along with various sulfide formation. The hydrothermal facies deposited above the vent complex are typically replaced at a later stage by higher temperature sulfide–silicate–carbonate assemblages. The deposits are often strongly zoned with Pb:Zn ratios decreasing away from the upflow zone. Sedimentary textures such as graded and cross-laminated beds, sulfide sedimentary breccias, slump structures, all indicate that most of the sulfides were deposited as sediments on the sea floor. They were then subjected to local redistribution and

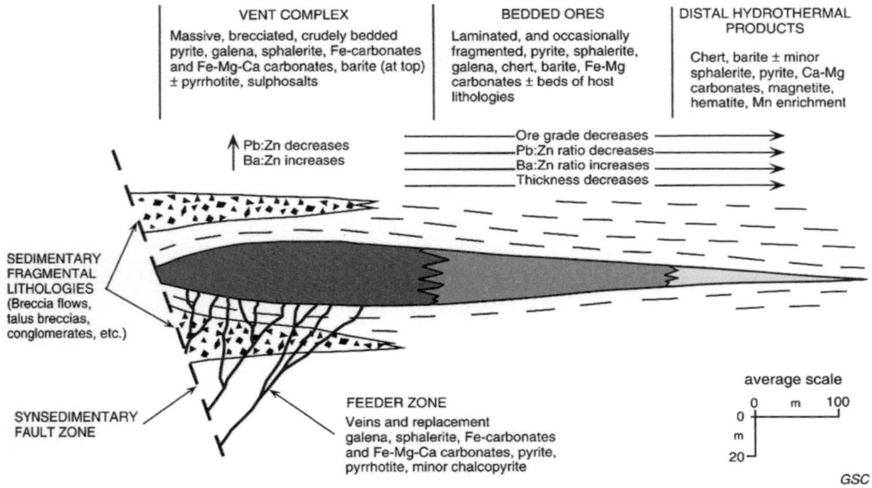

Fig. 2.46 Cross section through an idealized SEDEX deposit, showing the different facies of ores/sediments (after Lydon 2004)

recrystallization of mineral matter through subsequent diagenesis and metamorphism. Where the ore deposit has been metamorphosed and tectonized, the above picture will obviously be distorted and may have varying degrees of aberration.

The proximity of the sulfides to the fluid discharge sites defines whether an ore body is to be characterized as proximal or distal. The former has an alteration zone in the footwall along with fragmental (brecciated) sedimentary rocks. The distal ores form commonly in a brine pool depression away from the vent and are characterized by well-laminated sulfides interbedded with sediments that have a conspicuous absence of any wall rock alteration.

In summary, most SEDEX deposits form during post-rift reactivation of extensional structures accompanying the thermal subsidence or "sag" phase of a long-lived (100–200 My) basin (Fig. 2.47). They occur mostly within third-order basins near their down faulted margins. There is also a close spatiotemporal relationship between SEDEX deposits and sills, dykes, and related magmatic rocks in some instances. Magma injection into the lithosphere seems to have played an essential role in generating the heat source necessary for convection of metalliferous hydrothermal fluids and formation of SEDEX deposits. The nature of the sediment-fill in the basin plays a crucial role in SEDEX deposit formation. The syn-rift clastic sequence in the lower part of the basin acts as an *aquifer* while the reduced black shale and chert facies acts as an *aquitard*. The aquifer allows the fluids to carry metals and convect and slowly gets geopressured. Finally, the cap rock is ruptured and the mineralizing fluid is focused along active faults. As it

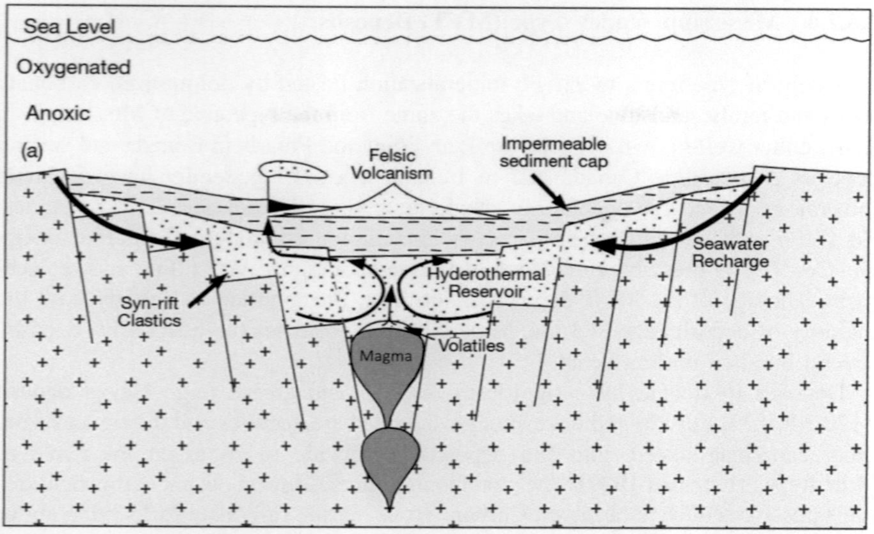

Fig. 2.47 SEDEX deposit model, based on the Bathurst Mining Camp, Canada. All the critical factors mentioned in the text are seen in this model (after Goodfellow 2004)

exhales on the sea floor, it mixes with the H_2S in the ocean water column. Based on available fluid inclusion data, morphology of stratiform sulfide bodies and vent textures, the hydrothermal fluids that vented into seawaters behaved either as buoyant plumes or bottom hugging brines, eventually producing vent proximal and distal deposits. Mineral assemblages, fluid inclusions, and stable isotope data indicate that many of the SEDEX deposits formed from Zn−Pb(± Cu, Fe, Ag, Au, Ba)-rich, low to moderate temperature, weakly to highly saline basinal fluids (Fig. 2.48) that were generated by reaction with sediments derived from continental crust and leached the metals from the Fe-oxide/-hydroxide coatings of grains in sandstones, lithic clasts, etc.

Fig. 2.48 Model for the generation of metalliferous brine and SEDEX deposit (after Lydon 2004)

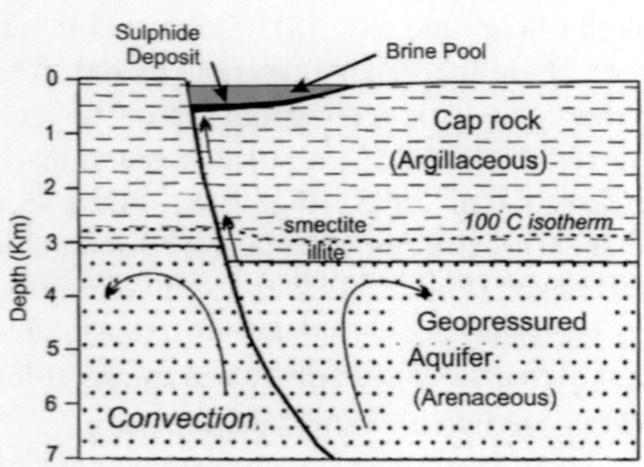

2.3.2.4 Mississippi Valley Type (MVT) Deposits

This deposit type refers to Zn–Pb mineralization hosted by dolomitized carbonate rocks and rarely sandstone and takes the name from the type area of Mississippi in USA. Other well-known deposits are Pine Point and Polaris in Canada and several deposits in maritime Canada and in Ireland. The ores generally have a simple mineralogy of coarse galena ± sphalerite and colloform sphalerite at times. Zn ≫ Pb in these deposits. The ores also contain the association of Ba–F–Cu–Ag–Ge–Co–V. The ores show evidence of low temperature cavity filling and replacement. The age of the MVT deposits range from Precambrian to Tertiary with the majority of deposits aged <1 Ga. Mineralization postdates the host rocks, even by several hundred million years.

Deposits in Ireland have significant reserves and grades (e.g., Navan deposit +120 Mt, 12% Zn–Pb) and have enough shared characteristics and differences from other carbonate-hosted zinc-lead deposits worldwide to be given the name of "Irish-type" (Hitzman 1999). The stratabound mineralization occurs in this field in a transgressive sequence of Lower Carboniferous marine carbonate rocks lying above a wedge of Upper Devonian continental red beds, similar to those seen in maritime Canada. The orebodies follow dislocation structures, such as, normal faults. Besides sphalerite and galena, Fe sulfides, and barite may be locally important. Textures of the ores are complex and variable; some indicate early diagenetic origin. The tectonic setting of these deposits indicates the extensional regime of a riftogenic basin margin and reactivation of basement faults. The Pb isotopes in the ores are radiogenic.

The MVT deposits are hosted by platform carbonates on stable cratons that postdate clastic wedge deposition. The dolomitization produces the necessary porosity-permeability in the host rocks for the movement of basinal fluids. Fluids are metal rich, saline Na–K–Ca–Cl rich brines and are commonly <150 °C, but may be hotter (<250 °C) in some cases, as in the Irish deposits. They are similar to basinal brines found in oil and gas districts. Mineralization is considered to have formed by basin dewatering, the reasons for which are often debated. Three distinct models are available for the explanation of the generation and movement of basinal fluids associated with MVT deposits. These are: (i) Topographic or gravity-driven fluid flow model; (ii) Sedimentary and tectonic compaction model; and (iii) Hydrothermal convection model. The first involves flushing of subsurface brines out of a sedimentary basin by groundwater flow from recharge areas in elevated regions of a foreland basin to discharge areas in regions of lower elevation (Garven et al. 1993) (Fig. 2.49). In the second, expulsion of basinal fluids through diagenesis and compaction and episodic fluid release from overpressured aquifers is envisaged (Fig. 2.50). The last invokes deep convection of hydrothermal brines due to buoyancy forces related to temperatures and salinity variations and producing long-lived flow systems that recycles solutions through the rock mass and brings about regional dolomitization.

There is a general consensus among many geologists that a continuum exists between VHMS and SEDEX deposits on one hand and between SEDEX and MVT

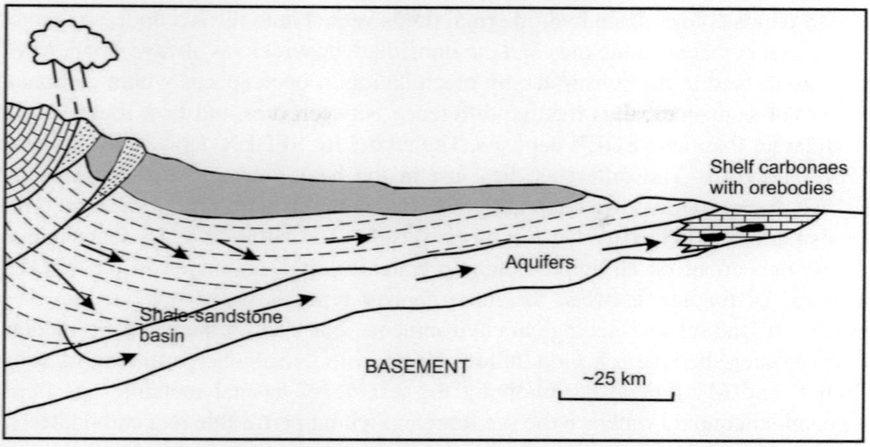

Fig. 2.49 Topography or gravity-driven fluid flow model for MVT deposits. Fluid flow initiated by the elevated thrust-faulted landmass which provides the hydraulic head (after Garven et al. 1993)

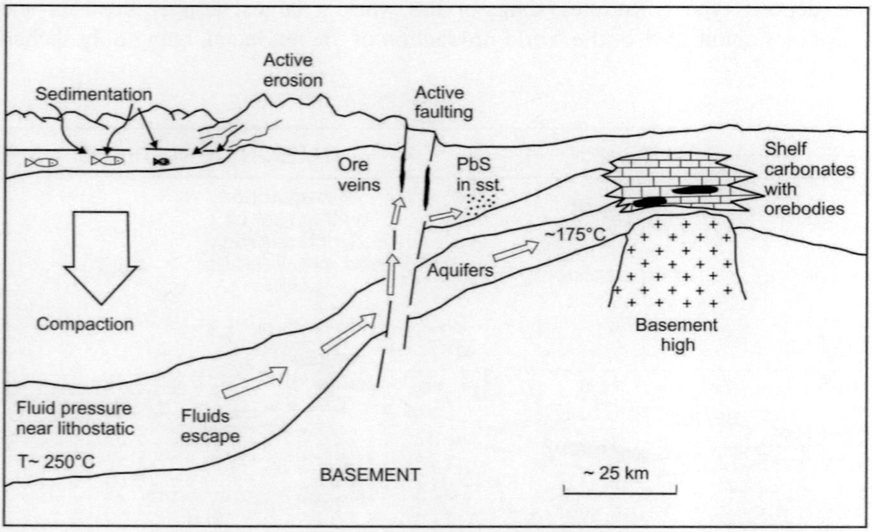

Fig. 2.50 Model for the formation of MVT deposits by expulsion of basinal fluids through diagenesis and compaction and episodic fluid release from overpressured aquifers (after Evans 1997)

(cf. Sangster 1990) on the other. Compared to most VHMS deposits, hydrothermal systems that generate SEDEX deposits are commonly long-lived, produce a temperature maximum quickly and decay slowly over millions of years. SEDEX and

VHMS deposits form from hydrothermal fluids vented onto the sea floor, so that the age difference between the ores and the immediate host rocks is always small. MVT deposits formed in the subsurface by precipitation in open spaces within carbonate platformal sequences; thus the age difference between ores and host rocks can be much larger than for SEDEX deposits. The model for SEDEX deposits is similar to VHMS deposits. The difference lies just in the heat source and lower T in the SEDEX formation, as discussed above. The metal association and alteration types are also different in the two deposit types. Besides, formation of VMS deposits may be less dependent on an ambient anoxic water column, compared to the SEDEX deposits. Distinction between the three deposit types is based not only on their geological features and geological environments, but also on their genetic models. Some workers believe (cf. Goodfellow 2004) in the cogenetic formation of some SEDEX and MVT deposits involving the mixing of basinal metalliferous fluids with ambient anoxic waters at the sea floor and within permeable reef carbonates, as shown in Fig. 2.51.

2.3.2.5 Sediment-Hosted Stratiform Copper (SSC) Deposits

This deposit type constitutes some of the world's largest copper deposits and contributes about 25% of the world production of the red metal, being only second

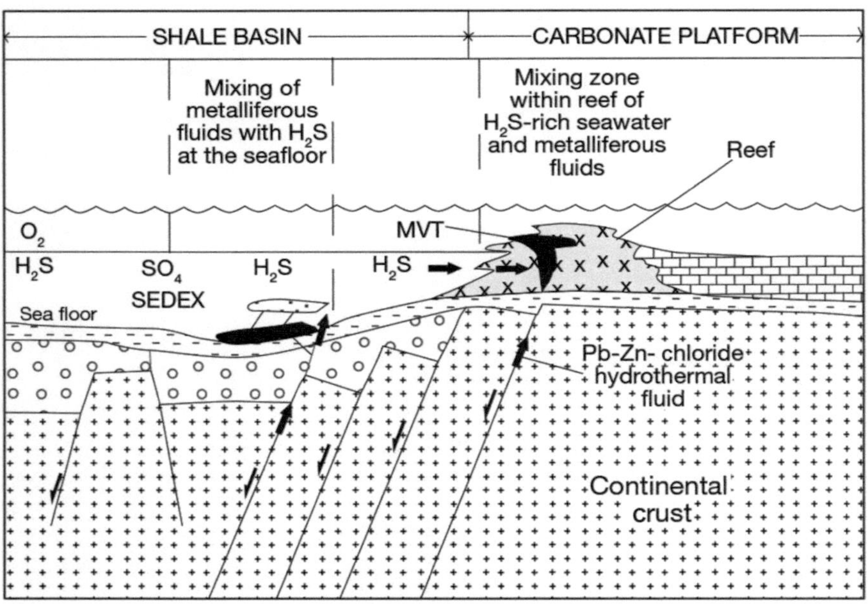

Fig. 2.51 Model for cogenetic formation of SEDEX and MVT deposits (after Goodfellow et al. 1993)

to porphyry copper deposits. Some deposits also produce large amount of Ag and a few are sources of Co, U, and other metals. The deposits are widespread, occurring in almost all geological eras from Early Proterozoic, postdating the first occurrence of undisputed red beds and *cyanophytes* (ca. 2.3–2.4 Ga), to Recent (Kirkham 1989). The most important and large deposits occur in Late Proterozoic and Late Paleozoic rocks which correlate closely with widespread desert sedimentary environment. Major part of the production comes from a few very large deposits and districts of the Central African Copper belt of Zambia and Zaire, Dzhezkazgan, and Udokan in erstwhile Soviet Union, and Lubin in Poland. White Pine, Michigan and Spar Lake, Montana are the significant producers in North America.

These deposits have several distinguishing features (Kirkham 1989). They contain disseminated, commonly zonally distributed sulfides that occur in reduced rocks near oxidation-reduction boundaries. The sequence of minerals from the oxidized to the reduced side comprises all or some of the following: hematite, native copper, chalcocite, bornite, chalcopyrite, galena, sphalerite, and pyrite. Mineral zones overlap upward and outward. In most localities, typical associated rock types are red beds and evaporites. Major deposits ("Kupferschiefer type") are located in anoxic sedimentary rocks immediately overlying typically red, continental clastic sedimentary rocks. The host rocks were characteristically deposited in arid to semi-arid areas within 20°–30° of the paleo-equator and in most areas are interbedded with evaporites.

Theories on the origin of sediment-hosted stratiform copper deposits revolve around two main issues: the source of their contained metal, sulfur, and other constituents and the manner of migration to and precipitation of these constituents in their depositional sites. The large stratiform Cu deposits appear to have formed in basins with unique conditions that allowed accumulation of large amounts of metal-bearing fluids, sufficient reduced sulfur, large amounts of reductants, and focusing of fluid movement into relatively small areas (Hitzman et al. 2010). General stratigraphic sequence in productive basins (Fig. 2.52) comprises a basal horizon of syn-rift red beds, often with mafic or bimodal volcanic rocks that serve as a source for rock-buffered oxidized fluids as well as a source of metals, particularly copper. This oxidized sedimentary package is overlain by marine to lacustrine sediments of sandstone, siltstone and shale, which may locally be organic-rich. This siliciclastic sequence grades upward to marine carbonates that contain a thick evaporite sequence. The upper portion of the basin contains shallow marine to continental siliciclastic sediments. The evaporate beds provide an effective top seal to the hydrologic system, whereas the basin edges themselves provide lateral containment. The total thickness of the sediments in the basin may range from several to >10 km. The stratiform Cu orebodies owe their origin to residual brines or brines from evaporate dissolution that undergo diagenetic infiltration and move downward into the basal oxidized red bed sequence. Geothermal heat and heat from igneous activity initiate convection of these highly saline fluids which leach metals from both the red beds and the basement. The oxidized metal-rich brines on their upward flow encounter organic-rich sediments which work as reductants and help to precipitate the copper sulfides. Fluids may also move up the faults in the basin

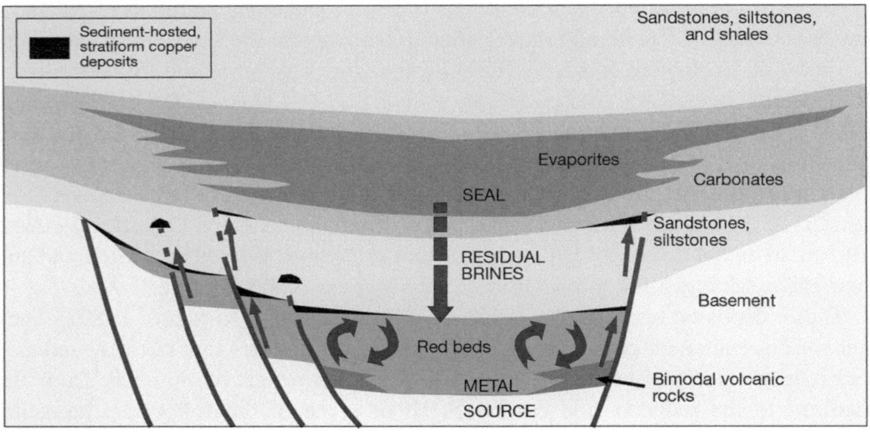

Fig. 2.52 Sketch of an intracratonic, hydrologically closed basin where stratiform copper deposits form (after Hitzman et al. 2010)

and precipitate sulfides where they encounter reductants, such as natural gas, petroleum, etc. However, the roles played by evaporites, red beds and anoxic host rocks demonstrate the importance of the environment of sedimentation and of diagenetic oxidation–reduction processes in the formation of this deposit type, which therefore, may also be classified under "Mineralization under sedimentary-diagenetic processes" (see below).

2.3.2.6 Metamorphism of Massive Sulfide Deposits

The relationship of ore deposits with metamorphism in a metamorphic belt may be casual or causal giving rise to *metamorphosed* or *metamorphogenic* deposits, respectively (Mookherjee 1976). It is generally noted by most ore geologists that the majority of stratabound VMS, SEDEX, SSC and MVT deposits, particularly in Precambrian terrains and/or in orogenic belts, are co-deformed and isofacially metamorphosed with their host rocks. The metamorphic record is better preserved in the associated rocks and the silicate gangue minerals, as the sulfides are less reactive and their thermal stability is generally inferior to the accompanying oxides and silicates. Ores show metamorphism in grades varying from greenschist to the granulite facies in the broad temperature range of 400–800 °C and at about 4–7 kb pressure. Some ores are also known to display evidence of polymetamorphism with the signatures of regional metamorphic event overprinted by a thermal imprint. In most situations, the orebodies may display morphological changes from their original, textural, and mineralogical changes and evidence of translocation if remobilized. Massive sulfide deposits of all kinds may retain textural features of deformation or get partially or wholly recrystallized while undergoing deformation and may turn into a metamorphic tectonite. Thus, distinguishing between primary, diagenetic, and subsequent deformation-recrystallization textures in such ores can be critical in defining the time of ore emplacement and developing appropriate genetic models. Metamorphism in middle to high grade does not generally obsure the original mineralogy of these deposits, even though some significant mineralogical changes may be produced. Pyrite may be converted to pyrrhotite by reaction with ferrous silicate minerals or by degassing of sulfur from the sulfide orebodies during metamorphism. During metamorphic retrogression, pyrrhotite may react to form pyrite and magnetite in minor quantities.

Metamorphosed deposits of these types are known in all continents (vide Spry et al. 2000), the best studied ones being in the Scandinavian Caledonides, Mt. Isa belt and Broken Hill deposit as well as Mt. Lyell and Roseberry, Tasmania, in Australia and deposits in the Cape province in South Africa–Namibia. In India, some of the best example of ore-host rock metamorphism is noted in the Paleoproterozoic sediment-hosted Pb–Zn sulfide deposits, and Neoproterozoic VMS deposits of Rajasthan, India (Deb 1979, 1990). Ore metamorphic features are also reported from Paleoproterozoic Khetri copper belt, Rajasthan (Sarkar and Dasgupta 1980) and Singhbhum copper belt, Jharkhand (Sarkar and Deb 1974).

While field studies, including detailed structural mapping help to identify the phases of deformation, petrographic criteria of crystallization–deformation relations in minerals in the host rocks and ores, along with conspicuous textural and mineralogical changes, and their compositional characteristics help to trace the tectono-thermal evolution of the mineralized belts (cf. Tiwary et al. 1998). Deformational structures in ores and ore bodies vary from folds (Fig. 2.53a), faults (including ductile shears) to compressional (flattening) and extensional stretching (with or without tearing). A fold structure in ores may be syn-deformational or post-deformational. In the first case, one will expect crestal thickening (in the base metal sulfide ores in particular) and may be, also axial cleavage in competent layers (Fig. 2.53b). These features will be absent if the fold-form is inherited through hydrothermal replacement of a compositionally suitable rock layer. Of course, deformational features may be recorded in a single grain to an entire ore body. Microfabrics indicating ductile deformation are slip bands, kinks, and translation twin lamellae in softer sulfides. The brittle vs ductile character of a mineral, ore or otherwise, is controlled by the mineral structure, ambient temperature and the rate of strain. Strong deformation in conjunction with enhanced temperature during dynamothermal metamorphism eventually leads grain growth and annealing. While softer sulfides are annealed and developed annealing twin lamellae and foam structure (Fig. 2.53c, d), the harder sulfides and some oxides underwent metablastic growth with flowage of sulfide–silicate schistosity and rare pressure shadows, and in some cases, intense fracturing during brittle deformation. Grain size generally increases with the rise in grade of metamorphism. But if the deformation outlasts recrystallization, then the deformational fabric with the grain size reduction may be conspicuous. The above two broad textural varieties may coexist in neighboring domains due to uneven strain-distribution. In extreme cases, a *durchbewegung structure* may develop in an ore which includes rootless fragments and rounded

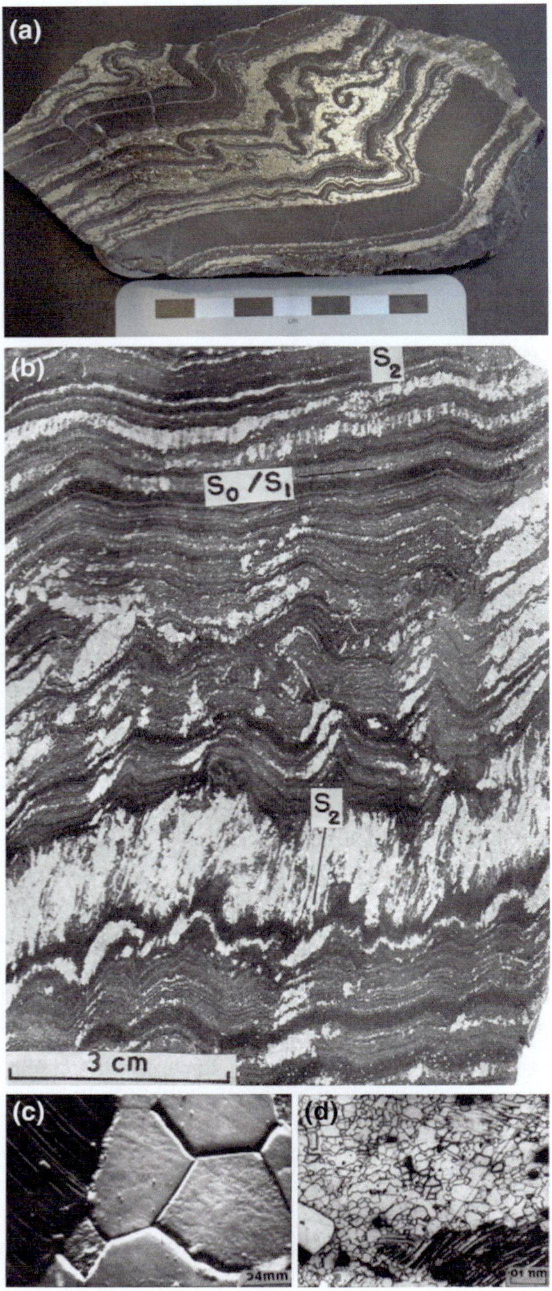
(a)
(b)
S2
S0/S1
S2
3 cm
(c)
(d)

◂**Fig. 2.53** **a** Intricate folding in carbonaceous chert and interlayered recrystallized pyrite bands in a sample from Balaria, Zawar ore deposit, Rajasthan (Photo: M. Deb). **b** Crenulated bands in pyrite-sphalerite-black chert rhythmite from Rajpura-Dariba deposit, Rajasthan. Note the transposed schistosity at high angles to the pyrite laminae and the subparallel orientation of pyrite boudins along it (after Deb and Bhattacharya 1980). **c** SEM photograph of foam structure in galena mass in Deri deposit, Rajasthan. Note the deformed twin lamellae in adjacent sphalerite (after Deb 1979). **d** Mosaic of recrystallized chalcopyrite grains in Ambaji deposit, Gujarat. Note relict translation gliding in patches of chalcopyrite (after Deb 1979)

clasts of competent components (silicate rocks and hard ore minerals) studded throughout an incompetent (sulfide) matrix. Such ores equate with high-strain domains of partitioned, non-coaxial rotational deformation arising through extreme heterogenous flow during a single event or separate overprinting events (Marshall and Gilligan 1989). Large-scale chemical mass-transfer from an ore deposit during metamorphism is not established.

2.3.2.7 Orogenic Gold Deposits

Gold mineralization in metamorphic tectonites, until recently, was generally referred to as "Lode gold" (Poulsen 1995), "Mesothermal gold" (Nesbitt et al. 1986), "Shear gold" (Laznicka 1993) or "Gold–quartz–carbonate lodes" (Kerrich and Fyfe 1981). It is clear that the nomenclature varies depending on which aspect of the mineralization is emphasized. Also it is now known that several deposits belonging to this broad group formed in P-T conditions surpassing the classical conditions of mesothermal regime. A term widely accepted now is "orogenic gold deposits" (Groves et al. 1998) which is defined as quartz-dominant lode gold systems formed in the compressional regimes of crustal accretion along convergent margins of continents (Fig. 2.54).

Giant orogenic gold deposits (>500 t Au) have been mined throughout the 20th century. These include Homestake, McIntyre-Hollinger, Kirkland-lake, Berezovkoe, Grass valley-Nevada City, Bendigo, and Kolar. Some continue to be important resource still, such as, Ashanti, Golden Mile/Superpit, Campbell-Red Lake, Morro Velho, Linglong, etc. Gold grades in this deposit type of ores are highly variable – ranging from as low as 1 g/t to often as much as 20–30 g/t. Figure 2.55 shows relative abundance of Au, Ag, and base metals in orogenic gold deposits in greenstone belts along with other common types of hydrothermal ore deposits for comparison (Poulsen et al. 2000).

Orogenic gold deposits share the following common features (Groves et al. 1998):

1. Orogenic gold deposits are quartz-dominant vein systems in which sulfides (mostly Fe sulfides) generally constitute $\leq$3–5% and carbonates $\leq$5–15%, with high Au: Ag ratio ($\sim$10:1 to 5:1)(Goldfarb et al. 2005). In the more common greenschist facies environment albite, white mica, chlorite, scheelite, and tourmaline are present as gangue in small but variable proportions. In the

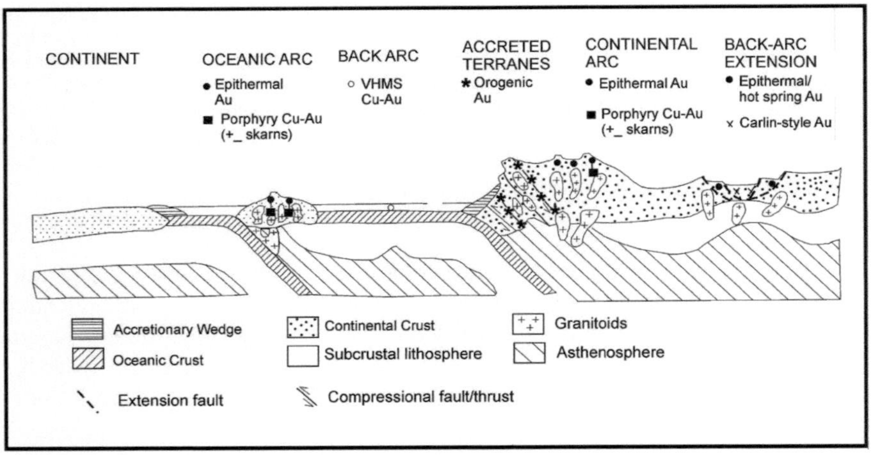

Fig. 2.54 Tectonic setting of different types of gold deposits, including orogenic gold deposits (after Groves et al. 1998)

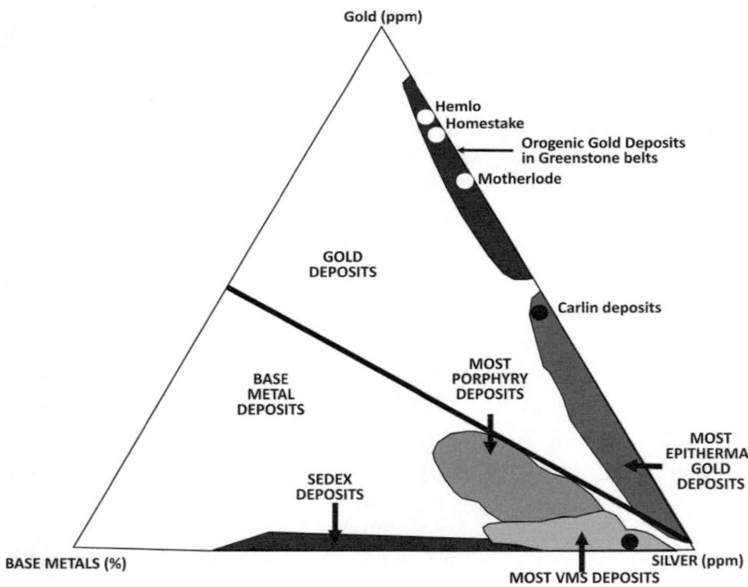

Fig. 2.55 Triangular plot of the relative abundance of Au (ppm), Ag (ppm), and base metals (%) in orogenic gold deposits in greenstone belts and other common types of hydrothermal ore deposits (after Poulsen et al. 2000)

amphibolites-granulite facies environment, hornblende, biotite, diopside, and even augite may be present in addition.

2. Orogenic gold deposits are mostly hosted in deformed metamorphic belts of all ages. They are most common in greenstone belts in Precambrian shield areas and in late Neoproterozoic-Phanerozoic sedimentary rock dominant fold belts. Syn-gold intrusions generally occur at, or within a few km of the deposits, but there are some notable exceptions (e.g., Golden Mile, Bendigo).
3. Metamorphic grade of host rocks is typically greenschist facies, but higher (e.g. Kolar) and lower (e.g., Donlin Creek) grades are also observed. Peak P-T conditions in the country rocks have been reached within a few million to tens of million years prior to ore formation for almost all deposits. Orogenic gold deposits in eastern Asia, in very high-grade metamorphic rocks, provide a globally unique example, where important deposits postdate metamorphism of their host rocks by billions of years; these are the only large Phanerozoic gold deposits known to be hosted in Precambrian cratons (Goldfarb et al. 2007).
4. The ore is epigenetic, hydrothermal in nature with characteristic albitization, sericitization, carbonatization, sulfidation, and tourmalinization of the wall rocks. Gold occurs mostly in veins but may also be present in the wall rocks.
5. Two major controls of gold mineralization of this type are structures and petrochemistry of the source and host rocks. First-order controlling structures are regional fault/shear zones in which the vein systems may extend for kms along strike, 1–2 km along dip with little change in mineralogy or even gold grade. A diversity of second-and third-order structures, from shear zones to fold hinges, is recognized as controlling the gold mineralization. Brittle-ductile structures locally host the ores. Hodgson (1989) proposed four types of structures for shear-related vein-type gold: shear fracture, extension fracture, hybrid (shear + extension) fracture, and fold-related structures. Robert and Poulsen (2001), on the other hand, identified four major vein types: fault-fill veins (mostly shear veins), extensional veins, stockworks , and breccias. The first is most common in these gold deposits, characterized by laminated quartz, slickenlines, and slip surfaces. Most fault-fill veins form lenticular mineralized structures in the central part of ductile or brittle-ductile shear zones, either parallel to or at a low angle with the shear zone boundary (Chattopadhyay 2010) (Fig. 2.56).
6. From a geochemical angle rocks rich in carbon or Fe^{2+} occurring along the path of hydrothermal fluid flow are conducive to gold deposition through redox reactions. Precambrian sulfide facies BIFs are often good hosts of gold mineralization.
7. The ores are interpreted to have formed from low salinity, near neutral H_2O–$CO_2 \pm CH_4$ fluids.
8. Emplacement of orebodies vis-à-vis the stage of metamorphism and the source of ore fluids and ore elements are other relevant issues in this context. Ore-forming fluids may be introduced during pre-peak/syn-peak stages of prograde metamorphism or during retrograde metamorphism.

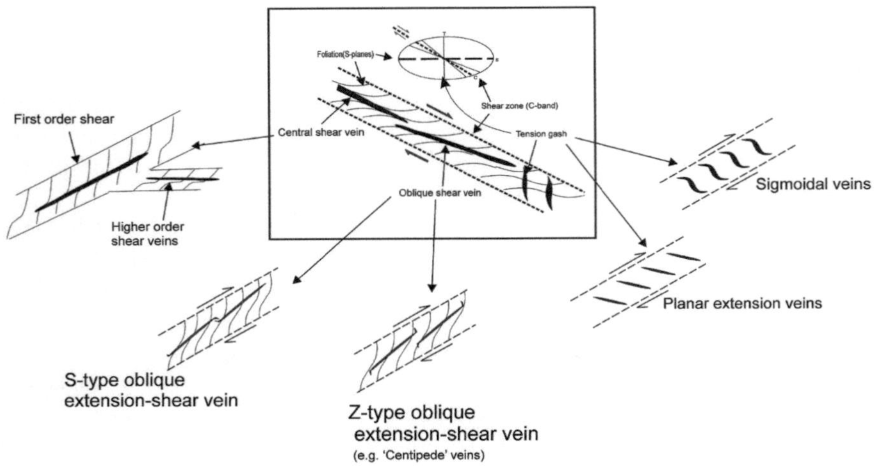

Fig. 2.56 Different types of shear and extension (or hybrid) veins associated with orogenic gold deposits (after Chattopadhyay 2010)

9. The general consensus regarding the potential ore sources for such deposits are metamorphic dehydration in a subducted slab, mantle devolatilization, magma crystallizing at depths, deep basinal fluids, or even the surface-derived fluids that penetrated deep into the crust (Goldfarb et al. 2005). Some deposits have been suggested to be multisourced and to have a multistage history of formation (cf. Deb 2014 and Chap. 5).
10. The gold resources in orogenic gold deposits show a secular concentration between 2.8–2.55, 2.1–1.75, 750–35 Ma (Fig. 2.57) which correlates well with continental growth, corresponding to the formation of Kenorland, Columbia, Gondwana/Pangea and perhaps a new super-continent, respectively. The gold deposits formed during growth of Rodinia appear to have been lost from geologic record due to erosion of Meso-Neoproterozoic blocks to their deep crustal levels that are below gold-favorable zones (Goldfarb et al. 2001; Groves et al. 2005).

2.3.2.8 Unconformity-Related Vein Complexes of Uranium

These deposits constitute approximately 33% of the world uranium resources and form due to geological and chemical changes in the hydrothermal solutions close to major unconformities. Uranium deposits that occur close to the early-middle Proterozoic unconformity or paleo-weathering surface and may be located above, below, or along it, are assigned to this group. Unlike the QPC deposits they are not hosted by pebbly rocks but are located at the base of a Proterozoic sandstone

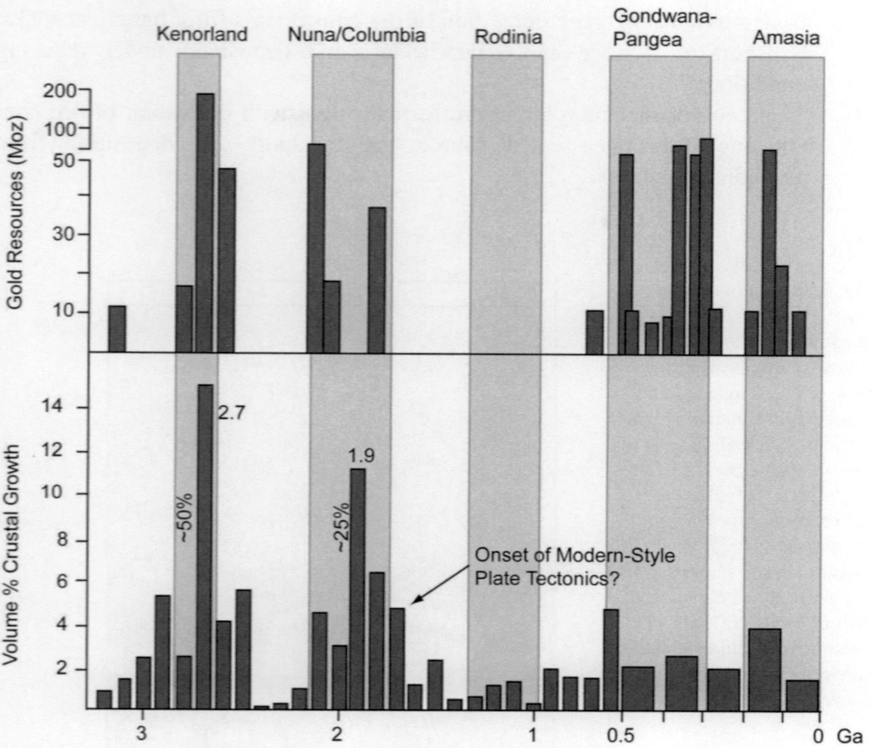

Fig. 2.57 Global resource versus age of orogenic gold deposits. Crustal growth through time and supercontinent assembly also shown (after Groves et al. 2005)

sequence, where they unconformably overlie pre-Middle Proterozoic metamorphic basement rocks which generally include graphitic pelitic units and underwent complex post-depositional transformation. The mineralization may be monotonous uranium-only or polymetallic, with Ni, Co, and As accompanying uranium mineralization in fair proportions along with some minor elements. The most well-known examples of this deposit type are in the Athabasca and Thelon basins of Canada and in the Pine Creek region of the Northern Territory, Australia. The most important deposits in the Athabasca basin in Saskatchewan, Canada are Cigar lake, Key lake, Rabbit lake, McArthur River, and Eagle Point while notable Australian deposits are Ranger I and III and Jabiluka I and II. The average grade range from 0.17% U in Ranger III to as high as 12.2% U in the main pod of the Cigar Lake deposit. The Australian deposits are lower grade but exhibit a larger ore tonnage.

Three conceptual genetic models have been proposed for the Canadian deposits of this type (Ruzicka 1995):

(i) Near-surface supergene derivation of ore constituents from basement rocks, transport by surface and groundwaters and deposition under reducing conditions.
(ii) Magmatic or metamorphic hydrothermal origin with derivation of ore constituents from deep-seated source and transport and deposition from ascending solutions.

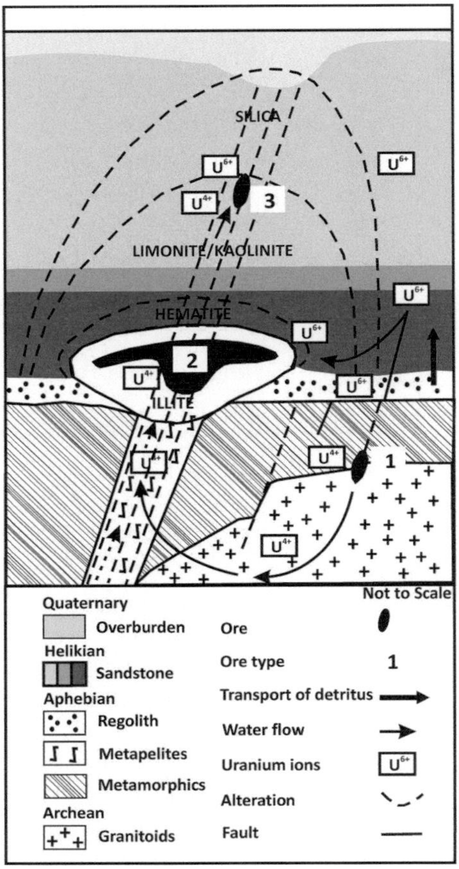

Fig. 2.58 A conceptual genetic model of unconformity-hosted uranium deposits. Arrows indicate flow paths of reduced and oxidized convective waters. Various styles of mineralization are: (*1*) High-grade polymetallic mineralization below the unconformity; (*2*) Medium-grade monometallic mineralization below the unconformity; (*3*) low grade monometallic mineralization above the unconformity (after Ruzicka 1995)

(iii) Diagenetic-hydrothermal origin with precipitation of uranium from local reductants (Fig. 2.58). Uranium was introduced in the geochemical cycle of granitic magmatism in the Late Archean or Early Proterozoic and was concentrated subsequently in sedimentary rocks and as pitchblende-brannerite deposits. Principal concentration of the ore elements took place during the pedogenesis of these basement rocks and diagenesis of the basinal sediments produced the metal reservoir. Subsequent tectono-thermal events involving the basin activated the hydrological system and caused convective cycling of fluids with mobilization/remobilization of the metals from the reservoir.

2.3.2.9 Sandstone-Hosted Uranium Deposits

This deposit type is also known as Colorado Plateau type or "roll-front" uranium deposit and takes its name from the type area of Colorado Plateau (Northrop and Goldhaber 1990) in Utah and New Mexico. They are also found in Wyoming, South Dakota and South Texas in U.S.A. Most deposits are Devonian or younger. Secondary roll-front deposits are mainly Tertiary. The deposits may be associated with sediment-hosted deposits of V and Cu in similar environment. These strata-bound deposits are characterized by microcrystalline uranium oxides, coffinite, pyrite deposited during diagenesis in localized organic-rich reducing environments within fine to medium-grained sandstone (feldspathic or tuffaceous) beds in the interstices of detrital quartz and feldspar grains. Mudstone or shale occurs above and/or below the host sandstone. The ore bodies are tabular, lensoid, or roll-like, i.e., sinuous in both plan and section, with the crescentic lens cutting across bedding. Replacement of wood or other carbonaceous material by ore constituents is observed. The ore is sometimes dispersed throughout the host rock. Ore grades vary from 0.1 to 1% U_3O_8. Oxidation of primary uraniferous minerals produces a variety of minerals, notably yellow carnotite in V-rich ores. The depositional environments are commonly continental-basin margins, fluvial channels, and braided stream deposits. Adjacent major uplifts provide favorable topographic conditions.

A common genetic model of roll-type uranium hinges on the paleohydrology of meteoric waters and the two valence states of uranium, U^{6+}, U^{4+}: U^{+6} being strongly soluble in oxidizing conditions while U^{+4} being relatively insoluble in reducing conditions. Three stages of U-collection to subsequent deposition in geological situations are recognized (Fig. 2.59): (1) Oxidative leaching of uranium from slightly enriched felsic rocks (in basement or clasts in host sequence). (2) Transportation by oxidizing surface and groundwaters through porous and permeable rocks. (3) Precipitation on reaching a redox boundary (Goldhaber et al. 1978). Roll-type deposits thus form when oxidized meteoric fluids containing uranyl carbonate complexes flow down a reduced aquifer which is highly permeable during mineralization, but is subsequently restricted by cementation and alteration. Redox reactions involved produce the uranium ores. Oxidized iron minerals characterize the rocks up dip while

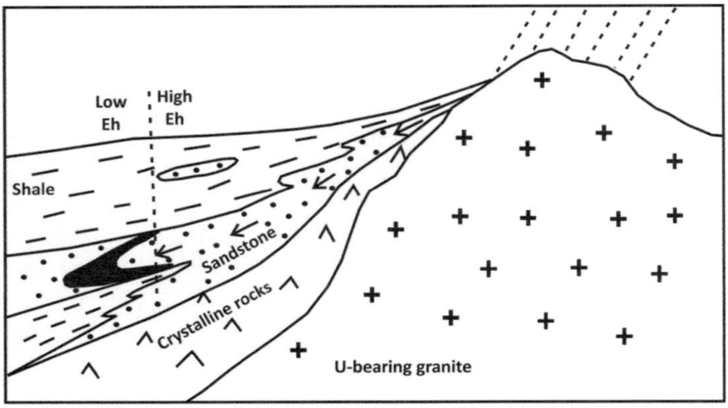

Fig. 2.59 A schematic model for the origin of roll-type uranium deposit (after Goldhaber et al. 1978)

reduced iron minerals are found in rocks down dip from the redox interface. Some uranium oxides may be redistributed by groundwater at the interface between reduced and oxidized ground. Felsic volcanic and plutonic rocks in the district are the common sources of uranium. In tabular ores, the source rocks for ore-related fluids are commonly in overlying or underlying mud flat facies sediments. Tabular ores form by fluid mixing at the interface between two low temperature meteoric fluids. The lower one, relatively stagnant, is a brine containing Na^+, Mg^{2+}, Ca^{2+} and Cl^-, and SO_4^{2-}. Another low-density flow containing such complexes as, $UO_2(CO_3)_2^{2-}$ and VO^+ move along the upper surficial zone of the lower unit, interact, and precipitate the ore (Northrop and Goldhaber 1990). Uranium oxide is produced by the breakdown of the uranyl complex, while the uranous ion and silica combine to form none too uncommon silicate U-ore mineral, coffinite.

2.4 Sedimentary(-Diagenetic) Processes

The ore deposits that form in the exogenous environment may have been transported from varying distances as terrigenous clastics into the basin or zone of deposition, and are termed *allochthonous* deposits. In contrast, there are other deposits of ore metals which form in the environment in which they are deposited and are termed *autochthonous* deposits (Evans 1997). The first group belong to the mechanically concentrated placer deposits (alluvial, beach, and eluvial) and their ancient counterparts, the quartz pebble conglomerate-type, paleo-placer deposits. The second group on the other hand, includes the various chemical (±biochemical) precipitates, such as the banded iron formations (BIFs), the sedimentary manganese deposits, stratiform/sedimentary copper deposits, to name the most important ones in economic terms. The sedimentary(-diagenetic) processes which mainly produce the second group of deposits may also have their imprint on some of the first group, such as the paleo-placer deposits, as discussed below. The residual deposits of Fe, Al, Ni, Mn, and Au and supergene deposits of Cu and Ag also develop in the exogenous environment but will be dealt with separately.

2.4.1 Placer and Paleo-Placer Formation

Mechanically concentrated placer deposits may contain many economically important resources, such as, gold, diamond, ilmenite, monazite, tin which are all of high-specific gravity and mostly of high hardness (barring gold), and are therefore resistant to attrition during winnowing and transportation by an aqueous medium. They are chemically stable as well. The source rocks which contain these economic minerals undergo weathering and the minerals are dislodged and subsequently transported over varying distances by water. The amount and distance of transportation are dependent on the quantum and velocity of the flow. Any physical feature which can bring about a drop in the rate of flow, such as, where the gradient flattens or at the inside of river meanders, a dike across the river, natural riffles on the floor of the river or stream, below rapids and falls, beneath boulders, in vegetation mats or where a tributary produces a confluence, can be responsible for the concentration of the economic heavy minerals by its separation from the lighter detritus, mainly silica sand. Such placer concentrations can get buried by later sediments and can occur along buried paleochannels. The most common aqueous environments for accumulation of placers are stream channels or beach and shelf regions. Stable coastal regions receive sediments from deeply weathered high-grade metamorphic terrains. The winnowing action of the surf brings about heavy mineral concentration, including that of gold and diamond (Sutherland 1982) if present in the source rocks, in raised, present and submerged beaches (Fig. 2.60). Both these environments are well represented in India, as detailed in a subsequent chapter. Well-known alluvial placer deposits of gold are recorded in Subarnarekha river in Jharkhand, Nilambur valley in Kerala, some tin deposits are found in Thailand and Malaysia whereas well-known beach placers are the ilmenite and monazite beach sands of the Malabar coast of India (Sarkar et al. 1995) and diamond deposits of Namibian coast. Russia produces about 200 t of placer gold per year.

2.4.1.1 Quartz Pebble Conglomerate (QPC) Deposits

Quartz pebble conglomerate (QPC) deposits, found in the Archean and Proterozoic sequences, are economically very important, being a major source of gold and a significant source of uranium. The world class example is the Witwatersrand gold fields in South Africa, which has a gold reserve of 52×10^3 t Au, accompanied by 533×10^3 t U, >3×10^3 t Ag, ~30 t PGE, and ~1000 t Hg. Other QPC deposits

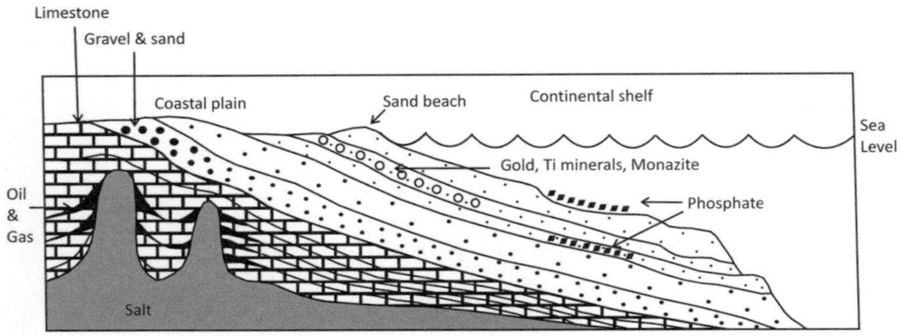

Fig. 2.60 An idealized section of a continental shelf showing the probable location of limestone, sand and gravel, phosphates (in outer portion of the shelf) and placer gold, Ti-minerals, monazite etc. (in the inner shelf). Salt domes may develop locally where previously deposited salt beds are deeply buried. This environment is conducive to oil and gas migration and entrapment (after Craig et al. 2001)

include Serra de Jacobina in Bahia state of Brazil, Dominion Reef, RSA, Tarkwa in Ghana. In the Blind River-Elliot Lake in Canada (with resource about 432×10^3 t U, 350×10^3 t Th and 72 t REE+Y), Witwatersrand and Jacobina deposits, uranium is extracted as a by-product of gold.

The Witwatersrand Goldfields in South Africa are located in the central portion of the Archean Kaapvaal craton within low-grade metamorphosed siliciclastic sediments of a Mesoarchean (2.9 Ga old) intracratonic basin. Seven major goldfields and a few smaller occurrences located around the northern and western margin of the 350 km-long Witwatersrand basin (Fig. 2.61) have so far yielded one-third of all gold ever produced on Earth and still remains the largest repository of gold resource ($\sim$30,000 tons still remaining) in the world. Each goldfield comprises one major and several minor reef horizons. The ore bodies are essentially stratiform, 1 cm to several meters in thickness, within fluvial to fluvio-deltaic, predominantly pyrite-rich (locally also uraninite-rich) quartz pebble conglomerates ("reefs"), polymictic conglomerate, carbon seam and pyritic quartzites, which overlie regional unconformity surfaces (Phillips and Law 2000; Frimmel 2005; Tucker et al. 2016).

Stratigraphically, the host Witwatersrand Supergroup overlies the +3.0 Ga basalts, volcaniclastics, and minor quartzites of the Dominion Group which represents the rifting phase of the granites and greenstones of the basement. The Witwatersrand Supergroup comprises a lower, dominantly marine West Rand Group (base fixed at 2.970 Ga) overlain by the mineralized fluvial sands, conglomerates and minor shales of the Central Rand Group. Subrounded pebble shapes (Fig. 2.62) suggest textural maturity and the oligomictic conglomerates are interpreted as channel-filled deposits (paleoplacers). A downstream decrease in grain size along with differentiation of heavy minerals indicates fluvial transport of ore minerals. The mineralized zone occurs at/close to the base of the host sequence. The

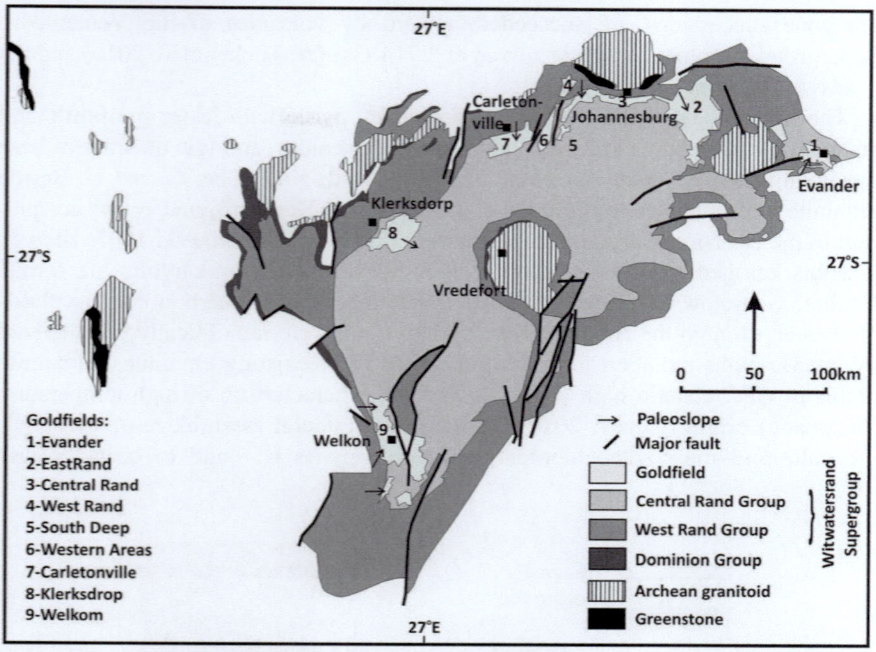

Fig. 2.61 Simplified geological map of the Witwatersrand basin, showing the location of the goldfields (after Frimmel 2005)

Fig. 2.62 A polished slab of a drill core from Witwatersrand deposit comprising subrounded to elliptical quartz and pyrite pebbles (Photo: M. Deb; sample from the collection of Late Prof. M. Schidlowski, MPI, Mainz, Germany)

ore zone successions are succeeded upward by volcanics of the Ventersdorp Supergroup (commencement age fixed at 2.714 Ga) (cf. Tucker et al. 2016) and the Transvaal sequence (Fig. 2.63).

The mineralogy of the reefs is dominated by pyrite with lesser pyrrhotite and arsenopyrite, widespread nickel, and cobalt sulfarsenides, and low quantity of base metal sulfides. The main elemental association with gold is Fe, C, and U. Beside uraninite and pitchblende, brannerite, thucholite (U-Kerogen), and pyrite are present in the ore zone. Carbon isotope studies (Hoefs and Schidlowski 1967) showed that this kerogen has a biogenic derivation possibly from prokaryotic life forms. Further, Schidlowski (1981) suggested that mobile hydrocarbon-bearing fluids permeated through the conglomeratic horizons and were radiolitically polymerized by intense alpha-radiation from accumulations of preexisting uraninites. Uraninite grains in QPC contain high (1–12 wt%) ThO_2, characteristic of high temperature (magmatic) origin (Cuney, 2010). A distinctive mineral assemblage of pyrophyllite–chloritoid–muscovite–chlorite–quartz–rutile–pyrite is found to occur around

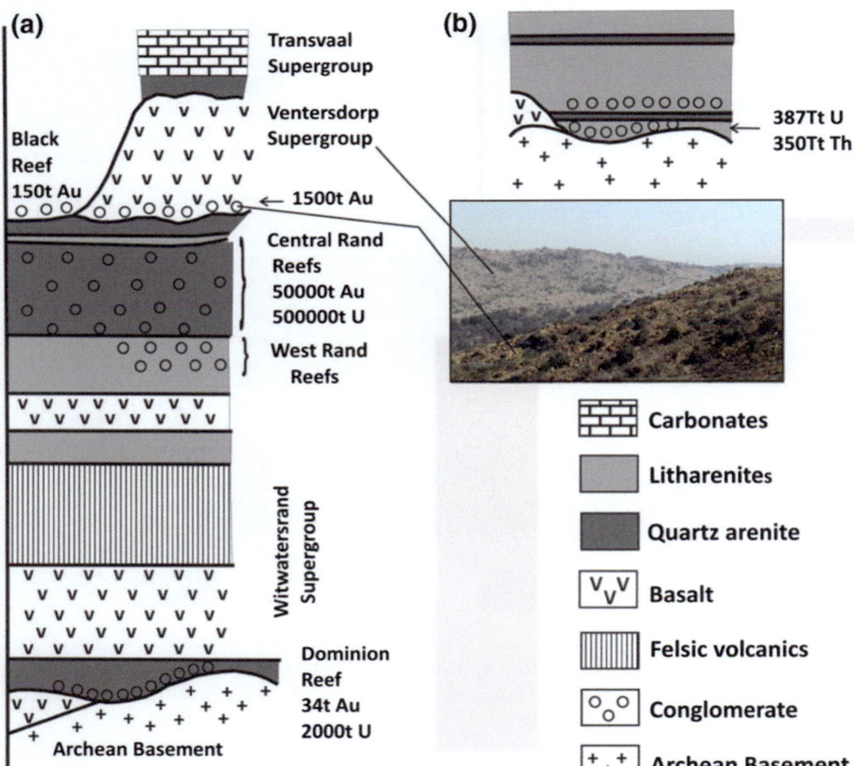

Fig. 2.63 Distribution of gold(-uranium) in **a** the basal sequence in Witwatersrand, RSA (*Inset* field photo by M. Deb) and **b** Elliot Lake, Canada (after Laznicka 1993)

the reefs in all the goldfields. This is considered as an alteration halo by some (Phillips and Law 2000) and denotes a metamorphic temperature between 300 and 400 °C.

The enormous, high-grade gold deposits of Witwatersrand clearly required three essential elements for their formation: source of the gold, concentrating mechanism, and preservation potential. While the primary source of gold is generally considered to be the greenstone belts and hydrothermally altered granites of the Archean Kaapvaal craton (Robb and Meyer 1990; Tucker et al. 2016), the gold mineralization at Witwatersrand has had its share of controversies revolving around the concentrating mechanism of gold, whether it was introduced as detrital particles during sedimentation (a placer theory), and underwent enhancement of concentration during burial, diagenesis, and subsequent metamorphism (a modified placer theory) or precipitated by epigenetic hydrothermal solutions moving along the sedimentary facies at a later stage of basin evolution (hydrothermal model). This spectrum of opinions is discussed briefly below.

In the early part of the history of Witwatersrand geology, since its discovery in the later part of ninteenth century, the *unmodified placer model* was popular. It envisaged that gold was transported from a source region into the basin in detrital form and deposited through sedimentary sorting processes (Fig. 2.64). It of course underwent burial but retained the original features of the depositional setting. This model however, lost favor by the beginning of twentieth century and saw a revival in the 1970s and 80s. Several studies, in more recent years (Minter et al. 1993; Minter 1999) inferred that the shape and/or composition of the gold grains reflect the source terrain. The accompanying rounded "buckshot" auriferous pyrite (Fig. 2.62) and uraninite grains were thought to be detrital, and presented strong evidence for primary sedimentary origin of gold in the anoxic atmospheric conditions presumably prevailing at about 2.85 Ga ago.

The *modified placer model* is based on the presence of secondary textures and typical hydrothermal minerals, silicates and sulfides, in the Witwatersrand reefs. In this model, gold and its associated minerals are considered as primarily detrital, partitioned into the coarse sedimentary fraction during transport and deposition and then remobilized after burial in scales of centimeters. Geochemical studies show an excellent correlation between carbon, gold, and uranium which is explained by trapping of the metals in algal mats (carbon seams). The placer models however, suffer from several weaknesses. A major problem is the difference between Witwatersrand reefs and modern placers. The reefs are planar whereas the modern placers are linear. A viable source area and large-scale alteration are the other controversial issues in this regard (Philips and Law 2000).

In the *hydrothermal replacement model*, stages of early basin formation, early diagenesis and generation of hydrocarbons, compression, migration of meteoric, and basin-derived fluids and gold mineralization are envisaged (Philips and Law 1997). A reduced, low salinity fluid is considered to have been introduced into the Witwatersrand Supergroup during metamorphism and tectonic evolution of the basin, resulting in widespread alteration. The fluid was channelized along structures, unconformity surfaces and primary stratification and the gold precipitation

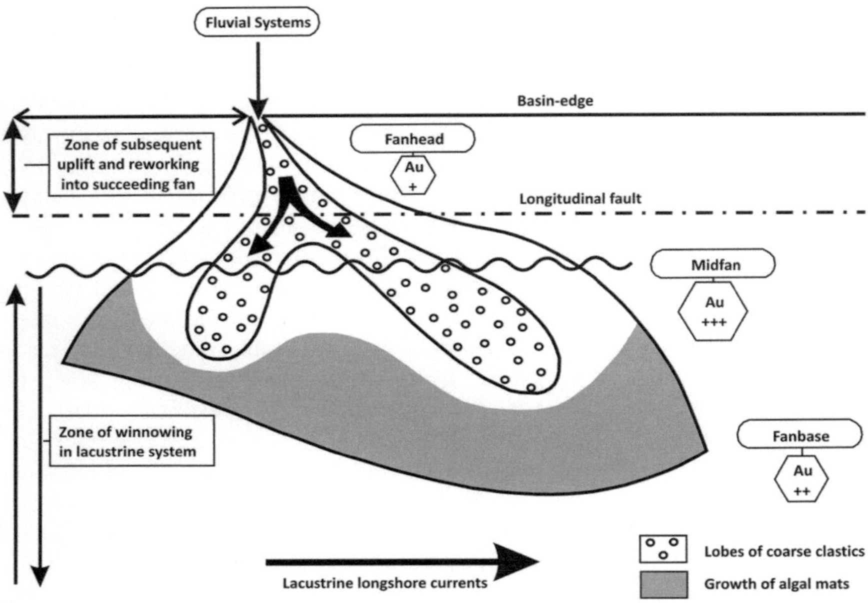

Fig. 2.64 Illustration of the placer modelplacer showing processes operating within a single fan and zones of gold concentration, being maximum in the mid-fan region. The distal portion of the fan is asymmetrical reflecting the effect of the long shore current. The algal mat is the inferred precursor of the carbon seams (after Pretorius 1976)

was controlled by reactions with carbon- or iron-bearing rocks above the unconformity surfaces. One hydrothermal model infers a separate origin for gold, uraninite, and hydrocarbons (Phillips and Law 2000), whereas another invokes a cogenetic origin (Barnicoat et al. 1997).

Agreement exists between the two viewpoints that the gold-bearing reefs of Witwatersrand show evidence of interaction with hydrothermal fluids, at least some of which brought about the precipitation of gold together with sulfide phases and bitumen. The basic difference between the two rests on whether the mineralization took place essentially during sediment deposition or it postdates sediment deposition. Recently, Large et al. (2013), based on a LA-ICP-MS study of pyrites, have concluded that both the competing models for the genesis of Witwatersrand gold are likely to be correct. They suggest that the hydrothermal event was widespread in the kilometer scale and involved basinal fluids that scavenged gold, arsenic, tellurium, and other trace elements from the gold-bearing sedimentary units in the Central Rand Group. Thus, the source of the gold in Witwatersrand deposits was both intra- and extra-basinal and involved multiple concentration processes. Some recent observations on the PGE content of gold in Witwatersrand deposits and identification of gold and uraninite grains which predate the commencement of Witwatersrand sedimentation are important in this context. The Re–Os ages of the gold are more than the sedimentation age (Kirk et al. 2002). Moreover, this gold has considerably higher Os-concentration and near mantle initial $^{187}Os/^{186}Os$ ratio (Frimmel et al. 2005). Minter (2006) determined U–Pb ages of 3.05 Ga for uraninite grains and a Re–Os isochron age of 2.99 Ga from pyrite grains in the Dominion and Witwatersrand conglomerate, having a depositional age of 2.97 Ga. These justify the belief that at least a part of the gold was detrital, while another part hydrothermal, derived by modification of the sediments during diagenesis-metamorphism or the hydrothermal fluids ingressed from below along structural discontinuities.

2.4.2 Banded Iron Formations and Related Ores

Within the broad group of chemically precipitated iron-rich sediments, banded iron formations are internally laminated stratigraphic units with at least 15% Fe. Band thickness varies from meters (macroband) through centimeters (mesobands) to millimeters (microbands). The interlayers are made principally of iron oxides and quartz/chert/jasper (cf. Figs. 3.1, 3.2 and 3.3), followed by carbonates. The oolitic chamosite–siderite–goethite–clay-rich rocks, commonly referred to as ironstones, are recognized as a distinctly separate type of iron-rich sediments. They formed in different environments than the cherty iron sediments and probably have a different origin and source of iron.

The banded iron formations are classified based on their general characteristics, depositional environments and the kinds of sedimentary rocks associated with them. Accordingly, there are two main types, the *Algoman type* and the *Lake Superior type* (Gross 1966). The former has limited resource of iron being relatively thin and discontinuous, traceable along strike for only a few tens of kilometers, shows intimate association with tholeiitic to calc-alkaline volcanic rocks and greywacke-type volcanoclastic sediments in deeper parts of basins or "eugeosynclinal" belts (Cloud 1973). Soft sediment slump and deformation structures are common, indicating their initial deposition as water-saturated chemical sediments in unstable, subsiding basins accompanied by extensive mafic volcanism. The mineralogy of the banded ores varies depending on the type of sedimentary facies (James 1954). The oxide facies are composed of magnetite $\gg$ haematite and their mixtures, deposited mostly as primary iron oxides. In general, these iron formations reflect highly reduced conditions of deposition from seawater-hydrothermal effusive systems. Silicate minerals in the silicate facies range from greenalite and minnesotaite to stilpnomelane, cummingtonite, and grunerite. Carbonate facies generally comprise siderite associated with magnetite or iron silicates but ankerite or ferroan dolomites are prevalent where carbonates are associated with hematite-rich facies. The sulfide facies of iron formations usually consist of fine grained pyritic carbonaceous mudstone with interlayered chert or siliceous shales. They exhibit rapid changes among the different facies. These iron formations often have a close spatial and genetic affiliation with other important metalliferous ores, notably with certain Ni sulfide deposits, auriferous lodes, and massive base metal (Cu, Zn) sulfide deposits (Hutchison 1981). They are widespread in Archean (>2.5 Ga) greenstone belts in the world's ancient cratons and reappear with lesser intensity in later ages. Examples are in the Superior province in northeastern Ontario, in the Yilgarn block of Western Australia, in the Kaapvaal craton of South Africa, and in the Baltic shield of northern Norway. In India, this type of BIF is mainly found in western Dharwar and in all the auriferous greenstone belts of eastern Dharwar craton.

Lake Superior type iron formations are, on the other hand, the main repository of iron resource in the world with the association of quartzites, dolomites, and black slates developed in shallow continental shelf environment. Epiclastic sediments are present but basic volcanic rocks in the ore districts are not as closely associated as in the Algoma type. These iron formations are commonly fine grained and oolitic, show lateral extent up to few hundred kilometers with great continuity. The thickness of these BIFs could measure in hundreds of meters. The conspicuous banding of the BIFs is due to alternation of cryptocrystalline silica/jaspilite and iron oxide bands, if the BIF is not metamorphosed above sub-greenschist facies. Otherwise, silica bands are quartzites. Oxide facies are predominant with hematite > magnetite, with the other facies rarely developed. They occur in a remarkably narrow time-stratigraphic interval of 2.0–2.5 billion years in age in virtually all of the earth's Precambrian shields. They do not reappear significantly in younger rocks. Examples of this BIF type (with synonyms, such as, taconite, jaspilite, itabirite) are numerous: Lake Superior and Labrador iron ranges of North America, deposits of the Hamersley basin in Western Australia, of the Transvaal Supergroup in South Africa, of Cerro Bolivar in Venezuela, of Quadrilátero Ferífero of Minas Gerais in Brazil, of Krivoi Rog in Ukraine. Deposits of this BIF-type are considered to have formed by deposition of iron and silica in colloidal size particles by chemical and biological processes in the oxygenated waters of a generally clastic-free shelf after upwelling currents brought reduced iron from iron-rich clastics from the continental slope (Drever 1974).

The two-fold classification scheme mentioned above is not applicable in every geological situation. For example, the BIFs with significant resource of iron in India are 3.1 or 2.6–2.7 Ga old and therefore not of Paleoproterozoic age. They generally do not have pillow lavas in close spatial relation as in case of the Algoma type, but mafic volcanic units, especially hydrovolcanics occur in the host sequence. Their mineralogy often is dominated by martite, secondary after magnetite. Thus, the Indian BIFs in many places have the characteristics of both the Algoman and Superior types of ores.

A third type of iron ores, the *Rapitan type*, has been described from the McKenzie Mountains of NW Canada from Neoproterozoic glaciogenic sediments (Klein and Buekes 1993). The host banded chert-hematite facies occurs in grabens and along fault scarps. The tectonic environments for the deposition of different iron formations mentioned above are illustrated in Fig. 2.65 (after Gross 1993).

The importance of iron being what it is today, the iron formations and the related ores have been in serious investigation during the last 100 years or so. Yet many of

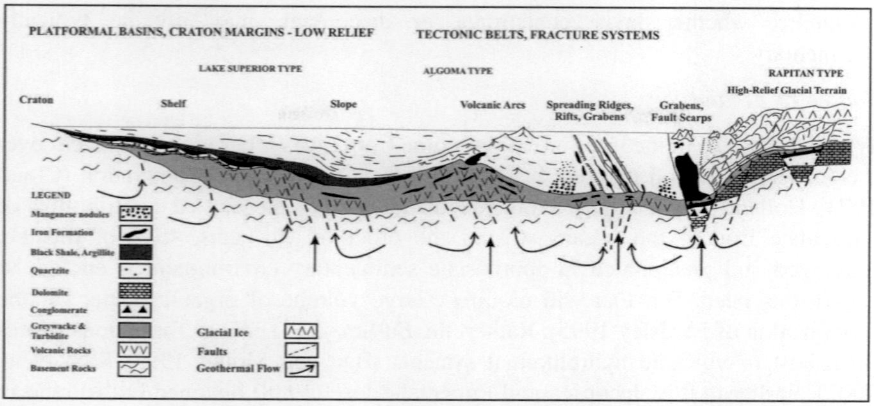

Fig. 2.65 Tectonic environments for the deposition of different iron formations (after Gross 1993)

the questions regarding these iron formations remain controversial. The moot questions concern:

(a) Ore petrography, including structure and textures.
(b) Sources of Fe and Si in the iron formations.
(c) Deposition of primary iron formation.
(d) Formation of ores.

Ore Petrography

BIFs and GIFs (Granular Iron Formations) are two petrogenetic varieties. BIFs were precipitated as chemical muds and the GIFs formed well-sorted chemical sands. The original mineral phases in the BIFs and GIFs are not yet incontrovertibly established, although some have reported the nano-spheric primary granules (120–200 nm) and microplaty hematite (Ahn and Busek 1990; Han 1988). Some interpreted magnetite to be early diagenetic (Johnson et al. 2008), while some others (Ayres 1972) held that most magnetite and hematite postdate burial stylolitization and even regional metamorphism (e.g., Morey 1999; Taylor et al. 2001). Both magnetite and hematite can form during diagenesis, but much of it is linked with Fe enrichment leading to Fe ore genesis (Bekker et al. 2010).

Much, but not all of banding and lamination in BIFs and GIFs is sedimentary. The mineralogy is generally not original. Layering in banded Fe-ores, on close inspection, is found not strictly parallel. Further, Fe-hydroxide particles, which believably were the initial Fe-components in these rocks, are highly reactive to dissolved silica in the system. This may be taken to imply that the silica component in the BIF was scavenged from seawater and later released during diagenesis when Fe(III)-hydroxides were transformed into stable Fe oxides. A part of the silica was possibly bound in clay. A reasonable implication of this view is that chert was not a direct chemical precipitate, as was generally believed until recently. Also laminated magnetite, whether layer concordant or discordant, may not be typically sedimentary.

Source of Fe and Si

Where did the Fe in the iron formations come from? An earlier view, persistent over a considerable period of time, was that iron was of continental derivation (Cloud 1973; Holland 1973, 1984). Holland concluded that the annual contribution of particulate iron to the oceans was of the order of 10^{15} gms. <1% of this Fe, mobilized and precipitated in appropriate sedimentary environment is enough to satisfy this need. But that will require a large volume of organic matter for the mobilization of Fe (Isley 1995). Rather, the $Eu/Sm_{CN} > 1$ in iron formations relates them best to volcanic hydrothermal systems (Bau and Moller 1993; Kato et al. 2002). Earlier to this, Jacobsen and Pimental-Klose (1988) obtained Fe/Nd ratio in the Archean/Early Proterozoic BIFs in the order of 10^5, like what is obtained in the modern hydrothermal systems. Some researchers are of the view that the deposition of large iron formations correspond closely in time with major mantle plume events. The Precambrian sea must have been saturated with volcanic silica [$Si(OH)_4$] in the absence of silica-mediating organisms. So there should not have been any difficulty for the supply of silica for the initial iron formations.

Primary Iron Deposition

Here also opinion varies. The principal views comprise the following:

(a) Oxidation of Fe(II) by cynobacterial O_2
(b) Metabolic Fe-oxidation
(c) UV photooxidation of Fe(II).

Cynobacterial Oxidation

This model invokes oxidation of dissolved Fe(II) with cyanobacteria-produced photosynthetic oxygen. These prokaryotic microbes likely flourished in the photic zones of near-coastal waters where Fe(II) and nutrients were rather easily available by upwelling of deep water, carrying hydrothermal inputs, with or without continental contributions.

$$CO_2 + H_2O\ (+\ \text{light}) \Rightarrow CH_2O\ (\text{organic molecule}) + O_2$$

Oxygen thus produced, was lethal to the biota that produced it until the latter did not have the O_2-mediating enzymes. In order to maintain the ecological balance, the oxygen produced had to be immediately removed from the system. This was done by oxidizing Fe(II) in solution. The process, Cloud (1973) suggested, was effective during the Early Proterozoic (2500–1900 Ma) time.

Metabolic Iron Oxidation

Metabolic ferrous iron oxidation is another interesting mechanism of iron formation deposition, a conclusion backed up by laboratory experiments with iron bacteria, as well as by modeling. This phototropic bacteria would have been capable of oxidizing enough Fe (II) to fully account for all the primary ferric iron deposited as BIF precursor sediment (Konhauser et al. 2002; Kappler et al. 2005).

U–V Photo-oxidation

This is an alternative to the biological models for Fe(II) oxidation, though suggested earlier (Cairns-Smith 1978). It proposes that the ferrous iron could have been photo-oxidized by the high flux of ultraviolet photons that would have reached Earth's surface prior to the rise of atmospheric oxygen and a protective ozone layer.

$$2\ Fe^{2+}_{(aq)} + 2\ H^{+} + hv \Rightarrow 2\ Fe^{3+}_{(aq)} + 2H_2O$$

The dissolved Fe(III) formed is substantially hydrolised as ferric oxy-hydroxide.

There are new suggestions too. Bekker et al (2010) emphasized the need for a view in which enhanced hydrothermal processes in Deep Ocean determined the ocean redox state independently of, or complementary to, atmospheric oxidation state (Fig. 2.66).

Origin of BIF (+GIF)-Related Fe Ores

Iron ores containing $\geq 55\%$ Fe, and associated with banded iron formations, particularly Lake Superior type, are commonly believed to have been secondarily derived from the latter with the help of oxidizing aqueous solutions. There are convincing field and laboratory evidence for it. There are however, a set of questions related to this. Did the process involve mainly the removal of silica or the addition of Fe? Was the aqueous fluid descending (supergene) or ascending (hypogene)? If ascending, was the aqueous fluid hot (hydrothermal) or cold? If ascending, was the water juvenile or modified meteoric? When did it take place in the history of the evolution of the iron formation? Researches to find the answers are still on. However, the present-day status of answers is as follows: the alteration process could be wholly supergene, wholly hypogene, hypogene with a supergene over-printing (Fig. 2.67) and supergene with a hydrothermal overprinting (Beukes 2002; Sarkar and Gupta 2005). A more recent study by Rasmussen et al. (2007) shows that hydrothermal iron ore formation from the BIF could be a long-lived, multistage process, spanning more than one billion years.

2.4.3 Sedimentary Manganese Deposits

Manganese deposits of diverse genetic types occur in the terrestrial geologic record. These deposits are of three lithologic associations: sedimentary rock-hosted, volcanic rock-hosted and karst-hosted, in order of predominance. Two genetic types of sedimentary rock-hosted deposits can also be identified: those with Mn derived via upwelling from oxygen-minimum zones and those formed on the margins of

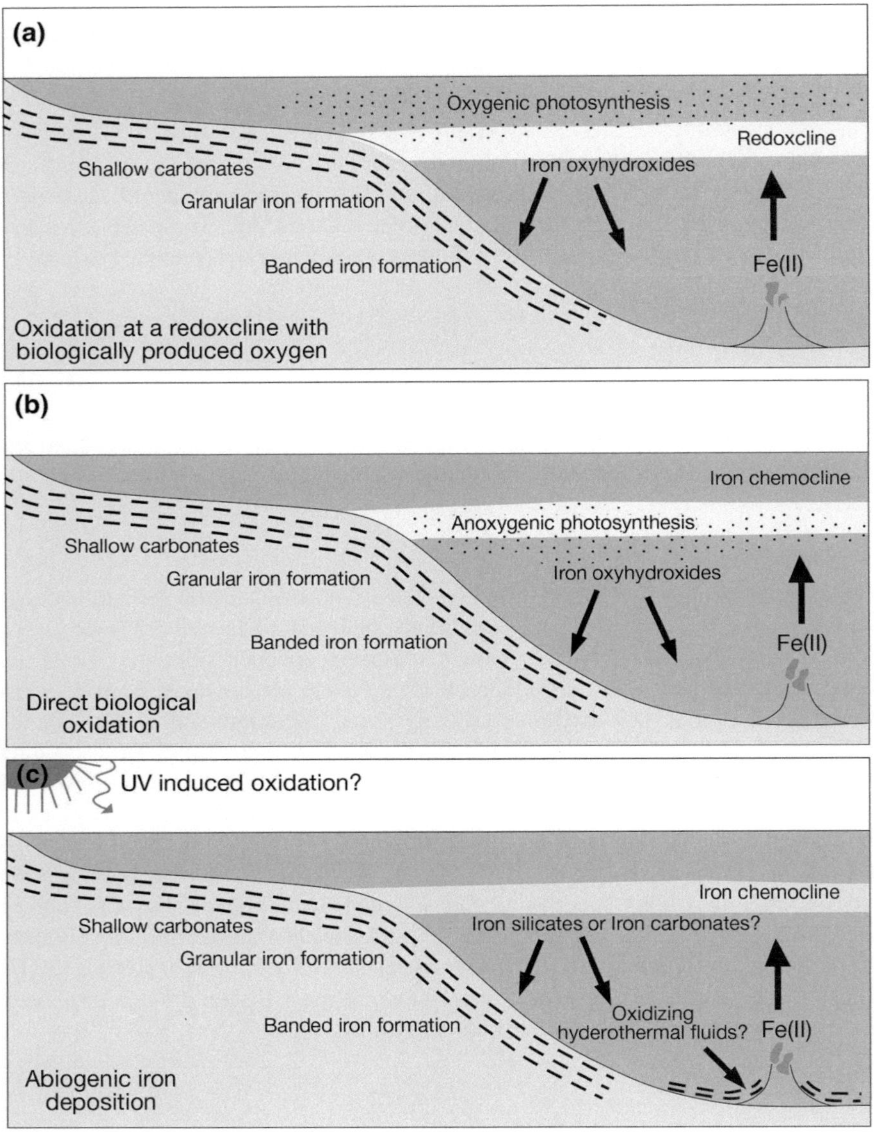

Fig. 2.66 Idealized models for the generation of Archean and Proterozoic iron formations (after Bekker et al. 2010)

euxinic basins. Most of the large tonnage deposits appear to form by the euxinic mechanism (Maynard 2010). The sedimentary rock-hosted manganese deposits easily outclass the other types with regard to size, spatial, and temporal distribution (Roy 1997). These ores occur in a variety of host rocks, mostly sedimentary, and

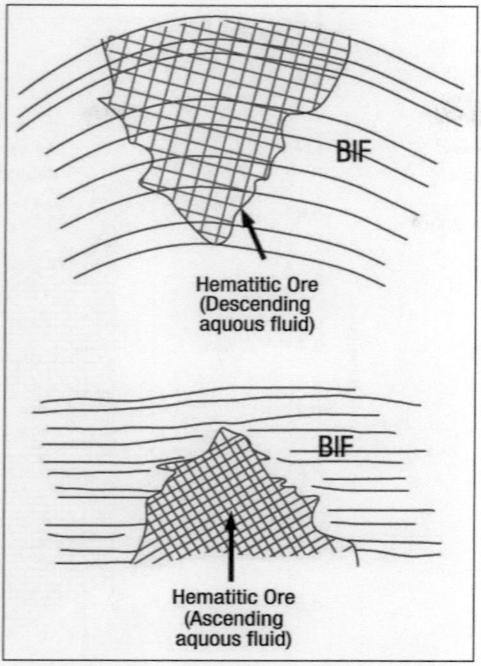

Fig. 2.67 Formation of Fe-ores from Banded Iron Formation by hypogene and supergene processes (after Sarkar and Gupta 2005)

rarely where volcanic rocks coexist. Some of these host rocks share a common genetic link with these deposits which form through a sequence of stages involving supply of the metal from a source or multiple sources, transport to a basin followed by direct deposition or concentration aided by early diagenesis. The geologic setting of most of these ancient manganese deposits indicates formation in shallow water basin-margin regimes. Manganese may be brought into the basin by endogenic hydrothermal solutions and by exogenic processes on the continent and coastal areas. Eh-pH of inorganic aqueous systems exert major controls on the solution and deposition of Mn as different species (Fig. 2.68). Further, presence of HCO^{3-}, SO_4^{2-}, HPO_4^{2-}, and organic matter may affect the behavior of Mn in exogenic conditions. Experimental studies have also indicated a much greater solubility of Mn compared to Fe at room T-P. Microbially mediated changes in the oxidation states of Mn leading to its dissolution and precipitation is also a distinct possibility.

Manganese deposits that form in the sedimentary environment can also be categorized as under:

1. Shallow marine manganese deposits with/without remobilization in subareal conditions
   a. Clastic-hosted deposits, e.g., Nikopol, Ukraine (Early Oligocene); Molango, Mexico (Late Jurassic); Groote Eylandt, Australia (Cretaceous), etc.

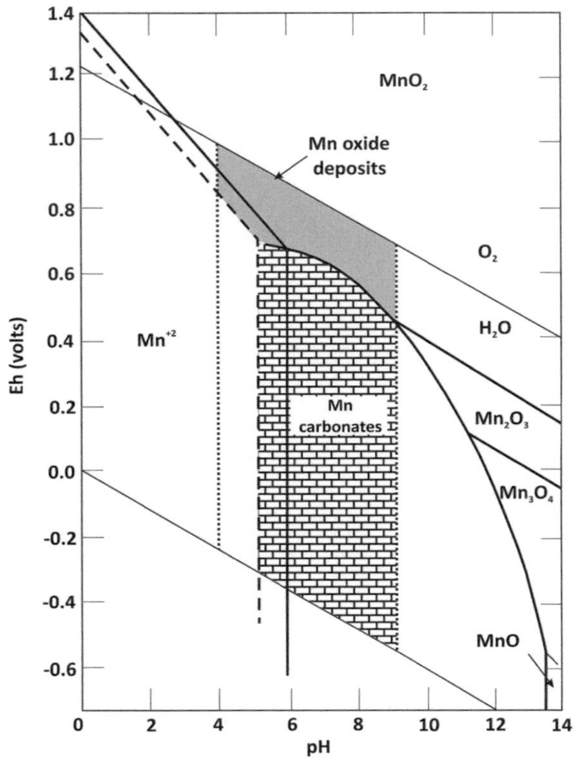

Fig. 2.68 Eh-pH diagram showing the stability of Mn oxides and Mn carbonates in natural waters. Dotted lines demarcate the boundary of natural waters. The *dashed* and the *bold lines* represent 10^{-4} and 10^{-6} M Mn^{2+}, respectively, (after Roy 2006)

b. Carbonate-hosted deposits, e.g., Úrkút, Hungary (Early Jurassic); Taojiang, China (Middle Ordovician); Imini-Tasdremt, Morocco (Cretaceous).
c. Banded iron formation (BIF)-hosted, e.g., ~2.2 Ga giant Kalahari Mn field and ~2.4 Ga Rooinekke deposit in South Africa (Paleoproterozoic), ~0.74 Ga Otjosondu deposit in Namibia.

2. Manganese deposits in black shales, e.g., Ulukent and Dodu, Turkey; large Early Proterozoic deposits in Francevillian succession in Gabon and in Birimian succession in West Africa, and the Kisenge district of DRC.
3. Manganese nodules and crust in deep marine environment.

Manganese deposits of the first two categories appeared in the Late Archean corresponding to development of oxygen oases in the otherwise reducing hydrosphere (Roy 1997). Examples in India are found in Chitradurga Group of western Dharwar, in the Eastern Ghat sequence in Kodur and Garbham and in the Iron Ore Group in Joda in Odisha. In the last occurrence, the Mn orebodies are constituted by pyrolusite, manganite, cryptomelane, braunite, interbedded with shale, and surficially modified by weathering. For all these Archean deposits there is no direct evidence of volcanic/hydrothermal input of manganese.

Large-scale deposition of manganese started from Early Proterozoic due to oxygenation of the atmosphere and development of stratified ocean, as explained below. The Kalahari Mn field with a potential resource of 13,600 million metric tons of ore (or 4200 Mt manganese metal), with Mn content between 20 and 48%, is the largest, representing 77% of the world's known land-based Mn resources. Here Mn oxide and Mn carbonate ores are interbedded with BIF in a cratonic shelf, constituting the 2.2 Ga old Hotazel Iron Formation in the Transvaal Supergroup (Beukes 1983). The Paleoproterozoic Sausar Group ($\sim$2.0 Ga) of central India hosts sedimentary manganese orebodies at Mansar, Chikla, Tirodi, etc. enclosed in metamorphic equivalents of limestone–shale–orthoquartzite assemblage indicating a shelf environment (Roy 1966, 1981). Presence of dolomite at the top (Bichua Fm) indicates further shallowing of the basin. Interbanded Mn oxide orebodies (represented by braunite, bixbyite, hollandite, jacobsite, hausmannite) and Mn silicate-oxide rocks (gondite) are hosted by metapelites and orthoquartzites (Mansar Fm) and less commonly occur as conformable lenses in carbonate rocks of older Lohangi Fm. The host sequence is complexly deformed and metamorphosed to grades ranging from low greenschist to upper amphibolite facies (Roy 1997).

There is a conspicuous gap in sedimentary rock-hosted Mn deposits between 1800 and 800 Ma (the "Boring Billion" period in Earth history, Mukherjee and Large 2015) that may correspond to a monotonous, low-oxygen ocean, but one without sulfidic deep water. Alternatively, Mn may have been precipitated entirely in the deep ocean, beneath a sulfidic oxygen minimum layer (Maynard 2010). Late Proterozoic Mn deposits are known in Jacadigo Group in Urucum, Brazil, in Damara sequence of Namibia, in the interglacial transgressive-regressive Datangpo sequence in China where Mn carbonates are hosted by black shales. The best Indian example is in the Penganga Group in the Godavari valley in Andhra Pradesh (Roy 1997). Sedimentary manganese deposits are more widespread, some examples of which are mentioned above. Very large deposits of unmetamorphosed Mn oxide and carbonate ores of Early Oligocene age occur at Chiatura (Georgia), and Nikopol and Bol'shoi Tokmak (Ukraine) in shallow marine intracratonic setting overlying the basement rocks of the Ukrainian shield. The Phanerozoic deposits are temporally coeval with stratified oceans, events of transgression and ocean anoxia (Force and Cannon 1988).

The geological–geochemical conditions of sedimentary manganese deposition through earth history demonstrates a primary redox control brought about by interplay of a variety of processes, mainly tectonic, along with the resultant climatic conditions prevailing in different periods (Roy 2006). Concentration of dissolved Mn in O_2-deficient seawater and/or sediment pore water through hydrothermal or terrigenous supply in stratified oceans and transfer of the dissolved metal to toxic continental shelves across the Mn^{2+}/Mn^{4+} redoxcline during transgression was responsible for deposition of Mn oxyhydroxide on shallow continental shelves (Fig. 2.69). When this initially precipitated Mn complex in oxic environment of stratified ocean got buried below the oxic-anoxic boundary of the water column, the metal was redissolved and when Mn^{2+} reached an optimum concentration level it reacted with organically derived bicarbonates in sediment pore water to produce

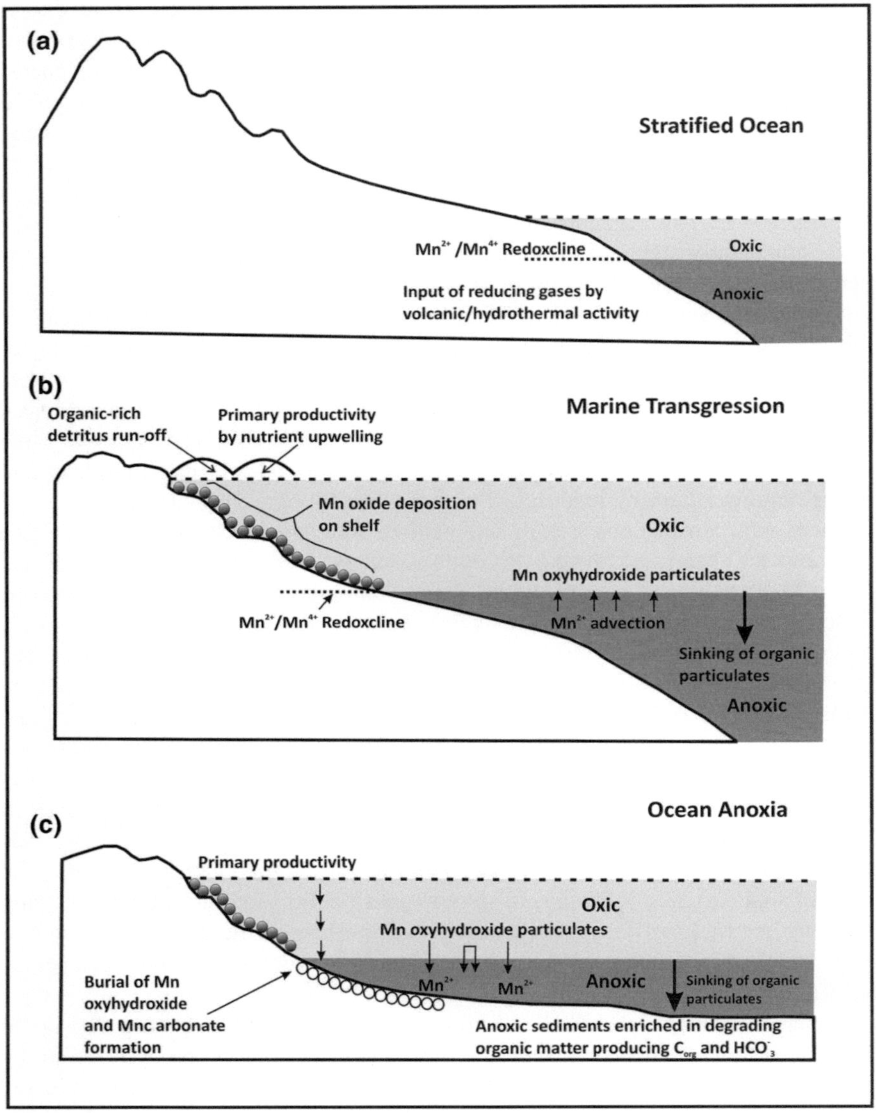

Fig. 2.69 Model representing sedimentary Mn deposition related to sea level changes (after Roy 2006)

Mn carbonates (Roy 2006). As noted in the earlier discussion, there is a strong concentration of Mn deposits in the Paleoproterozoic and a lesser occurrence in the Neoproterozoic, but, unlike Fe, there is an additional strong peak in the Oligocene. Therefore, it can be deduced that Mn is not controlled entirely by the level of oxygen in the earth's atmosphere. At each peak of Mn deposition, the associated ore deposits are concentrated in a few districts, suggesting a more local than global control on manganese metallogenesis (Maynard 2010).

Two major mineral deposit types occur commonly on present-day sea floor and ocean beds. These are the manganese nodules and crust, and metalliferous sediments, crusts, and mounds. They have been explored and studied extensively since they were discovered by the expedition of *HMS Challenger* between 1873 and 1876. Several useful reviews are available on the subject. The following summary is based on the publications of Cronan (1980); Roonwal (1986); Gross and Mcleod (1987). The manganese nodules and crust are widely distributed in deep ocean basins within 30° of the equator in the Pacific and Indian oceans, and in the south and southeastern Pacific, mainly at depths greater than 2000 m, below the carbonate compensation depth, where there is minimal detrital sedimentation and high biological productivity. Manganese nodules are commonly spheroidal to ellipsoidal, botroidal, discoidal, tabular, or faceted in shape, commonly 2–5 cm in their greatest dimension, and range from micronodules <1 mm in size to ones 1 m in diameter weighing a ton. Their growth rates average about 5 mm in a million years. Todorokite, birnessite, psilomelane, and hollandite are the most common of the large number of complex hydrous manganese oxide mineral phases present in the nodules. Iron in the nodule occurs as goethite, lepidocrocite, hematite, maghemite, and amorphous hydrated ferric iron oxide gel. The manganese nodules contain 1.7–3.5% combined Cu and Ni in minimum abundance of 10 kg/m^2 and with high Mn: Fe ratio. Estimates of potential resources of nodules in the world's oceans based on this minimum abundance and grade range from 14 to 99 billion tons.

Ferromanganese crusts containing 0.3–1% Co and occasionally platinum and deficient in Cu and Ni (in contrast to the nodules) are common at depths of 800–2500 m on seamounts older than 25 Ma in the Central Pacific. They also spread over thousands of square km on plateaus, rise and the eastern flank of the Mid-Atlantic ridge. Mineral occurrences in the Indian Ocean consist mainly of metalliferous sediments and sulfidic stockwork. These crusts were hydrogenetically deposited in areas of Oxygen Minimum Zone (OMZ) either from dissolved metals and/or oxide flocs suspended in the water column (Roy 1999).

The origin of the ferromanganese nodules involves the source of the metals and the mode of accretion of the oxyhydroxide phases. The source of metals in the nodules is attributed to a variety of processes and their combinations. These include discharge of hydrothermal fluids along active tectonic regimes, related to deep sea volcanism, or terrestrial weathering and transport by rivers into the sea. Secondary processes can also play significant role, e.g., leaching of metals from bottom sediments and volcanic rocks during diagenesis and lithification followed by transport and deposition by interstitial waters or by seawater and deposition at favorable sites. Two major mechanisms of nodule formation include precipitation, from enriched seawater around solid nuclei, of δMnO_2 with roughly equal Mn and Fe and relatively high content of Ce, Co, Pb, and Ti. The other mechanism is of diagenetic precipitation of todorokite with Mn > Fe and high content of Cu, Ni, Mo, and Zn. The role of microorganisms in the precipitation of Mn and Fe on the sea floor has also been emphasized by some workers.

2.5 Lateritic and Non-lateritic Residual Processes

Chemical weathering aided by water action and percolation leaves behind a host of economically important mineral resources as residual deposits near the surface. These include deposits of aluminum, iron, nickel, gold, and manganese. Chemical weathering proceeds in three distinct stages (Robb 2005): Dissolution, oxidation, and hydrolysis of the weathered rock material and removal of the soluble ions from the rock mass by percolating water; formation of new secondary minerals such as, clays, oxides, hydroxide, and carbonates; further breakdown of these secondary minerals and accumulation of hydroxides of iron and alumina in particular in the weathered zone.

2.5.1 Bauxites

Most common among the residual products is bauxite, a porous to pisolitic rock which could also be massive, nodular or earthy, comprising varying amounts of gibbsite, boehmite, diaspore, and kaolinite with subordinate amounts of silica, Fe oxides, and hydroxides and Ti oxides. Some bauxite ores are the source of gallium. Deposits typically occur on plateaus in tectonically stable areas. A large amount of scientific literature is available on this main economic resource for the Al metal, among which a mention may be made of Valeton (1972), Goudie (1973), Norton (1973), Maynard (1983) and Bardossy and Aleva (1990). Two major types of bauxites are distinguished in industry as well as in scientific literature. These are:

1. Lateritic bauxite
2. Karst bauxite.

Lateritic bauxite develops on or near the surficial parts of Al-silicate-rich rocks, such as granites, granite gneisses, basalts, arkoses, alkaline rocks, etc. as 85% of world's bauxite reserves belong to this type. The bauxite deposits on alumino-silicate rocks are commonly of the blanket type (Fig. 2.70) while the karst bauxites in impure carbonate rocks are irregular or sack formed. The most important stage in bauxite formation is Al-silicate breakdown in a weathering profile (Fig. 2.69):

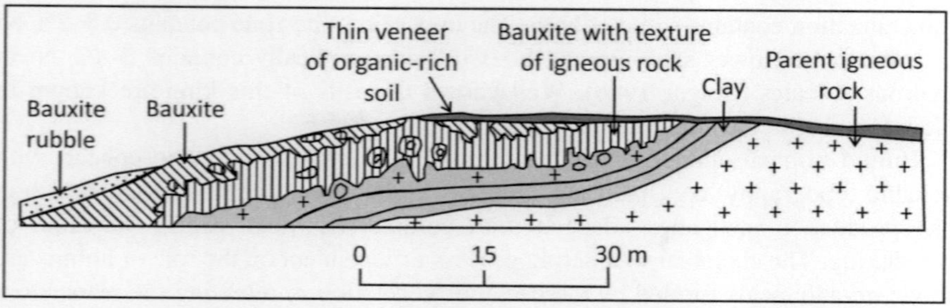

Fig. 2.70 A simplified section of a weathering profile on Alumino-silicate rocks, showing the formation of bauxite (after Craig et al. 2001)

$$2KAlSi_3O_8 + 2H^+_{aq} + 7\ H_2O \Rightarrow Al_2Si_2O_5(OH)_4 + 2\ K^+_{aq} + 4H_4SiO_{4aq}$$

(K-Feldspar) (Kaolinite)

$$Al_2Si_2O_5(OH)_4 + \ 5\ H_2O \Rightarrow Al_2O_3.3H_2O \ + 2H_4SiO_{4aq}$$

(Kaolinite) (Gibbsite)

Plagioclase feldspar and even muscovite mica may be involved in such a reaction if present in the protolith of the original rock (vide Sarkar and Gupta 2012). In karstic bauxite formation carbonate minerals dissolve out and the residue, dominated by clay minerals alter to bauxite mineral.

Bauxite accumulation can also be transported or detrital. Most deposits are Cenozoic in age. Largest resources are found in Australia, Brazil, Guinea, and India.

Several variables control the origin and quality of the residual product. These include: (1) climate (tropical with a mean temperature of 23–26 °C and annual rainfall of 1200–4000 mm); (2) topography (ancient planation surface dissected by valleys); (3) groundwater movement above the groundwater table; (4) relative rates of chemical and mechanical erosion; (5) type and amount of vegetation (tropical to savannah forest cover); (6) character of the bed rock; (7) lastly but most importantly, groundwater quality with respect to Eh, pH, and organic solutes.

Ni-rich in situ lateritic products are often found to develop on dunites and peridotites which have been uplifted and exposed to chemical weathering in warm, humid climates but protected by low rates of physical erosion. They are often associated with podiform chromite or serpentine-hosted asbestos deposits. The source rocks could be Precambrian to Tertiary in age, but the deposits form commonly by Cenozoic weathering. The weathered sequence from the top comprises: red, yellow, and brown limonitic and pisolitic soil; saprolites – continuous transition from soft saprolite below limonite zone, hard saprolite, and saprolitized peridotite to fresh peridotite. Boxwork of chalcedony and garnierite occurs near bed rock – weathered rock contact. Within such laterites, Ni-rich iron oxides are most common. Some deposits have a predominance of Ni silicates, such as, garnierite over quartz and goethite. The oxide and silicate ores are end members and most mineralization contains some of both. The upper limonite zone contains 0.5–2% Ni in iron oxides; lower saprolite and boxwork zone typically contains 2–4% Ni in hydrous silicates (Singer 1986). Well-known deposits of this kind are known in New Caledonia, Australia, USA (Oregon).

Humid tropical climate with abundant rainfall and vegetation, in concert with suitable topography, e.g., plateaus, drainage system, and parent rock composition can produce a manganese deposit with commercial yield within the zone of weathering. The chemical weathering process is dependent on the role of humic and fluvic organic acids formed by the decay of vegetation in releasing the manganese from the source rocks (Nicholson, 1992). In a Al–Fe–Mn triad, the solubility of Mn is maximum, as is its mobility. Thus during downward movement of Fe and Mn in supergene solutions, a change in Eh-pH may lead to precipitation of Fe in preference to Mn thereby bringing about a separation between the two elements. Where the weathered profile is of sufficient thickness, the upper zone is depleted in Mn which travels deeper and is reprecipitated in the lower zone (Roy 1981).

2.5.2 *Lateritic Gold*

Secondary gold enrichment in laterites is common in sulfidic deposits, though development of a fresh deposit of this type is not very common. Some important amongst them comprise Boddington-Hedges, (16.8 t/2.1 gm t^{-1}), and Gibson (8.5 t/1.6 gm t^{-1}) in Western Australia, Omai, Guiana (17 t/1.44 gm t^{-1}) and at a number of places in South America. Such deposit type prospects are also known in Nilambur region of northern Kerala (vide Sect. 3.5.5). These deposits occur in lateritic profile developed over greenstone-type mineralizations. Although gold in this situation could be largely secondary, primary gold may locally be present in considerable proportions. The secondary gold is generally of greater fineness (990) and may have forms varying from euhedral crystals through dendrites to irregular particles. The complexes that are believed to be capable of dissolving gold in such a situation and relocate it are chloride and thiosulfate ions, humic acid and may also be cyanide ions (Butt 1989).

Dissolution of gold from veins and lodes in near-surface environment are envisaged to take place by such reactions (Mann 1984):

$$4\ Au^o + 16\ Cl^- + 3\ O_2 + 12\ H^+ \rightarrow 4\ AuCl_4^- + 6H_2O$$

$$AuCl_4^- + 3\ Fe^{+2} + 6\ H_2O \rightarrow Au^o + 3\ FeOOH + 4\ Cl^- + 9\ H^+$$

Au^o precipitation from a solution in which gold exists as $AuCl_4^-$ complex may take place as a result of (1) lowering of Cl^- ion concentration; (2) by raising the pH, or (3) reduction of $AuCl_4^-$ ion by ferrous ion, as shown above. The last suggestion appears to be more realistic in view of the common association of gold with ferric hydroxide.

Mention of other enrichments, sometimes even reaching ore grades, are relevant in this context. These include gold contained in various surficial deposits and their cements, such as red brown hard pans, pedogenic calcrete, and secondary aluminosilicates. In some cases, gold is present in recently precipitated carbonates as well as in early formed pisolites (Butt 1989). Someone seeking explanations of gold nuggets, which have fascinated so many people for so long, may get an idea of the mechanism from the discussion made above. The accompanying diagram (Fig. 2.71) illustrates the enrichment and depletion of gold at Boddington-Hedges, Western Australia (Butt 1989).

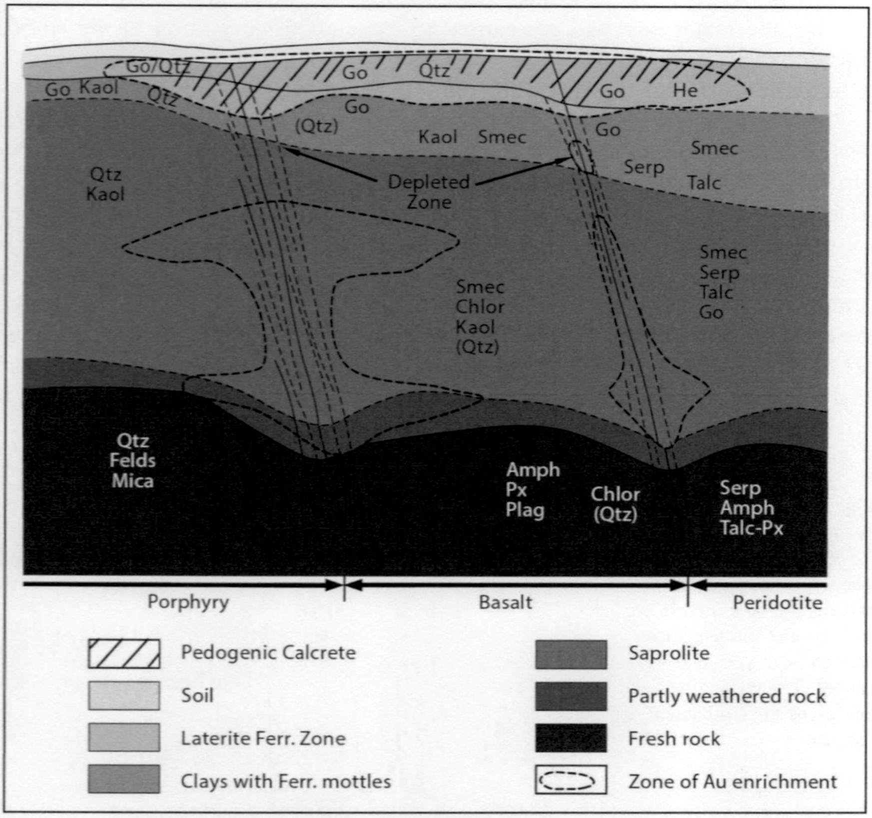

Fig. 2.71 Formation of supergene gold in deeply weathered lateritic profile in Western Australia (after Butt 1989)

2.5.3 *Supergene Oxidation and Enrichment (including "Exotic Ores")*

When a sulfidic ore body is brought close to the earth's surface through erosion, it will also get easily affected by the agents of chemical weathering. The sulfur is expected to get oxidized to sulfate ion, $SO_4^{=}$. The oxidation may go through the intermediate stages of SO_2 and free sulfur, but these products are rarely detected in the oxidized zone. The metals may be converted into insoluble compounds stable under surface conditions (e.g. oxides, carbonates, sulfates, silicates) or may be taken into solution. The dissolved metal will be slowly removed by the groundwater. Part of it may get deposited above the regional water table to form oxidized ores but the major part will get carried to the unoxidized portion of the sulfide deposit, where it will be precipitated by reaction with the primary sulfide minerals to form the supergene enriched sulfide ore (Fig. 2.72).

The weathering of sulfides, a complex natural process, thus results in: (a) metal ions getting into solution or into the form of an insoluble compound stable under surface condition; (b) conversion of the sulfur into sulfate ion and (c) production of relatively acid solution. We are particularly referring to the Cu-sulfide ores, as they are most susceptible to the above phenomena within the sulfide kingdom. This is because Cu has very low solubility of its sulfides and relatively high solubility of compounds it forms with the common anions in the oxidized zone. The stability of these phases is not only affected by higher redox potential (Eh), but also by lower pH or higher acidity. Some common reactions for acid production in geochemical environments are as follows:

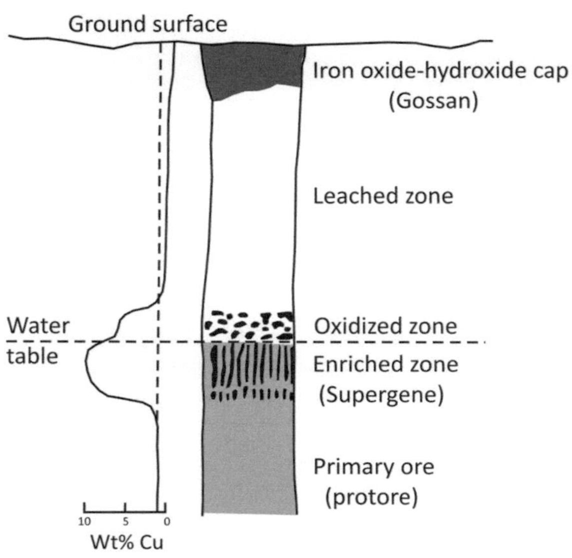

Fig. 2.72 A sketch of a near-surface section showing the formation of gossan, leached zone, oxidized ores, and supergene enriched ore, with special reference to a primary lean vein ore of copper. The dashed vertical line gives the Cu-content of the protore

$$4CuFeS_2 + 17\ O_2 + 10\ H_2O \rightarrow 4\ Fe(OH)_3 + Cu^{2+} + 8\ SO_4^{2-} + 8\ H^+$$
(Chalcopyrite) (Goethite)

$$2FeS_2 + 15/2\ O_2 + 4H_2O \rightarrow Fe_2O_3 + 4SO_4^{=} + 8\ H^+$$
(Pyrite) (Hematite)

The reactions mean that the oxidation of chalcopyrite liberates Cu^{2+} into the aqueous medium and leaves behind Fe oxide/hydroxide (regolith). Dissolution of pyrite often leaves behind polyhedral vacuoles, produced by the removal of Fe oxide/hydroxide. The material now obtained is a pock-marked goethitic/limonitic mass with some clay in most cases. It is called *gossan.*

Cu^{2+} leached from the top zone, i.e., gossan zone, percolates downward toward the groundwater surface, unless arrested by some specific reactions. Reactants available, the percolating fluids produce a complex mineralogy of Cu-carbonates, -silicates, -phosphates, -sulfates, -arsenates, as well as Cu oxide/hydroxide. A few partially altered grains of pyrite and chalcopyrite may still be there. The zone between the gossan and groundwater surface is called the *zone of oxidation*.

Below the groundwater table (surface) the environment is supposed to be reducing. The reactions there are like:

$$CuFeS_2 + 3\ Cu^{2+} \rightarrow 2\ Cu_2S + Fe^{2+}$$
(chalcopyrite) (chalcocite)

$$CuFeS_2 + Cu^{2+} \rightarrow 2\ CuS + Fe^{2+}$$
(chalcopyrite) (covellite)

$$14Cu^{++} + 5FeS_2 + 12H_2O \rightarrow 7Cu_2S + 5Fe^{++} + 3SO_4^{=} + 24H^+$$

Whether chalcocite or covellite will get preferentially deposited would depend on the Eh and pH at any particular location. Fe, Zn, and other metals displaced by Cu are carried away in solution and deposited later as limonite or smithsonite, etc. if the solution reaches an oxidizing or less acid environment. As the neo-formed phases, i.e., chalcocite and covellite are richer in Cu than chalcopyrite, which earlier dominated the ores below the groundwater surface, the metal-content, particularly of Cu increased and it is therefore called the *zone of secondary or supergene enrichment.* This zone gradually merges with the original ore or the *protore* (Fig. 2.72).

Silver also has necessary properties to show supergene enrichment whereas Pb, Zn, Ni, Co and Hg do not undergo such enrichment. Ag may be stuck in the oxidized zone as a chloride or may be reduced to the native metal if Eh remains low. However, solutions in the oxidized zone generally have low enough chloride ion and high enough Eh to permit considerable downward migration of Ag^+ to levels where very insoluble sulfide can precipitate. The simple sulfide argentite (Ag_2S) and the more complex sulfo-salt, proustite (Ag_3AsS_3) and pyrargyrite

(Ag_3SbS_3), are the common secondary minerals of silver in the zone of supergene enrichment.

Leaching, supergene enrichment and "exotic" deposit developments are interrelated, but not that if one of these phenomena developed well in one deposit, the other two will develop equally well there, or that all three will develop at all together. The "exotic deposits" of copper are rather less well known, although the mining of oxidic ore at Chuquicamata in northern Chile by the Incas and Spanish explorers took place hundreds of years ago. With breaks it was continuing until the middle of the last century when, however, the exotic deposit (now South Mine) was discovered beneath oxide tailing dumps, followed by systematic exploration and development. As Mote et al. (2001) defined, "generally exotic mineralization" occurs in paleo-drainage networks, leading away from principal porphyry deposits that have undergone supergene enrichment. Acidic, oxidizing copper-bearing fluids escaped the supergene enrichment system into the headwaters of the surrounding drainage network and flowed downhill to sites, where the precipitation of copper ore bodies occurred. More than 12 exotic copper deposits were reported from northern Chile. But what is not clear as yet is what caused the lateral escape of copper from the supergene system. Further, paleo-channel that may contain such a mineralization may be gravel covered and elude successful exploration.

Metals present, other than Cu in the exotic ores, are Mn, K, and Co. These are apparently contributed by the leached capping. When Mn is present in substantial proportions, the ore is referred to as *copper-wad*.

References

Ahn JH, Busek PR (1990) Hematite nanospheres of possible colloidal origin from a Precambrian banded iron formation. Science 250:111–113

Arndt NT, Lesher CM, Czamanske GK (2005) Mantle-derived magmas and magmatic Ni–Cu–PGE deposits. In: Hedenquist JW, Thompson JFH, Goldfarb RJ, Richards JP (eds) Econ Geol, 5–23

Ayres DE (1972) Genesis of iron-bearing minerals in banded iron formations mesobands in the Dales Gorge Member, Hammersley Group, Western Australia. Econ Geol 67:1214–1233

Bardossy GY, Aleva GJJ (1990) Lateritic bauxites. Developments in economic geology. Elsevier, Amsterdam, vol 27, 624 pp

Barnes HL (1979) Solubilities of ore minerals. In: Barnes HL (ed) Geochemistry of hydrothermal ore deposits, 2nd edn. Wiley, New York, pp 60–404

Barnes JW (1988) Ores and minerals. Introducing economic geology. Open University Press, Milton Keynes, 181 pp

Barnes S-J, Lightfoot PC (2005) Formation of magmatic nickel sulfides and processes affecting their copper and platinum group element contents. In: Hedenquist JW, Thompson JFH, Goldfarb RJ, Richards JP (eds) Econ Geol, 179–213

Barnes S-J, Maier WD (2002) Platinum-group element distributions in the Rustenberg layered suite of the Bushveld complex, South Africa. In: Cabri LJ (ed) The geology, geochemistry, mineralogy and mineral beneficiation of platinum-group elements. Can Inst Mining Metall Pet Spec vol 54:431–458

Barnicoat et al (1997) Hydrothermal gold mineralization in the Witwatersrand basin. Nature 386:820–824
Barrie CT, Hannington MD (1999) Classification of volcanic-associated massive sulfide deposits based on host-rock composition. Rev Econ Geol 8:2–11
Bau M, Moller P (1993) Rare earth element systematics of the chemically precipitated component in Early Precambrian iron formation and the evolution of the terrestrial atmosphere-hydrosphere-lithosphere system. Geochim Cosmochim Acta 57:2239–2249
Bekker A, Slack JF, Planavsky N, Krapez B, Hofman A, Konhauser KO, Rouxel OJ (2010) Iron formation: the sedimentary product of a complex interplay amongst mantle, tectonic, oceanic and biospheric processes. Econ Geol 105:467–508
Beukes NJ (1983) Paleoenvironmental setting of iron-formation in the depositional basin of the Transvaal Supergroup, South Africa. In: Trendall AF, Morris RC (eds) Iron formations: facts and problems. Elsevier, Amsterdam, pp 139–209
Beukes NJ, Gutzmer J, Mukhopadhyay, J (2002) The geology and genesis of high grade hematite iron deposits. Iron ore conference vol, Sept 2002, Perth, pp 23–29
Blevin PL, Chappell BW (1992) The role of magma sources, oxidation states and fractionation in determining the granite metallogeny of eastern Australia. Trans R Soc Edinburgh: Earth Sci 83:305–316
Brimhall GH (1991) The genesis of ores. Sci Am, 48–55
Burnham CW (1967) Hydrothermal fluids at the magmatic stage. In: Barnes HL (ed) Geochemistry of hydrothermal ore deposits. Holt, Rinehart and Winston Inc, New York, pp 34–76
Burnham CW (1979) Magmas and hydrothermal fluids. In: Barnes HL (ed) Geochemistry of hydrothermal ore deposits, 2nd edn. Wiley, New York, pp 71–136
Burnham CW (1997) Magmas and hydrothermal fluids. In: Barnes HL (ed) Geochemistry of hydrothermal ore deposits, 3rd edn. Wiley, New York, pp 63–123
Burnham CW, Ohmoto H (1980) Late-stage processes of felsic magmatism. Mining Geol, Spec Issue 8:1–11
Butt CRM (1989) Genesis of supergene gold deposits in the lateritic regolith of the Yilgarn block, Western Australia. Econ Geol Monogr 6:460–470
Cairns-Smith AG (1978) Precambrian solution photochemistry, inverse segregation and banded iron formations. Nature 76:807–808
Campbell IH, Naldrett AJ (1979) The influence of silicate: sulfide ratios on the geochemistry of magmatic sulfides. Econ Geol 74:1503–1506
Campbell IH, Naldrett AJ, Barnes SJ (1983) A model for the origin of the platinum group-rich sulfide horizons in the Bushveld and Stillwater complexes. J Petrol 24:133–165
Candela PA, Piccoli PM (1995) Model ore metal partitioning from melts into vapour/brine mixtures. In: Thompson JFH (ed) Magmas, fluid and ore deposits. Mineral Assoc Can Short Course Ser 23:101–127
Cathles LM, Erendi AHJ, Barrie T (1997) How long can a hydrothermal system be sustained by a single intrusive event? Econ Geol 92:766–771
Cawthorn RG (1996) Layered intrusions. Dev in Petrology 15, Elsevier, Amsterdam, 531 pp
Cawthorn RG, Barnes S-J, Balhaus C, Malitch KN (2005) Platinum group elements, chromium and vanadium deposits in mafic and ultramafic rocks. In: Hedenquist JW, Thompson JFH, Goldfarb RJ, Richards JP (eds) Econ Geol, 215–249
Cerný P (1991) Rare element granitic pegmatites Pt II: regional to global environments and petrogenesis. Geosci Can 18:68–81
Cerný P, Ercit TS (2005) Classification of granitic pegmatites. Can Min 43:2005–2026
Cerný P, Blevin PL, Cuney M, London D (2005) Granite-related ore deposits. In: Hedenquist JW, Thompson JFH, Goldfarb RJ, Richards JP (eds) Econ Geol, 337–370
Chattopadhyay A (2010) A review of the structural characteristics of orogenic gold deposits, with special reference to Indian gold fields. In: Deb M, Goldfarb RJ (eds) Gold metallogeny: India and beyond. Narosa Publ House, New Delhi, pp 123–153
Cloud P (1973) Paleoecological significance of banded iron formation. Econ Geol 68:113–1135

Collins WJ, Beams SD, White AJR, Chappell BW (1982) Nature and origin of A-type granites with particular reference to southeastern Australia. Contrib Mineral Pet 80:189–200

Cox SF (2005) Coupling between deformation, fluid pressures and fluid flow in ore producing hydrothermal systems at depth in the crust. In: Hedenquist JW, Thompson JFH, Goldfarb RJ, Richards JP (eds) Econ Geol, 39–75

Craig JR, Skinner BJ, Vaughan DI (2001) Resources of the earth: origin, use and environmental impact. 3rd edn. Prentice Hall, US, 520 pp

Cronan DS (1980) Under water minerals. Acadamic Press, London 350 pp

Cuney M (2010) Evolution of uranium fractionation processes through time: driving the secular variation of uranium deposit types. Econ Geol 105(3):553–569

Czamanske GK (1959) Sulfide solubility in aqueous solutions. Econ Geol 54:57–63

Deb M (1979) Polymetamorphism of ores in Precambrian stratiform massive sulphide deposits at Ambaji-Deri, western India. Miner Deposita 14:21–31

Deb M (1990) Regional metamorphism of sediment-hosted, conformable base metal sulfide deposits in the Aravalli-Delhi orogenic belt, NW India. In: Spry P, Bryndzia L (eds) Regional metamorphism of ore deposits. VSP, Netherlands, pp 117–140

Deb M (2014) Precambrian geodynamics and metallogeny of the Indian shield. Ore Geol Rev 57:1–28

Deb M, Bhattacharya AK (1980) Geological setting and conditions of metamorphism of Rajpura-Dariba polymetallic ore deposit, Rajasthan, India. Proceedings V IAGOD symposium 1, Schweiserbart, Stuttgart, pp 679–697

Detrick RS, Buhl P, Vera E, Mutter J, Orcutt J, Madsen J, Brocher T (1987) Multichannel seismic imaging of a crustal magma chamber along the East Pacific Rise. Nature 326:35–41

Dill HG (2010) The 'chessboard classification scheme of mineral deposits: mineralogy and geology from aluminum to zirconium. Earth Sci Rev 100:1–420

Drever JI (1974) Geochemical model for the origin of Precambrian banded iron formations. Geol Soc Am Bull 85:1099–1106

Einaudi MT, Meinert LD, Newberry RJ (1981) Skarn deposits. In: Skinner BJ (ed) Econ Geol 75th Anniv vol, 317–391

Eldrige CS, Barton PB, Ohmoto H (1983) Mineral textures and their bearing on the formation of the Kuroko ore bodies. In: Ohmoto H, Skinner BJ (eds) The Kuroko and related massive sulphide deposits. Econ Geol Monogr 5:241–281

Emmons WH (1936) Hypogene zoning in metalliferous lodes. Rept 1, 16th Int Geol Congr, pp 417–432

Evans AM (1997) An introduction to economic geology and its environmental impacts. Blackwell Sci Ltd, Oxford 364 pp

Force ER, Cannon WF (1988) Depositional model for shallow marine manganese deposits around black shale basins. Econ Geol 83:93–117

Franklin JM (1995) Volcanic-associated massive sulphide base metals. In: Eckstrand OR, Sinclair WD, Thorpe RI (eds) Geology of canadian mineral deposit types. Geol Surv Can, Geol Can 8:158–183

Franklin JM, Lydon JW, Sangster DF (1981) Volcanic-associated massive sulfide deposits. In: Skinner BJ (ed) Econ Geol, 485–627

Franklin JM, Gibson HL, Jonasson IR, Galley AO (2005) Volcanogenic massive sulfide deposits. In: Hedenquist JW, Thompson JFH, Goldfarb RJ, Richards JP (eds) Econ Geol, 523–560

Frimmel HE (2005) The case for a modified paleoplacer model for Witwatersrand gold. SEG Newslett 60:12–14

Frimmel HE, Groves DI, Kirk J, Ruiz J, Chesley J, Minter, WEL (2005) The formation and preservation of the Witwatersrand gold fields, the world's largest gold province. In: Hedenquist JW, Thompson JFH, Goldfarb RJ, Richards JP (eds) Econ Geol, 769–797

Garven G, Ge S, Person MA, Sverjensky DA (1993) Genesis of stratabound ore deposits in the mid-continent basins of North America. Amer J Sci 293:497–568

Goldfarb RJ, Groves DI, Gardoll S (2001) Orogenic gold and geologic time: a global synthesis. Ore Geol Rev 18:1–75

Goldfarb RJ, Baker T, Dubé B, Groves D, Hart C, Robert F, Gosselin P (2005) Distribution, character and genesis of gold deposits in metamorphic terrains. In: Hedenquist JW, Thompson JFH, Goldfarb RJ, Richards JP (eds) Econ Geol, 407–450

Goldfarb RJ, Hart C, Davis G, Groves D (2007) East Asian gold—deciphering the anomaly of Phanerozoic gold in Precambrian continents. Econ Geol 102:341–346

Goldhaber MB, Reynolds RL, Rye RO (1978) Origin of a south Texas roll-type uranium deposit. II: Sulfide petrology and sulfur isotope studies. Econ Geol 73:1690–1705

Goodfellow WD (2004) Geology, genesis and exploration of SEDEX deposits, with emphasis on the Selwyn basin, Canada. In: Deb M, Goodfellow WD (eds) Sediment-hosted lead-zinc sulphide deposits. Narosa Publ House, New Delhi, pp 24–99

Goodfellow WD, Franklin JM (1993) Geology, mineralogy and chemistry of sediment-hosted clastic massive sulphides in shallow cores, Middle valley, northern Juan de Fuca ridge. Econ Geol 88:2033–2064

Goodfellow WD, Lydon JW, Turner RJW (1993) Geology and genesis of stratiform sediment-hosted (SEDEX) zinc-lead-silver sulphide deposits. Miner Deposit Model GAC Spec Pap 40:201–252

Goudie A (1973) Duricrusts in tropical and subtropical landscapes. Oxford Press, UK, 174 pp

Gross GA (1966) Principal types of iron formations and derived ores. Can Min Met Bull 59 (646):150–153

Gross GA (1993) Industrial and genetic models for iron ores in iron-formations. Miner Deposit Model GAC Spec Pap 40:151–170

Gross GA, Mcleod CR (1987) Metallic minerals on the deep seabed. Geol Surv Can 86–21, 65

Groves D, Goldfarb RJ, Gebre-Mariam M, Hagemann SG, Robert F (1998) Orogenic gold deposits—a proposed classification in the context of their crustal distribution and relationship to other gold deposit types. Ore Geol Rev 13:7–27

Groves DI, Vielreicher RM, Goldfarb RJ, Condie KC (2005) Controls on the heterogenous distribution of mineral deposits through time. Geol Soc London Spec publ 248:71–101

Groves DI, Bierlein FP, Meinert LD, Hitzman MW (2010) Iron oxide–copper–gold (IOCG) deposits through earth history: implications for origin, lithospheric setting and distinction from other epigenetic iron oxide deposits. Econ Geol 105:641–653

Groves DI, Goldfarb RJ, Santosh M (2016) The conjunction of factors that lead to formation of giant gold provinces and deposits in non-arc settings. Geosci Front 7:303–314

Guilbert JM, Park CF (1986) The geology of ore deposits. W.H. Freeman and Co, 985 pp

Han TM (1988) Origin of magnetite in Precambrian iron formations of low metamorphic grade. In: Zachrisson E (ed) Proceedings of 7th quadrennial IAGOD symposium, Schweitzerbart, Stuttgart, pp 641–656

Hart CJR, Mair JL, Goldfarb RJ, Groves DI (2004) Source and redox controls on metallogenic variations in ore systems, Tombstone tungsten belt, Yokon, Canada. Trans R Soc Edinburgh, Earth Scs 95:339–356

Haynes D (2006) The Olympic Dam ore deposit discovery—a personal view. SEG Newslett 66: 1–15

Hedenquist JW, Arribas A Jr, Gonzales-Urien E (2000) Exploration for epithermal gold deposits. Rev in Econ Geol 13:245–277

Heinrich CA, Ryan CG, Mernagh TP, Eadington PJ (1992) Seggregation of ore metals between magmatic brine and vapour: a fluid inclusion study using PIXE microanalyses. Econ Geol 87:1566–1583

Henry CD, Elson HB, Macintosh WC, Heizler MT, Castor SB (1997) Brief duration of hydrothermal activity at Round Mountain, Nevada, determined from $^{40}Ar/^{39}Ar$ geochronology. Econ Geol 92:807–826

Hiemstra SA (1985) The distribution of some platinum group elements in the UG2 chromitite layer of the Bushveld complex. Econ Geol 80:944–957

Hitzman MW (1999) Characteristics and worldwide occurrence of Irish-type Zn–Pb–(Ag) deposits. In: Oliver H, Pongratz J, McGoldrick P (eds) Basin, fluid and Zn–Pb ores. CODES Spec Publ, pp 93–116

Hitzman MW, Oreskes N, Einaudi MT (1992) Geological characteristics and tectonic setting of Proterozoic iron oxide (Cu-U-Au-REE) deposits. Precambrian Res 58:241–287

Hitzman MW, Selly D, Bul S (2010) Formation of sedimentary rock-hosted stratiform copper deposits through Earth history. Econ Geol 105:627–639

Hodgson CJ (1989) The structure of shear-related, vein-type gold deposits: a review. Ore Geol Rev 4:231–273

Hoefs J, Schidlowski M (1967) Carbon isotope composition of carbonaceous matter from the Precambrian of the Witwatersrand System. Science 155:1096–1097

Holland HD (1973) The ocean: a possible source of iron in iron formations. Econ Geol 68: 1169–1172

Holland HD (1984) The chemical evolution of the atmosphere and oceans. Princeton Univ Press, Princeton 582 pp

Huston DL (1999) Stable isotopes and their significance for understanding the genesis of volcanic-hosted massive sulfide deposits: a review. Rev Econ Geol 8:151–180

Huston DL, Large RR (1989) A chemical model for the concentration of gold in volcanogenic massive sulfide deposits. Ore Geol Rev 4:171–200

Huston DL, Pehrsson S, Eglington BM, Zaw K (2010) Geology and metallogeny of volcanic-hosted massive sulfide deposits: variations through geologic time and with tectonic setting. Econ Geol 105:571–591

Hutchinson RW (1981) Mineral deposits as guides to supracrustal evolution. Am Geophys Union, Evolution Earth Geodyn 5:120–140

Hutchinson RW, Fyfe WS, Kerrich R (1980) Deep fluid penetration and ore deposition. Min Sci Eng 12(3):107–120

Institute of Mining Metallurgy (1985) High heat production (HHP) granites, hydrothermal circulation and ore genesis. In: Proceeding Conference, St Austell, Cornwall, England, 593 pp

Irvine TN (1977) Origin of chromitite layers in the Muscox intrusion and other stratiform intrusions: a new interpretation. Geology 5:273–277

Ishihara S (1981) The granitoid series and mineralization. In: Skinner BJ (ed) Econ Geol, 458-484

Isley AE (1995) Hydrothermal plumes and the delivery of iron to banded iron formation. J Geol 103:169–185

Jacobsen SB, Pimental-Klose MR (1988) A Nd isotope study of the Hamersley and Michipicoten banded iron formations: the source of REE and Fe in Archean oceans. Earth Planet Sci Lett 87:29–44

James HL (1954) Sedimentary facies of iron formation. Econ Geol 49:235–293

Jensen ML, Bateman AM (1981) Economic mineral deposits. Wiley, New York 593 pp

Johnson CM, Beard BL, Roden EE (2008) The iron isotope fingerprints of redox and biogeochemical cycling in the modern and ancient Earth. Ann Rev Earth Planet Sci 36:457–493

Kappler A, Pasquero C, Konhauser KO, Newman DK (2005) Deposition of banded iron formations by anoxygenic phototropic Fe (II)-oxidising bacteria. Geology 33:865–868

Kato Y, Kano T, Kunugiza K (2002) Negative Ce anomaly in the banded iron formations: evidence for the emergence of oxygenated deep sea at 2.9 to 2.7 Ga. Resour Geol 52:101–110

Keays RR, Jowitt SM (2013) The Avebury nickel deposit, Tasmania: a case study of an unconventional nickel deposit. Ore Geol Rev 52:4–17

Kerrich R, Fyfe WS (1981) The gold-carbonate association: source of CO_2 and CO_2 fixation reactions in Archean lode gold deposits. Chem Geol 33:265–294

Kerrich R, Goldfarb RJ, Richards JP (2005) Metallogenic provinces in an evolving dynamic framework. In: Hedenquist JW, Thompson JFH, Goldfarb RJ, Richards JP (eds) Econ Geol, 1097–1136

Kerrich R, Goldfarb RJ, Cline J, Leach D (2010) Metallogenic provinces of Laurentia in a superplume-supercontinent framework with a focus on gold. In: Deb M, Goldfarb RJ (eds) Gold metallogeny: India and beyond. Narosa Publ House, New Delhi, pp 1–29

Kirk J, Ruiz J, Chesley J, Walshe J, England G (2002) A major Archean gold and crust-forming event in Kaapvaal craton, South Africa. Science 297:1856–1858

Kirkham RV (1989) Distribution, setting, and genesis of sediment-hosted stratiform copper deposits. In: Boyle RW, Brown AC, Jefferson CW, Jowett EC, Kirkham RV (eds) Sediment-hosted stratiform copper deposits. Geol Assoc Can Spec 36:3–38

Klein C, Beukes NJ (1993) Sedimentology and geochemistry of the glaciogenic late Proterozoic Rapitan iron formations in Canada. Econ Geol 88:65–542

Konhauser KO, Hamade T, Morris RC, Ferris FG, Southam G, Raiswell R, Canfield D (2002) Could bacteria have formed the Precambrian banded iron formations. Geology 33:865–868

Large RR, McPhie J, Gemmell JB, Herrmann W, Davidson GJ (2001) The spectrum of ore deposit types, volcanic environments, alteration halos, and related exploration vectors in submarine volcanic successions: some examples from Australia. Econ Geol 96:913–938

Large RR, Meffre S, Burnett R, Guy B, Bull S, Gilbert S, Goemann K, Danyushevski L (2013) Evidence for an intrabasinal source and multiple concentration processes in the formation of the Carbon Leader Reef, Witwatersrand Supergroup, South Africa. Econ Geol 108(6): 1215–1241

Laznicka P (1993) Precambrian empirical metallogeny, Pt a & b. Elsevier, Amsterdam 1622 pp

Leach D et al (2005) Sediment-hosted lead-zinc deposits: a global perspective. In: Hedenquist JW, Thompson JFH, Goldfarb RJ, Richards JP (eds) Econ Geol, 561–608

Lesher CM (1989) Komatiite-hosted nickel sulfide deposits. Rev Econ Geol 4:45–101

Lesher CM, Groves DI (1986) Controls on the formation of komatiite-associated nickel–copper sulfide deposits. In: Friedrich GH (ed) Geology and metallogeny of copper deposits. Springer, Berlin, pp 43–62

Lindgren W (1933) Mineral Deposits. McGraw-Hill, New York 930 pp

London D (1995) Geochemical features of peraluminous granites, pegmatites and rhyolites as sources of lithophile metal deposits. Mineral Assoc Can Short Course Sr 23:175–202

London D (1997) Estimating abundances of volatiles and other mobile components in evolved silicate melts through mineral-melt equilibria. J Pet 38:1691–1706

London D (2005) Granitic pegmatites: an assessment of the current concepts and a direction for the future. Lithos 80:281–303

Lydon JW (1984) Volcanogenic massive sulfide deposits. Part 1: a descriptive model. Geosci Can 11:195–202

Lydon JW (1988) Ore deposits models#14. Volcanogenic massive sulfide deposits. Pt. 2. Genetic models. Geosci Can 15:43–65

Lydon JW (2004) Genetic models for Sullivan and other SEDEX deposits. In: Deb M, Goodfellow WD (eds) Sediment-hosted lead-zinc sulphide deposits. Narosa Publ House, New Delhi, pp 149–190

Maclean WH (1969) Liquidus phase relationships in the FeS–FeO–Fe_3O_4–SiO_2 system, and their application in geology. Econ Geol 64:84–865

Mann AW (1984) Mobility of gold and silver in lateritic weathering profile: some observations from Western Australia. Econ Geol 79:38–49

Marshall B, Gilligan LB (1989) Durchbewegung structure, piercement cusps, and piercement veins in massive sulfide deposits: formation and interpretation. Econ Geol 84:2311–2319

Matveev S, Ballhaus C (2002) Role of water in the origin of podiform chromite deposits. Earth Planet Sci Lett 203:235–243

Mavrogenes JA, O'Neill H (1999) The relative effects of pressure, temperature and oxygen fugacity on the solubility of sulfide in mafic magmas. Geochim Cosmochim Acta 63: 1173–1180

Maynard JB (1983) Geochemistry of sedimentary ore deposits. Springer, Heidelberg 305 pp

Maynard JB (2010) The chemistry of manganese ores through time: a signal of increasing diversity of Earth-surface environments. Econ Geol 105:535–552

Minter WEL (1999) Irrefutable detrital origin of Witwatersrand gold and evidence of eolian signatures. Econ Geol 94:665–670

Minter WEL (2006) The sedimentary setting of Witwatersrand placer mineral deposits in Archean atmosphere. In: Kesler SE, Ohmoto H (eds) Evolution of early Earth's atmosphere and biosphere—constraints from ore deposits. Geol Soc Am Mem 198:105–119

Minter WEL, Goedhart M, Knight J, Frimme HE (1993) Morphology of Witwatersrand gold grains from the Basal Reef: evidence for their detrital origin. Econ Geol 88:237–248

Mookherjee A (1976) Ores and metamorphism: temporal and genetic relationships. In: Wolf KH (ed) Handbook of Stratabound and Stratiform ore deposits, v 4. Elsevier, Amsterdam, pp 203–260

Mookherjee A (1999) Ore genesis—a holistic approach. Allied Publishers Pvt Ltd, New Delhi 657 pp

Mookherjee A (2000) Crustal fluids and formation of mineral deposits. The Vth Foundation lecture, Indian Geological Congress, Pune, pp 1–19

Morey GB (1999) High grade iron ore deposits of the Mesabi Range, Minnesota: products of a continental scale Proterozoic ground water flow system. Econ Geol 94:133–142

Mote TI, Becker TA, Renne P, Brimhall GH (2001) Chronology of Exotic Mineralization at El Salvador, Chile, by $^{40}Ar/^{39}Ar$ dating of copper wad and supergene alunite. Econ Geol 96: 351–366

Mukherjee I, Large RR (2015) Trends in nutrient supply, productivity and atmosphere oxygenation through the Boring Billion. AGU-GAC-MAC-CGU meeting, Montreal, Canada, Abstract PG12A-04

Mungall JE (1999) Why are large Ni–Cu deposits found only within plates? Geological Association of Canada, Mineral Association of Canada annual meeting, abstract with program 24, p 88

Mungall JE (2002) Kinetic controls on the partitioning of trace elements between silicate and sulfide liquids. J Pet 43:749–768

Murck BW, Campbell IH (1986) The effects of temperature, oxygen fugacity and melt composition on the behavier of chromium in basic and ultrabasic melts. Geochim Cosmochim Acta 50:1871–1887

Naldrett AJ (1989) Magmatic sulfide deposits. Oxford University Press, Oxford 186 pp

Naldrett AJ (2004) Magmatic sulfide deposits: geology, geochemistry and exploration. Springer, Berlin 727 pp

Naldrett AJ, MacDonald AJ (1980) Tectonic setting of some Ni–Cu sulfide ores: their importance in genesis and exploration. Geol Assoc Can Spec Pap 20:57–633

Nashlund HR, Henriquez F, Nystrom JO, Vivallo W, Dobbs, FM (2002) Magmatic iron ores and associated mineralization: examples from Chilean High Andes and coastal Cordillera. In: Potter TM (ed), Hydrothermal iron oxide-copper-gold and related deposits: a global perspective. PGC Publ, Australia, vol 2, pp 207–226

Nesbitt BE, Murowchic JB, Muehlenbachs K (1986) Dual origins of lode gold deposits in the Canadian Cordillera. Geology 14(6):50–506

Nicholson K (1992) Genetic types of manganese oxide deposits in Scotland: indicators of paleo-ocean spreading rate and a Devonian geochemical mobility boundary. Econ Geol 87:1301–1309

Northrop HR, Goldhaber MB (1990) Genesis of the tabular-type vanadium-uranium deposits of the Henry basin, Utah. Econ Geol 85:215–268

Norton SA (1973) Laterite and bauxite formation. Econ Geol 68:353–361

Norton D (1977) Fluid circulation in the Earth's crust. In: Heacock JD (ed) The Earth's crust. Am Geophy Union, Washington, Am Geophy Mono 20, pp 693–704

Pal N, Mishra B (2002) Alteration geochemistry and fluid inclusion characteristics of the greenstone-hosted gold deposit at Hutti, Eastern Dharwar Craton, India. Miner Deposita 37:722–736

Partington GA, Williams PJ (2000) Proterozoic lode gold and (Iron)–Copper–Gold deposits: a comparison of Australian and global examples. Rev Econ Geol 13:69–101

Partington GA, McNaughton NJ, Williams I (1995) A review of geology, mineralization and geochronology of the Greenbushes pegmatites, Western Australia. Econ Geol 90:616–635

Phillips GN, Law JDM (1997) Hydrothermal origin for Witwatersrand gold. SEG Newslett 31: 26–33

Phillips GN, Law JDM (2000) Witwatersrand goldfields: geology, genesis and exploration. Rev Econ Geol 13:439–500

Philpotts AR (1967) Origin of certain iron–titanium oxide and apatite rocks. Econ Geol 62:15–303

Pohl WL (2011) Economic geology, principles and practice. metals, minerals, coal and hydrocarbons—introduction to formation and sustainable exploitation of mineral deposits. Wiley-Blackwell, Sussex, p 663

Pollard PJ (1995) Geology of rare metal deposits: an introduction and overview. Econ Geol 90:489–494

Porter TM (ed) (2000) Hydrothermal iron oxide copper–gold and related deposits: a global perspective. Australian Mineral Foundation, Adelaide 349 pp

Porter TM (ed) (2002) Hydrothermal iron oxide copper-gold and related deposits: perspective 2. PGC Publ, Adelaide, vol 2, 367 pp

Poulsen KH (1995) Lode gold. In: Eckstrand OR, Sinclair WD, Thorpe RI (eds) Geology of Canadian mineral deposit types. Geological Survey of Canada, Geology of Canada vol 8, pp 323–328

Poulsen KH, Robert F, Dubé B (2000) Geologial classification of Canadian gold deposits. Geol Surv Can Bull 540:106

Pretorius DA (1976) The nature of the Witwatersrand gold–uranium deposits. In: Wolf, KH (ed) Handbook of stratabound and stratiform ore deposits, Amsterdam, Elsevier, vol 7, pp 29–88

Ramesh Babu PV (1999) Rare metals and rare earth pegmatites of central India. In: Mahadevan TM, Dhanaraju R (eds), Expl Res for Atom Miner 12:7–52

Rasmussen B, Fletcher IR, Muhling JR, Thorne WS, Broadbent GC (2007) Prolonged history of episodic fluid flow in giant hematite orebody: evidence from the U-Pb geochemistry of hydrothermal xenotime. Earth Planet Sci Lett 258:249–259

Reed MH (1997) Hydrothermal alteration and its relationship to ore fluid composition. In: Barnes HL (ed) Geochemistry of hydrothermal ore deposits. 3rd edn, Wiley, pp 303–366

Richards JP (2013) Giant ore deposits formed by optimal alignments and combination of geological processes. Nature Geosci 6(11):911–916

Richards JP (2016) Links between porphyry and IOCG deposits (SEG Thayer Lindsley lecture). Symp Mineral deposits and Ore-forming Processes. 35th International Geological Congress, Cape Town, South Africa

Richards JP, Mumin AH (2013) Magmatic-hydrothermal processes within an evolving Earth: iron oxide–copper–gold and porphyry Cu $\pm$ Mo $\pm$ Au deposits. Geology 41(7):767–770

Robb L (2005) Introduction to ore-forming processes. Blackwell Publ, UK 373 pp

Robb LJ, Meyer FM (1990) The nature of the Witwatersrand hinterland: conjectures on the source area problem. Econ Geol 85:511–536

Robert F, Poulsen KH (2001) Vein formation and deformation in greenstone gold deposits. Rev Econ Geol 14:111–155

Roedder E (1984) Fluid inclusions. Review in Mineral, vol 12, Mineralogical Society of America, 646 pp

Rona PA (1988) Hydrothermal mineralization at ocean ridges. Can Miner 26:431–465

Roonwal GS (1986) The Indian Ocean: exploitable mineral and petroleum resources. Springer, Berlin 198 pp

Roy S (1966) Syngenetic manganese formations of India. Jadavpur Univ Press, Calcutta 219 pp

Roy S (1981) Manganese deposits. Acadamic Press, London 458 pp

Roy S (1997) Genetic diversity of manganese deposition in the terrestrial geological record. In: Nicholson K, Hein, JR, Bühn B, Dasgupta S (eds) Manganese mineralization: geochemistry and mineralogy of terrestrial and marine deposits. Geol Soc London, Spec Publ 119: 5–27

Roy S (1999) Marine metallogenesis: new perspectives. In: Somayajulu BLK (ed) Ocean Science: trends and future directions. Ind Nat Sci Acad Akademia Books International, New Delhi, pp 129–150

Roy S (2006) Sedimentary manganese metallogenesis in response to the evolution of the Earth system. Earth Sci Rev 77:273–305

Ruzicka V (1995) Unconformity-associated uranium. In: Ekstrand OR, Sinclair WD, Thorpe RI (eds) Geology of Canadian Mineral Deposit Types. Geololgical Survey of Canada, Geology of Canada, No 8, pp 197–210
Sangster DF (1990) Mississippi valley type and SEDEX type lead-zinc deposits: a comparative examination. Trans Inst Mining Metall (Sec. B) 99: B21–B42
Sarkar SC, Dasgupta S (1980) Geological setting and transformation of sulfide deposits in the northern part of the Khetri copper belt, Rajasthan, India: an outline. Miner Deposita 15: 117–137
Sarkar SC, Deb M (1974) Metamorphism of the sulphides of the Singhbhum copper belt, India—the evidence from the ore fabric. Econ Geol 68:1282–1293
Sarkar SC, Gupta A (2005) Nature and origin of the iron ores of eastern India, a subject of scientific interest and industrial concern. In: Proceedings of SEM Mineral and Energy resources of eastern and northeastern India. MGMI, Kolkata, pp 79–101
Sarkar SC, Gupta A (2012) Crustal evolution and metallogeny in India. Cambridge University Press, Delhi 840 pp
Sarkar SC, Dwivedy KK, Das AK (1995) Rare earth deposits in India—an outline of their types, distribution, mineralogy, geochemistry and genesis. Glob Tectonics Metallogeny 5:53–61
Sawkins FJ (1990) Integrated tectonic-genetic model for volcanic-hosted massive sulfide deposits. Geology 18:1061–1064
Schidlowski M (1981) Uraniferous constituents of the Witwatersrand conglomerates: ore microscopic observations and implications for Witwatersrand metallogeny. In: Armstrong F (ed) Genesis of Uranium- and Gold-bearing Precambrian Quartz pebble conglomerates. US Geol Surv Prof 1161:1–29
Seedorff E, Barton MD (2004) Enigmatic origin of Carlin-type deposits: an amagmatic solution? SEG Newslett 59:14–16
Shanks WC III, Bischoff JL (1980) Geochemistry, sulfur isotope composition and accumulation rates of Red Sea geothermal deposits. Econ Geol 75:445–459
Shepherd TJ, Rankin AH, Alderton DH (1985) A practical guide to fluid inclusion studies. Blackie, Glasgow 235 pp
Sillitoe RH (1985) Ore related breccias in volcanoplutonic arcs. Econ Geol 80:1467–1514
Sillitoe RH (1995) Exploration and discovery of base and precious metal deposits in the Circum-Pacific region during the last 25 years. Resour Geol 19:119
Sillitoe RH (1996) Granites and metal deposits. Episodes 19(4):126–133
Sillitoe RH (2010) Porphyry copper systems. Econ Geol 105:3–41
Sillitoe RH, Burrows DR (2002) New field evidence bearing on the origin of the El Laco magnetite deposit, northern Chile. Econ Geol 97:1101–1109
Sillitoe RH, Mortensen JK (2010) Longevity of porphyry copper formation at Quellaveco, Peru. Econ Geol 105:1157–1162
Singer DA (1986) Descriptive model of lateritic nickel. In: Cox DP, Singer DA (eds) Mineral deposit models. US Geol Surv Bull 1693:379
Skinner BJ (1979) Many origins of hydrothermal deposits. In: Barnes HL (ed) Geochemistry of hydrothermal ore deposits, 2nd edn. Wiley, New York, pp 1–21
Skinner BJ, Peck DL (1969) An immiscible sulfide melt from Hawai. Econ Geol Monogr 4: 22–310
Solomon M, Groves DI (2000) The geology and origin of Australia's mineral deposits. Hobart Printing Author, Tasmania, 1002 pp
Spry PG, Marshall B, Vokes FM (2000). Metamorphosed and metamorphogenic ore deposits. Rev Econ Geol 11:310
Stolper E (1982) The speciation of water in silicate melts. Geochim Cosmochim Acta 46:2609–2620
Strong DF (1988) A review and model for granite-related mineral deposits. In: Taylor RP, Strong DF (eds) Recent advances in the geology of granite-related mineral deposits. Can Inst Mining Metall, Spec 39:424–445

Sutherland DG (1982) The transport and sorting of diamonds by fluvial and marine processes. Econ Geol 77:1613–1620

Taylor D, Dalstra HJ, Harding AE, Broadbent GC, Barley ME (2001) Genesis of high-grade hematite ore bodies of the Hamersley Province, Western Australia. Econ Geol 96:837–873

Tiwary A, Deb M, Cook NJ (1998) Use of pyrite microfabric as a key to tectono-thermal evolution of massive sulfide deposits - an example from Deri, southern Rajasthan. Ind Mineral Mag 62(2): 197–212

Trueman DL, Cerny' P (1982) Exploration for rare element granitic pegmatite. In: Cerny P (ed) Granitic pegmatite in science and industry. Mineral Assoc Can Short Course Handbook, vol 8, pp 463-493

Tucker RF, Viljoen RP, Viljoen MJ (2016) A review of the Witwatersrand basin—the world's greatest goldfield. Episodes 39(2):105–134

Urabe T (1987) Kuroko deposit modeling based on a magmatic-hydrothermal theory. Min Geol 37:159–176

Urabe T et al (1995) The effect of magmatic activity on hydrothermal venting along the superfast-spreading East Pacific Rise. Science 269:1092–1095

Valeton I (1972) Bauxites. Elsevier, Amsterdam, 226 pp

William-Jones AE, Heinrich CA (2005) Vapour transport of metals and the formation of magmatic hydrothermal ore deposits. Econ Geol 100:1287–1312

Williams PJ, Barton MD, Johnson DA, Fontbote L, de Haller A, Mark G, Oliver NHS, Marschik R (2005) Iron oxide copper-gold deposits: geology, space-time distribution and possible modes of origin: In: Hedenquist JW, Thompson JFH, Goldfarb RJ, Richards JP (eds) Econ Geol, 371–405

Wood SA, Sampson IM (1998) Solubility of ore minerals and complexation of ore metals in hydrothermal solutions. Rev Econ Geol 10:33–80

Zierenberg RA, Shanks WCIII, Bischoff JL (1984) Massive sulfide deposits at 21°N, East Pacific Rise: chemical compositions, stable isotopes and phase equilibria. Geol Soc Am Bull 95: 922–929

Zierenberg RA, Koski RA, Morton JL, Bouse RM (1993) Genesis of massive sulphide deposits on a sediment-covered spreading centre, Escanaba Trough, southern Gorda ridge. Econ Geol 88:2065–2094

Chapter 3
Metallic Mineral Deposits

Abstract Iron, and hence iron ores, have played a very important role in the history of human civilization, because of which the latter has been named "Iron Age". The principal types of iron ores are variously enriched hematite-dominated banded iron formation (Superior type) and variously modified magnetite-dominant banded iron formation (Algoma type). The Superior type deposits mainly formed during late Archean-early Proterozoic period. Algoma type is somewhat older. India's iron ore deposits, with an estimated reserve of 28 Gt, are concentrated in South India, Central India, and Eastern India. Mn ores, principally oxides to hydroxides, formed during separate events from early Proterozoic to recent (Mn-nodules) in South Africa, Brazil, Ghana, Gabon, India, and countries of previous USSR. Genetically the deposits are hydrothermal, sedimentary (-diagenetic), and of supergene enrichment. Chromium, obtained from chromite, is essentially magmatitic (orthomagmatic) in origin and chromite deposits are commonly associated with mafic–ultramafic rocks. Out of India's reserves of 139 Mt of chromite ores, >90% is located in eastern India and the rest in south India. Bulk of gold in nature forms in the "native state", which does not mean 100% purity in reality. Alloying with Ag (and Hg) is common. Most ores form from hydrothermal fluids of primary and secondary origin and as placers (and nuggets). Gold is reported from different parts of India, but the largest yield came from the Kolar gold field in Karnataka amounting to 800 t Au. Copper mineralization was recorded from the Singhbhum copper belt, occurring in association with altered volcanic rocks, now interpreted by some to represent an IOCG situation. Khetri copper deposits are associated with metasediments. Malanjkhand Cu(–Mo) deposit is associated with calcalkaline granitic rocks. SEDEX-type Pb–Zn deposits of Rampura-Agucha, Rajpura-Dariba, and Zawar belt were deposited in Paleoproterozoic sediments of Rajasthan. All these deposits are regionally metamorphosed in grades varying from granulite to upper greenschist facies. Total reserves of Pb–Zn ores were estimated to be 685.59 Mt in 2010. Mineralization of Sn, W, Nb, Ta, Li, Be, F, and REE occur in the pegmatites of Bastar-Malkangiri belt, Bihar mica belt and Rajasthan-Gujarat areas. Monazite in the beach sands of South India is a good source of REE. The same sands are also endowed with large Ti resource in the form of ilmenite and rutile. India is one of the five major Al-ore containing countries of

M. Deb and S.C. Sarkar, *Minerals and Allied Natural Resources and their Sustainable Development*, Springer Geology,
DOI 10.1007/978-981-10-4564-6_3

the world. Bauxite is of residual origin produced from Al-rich rocks. In India, the principal belts of bauxite are located along the East coast, Central India, and the West coast. Warm and wet climate with good drainage favor bauxitization.

Keywords Superior type iron ores · Algoma type iron ores · Kiruna type iron oxide deposits · Indian iron ores · Gonditic Mn ores · Mn nodules · Indian chromite ores · Greenstone-hosted gold in India · Singhbhum Cu–U deposits · Malanjkhand copper · Sediment-hosted Pb–Zn sulfide ores of Rajasthan · East coast bauxites

3.1 Introduction

In the last chapter, we have noted how mineral deposits form and transform through various geological processes. In the present chapter, we focus our attention on the mineral deposits from which we extract one or more metals (± nonmetals) economically or strategically. In the discussion, we have left out certain metals, such as, nickel and vanadium which are either absent or insignificantly present in India. These, however, have been discussed in moderate details in Chap. 2. Metallic mineral deposits discussed here include those of iron, manganese, chromite, gold, copper, lead–zinc, tin–tungsten, REE and aluminum. PGE mineralization has also been treated, though briefly. Ti mineralization finds mention in this chapter and also in Chap. 6.

3.2 Iron Ores

Iron played a crucial role in the development of human civilization right from the early days and in recognition thereof the historians have named one era in human civilization as the "Iron Age". Use of iron came later than copper and bronze in historical times, but it now forms the backbone of the modern industrial world. Whether it is in the field of construction, manufacture, transport – everywhere either iron or its improvised variant, steel, is the most sought after material. Iron and steel account for about 95% of all metals used by modern society. Till material for electronics became so important in contemporary world, steel held the sway. Older civilizations as well knew the use of steel. The Ashoka pillar of Delhi, constructed in fourth century BC, is an example par excellence of the achievements of ancients in the field of metallurgy. The pillar defies rusting even today. The old world, including India, was using low-grade near-surface ores with charcoal as the reductant. Large-scale iron manufacture started in early eighteenth century with commercial extraction of coal. World's crude steel production in 2013 was 1528 Mt. Of this, China is the major producer accounting for 50.2%, followed by Japan (7.1%), USA (5.6%), Russia (3.7%) and India and Republic of Korea, 4.3

and 4.25% respectively. Other countries, such as, Germany, Ukraine, Brazil, and Italy are not very negligible producers. Iron smelting is carried out by reducing iron oxides to iron metal by reaction with carbon monoxide gas, usually derived from coke.

The Clarke of Fe in the crust is 5.76 wt% and in sea water it is only 0.007 ppm (Wedepohl 1971). During magmatic differentiation, Fe is preferentially enriched in basic rocks, such as gabbro. In basalts bivalent iron predominates over trivalent. It is also enriched in Na- and fluid-enriched magmatic rocks as well as in hot brines emanating on the sea floor. In sedimentary rocks, trivalent iron predominates. The two valence states of Fe allow redox conditions of the depositional environments to be constrained, based on experimental Eh–pH data (Garrels and Christ 1965). Iron may get transported in the ferrous state by hydrothermal solutions or meteoric fluids. In ferric state under oxidizing conditions, Fe is soluble in acidic solutions ($pH < 4$) only. Ferric hydroxides are highly insoluble in supergene waters and precipitate in near-surface conditions to form laterites. Under arid conditions, expectedly the end-product of supergene alteration of Fe-silicates is however hematite and not goethite. Various acids, such as, tartaric acid, citric acid, salicylic acid, etc., are secreted by plant roots or are formed by decomposition of organic materials. They produce organometallic complexes which play an important role in the mobilization and fixation of iron in the pedosphere. Metastable Fe-bearing mineral phases form around submarine vents (black smokers) which gradually alter to more stable phases like goethite, hematite and pyrite.

We first discuss the general aspects of iron ores, based on world examples, followed by mineralization in India.

3.2.1 General Characteristics

3.2.1.1 Types of Deposits

Metamorphosed (volcano-) sedimentary iron deposits

1. Banded Iron Formation – Hematite-dominant and variously enriched (Superior type)
2. Banded Iron Formation – magnetite-dominant (Algoma type)

Sedimentary iron deposits

3. Continental fluvio-lacustrine (Clinton type) and Bog iron ores
4. Marine oolitic ironstones (Minette type)
5. Rapitan ores associated with glaciogenic sediments

Magmatism-related iron ores

6. Apatite-bearing iron oxide deposits (Kiruna type) in greenstone belts
7. Apatite-bearing iron oxide deposits in alkaline rocks
8. Fe–(Ti-V) deposits related to mafic intrusions
9. Contact metasomatic iron skarns (also Fe deposits in Ca-, Mg- and Na-skarns)
10. Volcanic-hosted Fe deposits (Lahn-Dill type)

Residual iron ores

11. Lateritic (Ferrites) and Karst iron ores

The cherty iron formations associated with quartzites ± dolomites and black slates in continental shelf environments are classified as the "Lake Superior type" (type 1). Volcanic rocks, represented by tholeiitic basalt, if present, are subordinate in the association. These are thinly banded cherty rocks with the iron-rich layers representing various sedimentary facies. Alternate or rhythmic banding of iron-rich and iron-poor cherty or jaspery layers, which normally range in thickness from a few mm to 1 m, is a prominent feature (Fig. 3.1). The lateral extent of the iron

Fig. 3.1 Rhythmic alternate banding of hematite and jasper in a monolithic BIF from Odisha, at *Shakti Sthal*, Indira Gandhi memorial, Delhi (Photo: M. Deb)

formations may be found to be in hundreds of km while the vertical extent may be in hundreds of meters. Fine granules and oolites of both chert and iron minerals are typical textural features. Iron oxide minerals are predominant with hematite > magnetite, suggesting a high Eh environment of deposition. Facies changes are gradual; they often have chemically graded boundary from BHJ (Banded hematite Jasper/chert) (Fig. 3.1) to BHS (Banded hematite shale) (Fig. 3.2).

The "Algoma type" cherty iron formations (type 2) are intimately associated with volcanic rocks and greywacke-type sediments in deeper basins of greenstone belts. Their continuity is limited, lateral extent is few km while the vertical thickness is in tens of meters. Texture is fine grained, but oxide facies is characterized by magnetite > hematite, suggesting a low Eh condition of deposition. Carbonate, sulfide, and silicate facies are developed abundantly. These iron formations are often associated with VMS deposits of Cu and Zn or with Ni deposits. Economic/subeconomic gold is locally associated with reduced (magnetite-rich) iron formation (Fig. 3.3).

Fig. 3.2 Exposure of banded hematite-shale at Gua mine, Odisha (Photo: S.C. Sarkar)

Fig. 3.3 Tightly folded banded magnetite (hematite) quartzite (originally chert) in Kolar gold field, Karnataka (Photo: M. Deb)

Both "Clinton" (type 3) and "Minette" (type 4) deposits are found in epiclastic rocks in littoral, shallow water estuarine-deltaic or lagoonal environment. They both have limited lateral extent, about 10 m of vertical thickness and are typically lenticular and discontinuous. They are both oolitic to pisolitic, deposited in high Eh conditions. While hematite is the main oxide mineral in Clinton ores, goethite is predominant in Minette. Sideritic carbonate facies is also often developed in both.

An iron formation in an unusual geologic setting (type 5) occurs along the Yukon and McKenzie district border in Northwestern Canada. It comprises a succession of jasper and blue hematite beds more than 150 m thick, near the base of the Rapitan Formation, a crudely stratified, poorly sorted conglomerate horizon at least 1500 m thick (Gross 1973). Differential compaction features, slump and scour and fill structures, intraformational breccias, etc., are all conspicuous throughout the iron formation. Occurrence of tuffaceous layers in the conglomerate and iron formation point to volcanic activity during the deposition of these rocks. These iron ores, with average Fe content of about 46%, are considered to be associated with glaciogenic sediments formed during major Neoproterozoic ice ages.

The Kiruna-type deposits (type 6) occur in volcano-sedimentary sequence of greenstone belts with keratophyres, quartz keratophyres, and granodiorite porphyry, members of sodic porphyry clan. Their mineral assemblage is distinctive: magnetite, hematite associated with F- or Cl-apatite and little titanium oxide. Thus, the host rocks of these ores are more acidic and more alkaline (Na-rich) and fluid-enriched than the Fe–Ti–V clan (type 8) which are related to mafic intrusions (Dill 2010). According to one view, the Kiruna type ores are on one side of a continuum while the IOCG mineralization of the "Olympic Dam type" is at the opposite end (Hitzman 2002).

Titano-magnetite-magnetite-apatite deposits or REE-apatite-Fe deposits (type 7) occur in a suite of rare alkaline magmatic rocks (ijolites, pyroxenites, melilitites, etc.) or in carbonatite–alkaline complex respectively. The contact metasomatic iron skarn deposits (type 9) are related to stocks of granitoid, gabbroid, or syenitic compositions intruded into calcareous sedimentary rocks. Iron in these can be in both non-sulfidic and sulfidic forms. The host skarn and alteration assemblages have variegated calc-silicate mineralogy, dominated by pyroxene and garnet. The stratiform, "Lahn-Dill type" (type 10) Fe deposits of central Europe are associated with volcanic rocks of spilite-diabase series and their volcano-clastic equivalents and display strong lateral facies change (Dill 2010). The siliceous ore with 38–45% Fe is considered to be the proximal vent facies of the exhalative ores. Some deposits are transitional into base metal deposits.

Continental Fe(–Mn) ores (type 11) represented by goethite-limonite-lepidocrocite are known to occur in karst cavities, channels, pockets in platformal limestones and in ferricretes. Lateritic iron ore deposits are of widespread occurrence in recent to near-recent tropical belts. They can have as much as 56% Fe and may grade into aluminous laterites or ferruginous bauxite.

3.2.1.2 Mineralogy and Petrography of the Ores

The principal ore minerals in any combination are hematite (α-Fe_2O_3), magnetite ($FeFe_2O_4$), maghemite (γ-Fe_2O_3), martite (secondary hematite after magnetite), goethite ($HFeO_2$). Other phases, siderite ($FeCO_3$) and pyrite (FeS_2) may occur with types 1–2 (see above) as subordinate phases. Even when pyrite occurs in larger quantities it is not used as an ore of iron. Magnetite is a cubic inverse spinel with 72.4% Fe in both ferrous and ferric forms. It converts to cubic maghemite with 69.9% Fe under oxidation. Many processes, such as, weathering, hydrothermal action, etc., can produce martite (secondary hematite) from magnetite. Hematite can also occur in the hydrous form, called hydrohematite or as flaky mass under metamorphism, called specularite. Mixture of iron hydroxides are commonly referred in the field as limonite.

Other phases, generally minor but sometimes in mineable quantities are greenalite (ferro-antigorite) [$Fe_6^{2+}Si_4O_{10}(OH)_8$] with 44% Fe, chamosite [$(Fe_4^{2+}Al_2)(Si_2Al_2)OH_{10}(OH)_8$] with 29% Fe and minnesotite [Fe-bearing talc-pyrophyllite]. They form if the Fe formation is metamorphosed to low-middle greenschist facies and give rise to biotite, stilplomelane, grunerite [Fe-amphibole], Mg-riebeckite, hornblende, actinolite in higher grades and modified composition of the protolith (Table 3.1).

The ore minerals from which iron is commonly extracted are hematite (Fe_2O_3), magnetite (Fe_3O_4), and goethite (FeOOH). The ore type 1, which is common in nature and in enriched state is preferred for iron extraction. It may have the following megascopic features and micro-structures: hard, massive to blocky, the latter often micro-laminated, platy, i.e., fragmental, cohesionless granular (blue dust), goethitic (pisolitic), canga (BIF/hard ore cemented by Fe^{3+}-oxides/hydroxides). Massive ore often contains neocrystals of hematite in a matrix hosting large grains of martite. Magnetite ores are first concentrated magnetically and in metallurgy, it needs oxidation of $Fe^{2+} \rightarrow Fe^{3+}$ before reduction to metallic Fe. Therefore, its exploitation is relatively expensive.

3.2.1.3 Distribution in Space

Fe ores, particularly of type 1, are found in all the five continents of the world. Some major regions with many deposits to name are Lake Superior (USA), Hamersley basin (Australia), Transvaal basin (South Africa), Labrador trough (Canada), Quadrilatero Ferrifero (Brazil), Krivoy Rog–Kursk (Ukraine). In India such ores occur in the Sandur belt (Karnataka), Dalli-Rajhara and Bailadila belts (Chhattisgarh), Singhbhum (Jharkhand) and the adjoining districts (Mayurbhanj and Keonjhar) in north Odisha. Ore deposits in Goa and western Maharashtra (Redi) are much smaller but they also belong to the Dharwar Group like the Sandur deposits. The type 2 deposits are associated with all greenstone belts in the

Table 3.1 Comparative mineral assemblages in diagenetic and metamorphic iron formations [P&T are estimates only. After French (1973) and Klein (1973)]

Probable original material	Diagenetic (primary)	Diagenetic (secondary)	Low grade metamorphic	High grade metamorphic
$P \leq 100$ b $T < 100$ °C	<1 kb 100–150 °C	1–2 kb 150–200 °C	2–5 kb 100–350 °C	5–11 kb 350–800 °C
1. $Fe(OH)_3$ $Fe(OH)_2$	Hematite Magnetite (?)	Hematite Magnetite	Hematite Magnetite	Hematite Magnetite
2. $FeCO_3$ (Fe, Mg, Ca)CO_3	Siderite Mackinawite	Siderite Ankerite	Siderite, ankerite, minesotaite, talc, stilpnomelane, riebeckite, pyrite	
3. Fe–Si gel (greenalite?)	Greenalite Chamosite	Pyrite, Greenalite (?) Chamosite Chlorite Minesotaite	Chamosite Quartz Chlorite Magnetite Minesotaite Stilpnomelane Riebeckite Crocidolite	Grunerite Ferro-actinolite Hercynite Hornblende Garnet Ortho- and clino-pyroxene Fayalite Quartz Magnetite
4. Volcanic glass and clay minerals	Montmorillonite Illite	Montmorillonite Illite Chlorite Stilplomelane (?)	Quartz Chlorite Stilplomelane Riebeckite	
5. Na–Si gel (magadiite)	Chert	Quartz Albite	Quartz Albite	Quartz Sodic Feldspar (?)

Canadian Shield and in parts of the Australian shield. Examples in the Archean cratons in India are the Gorumahisani and Badampahar deposits in Mayurbhanj district of Odisha and the iron ores of the Bababudan hill range with a large concentration at Kudremukh, besides the thin bands in most of the auriferous greenstone belts in the Dharwar craton.

Ironstones have been widespread in occurrence in west Europe and eastern USA. Clinton ores are found in Wisconsin, New York state, and Southeastern Quebec or in the south, in Kentucky and Alabama in Paleozoic (Cambrian to Silurian) rocks of the Appalachians. Minette-type ores occur in Jurassic rocks of contiguous regions in Luxembourg-France-Germany and Paleogene rocks in Egypt and Saudi Arabia.

Kiruna-type deposits are best represented in the Swedish Norrbotten district. The Kiirunavaara, Tuolluvaara, and Luossavaara orebodies have a cumulative reserve of over 1 billion tons of Fe. Similar deposits occur in Bafq, Chador Malu, and Chogart in east-central Iran. Iron oxide deposits in alkaline rocks are conspicuous in the Kola Peninsula of Russia and in Finland.

3.2.1.4 Distribution in Time

Banded iron oxide-silica ores (types 1–2) are well constrained by time in that bulk of type 1 BIF rocks developed during Neoarchean to Paleoproterozoic (2.7–1.8 Ga, with a peak between 2.4 and 2.2 Ga) and less prolific deposits during Meso- to Neoproterozoic. Derivation of hematitic ores from BIF could have taken place at an early stage of the tectono-thermal evolution of the iron formation or during Tertiary to recent or any combination thereof (Sarkar and Gupta 2012). Mafic volcanic-associated BIFs ("Algoma type") date back to Meso- to Neoarchean times (3.3–2.6 Ga). Clinton ores of the type area (Clinton, eastern USA) and Minette ores of type area (Alsace-Lorraine in France) are Ordovician to Devonian and Jurassic to Paleogene respectively, in age. Skarns are generally Proterozoic (Neo-) to Phanerozoic in age. V-Ti-bearing magnetitic ores within mafic intrusives were emplaced at various times during Archean-Proterozoic times. The Lahn-Dill type ores in central Europe are Devonian in age.

3.2.1.5 Tectonic Settings

Location of geologic entities, rocks and minerals in particular, are controlled by definite tectonic setting/settings (cf. Fig. 2.64). Type 1 deposits developed in Neoarchean to Paleoproterozoic platformal environments and type 2 deposits in greenstone belts. Clinton and Lorraine deposits formed in littoral environments of Phanerozoic platforms. Some ironstones (lateritic) formed in stable continental environments. Fluvio-lacustrine bog iron ore deposits developed under temperate and cold humid climatic conditions. Many intrusives hosting V-Ti-bearing magnetitic ores were emplaced in greenstone belts.

3.2.1.6 Origin of the Iron Ores

The problem of the origin of cherty iron formations has been dealt with in Chap. 2 (cf. Sect. 2.3.2) in terms of the source of Fe and Si, transport of Fe and Si, repetitive precipitation of Fe-dominant and Si-dominant layers and BIF deposition in particular time slots in Earth's history.

The Phanerozoic ironstones have no affiliation whatsoever with volcanism and also have no association with other metalliferous deposits. They have been probably formed by diagenetic replacement of calcareous oolite by hematite or siderite in contrast to the chemically precipitated BIFs. Microbial activity probably played a key role in Fe accumulation in the near-shore environments.

The genesis of Kiruna-type iron ores have been debated for a long time. Various suggestions include pyrometasomatic origin, exhalative origin with stacked mineralization of magnetite and chalcopyrite-pyrite, or through the action of saline hydrothermal fluids during basin subsidence and MORB-type volcanism in a failed rift. A more recent approach is to look at its origin in terms of the broad group of IOCG mineralization, as discussed in Chap. 2 (cf. Sect. 2.2.1.2).

Fe-rich duricrust or laterites developed on a variety of ferruginous bed rocks in subtropical to tropical zones. During humid phase, the Fe-bearing minerals are decomposed and alkalis and alkaline earths are dissolved and removed through percolation leaving behind a residue enriched in Fe, Al, Ni, Cr, Mn in various proportions. During drier periods, pH is lowered and silica is leached out producing a Fe–(Al)–(Mn) oxide–hydroxide-rich lateritic crust.

3.2.2 Iron Ore Deposits of India

Iron ore resources of India are fairly assuring. According to an Indian Bureau of Mines estimate (IBM 2012) India has got a total resource of 28 billion tons (Gt) of all grades (geological estimation), out of which 17.9 Gt belong to hematitic ore and the rest (10.6 Gt) belongs to magnetitic ores. In the hematite ore, 8093 Mt (45%) were under "reserve" category while the balance of 9299 Mt (55%) were of "resource" category. Thus the present estimate of reserve position does not give a complete picture as 55% of hematite resource still remains to be converted into reserves. Also the reserves are estimated at 60% Fe cut-off grade which is not realistic. A reassessment of iron ore reserve/resource at lower cut-off grades (45% Fe) is called for, keeping in mind the specific requirements of steel industry ($Al_2O_3 + SiO_2 \leq 5\%$). This will substantially increase the iron ore resource in the country. Hematitic ore distribution, statewise, is as follows: Odisha: 33%; Jharkhand: 28%; Chhattisgarh: 19%; Karnataka: 11%; Goa: 5%; others: 4%.

The domestic production of iron ores has seen a major dip since 2010–2011. From 208 Mt in 2010–11, it fell to 167 Mt in 2011–12 which may dip further to 120 Mt in near future. India's crude steel production capacity presently is 89 Mt, which is 4% of the total production of the world. Indian steel industry is in a position to compete globally in terms of product range, quality, and price. Presently it is the largest producer of sponge iron (20 Mt) in the world. According to a projection by Ministry of Steel, domestic steel production is slated to reach a target of 200 Mt by 2020. Of course it will depend on the internal political conditions, market, and external trade. As of now, it does not look impossible. With about 1.6 Mt iron ores required for the production of 1 Mt steel, this projected steel production will translate to 250–280 Mt of iron ore requirement per year.

Major iron ore fields in India are located in three regions

1. (a) Bababudan Hills, Chikmagalur district, Karnataka
   (b) Sandur belt comprising Donimalai, Ramanadurg, and Kumarswami ranges in the Bellary-Hospet region of Karnataka and adjoining region of Andhra Pradesh (Fig. 3.4).
2. Dalli-Rajhara, Rowghat and Bailadila region of Chhattisgarh, Central India (Fig. 3.5).

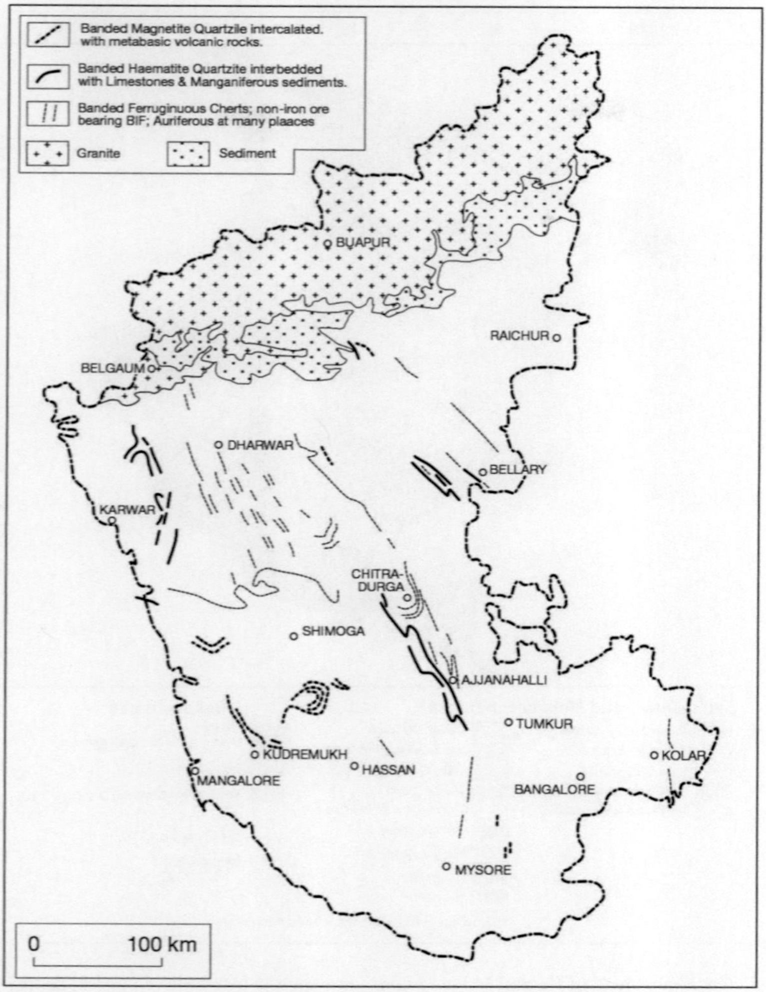

Fig. 3.4 Distribution of three varieties of banded iron ores in the state of Karnataka (after Radhakrishna 1996)

3. Gorumahisani-Badampahar (Mayurbhanj district, Odisha) and Noamundi-Thakurani (South Singhbhum district, Jharkhand), Joda-Barsua, Kalta, Kiriburu-Meghataburu-Gua-Bolani and Chiria (in Keonjhar, Bonai and Cuttack district, Odisha) (Fig. 3.6).

To these one may add the comparatively small deposits of Goa, about 30 in number, within the Bicholim Formation (Bicholim-Sirigao, Velguem-Pale, Cudnem-Surla, and Codli) with resource estimated at about 700 Mt. There are similar deposits in western Maharastra (Redi) as well as some magnetite-rich deposits in Haryana (e.g., Lohara, Mahendragarh district).

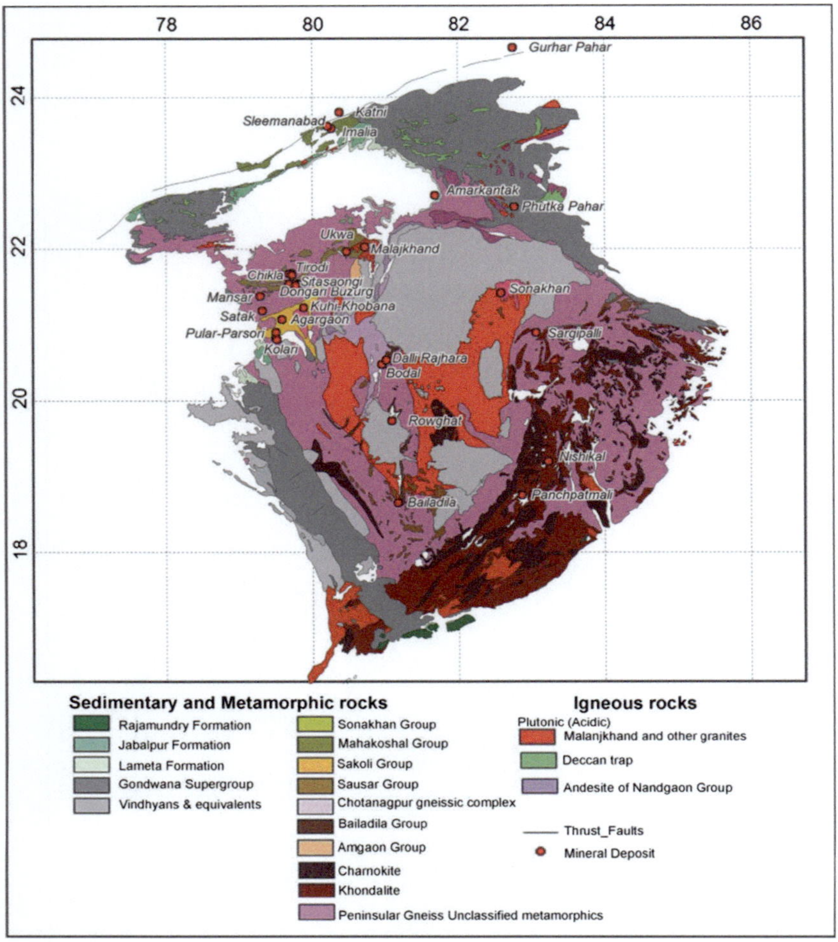

Fig. 3.5 Geological map of Central Indian craton, showing the locations of mineral deposits (after GSI 1998)

3.2.2.1 Iron Ores of South India

In Karnataka iron ores occur at four stratigraphic levels of the Dharwar Supergroup of rocks. The oldest, not too important from economic angle, are found in the Sargur Group within high-grade metamorphites at Maidur-Malavalli. The banded magnetite quartzite (BMQ) in the Bababudan Hills constitutes a horseshoe-shaped folded unit (Fig. 3.4), about 40 km in length and the most well-known deposit is at Kudremukh. The BIF is interbanded with 2.9 Ga old mafic volcanic rocks including

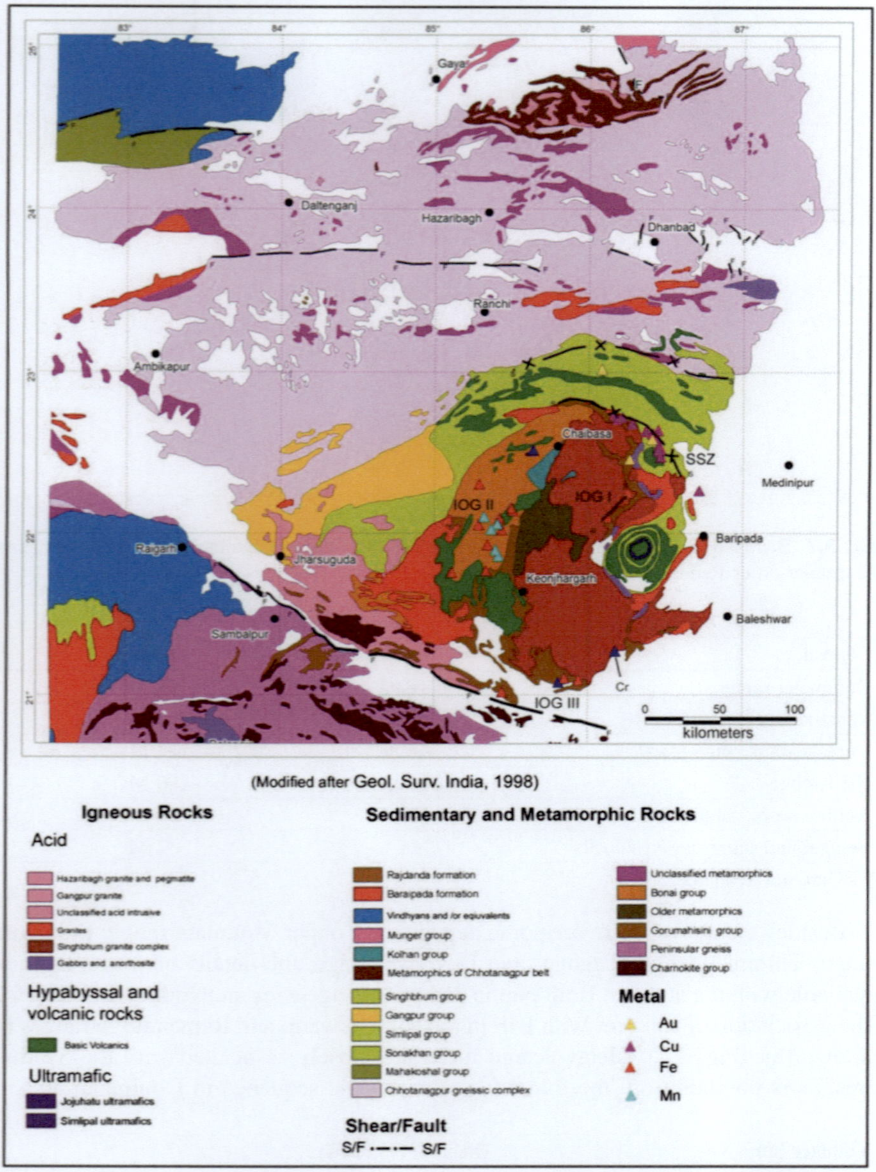

Fig. 3.6 Geological map of Singhbhum-North Odisha craton, showing the locations of mineral deposits (after GSI 1998)

possibly tuffaceous units. The whole association is metamorphosed to middle-upper greenschist facies. To extract the ore, the rock is crushed and the iron oxide (magnetite) is magnetically separated. The BIFs here resemble "Algoma type" and in Chikmagalur district itself its reserve has been estimated to be about 10 Gt (IBM 2005). The metal content hardly exceeds 40% Fe.

The Sandur belt is a repository of large reserves of high-grade hematite ores associated with BIF (Figs. 3.4 and 3.7). The major deposits in the belt are:

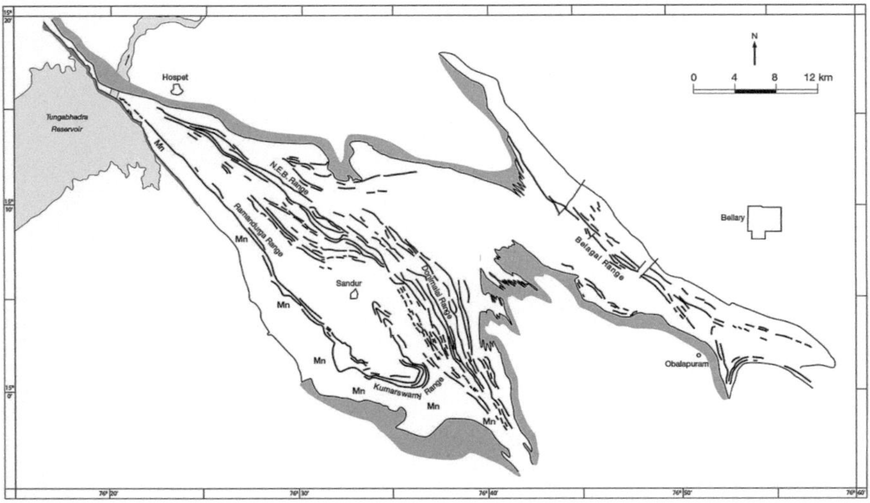

Fig. 3.7 Bands of iron ore and its association with Mn ores in Sandur schist belt, northern Karnataka (after Radhakrishna 1996)

Deposit	Fe-content	Reserve
Donimalai range	+64%	151 Mt
Ramandurga range	+62%	662 Mt
Kumarswami range (part)	+62%	182 Mt
NEB range	NA	280 Mt
Kumarswami range	NA	90 Mt
Belagal and other mining blocks	NA	32 Mt

NA Data not available

Besides the above, there are some deposits at Copper Mountain range, Ettinhatti range, Thimmappanguidi range, and Devadari range, the details of which are not available with the authors. Both public and private agencies are working in this belt. The association of Mn ores with BIF in the Kumarswami and Ramandurga ranges is noteworthy (Fig. 3.7). Metavolcanic rocks are closely associated with the Sandur ores, as is obvious from the following stratigraphic sequence in Donimalai area.

Younger intrusive	Diabase, pegmatite
Donimalai formation	Banded iron formation, shales and tuff
Krishnanagar formation	Metavolcanic rock (basalt andesite)

At Donimalai, ore bands vary from a few mm to several meters in thickness, the latter of course being rare. Fe bands in the BIF consist dominantly of hematite (martite), with relics of magnetite. Primary sedimentary structures such as compositional banding, ripple marks, cross laminations, slump structures, etc., are observed locally. Ore grade is high (>60%). Ores vary from lump to fines, hard to soft. Bulk of the ores has been epigenetically derived from BIF by descending meteoric water, but there are indications that a part could be early (Murthy and Chatterjee 1995; Sarkar and Gupta 2012).

Further up in the stratigraphy are the iron ores in carbonate association of the Chitradurga Group at Chikkanayakonahalli. Proved magnetite deposits are also confined to the high-grade terrains of Salem and North Arcot in Tamil Nadu.

The Fe formation of Goa and Maharastra are considered to be an extension of the Dharwar Group of rocks from northern Karnataka through Goa. The ore deposits of the region include iron, manganese, bauxite, and clay. In 1970, the total Fe ore in the region was estimated to be at 411 Mt with 81 Mt in lump forms with Fe content +60% and the rest occurring as fines. In four decades (2010), this estimate of total resource in Goa, of both hematite and magnetite ores, has gone up to 1141 Mt. The newly set up steel plants at Hazira (Gujarat) and Delvi (Maharastra) utilize the fine ores making them into pellets, using natural gas from Bombay High.

Besides the BIF-hosted iron ores, vanadium-bearing titaniferous magnetite deposits occur within gabbro-anorthosite members of ultramafic suites emplaced in the western Dharwar sequence. Such deposits occur around Masanikere (75° 54′–76° 5′ and 13° 50′–13° 55′) in the southeastern part of Shimoga schist belt and Tagdur-Nuggihalli area of Nuggihalli schist belt in Karnataka. These ores are associated with common sulfides like pyrite and chalcopyrite and are considered to be the result of magmatic differentiation with gravitative settling of magnetites (Pathan and Govindaiah 1986).

3.2.2.2 Iron Ores of Central India

Dalli-Rajhara deposits

These deposits have not been adequately discussed in the literature, although discovered by P.N. Bose way back in 1887. Serious exploration followed by exploitation started after India's independence.

The Dalli-Rajhara belt (cf. Fig. 3.5) is divided into two blocks or subsectors (Fig. 3.8):

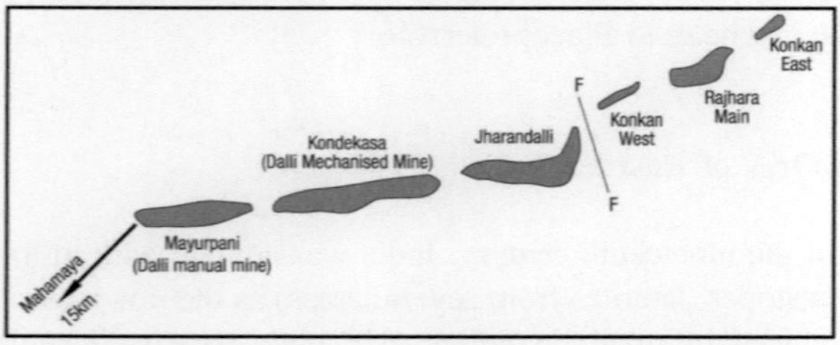

Fig. 3.8 A sketch of the disposition of iron ore mines in Dalli-Rajhara belt

Rajhara block combining Rajhara and the Konkan East and West on its flanks.
Dalli block comprising Jharondalli, Kondekasa (Dalli mechanized mine) and Mayurpani (Dalli manual mine)

The Rajhara main block contains mostly massive ores in the central and eastern parts and at depths in the eastern parts. At many other places, the ore is laminated. The massive ores of Rajhara contains +67% Fe. The Al_2O_3 content is generally low. The Mahamaya deposit being worked out through several mines is located 15 km away in the SW of Mayurpani. The ores in the Dalli-Rajhara mines group are being exploited for the Bhillai steel plant for more than half a century now and are fairly exhausted. The Rowghat deposits, occurring at a reasonable distance (200 km) (cf. Fig. 3.5) should be able to supply the necessary Fe ores to the plant.

Bailadila group of deposits

Initially discovered by P.N. Bose in the late nineteenth century (1889), these were later studied in some details by Crookshank (1963) and Chatterjee (1964). The Bailadila deposits (14 or more in number) occur in two ridges forming limbs of two synforms, the space between them representing an eroded antiform. Before the start of mining there in 1968, the estimated reserve was 1338.68 Mt. The balance reserve now should be around 1000 Mt. The ore grade is generally high (+67% Fe) and the $Al_2O_3 + SiO_2$, to the satisfaction of metallurgists is $\leq 2\%$. H_2O is low (<2%). These properties together with hard and tough character qualify them for use in both direct reduction route and in blast furnace. Bailadila offers one of the best iron ores in the world (vide Sarkar and Gupta (2012) for more details on the BIFs of Dalli-Rajhara and Bailadila belts).

Large ore bodies may be 2.5 km in length (strike) and a few hundred meters in thickness along the dip. The ore bodies may or may not be underlain by BIF. The mined ores may be massive, biscuity, soft laminated, blue dust, and lateritic. The bulk of the ores were formed at an early stage of the evolution of the sequence that contains them. Less importantly supergene process was imposed on it during the Tertiary-Recent times (Sarkar and Gupta 2012).

Bailadila deposits (cf. Fig. 3.5) occur at the southern end of the N–S trending 300 km long Bailadila Group, towards the northern end of which occur Dalli-Rajhara deposits. Inadequate radiometric data hardly constrain the Bailadila Group stratigraphically. But the general consensus is that it overlies the Bengpal Group (Crookshank 1963; Ramakrishnan 1990; Khan and Bhattacharya 1993). Its likely age is Neoarchean to Plaeoproterzoic.

3.2.2.3 Iron Ores of Eastern India

Until the end of the nineteenth century, India was content with utilizing ironstones (Gondwana ironstones, laterites from several areas) as the raw material for iron that was being produced in small quantities. BIF with associated hematite-dominant ores was discovered by P.N. Bose in the beginning of twentieth century.

He inspired Sir Jamshedji Tata to set up a modern steel plant. Tatas were forward-looking people and soon caught up with the idea. A steel plant came up at Kalimati, now called Jamshedpur, at the confluence of Kharkai and Subarnarekha rivers where it is working for a century, utilizing the coal from the nearby coal fields. Soon more iron ores were discovered from the Jamda-Koira valley, particularly along the limbs of the "horse shoe" structure (Fig. 3.9) of Jones (1934). Regionwise, the deposits are as follows:

Western Singhbhum dist., Jharkhand: Chiria, Gua, Noamundi, Kiriburu (Fig. 3.10), Meghatuburu

Mayurbhanj dist., Northeastern Odisha: Thakurani, Joda, Banspani, Malantoli, Bolani, Kalta, Barsua (Fig. 3.11), Gandhamardan, Daitari-Tomka

Keonjhar-Sundergarh-Cuttack dist., South-Central Odisha: Gorumahisani, Badampahar, Sulaipat.

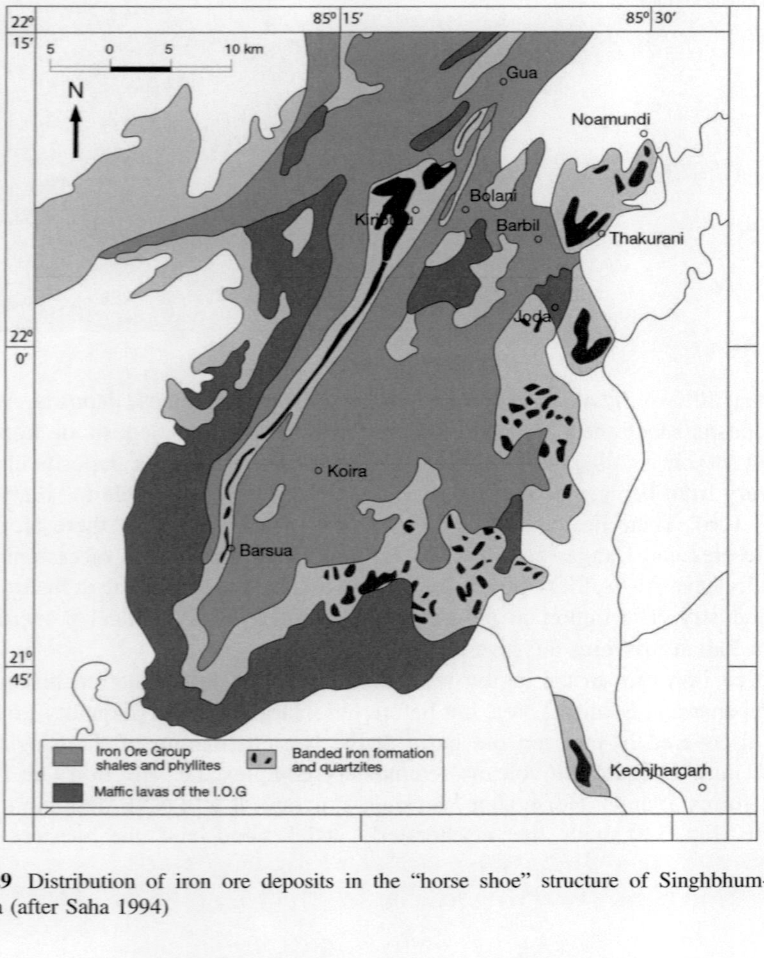

Fig. 3.9 Distribution of iron ore deposits in the "horse shoe" structure of Singhbhum-North Odisha (after Saha 1994)

Fig. 3.10 A view of Kiriburu iron ore mine, Jharkhand (Photo: S.C. Sarkar)

Fig. 3.11 A view of Barsua mine, Odisha (Photo: M. Deb)

About 50% of India's production of iron ores comes from these deposits. Among the deposits mentioned above, Chiria is the largest single deposit of iron ores (>2000 Mt). It is still poorly developed. In these eastern Indian deposits also the ores vary from being massive (Fig. 3.12a), through flaky or friable to "blue dust" (Fig. 3.12b). Their Fe content varies from 62 to 68%. Besides, there are some lateritic ores and Canga (56–60% Fe). Unfortunately the Fe ores of eastern India generally have Al_2O_3/SiO_2 ratio > 2, which, as such, is not welcome in the iron and steel industry. The important physical, mineralogical, and chemical properties of eastern Indian iron ores are given in Table 3.2.

Indeed iron ores in the region were first discovered in the Gorumahisani (and Badampahar and Sulaipat) area, but before long, large deposits of quality iron ores were discovered in the iron ore basin at the western margin of the Singhbhum granite massif. There the volcano-sedimentary complex, i.e., the Iron Ore Group (IOG) forms a large "Horse shoe" tectonic structure. It is a NNE-trending synclinorium (Fig. 3.9) with the northeasterly axial trend and the western limb

Fig. 3.12 **a** Hard, massive, blocky iron ore, Noamundi mine, Eastern ridge, Odisha. It is difficult to imagine this kind of ore to have formed by tertiary-recent supergene alteration of BIF. **b** A patch of unreplaced BIF (*upper left corner*) adjacent to transitional (laminated – blue dust) ore, Noamundi mine, West ridge (Photos: S.C. Sarkar)

Table 3.2 Common physical properties and chemical and mineralogical compositions of the Eastern Indian iron ores

Ore types	Important physical properties	Ore minerals	Gangue minerals	Fe-content
Hard or massive ores (HO)	Hard, massive to blocky, often micro-laminated. Steel gray in color	Hematite ≫ martite ± goethite	Quartz, phengite, kaolinite, talc, Al-phosphate	65–68%
Friable, flaky or laminated (including biscuit), "blue dust"	Cohesive, flaky, laminated, commonly fragmented	Hematite, martite	Kaolinite	62–66%

(continued)

Table 3.2 (continued)

Ore types	Important physical properties	Ore minerals	Gangue minerals	Fe-content
Blue dust (BD)	Fine (fractions of a mm) cohesionless, granular, deep bluish gray/black	Hematite with minor magnetite	Quartz	65–68%
Laterite-goethite ore (LGO)	Highly porous brown/yellow	Goethite, limonite	Kaolinite, rutile	56–58%
Canga	Pieces of BIF/hard ore and other rocks cemented by Fe oxide (hydroxide)	Hematite, goethite	Rock fragments and cryptocrystalline silica	58–60%

continueing towards southeast. Bulk of the iron ores of the region is located in the BIF within the two limbs. The eastern limb (normal) and the western limb apparently differ in certain aspects, such as, Fe concentration, orebody thickness, and nature of the overburden. Eastern limb orebodies are often richer and the eastern limb is steeper. These two features could be related in terms of percolation of water.

The ores of Badampahar-Gorumahisani-Sulaipat area are closer to Algoma type in character. The rest are so-called "Lake Superior" type and the ores formed from the BIF by hypogene, and more probably by hypogene + supergene processes (cf. Fig. 2.66). Relevant in the context is the fact that the eastern Indian BIFs are different being Archean in age, the oldest being 3.5 Ga (approximately).

3.3 Manganese Ores

Manganese has the atomic number 25 and atomic weight 54.938. It is an important metal for modern industries. It is used in metallurgy as well as in chemical industry. In steel industry it is used in small quantities (6–10 kg/ton) to remove the oxygen and sulfur. Manganese in steel renders it harder and tougher. Metallurgical use is made through several ferromanganese alloys. India now prefers export of manganese in the form of ferromanganese alloys to selling them in the raw state. Chemical uses of manganese need high degree of purity. The principal uses in the chemical industry are in the making of battery, paints and pigments, and in glass industry.

3.3.1 *General Characteristics*

3.3.1.1 Geochemistry and Mineralogy

Manganese occurs in four valence states: (II), (III), (IV) and (VI), Mn(III) being transient. In rocks and minerals however, Mn^{2+} and Mn^{4+} are more common. Higher valence states in nature are generally found in low temperature environments, characterized by higher oxygen fugacity. More than 50% of Mn minerals listed in the literature contain (OH) or H_2O. Average manganese content in mafic rocks (1000 ppm) is about 3 times larger than in granitic rocks (300 ppm). Average Mn content in sedimentary rocks is 0.056% or 560 ppm (Beus 1976).

Minerals common in continental deposits	Minerals in deep sea nodules
Pyrolusite (β-MnO_2)	Birnessite (γ-MnO_2)
Psilomelane-Cryptomelane [(Ba, K, Mn)O_2]	Todorokite [(Mn, Ba, Ca, Mg)$Mn_3O_7 \cdot H_2O$]
Manganite (γ-MnOOH)	Buserite (δ-MnO_2)
Hausmannite (Mn_3O_4)	Deep sea Mn nodules often contain Ni, Cu, Co, Pb, Mo, Tl, Zn. The first three metals may be economically exploitable one day
Braunite (Mn_2O_3)	
Bixbyite [$(Mn, Fe)_2O_3$]	
Nsutite $\left(Mn^{4+}{}_{0.85}O_{1.7}Mn^{2+}{}_{0.15}(OH)_{0.3}\right)$	
Manganosite (MnO)	
Rhodonite ($MnSiO_3$)	
Tephroite (Mn_2SiO_4)	
Rhodocrosite ($MnCO_3$)	
Spessartine (Mn-garnet) $\left[Mn^{2+}{}_3Al_2(SiO_4)_3\right]$	
Hubnerite [Mn$(WO)_4$], commonly occurring as a vein mineral is exploited for W rather than for Mn	

3.3.1.2 Types of Manganese Deposits

Manganese deposits in terrestrial record are mainly of three types: (i) Hydrothermal Mn deposits; (ii) Sedimentary Mn deposits; and (iii) Lateritoid Mn deposits, of which the second group is most important economically.

The hydrothermal deposits are common during Phanerozoic times. These deposits are small, often stratabound or irregular bodies or epithermal veins in a wide variety of host rocks. Some of these deposits have strong similarities with deposits that are being formed hydrothermally in the marine realm at or near spreading centers. Deposits of this type are known from Northern Apenine ophiolitic complex, Italy; Pindos geotectonic zone and at Othris in Greece; Olympic Peninsula, USA (Roy 1997).

Sedimentary Mn deposits outclass the other types with regard to size and spatial and temporal distribution. Such deposits formed in shallow water in the margins of stratified oceans above the redoxcline during sea level changes due to tectonism. World's largest Mn concentration of this type in a single geological sequence (ca. 2.4 Ga old Hotazel Formation, Postmarsburg Group, Transvaal Supergroup) took place, interbedded with BIF, during the "Kalahari Manganese Event" in South Africa. The platformal Mn deposits in Minas Gerais Supergroup in Brazil formed around the same time with similar characteristics. A geologically distinct Mn carbonate deposits formed in the intra-continental rift setting of the Lower Birimian greenstone sequence (ca. 2.3–2.0 Ga) in West Africa. Mn oxide deposits and BIF mark the glaciogenic Damara sequence (ca. 750–650 Ma) in Namibia. Giant deposits of diagenetic Mn carbonate formed during Late Jurassic at Molango, in Mexico while Mn oxide deposits formed in large shallow-shelf in Middle Cretaceous times in Groote Eylandt, Australia. Next only to Kalahari deposits, large-scale deposition of Mn oxide and carbonate ores took place in shallow water shelves during Early Oligocene times in Chiatura, Georgia, Nikopol in Ukraine and Northern Ural, Russia.

Supergene manganese enrichment in continental weathering profiles was dictated at various locations by climate, topography, and prevailing drainage system. Best-known deposits of this type with high-grade Mn ores are located in Brazil, Gabon, Ghana, Ivory Coast, and Mexico.

3.3.1.3 Distribution of Manganese in Space and Time

The distribution of Mn deposit is rather secular. There are only a few not-too-large Mn deposits that may be related to the initial deposition in the Archean time. Of the six deposits of initial Archean age, four are located in India (Tumkur-Chitradurga, Sandur, Jamda-Koira valley and Eastern Ghats). The other deposits of this age are found in Brazil. The Paleoproterozoic age is suggested for a few large and super-large deposits. These include deposits in South Africa (Kalahari), Brazil (Minas Gerais, Urukum), Ghana (Dagwin), Gabon and India (Sausar belt, Central India; Gangpur Group, Odisha). Mesoproterozoic is virtually barren of Mn deposits. Mn-mineralization came back in the Neoproterozoic but in glaciogenic sequences in Brazil, China, and Namibia. The only exception is the Penganga Group deposits in Andhra Pradesh, India which were formed in a shelf sequence. Mn deposits again became important during the Phanerozoic. However, the Paleozoic deposits in China and Kazakstan are rather small; deposits become larger in the Mesozoic. Giant-size land based deposits formed during the early Oligocene in Georgia (Chiatura), Ukraine (Nikopol), west Siberia, Middle Volga, and the south Urals in the CIS (former USSR). The USA has Mn deposits at a number of places, but none is large and at present there is no Mn mining in that country. The theater of Mn deposition soon shifted to the oceanic realms in the form of Mn nodules and crust, the former being best developed in the Pacific Ocean.

3.3.1.4 Tectonic Settings of Manganese Deposits

Preferred tectonic environments for sedimentary (-diagenetic) manganese deposition are basin margins – shallow shelves to deeper shelves. Mn nodules and crusts, of course, develop in deeper parts of basins. Changes in climate and sea level, as we shall see later, are indirectly related to tectonics and directly influenced manganese deposition through control of oxygen regime by causing transgression/regression of the oceans (cf. Fig. 2.69).

3.3.1.5 Origin of Manganese Deposits

Manganese may be supplied to basins by endogenic hydrothermal solutions and from continental rocks by weathering, erosion, and transport in solution of the metal. Of the continental rocks, basic rocks are expected to have been the best provider of the metal. Various experimental studies on Mn–Fe–H_2O system and field observations support the conclusion that Eh–pH are the main controls in Mn and Fe deposition (Fig. 3.13).

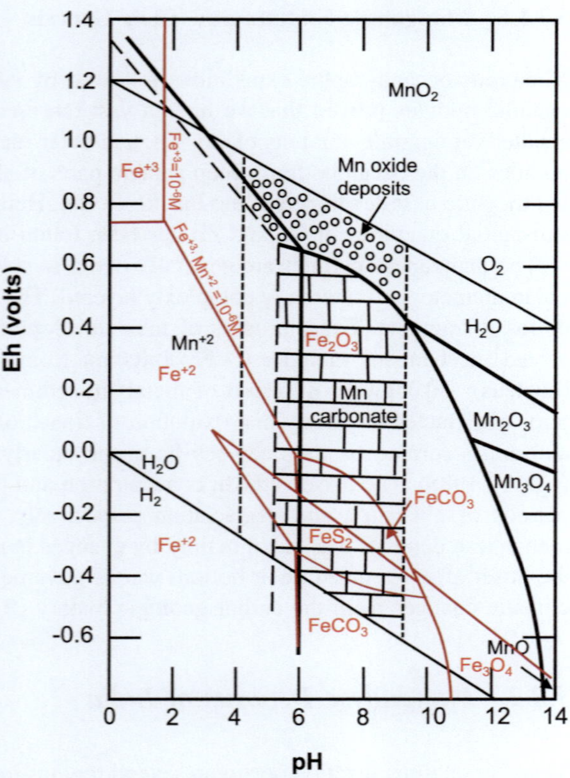

Fig. 3.13 Stability fields of common iron minerals (shown in *red*) (after Garrels and Christ 1965), superimposed on those of common Mn minerals (vide Fig. 2.68 for details of Mn stability) in a Eh-pH diagram. Total activity of dissolved carbonate, 1M, of dissolved sulfur, 10–6M

The initial solid phases in equilibrium with sea water are Mn_3O_4 and γ-MnOOH. Owing to the transient nature of Mn(III) it is ultimately Mn^{2+}–Mn^{4+} redox pair that effectively controls the precipitation of Mn from the dissolved state (Roy 2006) and there is no reason why this was not the case in the past. In this context, the oxygen in the atmosphere and the interacting hydrosphere played an important role.

In an aqueous solution characterized by low Eh, both Fe(II) and Mn(II) are soluble. That keeps hydrothermally produced Fe(II) and Mn(II) dissolved in oxygen-starved basins. With the rise of oxygen potential Fe(II) is precipitated as Fe(III), but Mn still remains in solution until Eh rises further. Therefore, in an uninterrupted sedimentary sequence in the upper levels where Eh commonly rises, we expect precipitation of Mn oxides in a system containing both Fe(II) and Mn(II) (Fig. 3.13).

During the Archean time, the world was generally anoxic. That explains the scant availability of Mn ores in the rocks (possible erosion not ignored). Local accumulation is difficult to explain. It is possible that high UV-dependence of manganese photo-oxidation in the presence of Fe(II) may explain the situation (Anbar and Holland 1992). Local precipitation of Mn ores could have developed in "oxygen oasis" (Kasting 1993). After oxy-atmoversion, transgression and regression played important roles in controlling Eh-regimes in platformal regions.

3.3.1.6 Manganese Nodules and Their Genesis

Numerous oceanographic expeditions followed by laboratory investigations of the metallic nodules proved that we have a vast resource of Mn, Fe, Cu, and Ni and smaller yet tangible amounts of Co, Zn, and other metals in the (Ferro-)manganese nodules on the ocean bottom. Deep pelagic parts of all major oceans contain them, but they are best developed in the Pacific Ocean. Here up to 2 wt% (Cu + Ni) with substantial quantities of Co and Zn has been found locally.

The manganese nodules are generally ovoid to spheroidal in shape, up to a few cm in diameter and internally complexly layered. The growth may be around a rock or fossil fragment. Exact process of their development is not known. Some suggested mechanisms include: (i) Precipitation from sea water around a substrate (nucleus); (ii) Upward diffusion of metals in sediments through pore-water transport; (iii) Bacterial activity and oxidation of transition metals. Manganese nodules and crusts formed on the deep sea floor particularly since the late Miocene time (Glasby 1988). The dissolved Mn concentration and its deposition corresponded to periods of glaciation and deglaciation respectively. Formation of such deep-sea manganese deposits was aided mainly by changes in ocean circulation patterns and the onset of oxygenated polar bottom waters governed by interrelated tectonic and climatic changes as in the earlier geologic history (Roy 2006).

3.3.2 Manganese Deposits of India

Manganese mineralization occurs in several regions in India. These are the Dharwar region (Chitradurga-Tumkur, Shimoga, Goa, and Sandur areas), Eastern Ghats (both in parts of Odisha and Andhra Pradesh), Central India (Maharashtra and Madhya Pradesh), and in Eastern India (Kendujhar or Keonjhar district and Goriajhar in Sundergarh district of Odisha) – the most important of them being the Central Indian region. The inferred reserves of all categories of ores in the country are 142 Mt (Mulay 2001).

3.3.2.1 Dharwar Region

Manganese mineralization originally took place during the Late Archean (2.6–2.7 Ma) and is associated with phyllites, carbonates, and ferruginous cherts, interlayered with layers of mafic volcanic rocks in the Chitradurga-Tumkur and Kumsi-Hornhalli areas. They usually occupy the lower parts of the Chitradurga Group. In North Kanara and Goa, the mineralization took place at an upper horizon. Mn ore minerals are of low temperature varieties. Opinions on the origin of these deposits vary from sedimentary-diagenetic (Naganna 1977) to supergene concentration from manganiferous sediments (Fermor 1909; Roy 1966). Krishna Rao et al. (1982) divided these ores into the following genetic types:

1. Sedimentary oolitic ores
2. Reworked sedimentary oolitic ores
3. Cavity filling replacement types
4. Float ore

The above classification does not make a clear distinction and "float ore" has no genetic connotation, except that they are scattered on the surface as pieces of different sizes and shapes. The presence of stromatolites along with relict layering confirms that the original deposition was sedimentary-diagenetic, formed in the marginal shallow water part of a basin. The Mn oxide ores were metamorphosed to low grade (Roy 1981).

The Mn-bearing sequence at Sandur also contains stromatolites. The basalt andesite volcanic suite with BIF is overlain by phyllitic rocks containing bedded Mn ores (Fig. 3.7). Here also an Iron formation, i.e., Donimalai Formation overlies the Mn-bearing Deogiri Formation. Mn, like BIF of the belt, could be related to the mafic volcanism.

3.3.2.2 Penganga-Godavari (P-G) Valley

Manganese mineralization, though not a large one, occurs in the P-G valley, included in the Chanda Formation of the Neoproterozoic Penganga Group in Andhra Pradesh. Ore-bearing intervals occur in two stratigraphic horizons. Bedded chert, associated with Mn oxides is enclosed in a thick limestone sequence (Bandyaopadhyay 1996) (Fig. 3.14). Mukhopadhyay et al. (1997) concluded that the Mn oxide deposition took place at the base of the slope of a distally steepened deep water ramp. Gutzmar and Beukes (1998, 2000) discovered the Mn carbonate rhodochrosite and kutnahorite below the Mn horizon and concluded that these

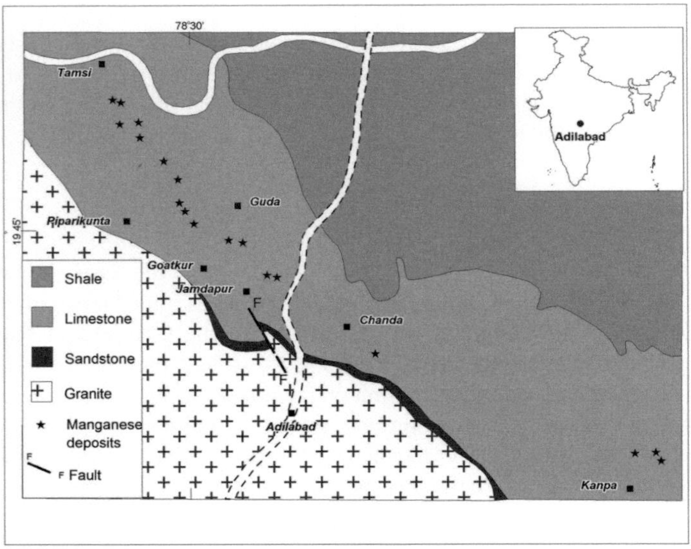

Fig. 3.14 Geological map showing distribution of Mn ore deposits near Adilabad in Pranhita-Godavari valley (after Bandyopadhyay 1996)

minerals on oxidation produced the ore. The two models are far too different to be reconciled. No volcanic rocks have been reported by any of the groups. Ores are not metamorphosed and there is no evidence of glaciation.

3.3.2.3 Eastern Ghat Belt

Manganese mineralization in the Eastern Ghats occurs over a total distance of ~200 km, spreading over both the states of Odisha and Andhra Pradesh. In the Odisha sector the important sites are at Boirani–Kodala (Ganjam district), Kanaital (Bolangir district), and Ambadala and Nishikal (Rayagada district) (cf. Fig. 3.5). Originally deposited ≥2.8 Ga ago, the ores now occur as bedded deposits co-deformed and co-metamorphosed (at amphibolites/granulite facies) with the host rocks. Near-surface supergene modifications are also conspicuous.

Some information on East Coast Mn deposits of Odisha is as follows:

1. Kutinga–Nishikal zone: 12.5 Mt; average grade 29% Mn; 11.5% Fe; 0.3% P; cut-off grade 20% Mn.
2. Ambadala zone: Mn-content of ore up to 26%.
3. Kanaital–Uchhabapalli zone: 6.5 Mt; average grade 27.4% Mn, 14% Fe and 0.3% P.

Mn ores occur in Kodur, Garividi, and Garbham in Andhra Pradesh. Average grade of Mn ores from Srikakulam district, Andhra Pradesh is 28–40% Mn, 5–12%

Fe, and 0.3% P. The rock name Kodurite is no more in vogue. Here the sedimentary-diagenetic manganese ores, metamorphosed to amphibolite-granulite facies, exists along with its supergene alteration products. They are enclosed by calc-silicate rocks and pelitic granulites.

3.3.2.4 Sausar Belt, Central India

The Sausar belt, the largest repository of manganese ores in the country, is about 200 km long (Fig. 3.15). Within the Sausar Group the Mn-bearing rocks and ores are enclosed within the pelitic Mansar Formation and calcareous Lohangi Formation in more than one stratigraphic unit. Lower Mansar Formation is the better container of the ores (Fig. 3.16). There are more than 200 Mn ore deposits in this belt, spread over four districts of Maharashtra (Nagpur and Bhandara districts) and Madhya Pradesh (Chindwara and Balaghat districts).

"Proved" and "probable" reserve of 17 major deposits is about 20 Mt. Dongri Buzurg (6.22 Mt) and Chikla (4.71 Mt) are the largest deposits in the country. Other important deposits are at Mansar and Tirodi. Average Mn ore in this belt contains 41.50% Mn, 10.26% SiO_2, 4.68% Fe, and 0.18% P. Phosphorous has a tendency to increase with the increase in the Mn-content of the ore. Mn-bearing rocks are a part of the Sausar sequence and the whole complex has been metamorphosed up to upper amphibolite facies.

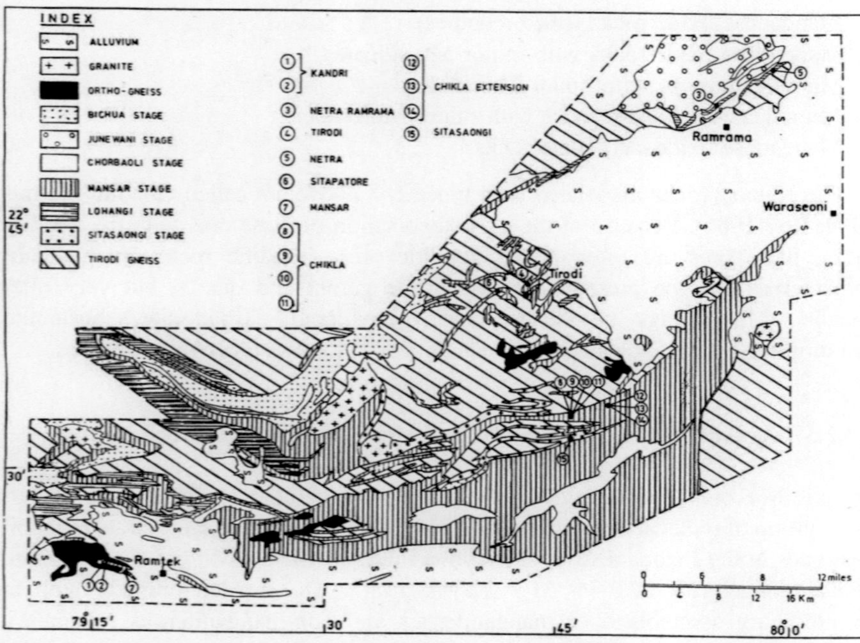

Fig. 3.15 Geological map showing distribution of Mn ore deposits in Sausar Group of rocks in central India (after Narayanaswami 1962)

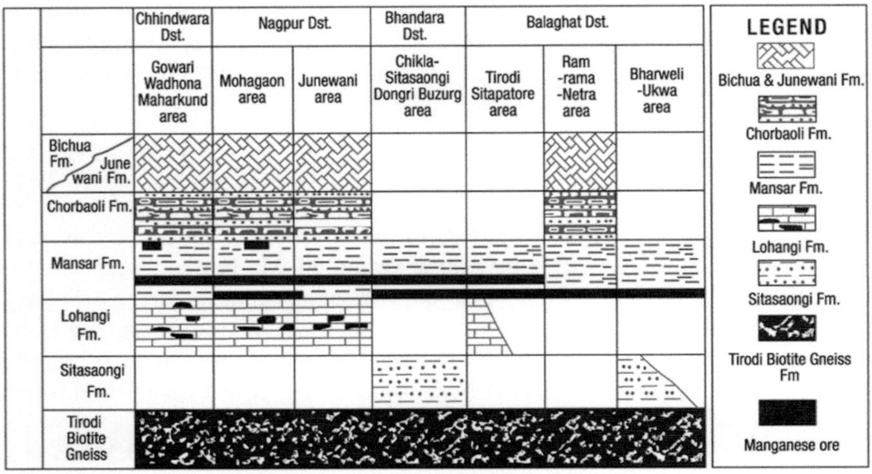

Fig. 3.16 Synoptic diagram showing stratigraphic positions of the manganese ore horizons in the manganese ore belts of Madhya Pradesh and Maharashtra (thickness and lateral extent not to scale) (after Roy 1966)

Manganese-rich rocks in the Sausar Group have been distinguished into five broad groups (Dasgupta et al. 1990, 1992):

1. Manganese oxide rocks (Mn orebodies).
2. Manganese oxide rocks with minor Mn-silicates.
3. Mn silicate rocks with minor Mn oxide.
4. Mn silicate–carbonate rocks with minor Mn oxides.
5. Manganese oxide–carbonate rocks.

The country rocks associated with types 1–3 above are called Gondite (Fermor 1909; Roy 1966). Because of the close association of these ores with the gonditic rocks, these ores are often called gonditic ores. Gonditic rocks are generally characterized by the presence of spessartine garnet and quartz, but very often contain Mn-bearing pyroxenes and pyroxenoids (rhodonite, bustamite, Mn-diopside), amphiboles (tirodite, winchite), and a few other minor phases.

3.3.2.5 Aravalli Supergroup and Equivalents in West-Central India

Several Mn ore deposits are known in the southernmost part of the Aravalli Supergroup of rocks in the Banswara district of Rajasthan. The same strata probably continues in the Jhabua district of Madhya Pradesh where a Mn ore deposit occurs at Kajlidongri (22° 57′; 74° 31′) (Nayak 1980). The bedded orebody with an extraordinary assemblage of manganiferous and non-manganiferous minerals is hosted by quartzite rocks in a suite of quartzites, phyllites, and gonditic rocks. The ores and the host rocks are regionally metamorphosed to a low grade.

Fig. 3.17 View of semi-mechanized Bamebari Mn mine of Tata Steel Ltd., near Joda, dist Keonjhar, Odisha (Photo: S.C. Sarkar)

3.3.2.6 Eastern India

In eastern India manganese deposits occur in Keonjhar district and in Goriajhar (Sundergarh district, Odisha). In Keonjhar much of the mineralization occurs along the eastern limb of the "horse shoe" synclinorium and also along its core. This mineralized zone containing both the iron and manganese mineralizations is often referred to as the Jamda-Koira valley (cf. Fig. 3.9). The main occurrences are at Joda (Fig. 3.17), Kalimati, Gurda, Phagua, and Mahulsukha. It will be interesting to note that here also Fe and Mn mineralization occur in two litho-units of an Archean volcano-sedimentary stratigraphic sequence, i.e., the Iron Ore Group (IOG) as presented below (Table 3.3).

Manganese ores in this region occur in two principal modes: stratabound (including stratiform) and lateritoid. Cryptomelane, pyrolusite, lithophorite are the principal ore minerals. Hematite, goethite, and kaolinite are common associates. Mafic volcanism appears to be responsible for Mn, while its concentration as stratabound to lateritoid ores could have taken place by diagenesis and supergene fluid activities (Sarkar and Gupta 2012).

In the Goriajhar–Monomundra area of Sundergarh district, Odisha, Mn ores occur in a gonditic association and has been subjected to exploitation for quite some time now.

Table 3.3 Stratigraphic sequence of the IOG rocks in the Jamda-Koira valley, showing stratigraphic dispositions of the manganiferous and iron formations (after Patel et al. 2005)

Formations	Rock units
Upper volcanic	Mafic lavas and tuffs
Ferruginous shale formation	Shales of various denominations that include tuffaceous matter to BIF with derivative iron ores
Manganiferous formation	Cherty member: massive chert grading to dolomite shale member: Mn-bearing shale with Mn ores
Banded iron formation	Laminated banded hematite jasper (BHJ) and banded hematite quartzite (BHQ) with Fe ores
Volcanic formation	Upper member: Tuffaceous shale Lower member: Mafic lava
Basal formation	Sandstone, grit, conglomerate

3.4 Chromite Deposits

Chromium is used in metallurgy (in special quality steel manufacture), furnace lining and in chemical industry including tannery. Average contents of Cr and PGE in mafic and ultramafic magmas are 200–1000 ppm and 0.2–50 ppb respectively (Crocket 2002). Currently chromite ores grade 25–35% Cr and the PGE ores 2–20 ppm (2–20 g/t). In the present-day world, exploitable reserve of Cr is believed to be of the order of >1 billion tons and PGE around 77 thousand tons. Over 45% of Cr is produced by the Union of South Africa and the rest is produced by Kazakhstan, Turkey, India, and Albania. ~80% of Pt and 40% of Pd are produced by the Union of South Africa. Several other countries such as, Russia, China, Finland, and Zimbabwe also produce these metals but in a small way. India however, is not in the list and therefore virtually all of PGE consumed in the country are imported from outside.

Chromite deposits are associated with ultramafic and mafic rocks (including anorthosite, rarely, as in Bushveld). Chromite deposits are mainly divided into two types, depending on their lithotectonic association which have some conspicuous features. These are

1. Layered stratiform intrusions (Bushveld type)
2. Ophiolites (Alpine type)

The layered intrusions, for example, the Bushveld and the Great Dyke (Union of South Africa) and Kemi (Finland) as a whole contain much more chromite than the ophilite bodies in Russia, Turkey, and Kazakhstan. The layered chromitite[1] bodies, occurring in tectonically stable areas, are mm to m scale in thickness and may be extensive over kilometers. All major stratiform chromite deposits are Late Archean to Paleoproterozoic in age. Podiform chromite[2] in ophiolites occurs in supra-subduction zone (Zhou and Robinson 1997). This variety of chromite is generally refractory in nature, rich in Al_2O_3 (>20% Al_2O_3). It constitutes the refractory grade chromite ores with higher Cr/Fe values. Ophiolite complexes are representative of the uppermost mantle as well as oceanic crust, emplaced by subduction tectonics. The complex is characterized by deformed harzburgite and peridototic mantle overlain by less deformed layered dunite, gabbro, sheeted dykes, and basaltic lavas.

The principal chromium mineral is chromite ($FeO{\cdot}Cr_2O_3$) with a spinel structure. However, a general formula should be (Mg, Fe^{2+}) (Cr, Al, Fe^{3+})$_2O_4$ as it commonly contains Mg, Al, Fe^{3+}, besides Fe^{2+} and Cr^{3+}. Cr in hexavalent state (Cr^{VI}) is toxic and hence an environmental hazard.

---

[1]A rock consisting essentially of chromite cumulates with the chromite grain interstices occupied by serpentine veneers.

[2]Characterized by chromite 'nodules', set in serpentinized dunite or chromite-bearing dunite.

3.4.1 Origin

Origin of chromite ores is believed to be related to ultramafic–mafic magmatism because of their close field association, chemical compatibility, ore petrography, and results of experimental studies, as outlined in Chap. 2, under "Essentially Magmatic Processes".

3.4.2 Chromite Mineralization in India

India had a moderate reserve of 179 Mt of chromite ores, including the mined out part. The present remaining reserve is 66 Mt in 2010. More than 90% of this reserve occurs in eastern India, particularly in Odisha. India's chromite production in 2010 was a little over 4 million tons. A small production comes from Karnataka.

The three fields of chromite mineralization in eastern India (cf. Fig. 3.6) are

1. Jojohatu-Hatgamaria area (West Singhbhum), Jharkhand.
2. Sukinda (Jajpur-Dhenkanol district), Odisha.
3. Baula-Nuasahi (Keonjhar district), Odisha.

Of the three fields mentioned above, the Jojohatu field has been the weakest and the mineable part has already been extracted. The ultramafic bodies, generally composed of rhythmically layered dunite-chromite, clot peridotite and pyroxenite, are co-folded with the country rocks.

Sukinda is a fairly large field of chromite mineralization, being extensive for about 25 km in strike and little less than 2 km in plan width. Surface exposures of primary rocks and ores are poor due to surficial weathering. Interpretation of subsurface geological–geophysical data suggested a package of ultramafic intrusive and basic effusive, occurring as a folded sequence (Fig. 3.18) (Basu et al. 1997).

The chromitite seams at Sukinda occur as disconnected bands and lenses within serpentinized dunite. The tabular orebodies, six in number and concordant with the

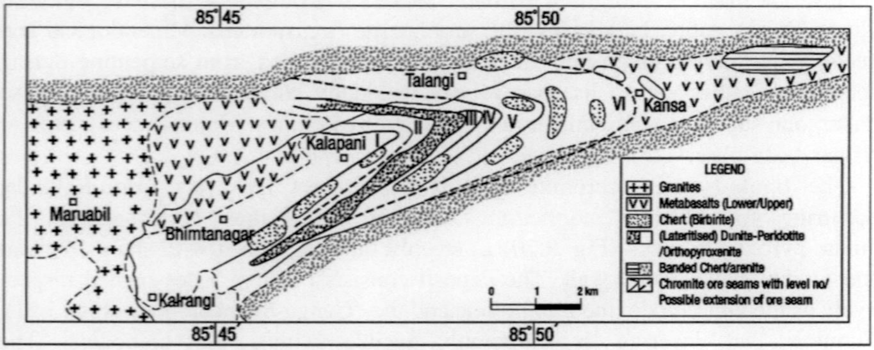

Fig. 3.18 Inferred geological map of the Sukinda ore field, Odisha (after Basu et al. 1997)

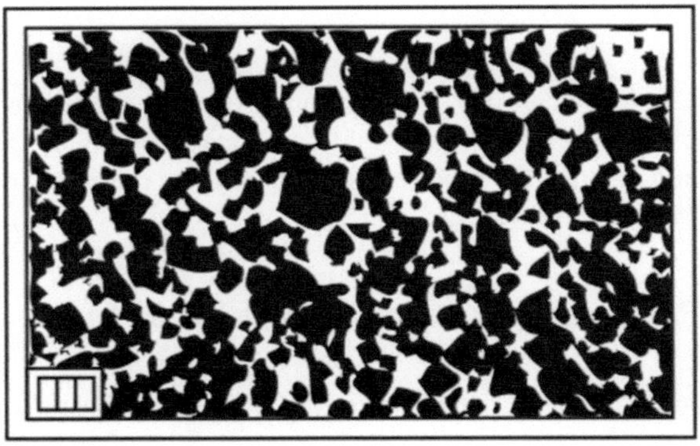

Fig. 3.19 Cumulus texture of chromite (*black*) with interstitial serpentinite (*white*) – a transmitted light photomicrograph, Sukinda. The scale represents 1 mm (after Chakraborty and Chakraborty 1984)

folded host rocks, may be 3 km in length (straightened strike) and up to 20 m in thickness. The chromitite is characterized by cumulus texture of chromite with interstitial serpentine (Fig. 3.19). The total ore reserve at the beginning, down to 250 m depth, was 112 Mt with an ore grade of >42% Cr_2O_3. Sukinda, India's largest chromite deposit is much smaller than the super large deposit of Bushveld (South Africa) and also compared to the chromite deposits in the Great Dyke of Zimbabwe.

Nickeliferous limonite occurs over a wide area above the serpentinized ultramafic rocks of Sukinda valley (Chakraborty and Chakraborty 1976). It is overlain by a thin mantle of laterite and is underlain successively by silicified ultramafics and serpentinites with layers of chromitite. The nickeliferous limonite band extends from 1.22 up to 7.0 m on an average, below the laterite cover. NiO content of limonite varies from 0.32 to 1.17%, in primary ultramafic rocks from 0.44 to 0.75% and in the chromitite from 0.13 to 0.40%. These authors concluded that Ni was fixed in the lattice positions of olivine and chromite respectively. It was released during residual weathering of these minerals and precipitated in the limonite profile in an acid environment. Later, Som and Joshi (2002) studied three vertical profiles of Sukinda chromiferous ultramafic body at Kansa, Kaliapani, and Kalarangi in details. The profiles comprised partially altered serpentinite at the base followed upward by saprolite, laterite (in situ) and laterite (transported). Mineralogical and geochemical investigations revealed that Ni was released from serpentine during chemical weathering and has partly been residually enriched in goethite as also undergone supergene enrichment in Fe and Mn hydroxides. Major concentration of Ni has been observed in saprolite.

The Baula-Nuasahi chromite field is located not far away from Sukinda, occurring towards the northeast. Chromite ore bodies occur within the dunite-pyroxenite rocks (Fig. 3.20) as steeply dipping bodies with sharp footwall and gradational hanging wall. The deposit consists of three lodes named respectively as the Durga lode, the Laxmi lode and the "Ganga-Shankar" lode (Fig. 3.21). Compositional layering is commonly visible within the orebodies. The dunite-peridotite is intruded by a gabbro-anorthosite complex.

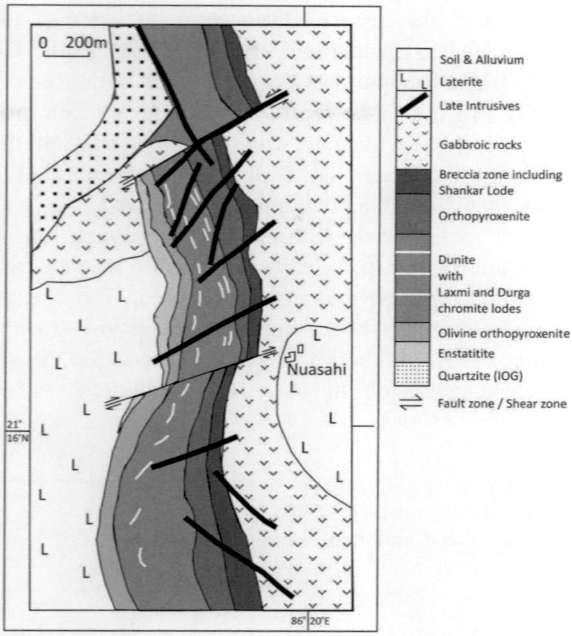

Fig. 3.20 Geological map of Baula-Nuasahi area, Odisha (after Basu et al. 1997)

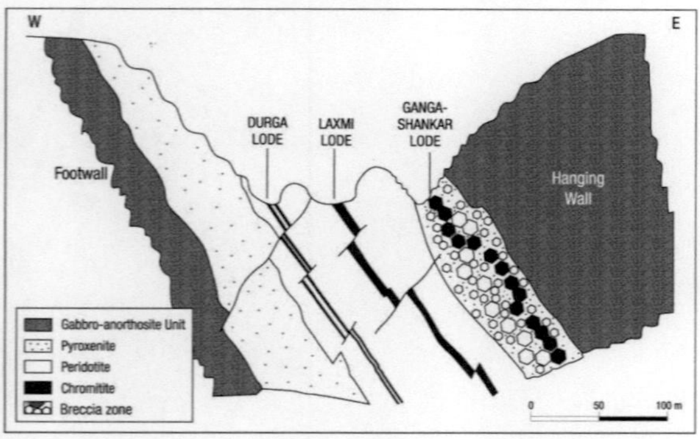

Fig. 3.21 A geological section across Baula mine, showing disposition of the chromite lodes. PGM mineralization associated with the Ganga-Shankar lode system is hosted by the breccias zone (after Auge et al. 2002)

There are several studies on the genesis of Baula-Nuasahi and Sukinda chromite (cf. Sarkar and Gupta 2012 and references therein). Latest among these is Mondal et al. (2006). According to these authors the chromitites of Nuasahi and Sukinda massifs were crystallized from mantle-derived boninitic magma, produced through a second stage melting of depleted upper mantle in the presence of fluids generated by a subducted slab in a suprasubduction zone setting. Evidence of "mixing" is absent in Nuasahi-Sukinda magmatic complexes (Mondal et al. 2006).

Chromite deposits in Karnataka occur mainly in Hassan and Mysore districts, confined within Meso-Archean Nuggihalli and Sargur schist belts, respectively. In the former, the deposits from south to north of the belt are: Jambur, Tagdur, Ranganahalli, Bhaktarhalli, Byrapur, Aladahalli, Gobbalahalli, and Belagumba (Fig. 3.22). Byrapur has been the chief producer of chromite ore with 48–52% Cr_2O_3. All the other deposits in Nuggihalli belt are of low grade with 27–40% Cr_2O_3. In Mysore district, within Sargur schist belt, the larger chromite deposits are at Sinduvalli, Thalur, and Urdbur (Nijagunappa et al. 1986). Estimated ore reserves in the state, including that already mined out is only about 3 Mt. The plutonic suite of host rocks for the chromitite bodies comprises altered dunite and peridotite, while anorthosite, pyroxenite, and gabbro host the associated magnetite bands. The ores are deformed and metamorphosed. The chemistry of chromite from these deposits correlates better with stratified deposits (Mukherjee et al. 2010; Sarkar and Gupta 2012). Mukherjee et al. (2012) modeled the geodynamic setting of Western Dharwar craton during chromite metallogeny. According to them, TTGs in the

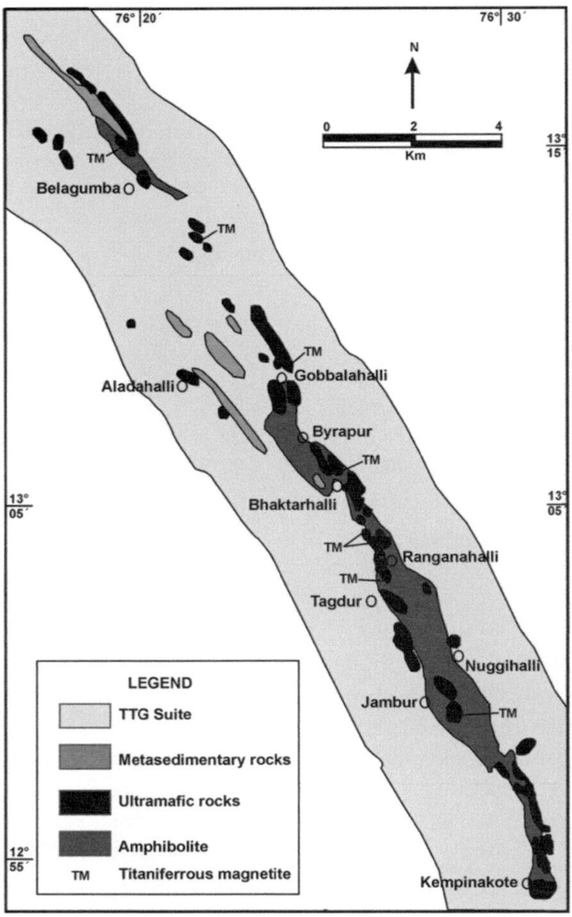

Fig. 3.22 Geological map of Nuggihalli schist belt, Karnataka (after Jafri et al. 1983)

region formed by the melting of the subducting oceanic crust with simultaneous formation of komatiites by hydrous melting of the depleted mantle wedge. Isotopic and geochemical data indicate depleted mantle as the source of the komatiite-derived sill-like mafic–ultramafic complexes that host the chromite bodies whose metallogeny was probably related to the amalgamation stage of a supercontinent assembly.

3.5 Ores of Platinum Group Elements (PGE)

PGE includes platinum (Pt), palladium (Pd), rhodium (Rh), ruthenium (Ru), iridium (Ir) and osmium (Os). They are very rare metals, having a potential world reserve of 77,000 tons. Of these Pt and Pd are the principal members. PGE is divided into two subgroups: PPGE (Pd, Pt and Rh) and IPGE (Ru, Ir, Os). Platinum group elements are very chalcophile ($D_{PGE}^{sulfide-silicate}$ ~ 104–105, Ballhaus et al. 1994; Fleet et al. 1996) and very strongly siderophile ($D_{PGE}^{Metal-Silicate}$ ~ 105–106; O'Neil et al. 1995). Copper is also chalcophile and hosted mainly in sulfides. Nickel is both chalcophile and lithophile, being partitioned between sulfides and olivine.

World production of PGE was 492,000 kg in 2011 (BGS 2015). South Africa is the producer of 80% Pt and 40% Pd. Russia produces the rest of the remaining Pd.

Major PGE mineralization occurs as

1. Reefs in layered mafic–ultramafic intrusions
2. Major Ni–Cu sulfide deposits
3. Placers

Reefs, as stratiform bodies, occur in the middle of layered intrusions and are thin compared to their lateral extent. PGE are associated with sulfides and chromites in the reefs. Ni and Cu are usually minor by-products. PGE may, in these bodies, be concentrated in the base, rarely though. Ni–Cu sulfide deposits of Norilsk, Siberia, Russia are richer in PGE, compared to most other Ni–Cu sulfide deposits. Placer deposits have been located at places, but in totality they contribute a small portion to the reserve. PGE-containing minerals are many. But some of them are difficult to identify because of size and compositional complexity.

A number of mines produce PGE as the principal commodity or as by-products. Three important deposits amongst them are the Bushveld Complex of South Africa, the Great Dyke of Zimbabwe, and the Norilsk deposit of Russia. A mention may be made of the only known PGE prospect in India at Boula-Nuasahi.

In the Bushveld Complex, the known largest repository of PGE in the world, every layer of chromitite is abnormally rich in PGE (1–3 g/ton), but they still do not make the grade everywhere. Three units that make the economic grade in Bushveld Complex with respect to PGE are the Merensky Reef, UG2 chromitite and a basal unit in the northern limb called Platreef. The PGE-rich Merensky Reef (cf. Fig. 2.2) lies at the base of cyclic units that contain thin basal chromitite layers. At

Rustenburg mine the basal chromite (<1 cm) rests on an anorthosite layer and is overlain by a 30–90 cm thick pegmatitic feldspathic pyroxenite. In thin chromitite layers grades can reach 50 g/ton (Pt + Pd + Rh + Au). Merensky Reef contains about 3% sulfides (pyrrhotite, chalcopyrite and pentlandite) and there is a good correlation between all the PGE and Ni, Cu, and S. The UG2 chromitite layer is between 60 and 120 cm thick. Pt + Pd + Rh + Au average 5 g/ton, with a suggestion that thinner the layer, higher the grade. Here also Pt, Pd, Cu, Ni and S show direct correlation. Platreef has a different geology. At one or more different levels sulfide-bearing PGE mineralization is present. Average grade of Pt + Pd + Rh + Au may be ≥ 3 g/ton for a thickness of several tens of meters. Cu and Ni are higher than in the Merensky Reef and correlate with S but not the PGE.

PGE (and Cr and V) are found in deposits that are generally derived from the mantle by partial melting which produce basic to ultrabasic magma. But there may be a rare exception too, in the example of Sudbury Igneous Complex (SIC), Ontario, Canada. Here the ore-bearing igneous complex bears the shape of a lopolith and consists of norite to granophyres with the Ni–Cu sulfides concentrated in the footwall zone. As there is no evidence of mantle parentage of any rock of the complex and there are evidence of shatter cone, high pressure mineralogy formed in the adjacent rocks, the SIC is believed to have formed by meteoric impact within the crustal rocks.

3.5.1 PGE Mineralization in India

The Nuasahi ultramafic complex in Odisha, mentioned above, contains high concentration of platinum group elements (Pt up to 9.6 ppm and Pd up to 17.9 ppm) and Ag (up to 6.6 ppm) in a chromiferous sulfide-rich breccias zone (Mondal et al. 2001) (Figs. 3.20 and 3.21). This NNW–SSE-trending zone is $\sim$10–15 m thick and occurs on the hanging wall side of the orthopyroxenite unit in the eastern part of the complex. The zone comprises heterogenous assemblage of differently oriented angular to subrounded fragments of massive, spotted, and banded chromitite of the Shankar lode and its host ultramafic rocks, represented mainly by serpentinite. The matrix is a complex assemblage of chromite, magnetite-ferrian chromite, orthopyroxee, amphibole, biotite, talc, carbonates, and sulfides. The sulfide mineralization is restricted to the matrix and in the margins of chromitite fragments. Auge et al. (2002) reported 18 PGM species from the breccias zone and the alluvial concentrates. Sudburyite is the most common and occurs preferentially within coarse grains of chromite. Michenerite occurs with palladian melonite. Irarsite-hollingworthite solid solution forms inclusions in cobaltite-gersdorffite grains occurring within chalcopyrite. The mineralization can be ascribed to magmatic processes (Mondal et al. 2001). A hybrid magma, formed by the partial incorporation of serpentinized and chromiferous ultramafic rocks and chromitite in the younger gabbroic magma, was emplaced within a preexisting breccias zone, which itself developed during brittle deformation along a NNW-trending shear zone. PGM mineralization has also been reported from several places in Dharwar and Central Indian cratons. But all of them remain unworkable till date.

3.6 Gold

Gold is known to have been mined in substantial amount for at least 6000 years. Egypt was probably the first place where gold objects first appeared as early as 4000 BC. Over the centuries, it has been the principal medium of international monetary exchange and its physical properties and common occurrence as native metal made it prized for jewelry and other adornments. Gold is still the backbone of international finance and one commodity whose value has appreciated steadily and steeply since 1971 (Fig. 3.23). Gold prices worldwide are soaring (in August 2014, it was at US$1294/oz or about US$45/g). As world economies reduce their dollar reserves, gold is regaining favor as a repository of wealth and a convenient, reliable, and stable medium of exchange. China and India, huge exporters and holders of US currency, are both avid buyers of gold. In these countries, the ever-increasing affluent middle-class consumes a large portion of the gold in the market via jewelry. To some extent wealth flows away from the dollar into gold. Ultimately these two fastest growing economies of China and India may set the standard and history as to which asset will measure wealth and so far gold comes well ahead.

In India gold (22 Ct) is selling at about Rs. 30,000 per 10 g and the consumption remains the highest in the world, close to that of China and followed by Japan. The

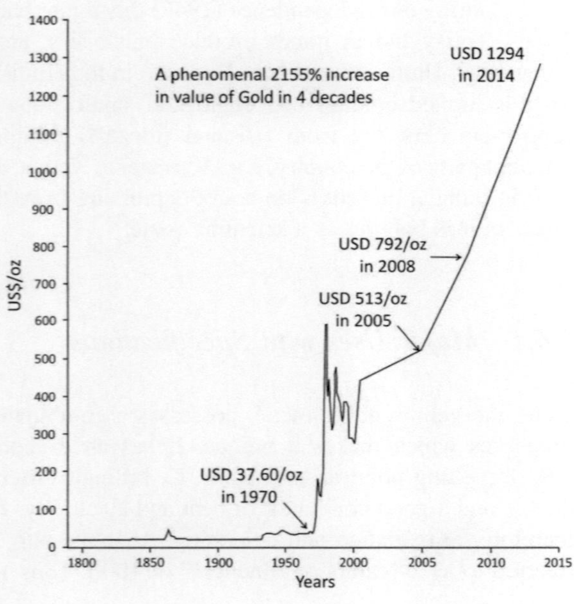

Fig. 3.23 Graph showing the growth of gold value in dollars in the last 200 years

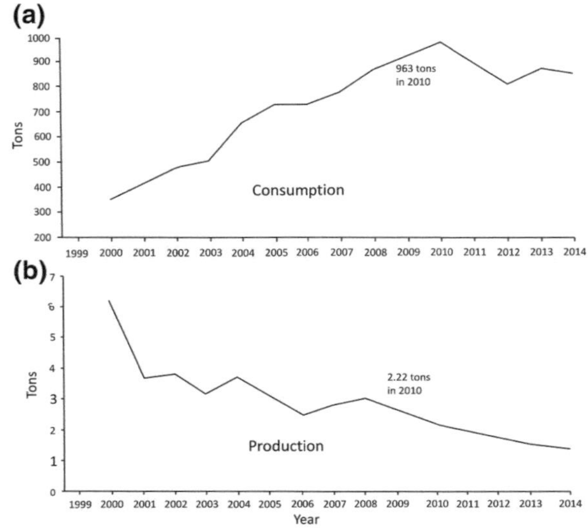

Fig. 3.24 Graphs showing **a** the consumption and **b** production of gold between 2000 and 2014 in India

consumption figure reached near the 1000 ton mark for the first time in 2010 (Fig. 3.24a), according to the World Gold Council. The contrast with China with regard to production is glaring: While India produces only 2–3 tons of gold per annum (Fig. 3.24b) from primary ores, China produced 270 tons of gold in 2007, keeping pace with its rising consumption. India which has the highest consumption of gold in the world showed a steep decline of production in the past decades. Around this time 100 years ago India produced about 20 tons of gold from its mines. During our Independence (1947) this figure had already plummeted to about 5 tons. Today Indian mines produce much less, around 3 tons from only three mines (e.g., Hutti, Uti and Hira Budinni) in the Hutti-Muski schist belt, Karnataka. To this limited amount one could add small amounts of by-product gold from copper smelters and from artisanal (illegal?) mining of alluvial placer gold in different parts of the country, e.g., Karnataka, Kerala, and Jharkhand. The picture of gold in India at present is no doubt depressing from the resource point of view, as much as it is baffling as a scientific issue.

3.6.1 Major Uses with Specifications

Gold, the yellow noble metal, possesses a combination of chemical and physical properties which makes it unique. It has an exceptionally high specific gravity (19.3), melting point around 1060 °C, brilliant luster, is extremely malleable and ductile and a good conductor of heat and electricity. It is resistant to corrosion and therefore, to oxidation and tarnishing. Absolute pure gold, not found in nature, is referred to as 24 carats or "fineness" of 1000. Thus 1 carat denotes 41.67 parts of gold per 1000 parts of total metal or 4.167% gold. In jewelry gold is commonly of 22 carats, the rest 2 carats being that of copper. Other than its major use in jewelry, gold also finds use in ayurvedic medicine, dentistry, and laboratory instruments. In recent years, it has achieved wide industrial use in electronics and computer industry. Gold is used for fabrication of semi-conductors, printed circuits, connectors, and other micro-components for computers and space applications. Gold has no entirely suitable substitute. Platinum is a substitute in jewelry, but is costlier and lacks the appeal and color of gold. Silver is not sufficiently tarnish-resistant and somewhat less ductile.[3]

3.6.2 Geochemistry

The crustal abundance of gold is 4×10^{-3} mg/kg in Earth's crust and 4×10^{-6} mg/L in Earth's oceans (Institute MI 2011). Gold is more abundant in mafic than in felsic igneous rocks. The iron meteorites called siderites contain as much as 8 ppm of gold, but the average ranges from 0.67 to 1.3 ppm or 150–300 times as much as crustal values.

Gold deposits are however found more commonly in felsic or intermediate igneous rocks than within mafic or ultramafic types. Gold is chemically inert in natural environments, at the earth's surface. During weathering and decomposition of rocks it is hardly affected. Thus, it gets residually concentrated at the weathering site or transported in the native state and concentrated locally along with other resistant minerals in placer deposits.

In ore deposits, gold is commonly associated with pyrite and/or arsenopyrite and other sulfide minerals of Ag or Cu although it does not form a separate sulfide phase. All gold ores contain some Ag. On the other hand, ores predominantly of Ag contain appreciable quantities of Au. Cu-ores particularly of the VMS and porphyry type produce the bulk of by-product gold.

It is clear that gold as an element occurs in a wide variety of ore deposit types. Thus, gold deposits are identified, based mainly on three main factors (cf. Poulsen et al. 2000): (1) *Concentration of gold in a deposit*: while grade of gold in different deposit types can be quite variable (Fig. 3.25), majority of deposits that are mined presently have grades exceeding 1 g/t of ore. (2) *Total amount of gold in a deposit*: The product of gold concentration and ore tonnage determines the total amount of gold in a deposit (Fig. 3.25). The total tonnage may be calculated using low tonnage and high grade, or high tonnage and low grade, or any intermediate combination. The total tonnage may thus range from 1 to 10,000 t of gold for economically viable deposits. (3). *Relative amount of gold in a deposit*: In a variety of gold deposit types (cf. Tables 1.5 and 3.4), a compositional continuum exists between Au and other metals, e.g., Ag, Cu, Mo, Zn, Pb. One defining characteristic of "gold deposits" therefore, is such: in a "gold deposit" in which other economic elements are also present, the worth of the mineable gold should exceed the worth

[3]'Fineness of gold' in an Au-alloy is 10× its percentage in the alloy

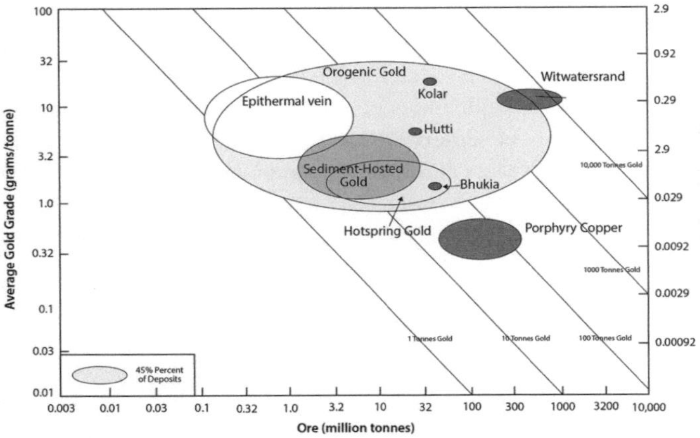

Fig. 3.25 Grade–tonnage relationship in different gold deposit types. Note the position of Kolar, Hutti and Bhukia deposits in India (after Singer 1993)

of the other elements taken together (cf. Fig. 2.54). Gold deposits also show considerable variation in Ag:Au ratio.

3.6.3 Mineralogy

In the natural state, gold occurs as native metal which however, is rarely pure. It is invariably found alloyed with some proportion of silver or copper or iron, or with bismuth and mercury. When the proportion of silver is considerable (15–45%), the alloy is called *electrum*. Other minerals of gold are tellurides, composed of Au or Au + Ag combined with Te. The minerals are calaverite ($AuTe_2$) and sylvanite [(Au, Ag)Te_2]. 95% of gold in nature occur in the native state or alloys.

3.6.4 Different Mineral Deposit Types

The classification of gold deposit types is fraught with some inherent problems as it can be approached through genetic, geochemical, tectonic, or economic point of view (cf. Emmons 1937; Boyle 1979; Cox and Singer 1986; Bache 1987). An easy to use, though not quite exhaustive, lithotectonic classification of gold deposits is presented in Table 1.5. The attempt to classify gold deposits is also hampered by: (1) Several large gold deposits carrying the signatures of two or more ore-forming systems, giving rise to a hybrid or composite look. (2) Different geological environments also superimpose on each other, e.g., continental arc may superimpose on older accreted volcanic arc, making it difficult to decide with which environment the deposit is associated. (3) Geological attributes used to distinguish between deposit types get blurred where the deposits are deformed and metamorphosed, particularly in Precambrian terranes (cf. Poulsen et al. 2000).

In spite of these inherent problems, most geologists would agree that classification schemes have a high predictive value for discovery of new deposits. There are three unifying aspects to all classification schemes (cf. Poulsen et al. 2000): First, there are a finite number of well-studied gold deposit types, number varying between 8 and 20. Second, some deposit types, e.g., skarns, paleoplacer, etc., are recognized in all schemes, but may differ in name only; third, to each such deposit type, there is a model attached which lists the main attributes, a perceived mode of genesis and also specifies a particular geologic setting. 17 commonly recognized gold deposit types are listed in Table 3.4 with a mention of their salient features, and world, including Indian examples.

Table 3.4 Commonly recognized gold deposit types (after Poulsen et al. 2000)

I. **Intrusion-centered, large, district-scale hydrothermal system, genetically linked to the deposits**

Deposits having spatial and genetic relationship with igneous intrusions. The intrusions related deposits are further subdivided (cf. Sillitoe 1991; Poulsen et al. 2000) as:

1. **Porphyry gold and gold-copper deposits**: associated with multidirectional stockwork mineralization and K-silicate alteration adjacent to small composite stocks of diorite to syenite in composition
2. **Skarn Au**: replacement deposits in skarn and carbonatized rocks adjacent to diorite to granodiorite
3. **Breccia pipe**: ore as disseminations in funnel-shaped magmatic-hydrothermal, hydromagnatic and collapse breccia
4. **Carbonate replacement**: semi-massive to massive sulfide bodies (mainly pyritic) in silicified limestone. Represents an end member of more typical gold-poor Pb–Zn mantos
5. **High sulfidation epithermal veins**: also disseminations and replacement bodies of high sulfur assemblages within vuggy silica-rich rocks. Argillic alteration with hypogene alunite
6. **Batholith-associated quartz veins (Korean-type)**: quartz vein swarms in and adjacent to granitoid batholiths. Fault-controlled mineralization

II. **Linked to magmatic-hydrothermal systems (exact link less well understood)**

7. **Low sulfidation Epithermal veins (volcanic-associated)**: formed at low temperatures (less than 200 °C) and at shallow depth (Lindgren 1933). Crustiform colloform chalcedonic quartz veins with adularia-calcite-sericite
8. **Hot spring deposits**: downward transition to 7 above. Siliceous sinter cap and geyserites. Mercury and native sulfur mineralization
9. **Sediment-hosted micron gold (Carlin-type)**: sediment-hosted, very finely disseminated Au–Ag deposits. Stratabound low sulfide replacement of carbonate rocks. Structurally and stratigraphically controlled zones of silicification and brecciation
10. **Gold–copper sulfide-rich veins**: quartz-sulfide veins (>20% sulfides) occurring in clusters. Au generally < Ag
11. **Submarine Au-rich massive sulfides**: gold (in ppm) exceeds combined Cu, Pb, Zn (in percent). Felsic volcanic rocks and sub-volcanic intrusions common
12. **Iron Oxide–Copper–Gold (IOCG):** structurally controlled, epigenetic deposits, formed in extensional environments within co-eval anorogenic granitoids or older volcano-sedimentary metamorphic rocks

(continued)

III. **Other Au deposit types**

13. **Iron-formation-hosted veins and disseminations**: Stratabound orebodies within thick sequences of iron-rich metasediments or BIF and volcaniclastics. Also in derived laterites. Syntectonic quartz veins with gold commonly concentrated in vein selvedges
14. **Paleo-placer deposits**: formed by the mechanical accumulation of economic important, detrital and resistate minerals. Pyrite-bearing quartz pebble conglomerate and quartz arenite
15. **Greenstone-hosted lode gold deposits**: epigenetic fault- and shear zone-related quartz-carbonate vein deposits of gold in metamorphic terranes (Kerrich 1993). (Also termed as orogenic gold deposits, shear gold deposits, gold-only deposits)
16. **Turbidite-hosted quartz-carbonate veins (Bendigo type)**: fold and fault-controlled quartz-carbonate veins with arsenopyrite being the dominant sulfides
17. **Non-carbonate-hosted (e.g. black shale) stockwork and disseminations**: along faults, permeable units and lithological contacts in a variety of host rocks

Examples of deposit types

# (as above)	World	Indian
1	Grasberg (Indonesia), Fort Knox (Alaska)	Malanjkhand?
2	Suan (Korea), Red Dome (Australia), Fortitude (Nevada)	Bhukia?
3	Creeple creek (Colorado), Kidston (Australia)	
4	Ruby Hill (Nevada), Mammoth (Utah)	Imalia?
5	Gold Goldfield (Nevada), El Indio (Chile), Nansastu (Japan), Yanacocha (Peru)	
6	Chenoan (Korea), Linglong (China)	Jonnagiri
7	Creede (Colorado), Hishikari (Japan)	
8	McLaughlin (California), Champagne Pool (New Zealand)	
9	Carlin (Nevada), Golden Reward (S. Dakota), Guizhao (China)	Imalia?
10	Tennant Creek (Australia)	Bhukia?
11	Boliden (Sweden), Mt. Lyell, Mt. Morgan (Australia)	
12	Olympic Dam (Australia)	
13	Homestake (S. Dakota), Cuiaba (Brazil)	Ajjanahalli
14	Witwatersrand (S. Africa), Tarkwa (Ghana), Jacobina (Brazil)	
15	Mother Lode (California)	Kolar, Hutti, Ramagiri etc.
16	Victoria Goldfields (Australia), Ashanti (Ghana)	Gurhar Pahar, Gulaldih
17	Muruntau (Uzbeckistan), Andacollo (Chile)	

3.6.5 Types Represented in India

As is now generally accepted, there are two main types of environments in two different geological periods in which gold is found in India (Fig. 3.26): (a) Neoarchean granite-greenstone belts of Dharwar craton, South India and their

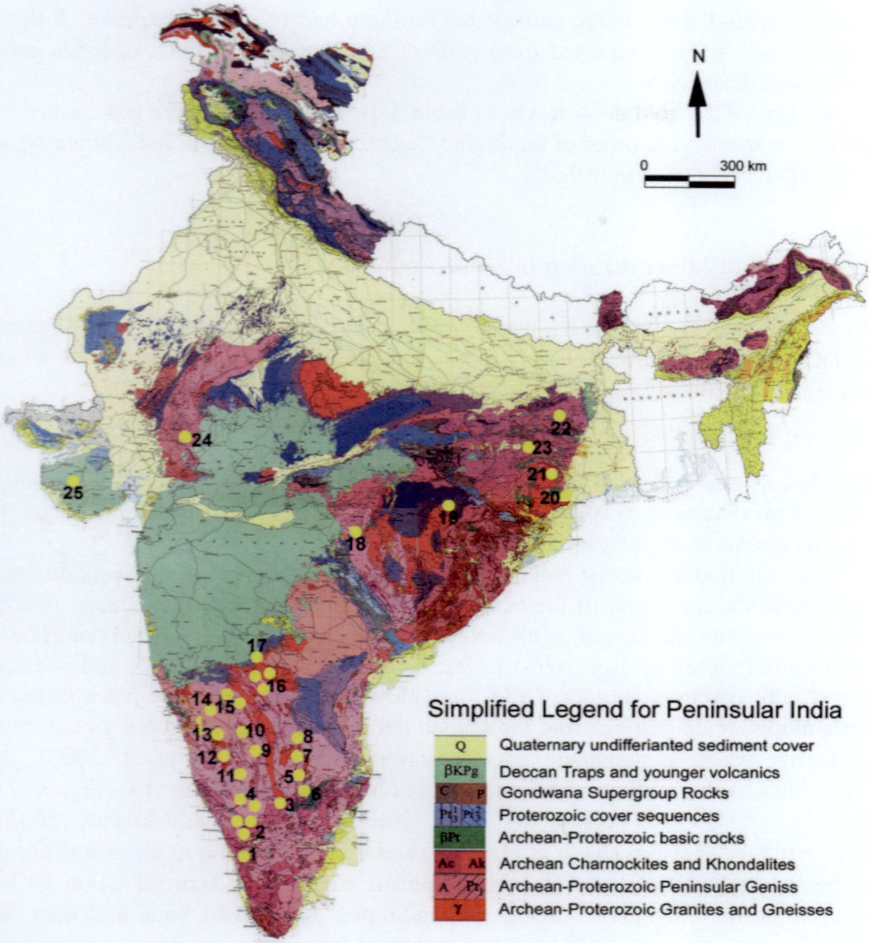

Fig. 3.26 Location of gold deposits and prospects on a geological map of India. *1* Allapadi, *2* Wynad, *3* Kavudahalli–Porsedyke, *4* Amble–Woolagiri, *5* Kolar, *6* Mallapakonda-Chigaragunta, *7* Ramagiri, *8* Penkacharla, *9* Bellara, *10* Ajjanahalli, *11* Kempinkote, *12* Honnuhalli, *13* Kudrekondo–Palavanahalli, *14* Chinmulgund–Haveri, *15* Gadag, *16* Hutti, *17* Manglur, *18* Bhiwapur–Chanderi, *19* Sonakhan, *20* Talkot–Benapal, *21* Kunderkocha, *22* Lawa–Mysera, *23* Parasi, *24* Bhukia, *25* Alech hill (after Sarkar 2010)

variants in the granulite terrain further to the south. Sonakhan in central Indian craton (cf. Venkatesh 2001), though of Paleoproterozoic age and Kunderkocha in the Archean Singhbhum craton of eastern India are probably the only two examples where this type of auriferous environment is found outside the Dharwar craton. (b) Paleo- to Mesoproterozoic metasedimentary zones hosting Au-sulfide mineralization in carbonate and argillite facies of rocks in varied tectonic settings in western, central, and eastern parts of the northern half of peninsular India. A third type, lateritic gold, is reported from parts of Dharwar craton and is of lesser economic significance.

In terms of the gold deposit types (Table 3.4), # 15, 6, and 13 are represented in the Dharwar craton in order of importance, and 2 or 10 in western India while 16, 4 or 9 and 1 occur in central India.

3.6.5.1 Gold Mineralization in South India

South India is divided in the geological context into Western Dharwar craton (WDC), Eastern Dharwar craton (EDC), and Southern Granulite province. Gold mineralization in each is described below.

Western Dharwar

The **Kempinkote** deposit at the southern part of West Dharwar, now virtually mined out, occurs in hornblende- and biotite-bearing quartzo-felspathic rocks of varying extent in a greenstone belt.

In the Chitradurga schist belt, the **Gadag deposits** (Mysore, Kabuliyatkatti etc.) occur near the north end of the belt and at Hosur, in a rock association of basalt, basaltic andesite, clastic sediments and chemical sediments like BIF and dolomite – all metamorphosed to greenschist facies. Gold occurs in quartz reefs, and is associated with common sulfides. Gold mineralization in the belt took place through metamorphogenic fluids during the second deformation phase in the region except at Hosur, where a magmatic contribution is indicated (Deb et al. 2008). At **Ajjanhalli** in the southern part, mineralization is associated with BIF (Fig. 3.27). H_2O–CO_2–CH_4–NaCl containing fluid inclusions (Pal and Mishra 2003), $\delta^{18}0 = 6.5–8.5$ per mil (Kolb et al. 2004) and the deformation and mineralization corresponding to the peak of metamorphism suggest the mineralization to be syntectonic (or orogenic). Modern mining started a few years back and then the operation stopped, reportedly due to a technical flaw.

Eastern Dharwar

Eastern Dharwar Craton (EDC) has at least two world class ore systems in the now-inactive Kolar deposit (since the late nineties), which has produced about 800 t of the yellow metal in last 100 years of its operation, and the Hutti group of deposits (Hutti, Uti, Hira Buddini) with a total reserve of ≥ 120 t of Au.

Kolar Belt

The Kolar schist belt is a N–S trending roughly $80 \times (4 - 8)\ \text{km}^2$ greenstone belt (Fig. 3.28). BIF interlayered with graphitic sulfidic schists occur near the western margin. This is followed to the east by massive to schistose amphibolite, and a complex rock unit called Champion Gneiss. Between Champion Gneiss and the Peninsular Gneiss (granodiorite-tonalite) again occurs an amphibolite unit.

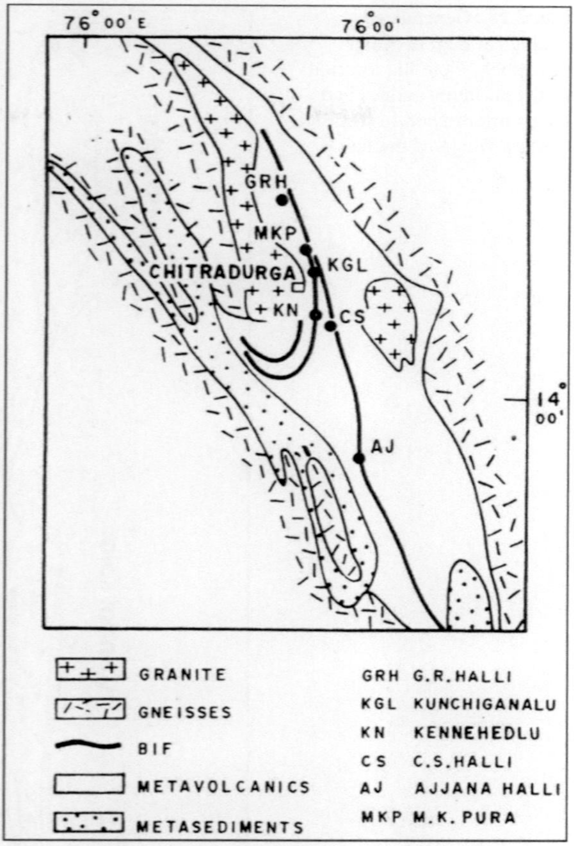

Fig. 3.27 Generalized map of a part of the Chitradurga schist belt, showing the disposition of the BIF bands and the location of gold prospects

High-Mg basalt is also present besides OFB and LKT that are dominant. According to Krogstadt et al. (1989) and Rajamani (1996) the belt consists of rocks (tholeiite and komatiite) that mark the collision and suture of two discrete Archean gneissic terranes. Metamorphism corresponds to amphibolite facies. Granitic pegmatites and dolerite dikes are conspicuous in Kolar, particularly in underground workings.

The regional structure of the Kolar field is generally believed to be a compressed syncline (cf. Narayanswami et al. 1960), but some other workers (e.g., Hamilton and Hodgson 1986) doubt this interpretation. The greenstone belt is characterized by a few N–S trending shear zones and NNW (–NW) – and NNE-trending regional faults. The control of the first set on mineralization is conspicuous, but that of the others is contentious. Champion reef was the main repository of gold in Kolar contributing >95% of gold (~770 t) to the total production. It extends along the strike for more than 8 km and along the dip, >3 km. It consists of >10 large

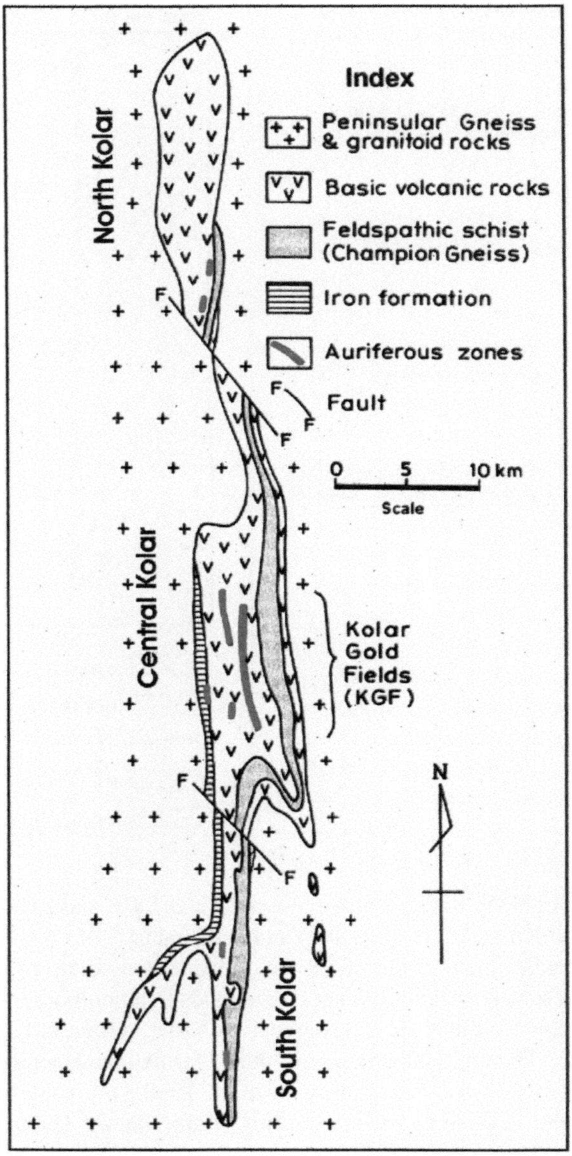

Fig. 3.28 Generalized geological map of Kolar schist belt. Note the location of the auriferous zones and the iron formation in the western fringe of the belt

orebodies arranged in an en echelon pattern. Ore shoot in the Champion lode plunge northerly with a pitch of 40°–50°, the angle increasing below ~2 km. In the Oriental lode to the west, the ore shoots pitch in the opposite direction. Gold ores in Kolar are divisible into three types

1. Gold-quartz type: seen best in the Champion reef; occurs also in Mundy, Muscom, and Chigargunta.
2. Gold-sulfide-quartz type: characteristic of the Mc Taggart and Oriental ore zones in the west.
3. Gold-bearing sulfides, commonly banded: seen at Malappakonda, a relatively small occurrence.

The ore-bearing vein quartz, with prominent marks of ductile-brittle deformation appears to have been multiply introduced and deformed. Gold in the Champion reef occurs along with a large number of minerals, both rock-forming and ore-constituents. Fineness of gold varies from 953.7 to 757.7 and the Ag-content may go up to 23 wt%. Ag-content decreases downward and the Hg-content increases (max. 2.56 wt%) in that direction. Scheelite occurs in the upper 1.5 km. Other elements participating in different minerals are Fe, Pb, Zn, Cn, Ni, Te, Bi, As, Sb, S, and Cl (Genkin et al. 1988). The fluid composition in inclusions from quartz in the Champion reef is represented by the system, $H_2O–CO_2 \pm CH_4$ and the salinity varies from 3–5 (early) to 7–12 wt% NaCl equiv. Quartz-sulfide ores in the western sector is characterized by a large content of Fe-sulfides (20–30% modal) and a little gold and tungsten. Galena in the ores is enriched in radiogenic Pb (Chernyshev and Safonov 1988; Balakrishnan et al. 1990). Pb–Pb model age of galena deposition in the gold ores of Kolar is ~2.6 Ga (Chernyshev and Safonov 1988).

Hydrothermal alteration related to gold mineralization in the Champion reef is interesting. Biotite is the characteristic mineral in the outer zone, while in the inner zone developed neo-amphibole, quartz, feldspar, and in the immediate contact with quartz veins, garnet, and diopside. This alteration pattern is not a constant feature though. Pyroxene could be represented by not only diopside, but also augite.

Ore genesis in the Kolar belt has been studied by a large number of individuals and groups. We refer to a few relatively recent ones. Siva Siddaiah and Rajamani (1989) proposed two different models. The Champion Reef, according to them, formed from hydrothermal fluids derived from granitic intrusions (2630–2550 Ma) and the sulfidic western lodes represent interflow sediments between volcanic units. To Rajamani later (Rajamani 1996), the ore fluids appeared to be of metamorphic origin, being low salinity, CO_2 – rich water vapor at t = 400–450 °C carrying high contents of Na and K. Hamilton and Hodgson (1986) suggested that the ore-forming hydrotherms were derived from granitic magmas emplaced in the neighborhood. Safonov et al. (1988) critically discussed the problem and chose the following observations to base their conclusion: (1) SiO_2 – dominant ore veins, with poor development of Fe-sulfides in the wall rocks; (2) Notable increase of CO_2, H_2O, S, F, and B in vein walls; (3) post-peak metamorphic wall rock alterations, but still at high temperatures; (4) High Hg-contents in ores and wall rocks compared to mafic rocks (2.18–1.4 wt%); (5) Small variations in $\delta^{34}S$ in ore-associated sulfides and generally low values of $\delta^{18}O$ (<+10 per mil); (6) High-content of radiogenic lead in associated galenas. The authors concluded that the gold mineralization proper was superimposed on already deposited vein quartz and sulfide minerals, gold and the accompanying elements having been provided by a granitic magma chamber, "influenced by mantle processes". Sarkar (1988) suggested the mineralization to be of metamorphic origin and later (Sarkar 2000) updated the mineralization to that of orogenic gold. Mishra and Panigrahi (1999) invoked gold deposition by mixing of two fluids: a low T (≤ 200 °C) – low salinity (<7 wt% NaCl equiv.) fluid with a high T (≥ 400 °C) – high salinity (≥ 50 wt% NaCl equiv.) fluid. But what is the relationship of the ores with these high salinity fluids? The latter may surely be derived from the contemporaneous granite mentioned above. The age of mineralization is about 2550 Ma (Au-sulfide-telluride sample from 108th level of the Champion lode, using Pb–Pb method and a Proterozoic model; Ralph Thorpe GSC 2008 Open File Report) which overlaps with the age of granodiorite near the ore zone (2551 ± 2.5 Ma) but is 140–180 Ma later than greenstone volcanism (cf. Ram Mohan and Sarma 2010).

Hutti-Maski Belt

The hook-shaped 65 km long Hutti-Maski gold mineralized belt of Northern Karnataka (Fig. 3.29) contains a number of sites of gold mineralization, of which Hutti, the major Indian working gold mine at present, is the most important. It has

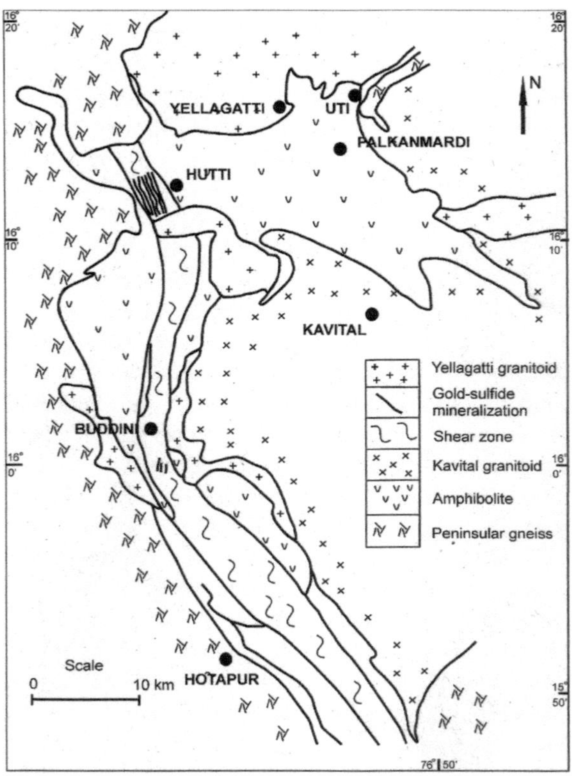

Fig. 3.29 Generalized geological map of Hutti-Maski schist belt, Northern Karnataka

so far produced ~45 tons of gold and hopefully still contains as much reserves at Hutti, where the ore zone comprises six lode systems.

The host rocks in the belt are principally Fe-tholeiites with subordinate clastic and chemical sediments. A dacite body is exposed underground. General enrichment of LIL elements and slight enrichment in LREE are more suggestive of marginal basin basalts (MBB) rather than Mid-ocean ridge basalts (MORB). Grade of metamorphism is greenschist-amphibolite facies transition (Riyaz Ulla et al. 1996). Western boundary of the belt lies against Peninsular Gneiss, while the younger granitoids occur in the north, south and east. A granitoid pebble at the base of the sequence yielded a SHRIMP weighted mean zircon age of 2576 ± 12 Ma, which is interpreted to be the maximum age of the belt (Vasudev et al. 2000). Three phases of deformation (D1–D3) produced folds and related axial shears that control the mineralization. The first major folding (-shearing) produced the foliation, quartz veins, distal chlorite-sericite and proximal biotite-plagioclase alterations at ca. 530 °C/3 kb. In the next major tectonic phase laminated quartz veins, inner chlorite-K feldspar developed at 300–400 °C/<3 kb (cf. Kolb et al. 2004). There is an increase of SiO_2, Al_2O_3, Na_2O, S, H_2O, CO_2 and decrease of Fe_2O_3, CaO, TiO_2, and MnO_2 in the ore zone.

Gold occurs generally with such sulfides as arsenopyrites, pyrite, pyrrhotite, chalcopyrite, sphalerite, galena, and a little magnetite. Scheelite is ubiquitous. Fluid inclusions yielded a salinity of 13.5% NaCl equivalent. Mineralization followed a retrograde P-T path. The age of first-stage gold mineralization, at 2547 ± 10 Ma, represented by hydrothermal monazite from the wall rocks of Hutti deposit (Sarma et al. 2008), followed by the second stage 10 Ma later (Rogers et al. 2013) post-dates greenstone formation by about 40–50 My but overlaps with regional magmatism (Yallagatti radiogenic granite: 2532 ± 3 Ma; Anand et al. 2005).

Kolb et al. (2004) assigned a typical orogenic gold status to the Hutti mineralization. Recently, Rogers et al. (2013) have proposed a two-stage model of ore genesis at Hutti, based on stable isotope studies. The first stage is represented by a typical orogenic gold mineralization (grade < 5 g/t) in a compressional setting with a low salinity metamorphic fluid source. The second stage produced high grade (up to 1 kg/t) auriferous zones in a transcurrent tectonic setting through a mixture of magmatic fluid and isotopically light formation fluids contemporaneous with the highly radiogenic Yelagatti granite intrusion. This inference is supported by the earlier work of Krienitz et al. (2008) in Hira Buddini mine where a bimodal distribution of $\delta^{11}B$ of tourmaline was identified which require an origin from at least two isotopically distinct fluid sources that subsequently mixed. The lighter fluid (+1‰) had the B released from metamorphic devolatilization of greenstone lithologies whereas the ^{11}B-rich fluid is attributed to degassing of I-type granite that intruded the greenstone sequence, providing the heat and fluids to the hydrothermal system.

Ramagiri–Penkacherla Belt

The Ramagiri–Penkacherla belt, ~100 km long and about 10 km wide, occurs in Andhra Pradesh. The belt has a very impersistent history of mining.

Gold mineralization was concentrated in a system of five lodes, Yeppamana and the main lodes among them being better developed. The ore shoots, best observed in the Yeppamana mine (operative till 2001), are folded with the axes plunging 55°–60° toward N. The host is a metamorphosed (upper greenschist facies) tholeiite, with sub-ordinate BIF. Gold occurs in quartz-carbonate veins with pyrite, arsenopyrite and chalcopyrite. $\delta^{34}S$ in pyrite ranges between +1 and +3 per mil. Elements such as As, Cd, Cu, Pb, Zn, and Sb have been added along with Au, Ag (Sarkar 1988; Manikyamba et al. 2004). Not much data are available to constrain a rigorous genetic model for this deposit.

Jonnagiri Belt

It is a rather small, $25 \times 0.5 - 5\ km^2$, dolphin-shaped schist belt in Ananthapur and Kurnool districts of Andhra Pradesh (Fig. 3.30). The rocks composing the belt are mafic and felsic volcanics metamorphosed to upper greenschist/lower amphibolite facies (~5 kb/500 °C; Chinnaswami and Mishra 2013). Three phases of deformation affected the rocks, like most other greenstone belts, producing folds, cleavages, and sheer zones. Late granitoid intrusives ranging in composition from granite to tonalite are present.

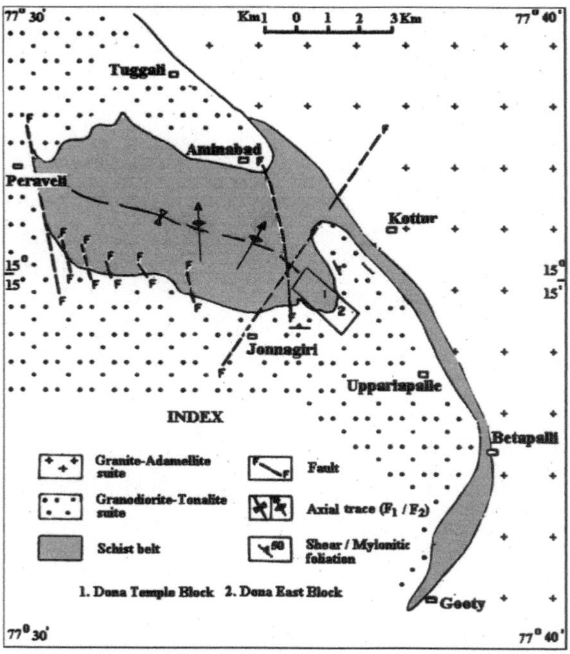

Fig. 3.30 Generalized geological map of Jonnagiri schist belt, Andhra Pradesh

Mineralization in quartz-rich veins follows shear zones that traverse both schistose and granitoid rocks. The mineralization is fairly rich, but the reserves are small and discontinuous. According to Chinnaswami and Mishra (2013), Au-precipitation took place as a consequence of post-peak metamorphism, fluid phase separation accompanied by fluid-rock interaction and decrease of fO_2 in the fluid.

While summarizing the data on gold mineralization in South India, as presented above, one inevitably notes that Hutti remains presently the only producing mine in the country although favorable ages of rocks, lithologies, structures, and metamorphic setting are all found in the EDC. The argument that middle to upper greenschist facies metamorphism is most conducive for orogenic gold endowment (Groves et al. 2003), and gold metallogeny in EDC is mostly in amphibolite facies rocks, is not quite convincing as the two biggest gold endowments in EDC are in higher grade rocks, with the Kolar deposit located close to the border of granulite facies terrain. Rather, the answer seems to lie in the multistage history of gold metallogeny where a superimposition of a magmatic process on the metamorphic fluid focusing along a structural conduit (Fig. 3.31) can increase the economic potentiality of an individual deposit manifold, like in Hutti and Kolar (Deb 2014).

Southern Granulite Province

The Southern granulite province contains gold mineralization but fewer in number and much poorer in quality. We mention in this context the Wynad-Nilambur region in North Kerala where gold occurs in (1) Primary quartz-rich veins, (2) Lateritic gold, and (3) Placer gold. The mineralized zone is located in the western extension of the Bhavani shear zone and the auriferous quartz veins are syn- to post-tectonic with respect to shearing. Srikantappa (2001), based on fluid inclusion studies, concluded that gold deposition here took place at ~2 kb and 300–400 °C. Results of stable isotope studies led Santosh et al. (1995) to suggest a mantle source for the ore-bearing fluids for the Nilambur-Wynad mineralization.

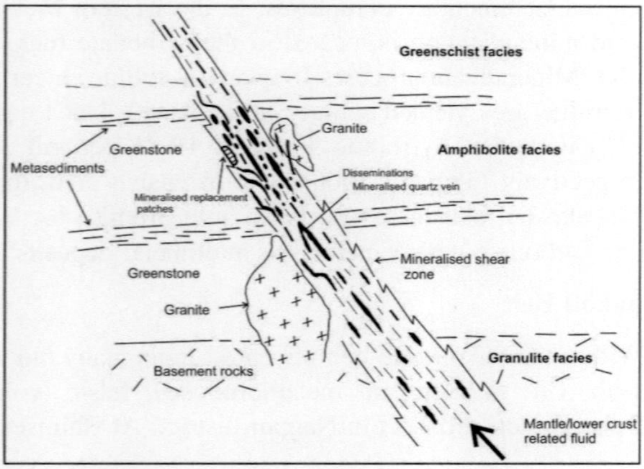

Fig. 3.31 Sketch of superimposed, multistage metamorphic-magmatic model of enhanced orogenic gold concentration (after Deb 2014)

Attapadi Valley, along Tamil Nadu–Kerala border, also located at the western extension of the Bhavani shear zone, is another area where gold occurs in sulfidic quartz veins in amphibolite–granulite facies rocks.

Gold has been identified in highly ferruginous and nodular laterites of Nilambur area in northern Kerala with good development of weathering profile at Maruda, Kappil, Arippamadu, and Thannikadavu. At Maruda, the oxidized zone contains high concentration of gold as small grains or nuggets, closely associated with iron oxide/hydroxide matrix. Omana and Santosh (1996) suggest dissolution of gold in an acidic environment and reprecipitation during weathering and lateritization.

Other Places in South India

Gold mineralization is reported from several other places, mostly greenstone belts, in Southern India (Fig. 3.26).

3.6.5.2 Gold Mineralization in Central India

Occurrence of gold is known from a number of places in Central India. The better known ones, however, are located in the Mahakoshal belt (Sonbhadra and Sidhi districts, UP and MP), Pular-Parsori, and Bhimsain Killa in Sakoli fold belt (SFB), MP. None, however, has been proven to be a viable proposition for exploitation.

Mahakoshal Belt

The ENE–WSW to E–W trending Mahakoshal belt extends for a 600×20 km^2 area, from SW of Jabalpur to the Palamau district of Jharkhand. Lithologically, the belt consists of quartzite, carbonates, pelites, cherts, BIF, greywackes, mafic volcanics, ultramafic to felsic intrusives. Some workers have called it a greenstone complex, developed in a continental rift (Nair et al. 1995; Jain et al. 1995). It is a riftogenic basin alright, but one needs to be extremely liberal to call it a greenstone belt. The eastern part of the Mahakoshal belt containing relatively better mineralization, hosts such prospects as Gurhar Pahar, Gulaldih, and Sona Pahari in volcani-sedimentary complexes. In the western Mahakoshal the only mentionable gold mineralization is located in the carbonate rocks of Imalia, Sleemanabad district. Mineralization occurs in quartz + sulfide $\pm$ carbonate veins. Fluid inclusions from the ores yielded salinity of 2–14 wt% NaCl equivalent, 250−300 °C/3.3 kb, δ^{34}S in pyrite, pyrrhotite +7.81 to +9.81 per mil and +9.16 to +14.45 per mil respectively (Tripathi 2008). The extensive platformal carbonates in the western Mahakoshal belt in central India hold promise for both Carlin-type gold deposits and carbonate-replacement type multimetal deposits.

Sakoli Belt

Within the Sakoli fold belt of central India many old workings of gold occur, along with Cu sulfides, in metamorphosed felsic volcanic and mica schists at Pular-Parsori prospect in Nagpur district. At Bhimsen Killa Pahar prospect, 25 km

NE of Pular, a 900 m-long brecciated cherty zone is auriferous. It also occurs within felsic volcanic rocks interbedded with andalusite-staurolite-bearing mica schist. Ore genesis is controversial; while Bandopadhyay et al. (1992) suggested a volcanogenic origin followed by tectonothermal mobilization, Chattopadhyay and Saha (1998) suggested hydrothermal mineralization related to acid magmatism.

Other Occurrences

At Sonakhan, gold occurs along with metachert and silicified brecciated chert in a small greenstone belt. The deposit may not prove unworthy on detailed prospecting (Venkatesh 2001).

The Kotri-Dongargarh belt is presently rated still lower in terms of prospectivity for gold. Sonadihi deposit is worth mentioning in this context.

The 2.5 Ga porphyry-style copper deposit at Malanjkhand contains about 30 t (<1 ppm Au) of the yellow metal.

3.6.5.3 Gold Mineralization in Western India

Mentionable occurrences of gold in Western India are located at Bhukia-Jagpura (Banswara district) (Fig. 3.32) and Hinglaz Mata-Bharkundi (Dongargarh district), SE Rajasthan. Once there was a flourishing gold-mining industry in this region, as evidenced by extensive ancient workings. Gold occurs, associated with sulfides in amphibole-bearing dolomitic marbles, calc-silicates, mica, schists, quartzite, and amphibolites of Delwara Formation in Lower Aravalli Supergroup. The rocks are metamorphosed to green schist-lower amphibolite facies. The ore veins are remobilized and structurally controlled (Deol et al. 2014). Wall rock alterations include chloritization, biotitization, sericitization, silicification, and albitization (cf. Grover and Verma 1993; Verma and Golani 1999). Golani (2008) visualizes a mantle-flux in the gold-bearing carbonates of this area. Deol et al. (2010) propose that sulfides were deposited by layer-parallel influx of ore-bearing hydrothermal fluids along S_1-surfaces in the host rock during the first deformation. The syn-kinematic granite bodies near the ore zone acted as the heat source for the hydrothermal system and possibly contributed, at least some fluid. Reportedly a joint venture by an Australian company, M/s Indo Gold and M/s Metal Mining India, established a significant reserve (38.5 Mt) of low-grade ore at Bhukia–Jagpura area.

Gold is also reported from the Newania Carbonatite, Udaipur district (Golani and Pandit 1999).

3.6.5.4 Gold Mineralization in Eastern India

A couple of areas in Eastern India bear evidence of gold mining in the past and quite a few of them have been further explored in the recent past, but none turned out to be a big success. In this context, the name of the Kunderkocha deposit in Eastern Singhbhum should come up first. It is located within the Iron Ore group

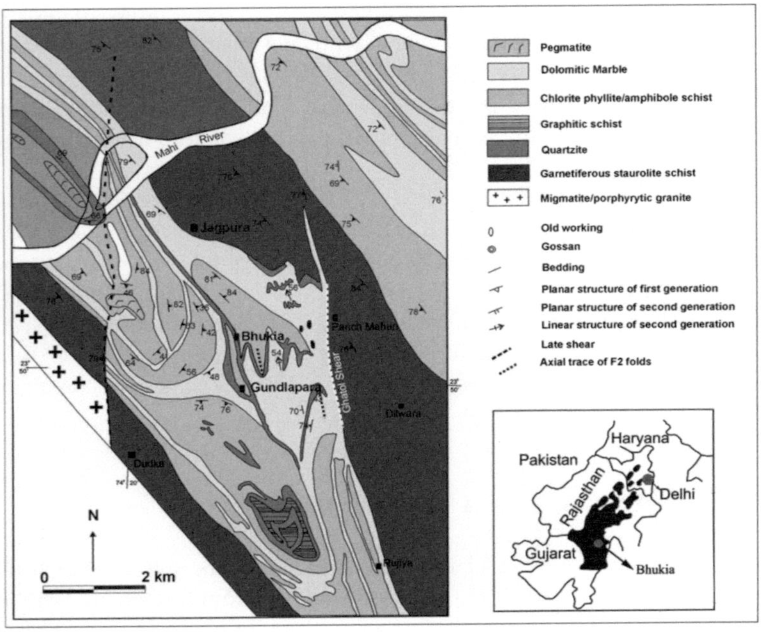

Fig. 3.32 Geological map of Bhukia-Jagpura area, South Rajasthan. *Inset* shows the location of Bhukia in the Aravalli-Delhi orogenic belt (*black*) (after Mineral belt map of GSI)

(IOG) rocks in the Badampahar-Gorumahisani belt. At Kunderkocha small-scale mining went on for long and then with a break of some years the operation has been resumed under a changed ownership. The mineralization is in smoky and bluish quartz venis, traversing black chert, graphitic phyllite, and tale-chlorite schists. The mineralization has not been proven to be large and continuous, but locally the mineralization is extraordinarily rich (up to 1300 g/t: Ziauddin and Narayanswami 1974).

Insignificant contents of gold are reported from the basal member (QPC) of the Dhanjori Group of rocks at several places.

Gold occurs along with the basic volcanic-associated copper ores of the Singhbhum belt. Cu-concentrate from Mosaboni Mines assayed 3 ppm and those from the Rakha and Surda mines, 9–10 and 7 ppm respectively.

Sonapet, as the name indicates, was locally known for its gold occurrences. The area is located south of the Dalma range near its western end. Here gold-bearing blue-dark gray quartz veins, emplaced in laminated magnetite-sericite-biotite schists and hornblende schists, locally showed very high (63 g/t) gold-content (Ziauddin and Narayanswami 1974). But the mineralization is impersistent. Same is the situation at Pahardiha, further south.

The situation further brightens up somewhat, north of the Dalmas. The Lawa gold deposit, NW of Chandil, and now mined out and under Subarnarekha reservoir, was hosted by quartzite and pelite in a sequence of pelites, psammites, psammopelitic rocks, occasionally interlayered with carbonaceous beds, amphibolites, and felsic volcanic rocks. Three subparallel ore zones are spread over a strike length of 1 km. The northern or the footwall ore body is defined by pyrrhotite-pyrite-rich sulfidic assemblage sans gold, whereas the centrally occurring ore zone gold was associated with Cu-rich sulfides. The pyrrhoite-rich ore bodies are lensoid in shape. In the opinion of Gupta (1975), the pyrrhotite-rich mineralization was possibly syngenetic, while the Cu–Au mineralization was caused by a later hydrothermal phase.

The most recent and that too prospective, find of gold minealization in this part of the country is located in the Lungtre-Parasi-Sindauri region north of the Dalma range (Singh et al. 2005). It is often briefly referred to as Parasi prospect. The host rock association consists of magnetite-biotite-quartz-sericite schists, phyllites and quartz schists, ferruginous quartzite, felsic tuff, rhyolite, acid agglomerates, actinolite-tremolite schists, metabasics, and siliceous carbonates. Sheared quartzose schists, traversed by quartz and quartz-calcite veins and phyllites with quartz ribbons are the major hosts for gold mineralization (up to 10 g/t). Prospecting activities are still on. Preliminary estimates reportedly suggest a total reserve of 4 t gold here. The distribution of gold deposits and prospects in eastern India in space and time is presented in Fig. 3.33.

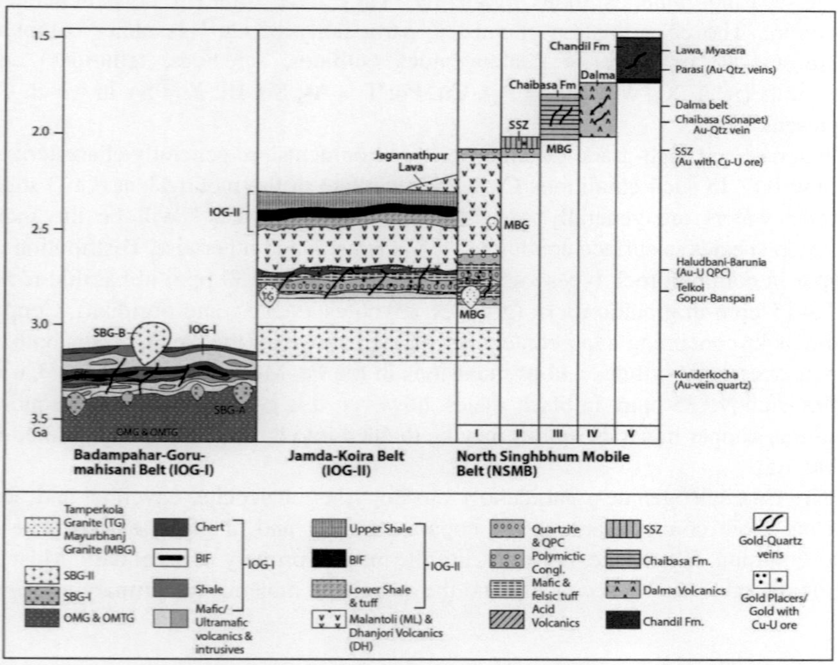

Fig. 3.33 Space-time diagram of Singhbhum region showing the position of various gold deposits and prospects (after Gupta 2010)

3.7 Copper Ores

Copper drew the attention of human society long back in its history of civilization, even when the use of iron was not known. Its greatest use in electrical power generation, distribution, and making electrical appliances in different industries came rather recently. The word copper is apparently derived from the Latin word "cuprum", that was mined by the Romans in the island of Cyprus in the Mediterranean Sea. Some of its uses were known in India in the pre-Christian era as evidenced by the profuse mention of its use in Kautilya's *Arthasastra*. The two widely used alloys of copper are brass (Cu + Zn) and bronze (Cu + Sn).

3.7.1 Geochemistry and Mineralogy

Atomic weight of copper is 63.54 and atomic number is 29. It is a reddish, malleable, and ductile metal with a melting point of 1084 °C and specific gravity of 8.95. A good conductor of heat and electricity, copper makes alloys with numerous metals, besides Zn and Sn.

Copper occurs in the valency states Cu^0, Cu^+, Cu^{2+}, the last two participating in copper-bearing minerals more frequently. They have different crystal-chemical behaviors. The co-ordination chemistry, particularly of Cu^{2+} is rather complex. There are 80 or more of chalcogenides (sulfides, selenides, tellurides) and sulfo-salts ($A_mT_nX_p$, where A = Ag, Cu, Pb; T = As, Sb, Bi; X = S), in which Cu is present.

Magmas and high-grade metamorphic environments are generally characterized by low fO_2. In such conditions Cu will precipitate in the monovalent (Cu^+) state. Surface waters are generally oxygenated. As a result, Cu^{2+} will be the more common species in surface conditions, if not constrained otherwise. Distribution of copper in common rock types vary widely, from 120 to 130 ppm in basaltic rocks to 10–15 ppm in granitic rocks (granites, rhyolites, dacites, and obsidian). Copper in the rocks containing a fair content (mentioned above) of the metal, occurs both as minute accessory sulfides and as inclusions in the Fe–Mg silicate structures. Cu in shales is above 35 ppm. In black shales, however, it is generally many times more. Common copper minerals in ores may be divided into three groups, as presented in Table 3.5.

The Cu-chalcogenides, particularly chalcopyrite, chalcocite, covellite, and bornite are more common among the copper minerals and of these, chalcopyrite is most common. Chalcocite, covellite, bornite may be primary or secondary. Most of the oxidic minerals are secondary and the associated minerals are primary in origin

Table 3.5 Common copper minerals in ores

Chalcogenides[a]	Oxidic and native minerals	Associated minerals
Chalcopyrite, $CuFeS_2$	Cuprite, Cu_2O	Pyrite, FeS_2; Pyrrhotite, $Fe_{1-x}S$; Sphalerite, ZnS; Galena, PbS; Molybdenite, MoS_2; Magnetite, $FeFe_2O_4$; Hematite, Fe_2O_3, Goethite, γFeOOH; Jarosite, $KFe_3(SO_4)_2(OH)_6$
Chalcocite, Cu_2S	Tenorite, CuO	
Covellite, CuS	Antlerite, $CuSO_4{\cdot}2Cu(OH)_2$	
Bornite, Cu_5FeS_4	Brochantite, $CuSO_4{\cdot}3Cu(OH)_2$	
Digenite, $Cu_{9-x}S_5$	Atacamite, $CuCl(OH)_3$	
Energite, Cu_3AsS_4	Crysocolla, $Cu_2Si_2O_6{\cdot}nH_2O$	
Tennantite, $Cu_8As_2S_7$	Delafossite, $Cu_2O{\cdot}Fe_2O_3$	
Tetrahedrite, $Cu_8Sb_2S_7$	Native copper, Cu	

[a]cf. Appendix A

excepting goethite and may be hematite. Gangue minerals vary depending on the types of ores discussed below.

3.7.2 *Principal Types of Copper Ore Deposits*

Copper deposits vary in geology, geochemistry, mineralogy, and time and space of emplacement. We can divide them into the following types, which have been discussed in details in Chap. 2 in terms of their genesis:

1. Mafic volcanic-associated Cu deposits. Example: Cyprus Cu deposits; Lake Superior Cu deposits
2. Porphyry Cu deposits. Example: Andean deposits
3. Sediment-hosted stratiform Cu deposits. Example: Cu belt deposits of central Africa, Udokan deposit, Russia
4. Polymetallic deposits. Example: Kuroko (Japan) and Buchan (Canada) deposits
5. Iron oxide–copper–gold (IOCG) type ores. Example: Many in Australia, South America but not confirmed as a genetic model
6. "Exotic" deposits, including supergene enrichment. Example: Chuquicamata mine and El Salvador mine, Chile.

3.7.3 Copper Mineralization in India

The principal copper belts/districts in India are

1. Singhbhum copper (–uranium) belt, Jharkhand (cf. Fig. 3.6)
2. Malanjkhand deposit, Madhya Pradesh (cf. Fig. 3.5)
3. Khetri copper belt, Rajasthan (cf. Figs. 3.37 and 3.38)
4. Ingaldhal copper deposit, Karnataka (cf. Fig. 3.39)

1. *Singhbhum copper (–uranium) belt*

Copper mineralization in Singhbhum is distributed along an arcuate belt of 120 km, stretching from Duarpuram in the west to Baharagora in the east, with the known copper deposits being located mainly in the eastern half (Fig. 3.34). In this sector, the known deposits are located at Baharagora, Badia-Mosaboni, Dhobani-Chirudih-Samaidih, Pathargora, Surda, Kendadih-Chapri, Roam–Rakha Mines-Tamapahar, Ramchandrapahar-Byanbil-Nandup, and Turamdih-Dadkidih. Mining in most of them took place for different periods of time during the last century. Presently their status is in Table 3.6.

Now that the price of copper has gone high, the Government has a second thought of resuming mining in most of them, and in fact, the Surda mine has resumed production in 2007. Mining in several other mines may soon restart.

Singhbhum copper belt broadly coincides with a regional dislocation/shear zone and the mineralization is associated with several rock types, starting with amphibolites through chlorite schist to magnetite-chlorite-quartz schists and albite schists–albitite. Ores were differentially subjected to deformation and

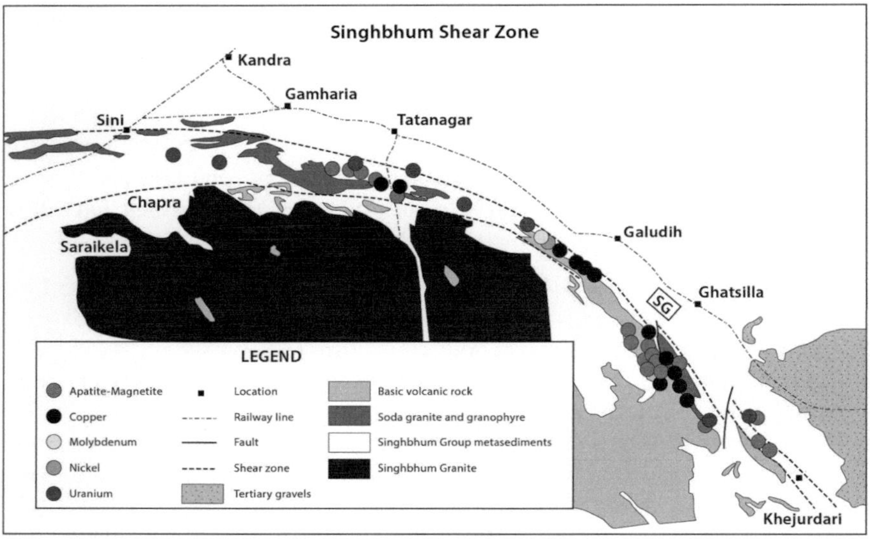

Fig. 3.34 Disposition of the Singhbhum shear zone in eastern India showing the location of different Cu, U and Apatite-Magnetite deposits and prospects (after Sarkar 1984)

Table 3.6 Status of copper mines, Singhbhum copper (–uranium) belt

	Mosaboni-Badia	Dhobani	Pathargora	Surda	Kendadih	Roam-Rakha Mines
Year of opening	1928	1934	1957	1956	1973	1978
Year of closing	1998	1953	2000	2003	2000	2001
Remaining reserves (Mt)	19.80 (1.7% Cu)	1.4 (2.4% Cu)	3.58 (1.32% Cu)	26.09 (1.2% Cu)	15.00 (1.73% Cu)	85.36 (1.32% Cu)

metamorphism (Sarkar and Deb 1974). They contain more than 40 common and uncommon ore minerals in them (Sarkar 1984). Economic metals extracted from them are Cu, Ni, Te, Bi, Se, Au, and Ag.

The origin of the copper ores of Singhbhum is controversial. Opinions vary from magmatic hydrothermal (Dunn 1937), albitization-related (Banerji 1962), modified volcanogenic (Sarkar 1984) to IOCG (Pal et al. 2009). There is obviously no tell-tale story about ore genesis along the Singhbhum belt. Points in favor and against the IOCG model have been discussed by Sarkar and Gupta (2012). Study in this regard needs to be pursued.

2. *Porphyry copper/granite associated copper ores of Malanjkhand, Central India*

The Malanjkhand deposit of copper (–molybdenum) is a chance find in Balaghat district, Madhya Pradesh, Central India. The primary report however gathered dust until the Geological Survey of India took a serious look at it sometime in the middle of the last century. By 1968, the deposit was proven. The mine was made operative by 1982. Presently the open pit mine has a capacity of 2 million TPA. According to Hindustan Copper Ltd. website (www.hindustancopper.com), the proved reserve of copper ores at the Malanjkhand deposit stands at 221 Mt with an average grade of 1.31% Cu. This information probably refers up to a depth of −300 mRL. Malanjkhand produced 2,233,523 t (1% Cu) ores during 2010–2011.

The Malanjkhand ores occur along a shear zone within the Malanjkhand Granite, a TTG massif, temporally located along the Proterozoic – Archean boundary ($\sim$2490 Ma) and intrusive into the Archean Nandgaon Group. The granite is overlain by the Chilpi Ghat Group with an unconformity in between (Fig. 3.35). The massif is polyphase, the last one being a microgranite, post-mineralization in age. The granitic rocks belong to the "Magnetite series" and to the "I-type". The principal ore minerals are chalcopyrite and pyrite with subordinate magnetite and minor/accessory sphalerite and molybdenite. The trace phases include rutile, cassiterite, galena, hessite/stuatzite, empressite, clausthalite, and native gold. Gold commonly occurs as minute grains (up to 3 ppm) occurring within chalcopyrite grains. Deformation and re-crystallization are common in ores. The major phases in the secondary ores are Fe-oxyhydroxides, chalcopyrite and malachite and a host of

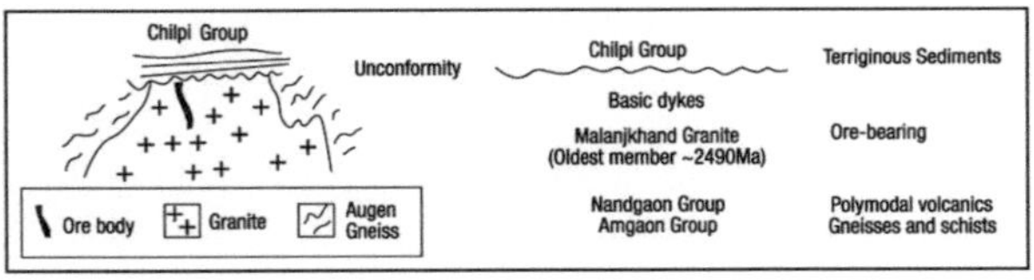

Fig. 3.35 Sketch showing the stratigraphic position of Malanjkhand granite and associated Cu–Mo orebody

minor phases (Sarkar et al. 1996 and references therein). $\delta^{34}S$ in chalcopyrite and pyrite cluster around zero.

On the plan the ore zone is elongate though not straight and show penetration by aplite/micrograinte and diabase (Fig. 3.36). Wall rock alteration in and around the ore zone in Malanjkhand comprise (1) silicification, (2) Na/K-replacement of andesine feldspar, (3) replacement of hornblende by epidote or chlorite, (4) increase in Mg/Fe ratio in biotite and (5) appearance of hematite dusts, seen in quartz and feldspar.

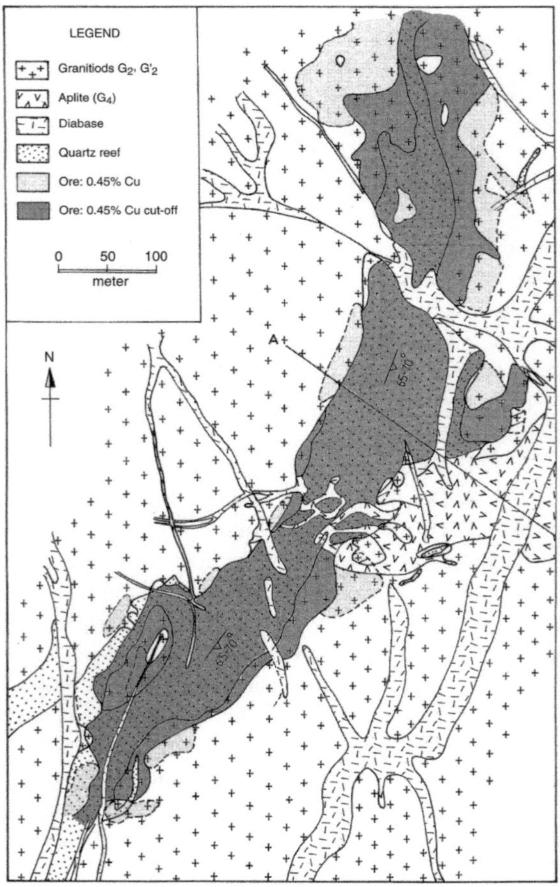

Fig. 3.36 Geological map of Malanjkhand deposit (after HCL)

Molybdenite, the second important ore mineral in the ore occurs as: (1) clusters and clots in veins and veinlets and disseminations in nearby host rock; (2) disseminations and minor clusters in the copper ores and (3) late stage transverse veins, fewer but more persistent than (1). A five-point chronometric Re–Os isochron study of Malanjkhand molybdenite belonging to type 1 above, gave a model age of mineralization to be 2489 ± 1.4 Ma (Stein et al. 2004).

The initial mineralization of Malanjkhand ores is hydrothermal as is suggested by the typical wall rock alteration, structural control, $\delta^{34}S$ values and occurrence within a granitoid complex. The entire mineralogenetic history of Malanjkhand deposit may be enumerated as follows:

1. Early shearing accompanied by some hydrothermal alterations
2. Emplacement of ore-free/ore-poor quartz veins
3. Emplacement of molybdenite-bearing veins and clusters
4. Main ore genetic event
5. A late phase of molybdenite vein emplacement

All the above features are characteristics of porphyry copper deposits. Only one feature of the typical porphyry copper system is not conspicuous, i.e., the stockwork. It could be explained by erosion, as the top of the orebody was eroded long before mining and covered by the Chilpighat sediments in a large part. Alternatively, enough fluid pressure did not develop to produce shattering, probably due to prolonged shearing (vide also Sarkar and Gupta 2012).

3. *Khetri copper belt and the Kho-Dariba deposit in Rajasthan, Western India*

Although both the Khetri copper belt (KCB) and Kho-Dariba (KD) deposit (Fig. 3.37) bear evidence of ancient mining, a vigorous modern exploration programme was taken up by the Government agencies only after independence in 1947. It is said that the KD was the captive copper mine for the Moghul emperors of Delhi. As a result of modern exploration, a total reserve (mine + residual tonnage) of 83 Mt of ores with a Cu-content of 0.88–1.5% was estimated at the KCB. KD was a small deposit and much exploited. Hindusthan Copper Limited (HCL) mined out the remaining 0.5 Mt copper ores (@ 1.3% Cu). The Kho-Dariba deposit is hosted by carbonaceous phyllite (+quartzite). Structural controls, including axial zone controls of a part of the orebody is interesting.

The major deposits in KCB in the North Delhi Fold Belt, north to south, are: Banwas, Madan Kudan, Kolihan, and Chandmari, with a few smaller deposits (Akwali, Satkui, Dholamala) occurring further south. The whole mineralized belt is about 80 km long. Location and disposition of the ore deposits along KCB are controlled by broad structures (Fig. 3.38).

The host rocks at Madan Kudan and Kolihan are both garnetiferous chlorite schist and banded amphibolite-quartzite. At Chandmari it is garnetiferous chlorite schist alone. Carbonaceous phyllite is present in the ore zone in the southern part of the belt. Cordierite-anthophyllite-cummingtonite rocks of Madan Kudan area do not constitute immediate wall rocks of any ore body. These pelitic metasediments are

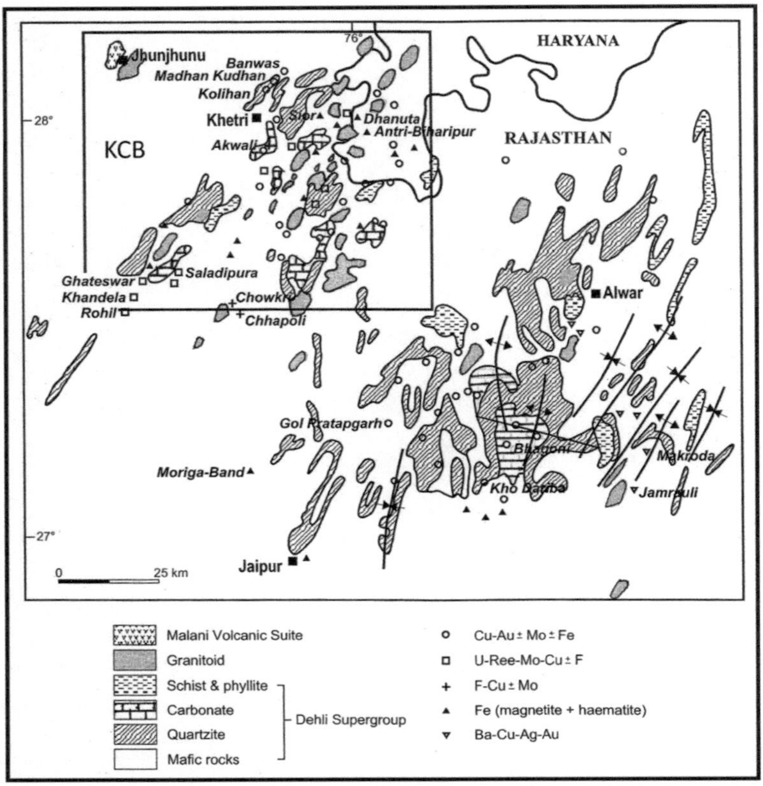

Fig. 3.37 Regional geological map of northern Rajasthan and Haryana showing the disposition of Khetri copper belt (KCB) and Kho-Dariba deposit (after Knight et al. 2002)

metamorphosed to upper greenschist to lower amphibolite facies. The demonstrated reserve by HCL in the Madan Kudan (Khetri mine) is 26 Mt with 1.13% Cu; in Banwas mine is 25.02 Mt with 1.69% Cu; in Kolihan mine is 20.64 Mt with 1.35% Cu and in the Kolihan-Chandmari intervening block the estimated reserve is 12.10 Mt with 1.03% Cu. During 2010–2011 Khetri and Kolihan mines produced 481,966 t (0.97% Cu) and 489,654 t (0.965% Cu) ores respectively. Mineralization was sedimentary-diagenetic, affected during post-depositional dynamothermal modifications (Sarkar and Dasgupta 1980).

4. *Ingaldhal copper deposit, Karnataka*

Copper mineralization in Chitradurga Group of Dharwar Supergroup is represented mainly by the Ingaldhal deposit, located in the 40 km long Ingaldhal-Kunchiganahalu belt (Fig. 3.39). This small deposit has a reserve of 1 million ton (@ 1.4% Cu) in two modes of mineralization: stratiform and

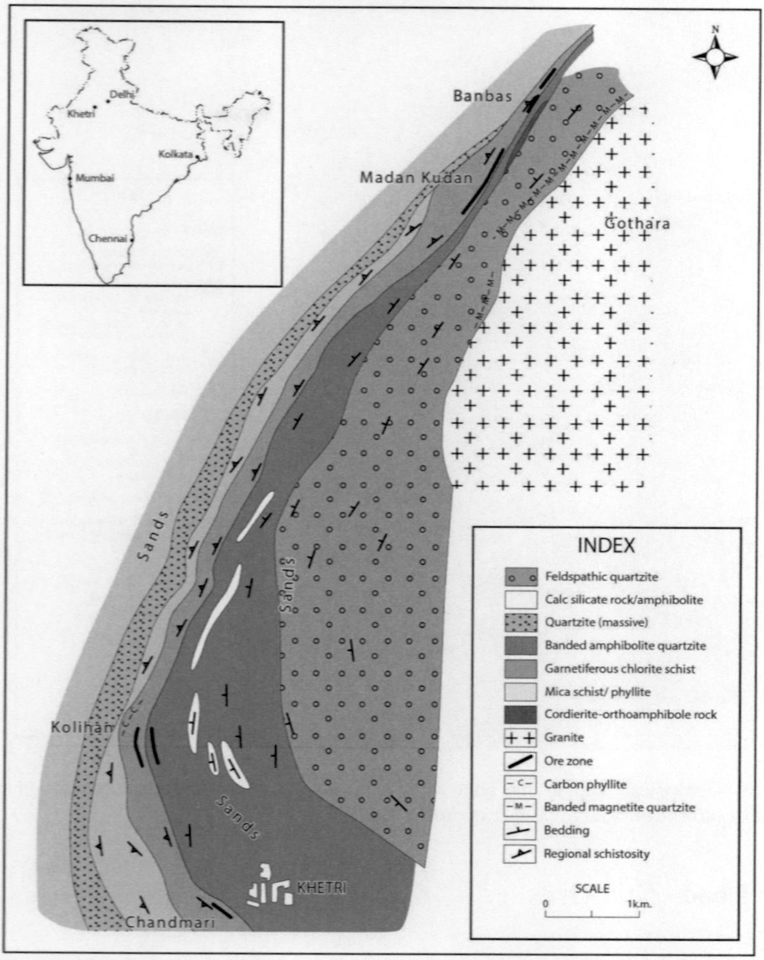

Fig. 3.38 Geological map of northern part of Khetri copper belt (after Sarkar and Dasgupta 1980)

vein-type. The stratiform ores occur as banded sulfide-chert body, extending for 7 km along strike. The ores comprise pyrite, pyrrhotite, magnetite mainly, with chalcopyrite and sphalerite as minor phases hosted by pyritiferous carbonaceous phyllite. The vein ores are lensoid bodies extending few hundred meters along strike within altered basaltic rocks. Chalcopyrite is the main constituent of these ores followed by pyrite and pyrrhotite. The ores have been metamorphosed to upper greenschist facies. Most workers agree that these ores belong to the "volcanogenic type" (Anantha Iyer and Vasudev 1985; Sarkar 1988).

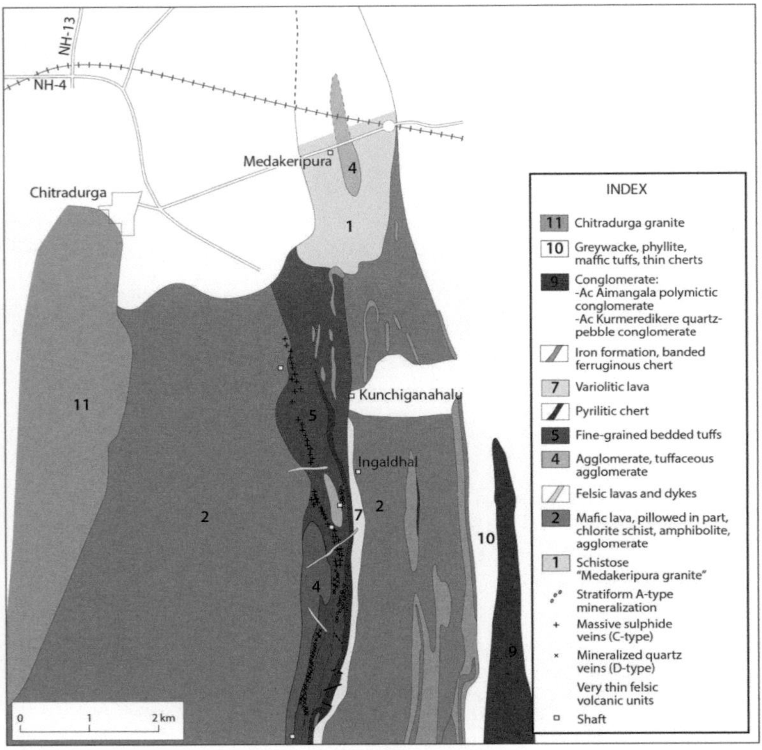

Fig. 3.39 Geological map of the area around Ingaldhal copper deposit, Chitradurga district, Karnataka (after Chitradurga copper company Ltd.)

3.8 Lead–Zinc Ores

Lead has been in human use since long. Romans used lead pipes for water transport and those who could afford, used lead utensils for daily use. There is a conjecture that the decline of the Roman Empire was due to lead poisoning among the aristocracy for their unrestricted use of lead. Many other ancient civilizations including India and China, knew its use sometime earlier than the beginning of the Christian era. Zinc, a common associate of lead, established its identity in early sixteenth century (1520 AD). However, some societies used it for making brass. India played a great role in establishing zinc metallurgy (Gandhi 2000 and references therein).

3.8.1 *Essentials of Lead and Zinc Metals and Their Mineralogy*

Lead and zinc are important base metals next only to copper. Atomic number and atomic weight of lead and zinc are: 82 and 207.2 and 30 and 65.37 respectively and yet in nature (ore deposits) they have close association.

Lead is unusually soft and scratchable by finger nail. It has two oxidation or valence states, plumbous (Pb II), and plumbic (Pb IV). Ag is a common impurity in galena and hence it passes on to metallic lead. It is extracted either by Parkes process or by electrolytic process. Lead has four isotopes: ^{204}Pb, ^{206}Pb, ^{207}Pb, and ^{208}Pb. Of these ^{204}Pb is primitive or non-radiogenic, while the others are radiogenic having formed from disintegration of uranium or thorium. Pb isotope ratios help in determining the age of Pb-ore (galena) in a deposit.

Zinc is a moderately hard metal with a silvery look. It has a single oxidation state (Zn II). Zinc occurs in many natural compounds, i.e., in minerals, the most important of which is zincblende or sphalerite (ZnS). The position of zinc in sphalerite structure may be occupied by Fe, Mn, or Cd up to several atomic per cents (Fe up to 20 at.%). Au may also be present as an economic by-product (Emsbo 2000). Both galena and sphalerite in pure state are diamagnetic semi-conductors. The presence of transition metal impurities in sphalerite profoundly affects the magnetic and electrical properties though. Metamorphism of Pb–Zn ores is common.

Lead is used in storage batteries, cable-carrying electric wires, solder, ammunition, containers of nuclear waste. Zinc is used in dry cell and in galvanizing iron and steel. A zinc–copper alloy, brass, is in wide use. A list of Pb and Zn minerals is given below

Lead minerals	Zinc minerals
Galena (PbS)	Zincblende, Sphalerite (ZnS)
Cerrusite ($PbCO_3$)	Wurtzite (ZnS)
Anglesite ($PbSO_4$)	Smithsonite ($ZnCO_3$)
Clausthalite (PbSe)	Zincite (ZnO)
Altaite (PbTe)	Willemite ($ZnSiO_4$)
	Franklinite ($ZnFe_2O_4$)
	Gahnite ($ZnAl_2O_4$)

Galena is the most dominant mineral amongst the Pb-minerals and sphalerite the most important Zn-mineral, except in not-too-large non-sulphide zinc deposits at few places (Hitzman et al. 2003). Wurtzite is actually sulfur-deficient relative to sphalerite ($\sim$1 atom per cent) (Scott and Barnes 1972) and not a high temperature variant of sphalerite as was earlier believed.

3.8.2 Types of Pb–Zn Deposits and Their Origin

Types of Pb–Zn deposits accepted presently include SEDEX (Sedimentary exhalative type), MVT (Mississippi valley type), VMS (Volcanogenic massive sulfide type), Irish Midland type, Broken Hill type and Skarn type. It should be mentioned here that these divisions are not very clear cut always. Generally some features are shared among themselves and locally some more features are overlapping. Of the above, except VMS and Skarn types, all others are sedimentary rock affiliated.

The SEDEX and MVT are the two major groups. Carbonate-hosted Pb–Zn deposits of Irish Midlands are distinguished from the MVT by some workers (Wilkinson 2003 and references therein). Broken Hill (Australia) type deposits are often assigned to the SEDEX type. Indeed they have many similarities except the fact that these have more Ag and Pb compared to most SEDEX deposits.

The origins of these deposit types as well as their distribution in space and time is discussed in Chap. 2.

3.8.3 Pb–Zn Mineralization in India

Bulk of the Pb–Zn mineralization discovered in India, whether in the ancient times or in the recent past, is all located in northern India, in the states of Rajasthan and Gujarat (Fig. 3.40). Smaller deposits are known from eastern India: Sargipalli in Odisha (Fig. 3.35), and Agnigundala area (vide Fig. 4.8) in Andhra Pradesh and much smaller deposits in a few places in the Himalaya, in the states of Sikkim (e.g., Rangpo, Dikchu) and Uttarakhand (e.g., Askote). Total resource of Pb–Zn ores, as on 1.4.2010 (Indian Bureau of Mines 2012) and as per UNFC system is estimated at 685.59 Mt. Of these 108.98 Mt (16%) fall in "Reserve" category while the balance 576.61 Mt (84%) are classified as "remaining resource". Resource of ore according to ore grade is as follows:

+10% Pb + Zn	139.85 Mt
5–10% Pb + Zn	252.30 Mt
<5% Pb + Zn	293.44 Mt

The state of Rajasthan, which covers the major part of the northwestern Indian shield, is endowed with substantial reserve of base metal sulfides, particularly of Pb and Zn. Rajasthan's share of Pb–Zn resource is 607.53 Mt which amounts to 88.61% of national resource. The details of the reserve/resource base in Rajasthan, according to M/S Vedanta are shown in Table 3.7. By and large, the Indian Pb–Zn deposits are comparatively richer in zinc with an insignificant lead content.

Presently, economically viable Pb–Zn deposits in India are only three: Rampura-Agucha, Dariba-Rajpura-Sindeswar Khurd and Zawar group of deposits, all in the state of Rajasthan. While the first two have characteristics of a SEDEX

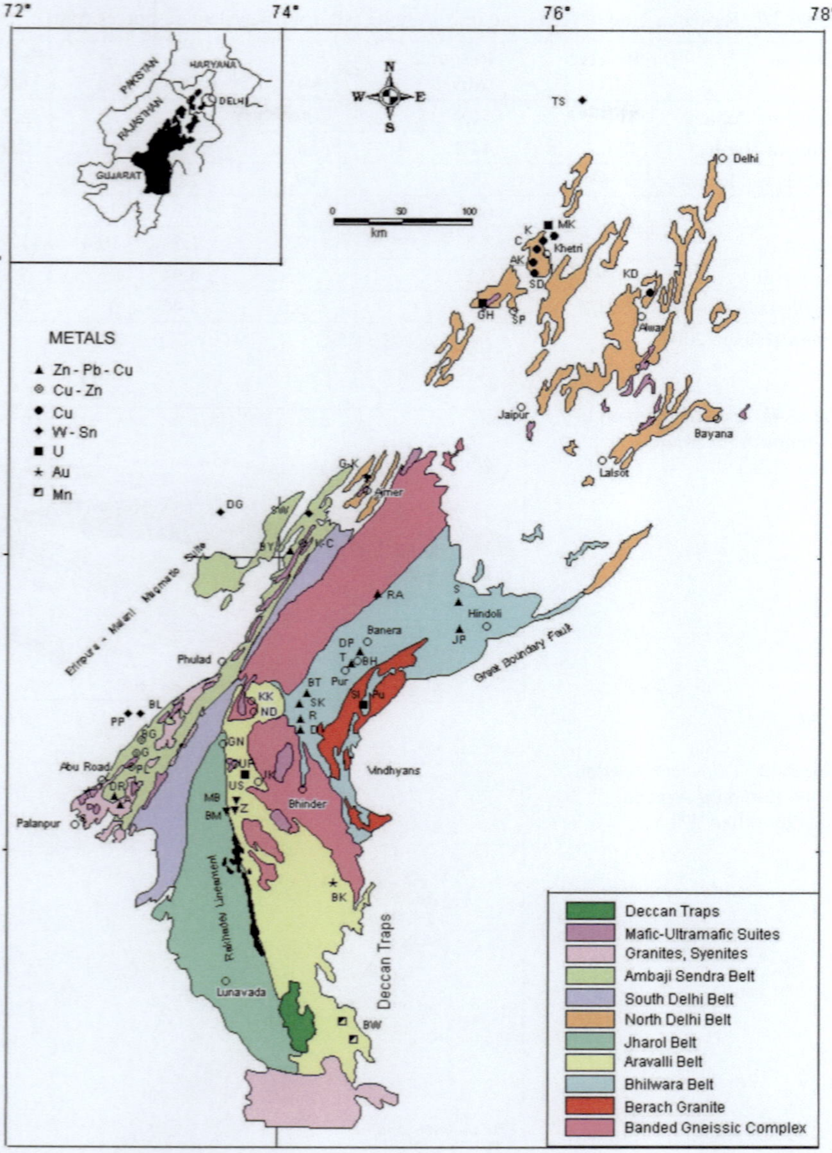

Fig. 3.40 Geological map of Aravalli-Delhi orogenic belt, showing the locations of mineral deposits (after Deb and Thorpe 2004)

type deposit, the third deposit is more akin to a MVT type. A recent addition to this list of exploitable deposits is Kayar, near Ajmer (cf. Fig. 3.40).

The *Rampura-Agucha deposit* in Bhilwara district is represented by a single lensoid orebody, 1.5 km in strike length (Fig. 3.41) and about 60 m in width and

Table 3.7 Resource base of lead–zinc ores in India (after www.vedantaresources.com)

Deposit	Reserve (Mt)	Resource (Mt)	Total (Mt)	% Pb	% Zn	% TMC
Rampura-Agucha	57.5	51.9	109.4	1.8	13.7	15.5
Rajpura-Dariba	10.0	44.3	54.3	1.6	6.4	8.0
Sindeswar Khurd	20.4	78.7	99.1	2.6	4.6	7.2
Zawar	9.9	68.5	78.4	1.9	3.8	5.7
Kayar	6.2	1.5	7.7	1.5	10.4	11.9
Sargipalli[a]	4.30	0.12	4.42	5.94	0	5.94
Agnigundala[a]	3.07	0	3.07	5.46	0	5.46

[a]From Haldar (2004)

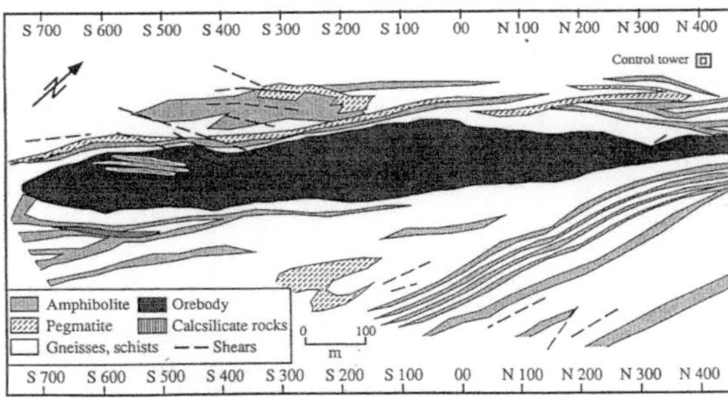

Fig. 3.41 Geological map of Rampura-Agucha orebody (after HZL)

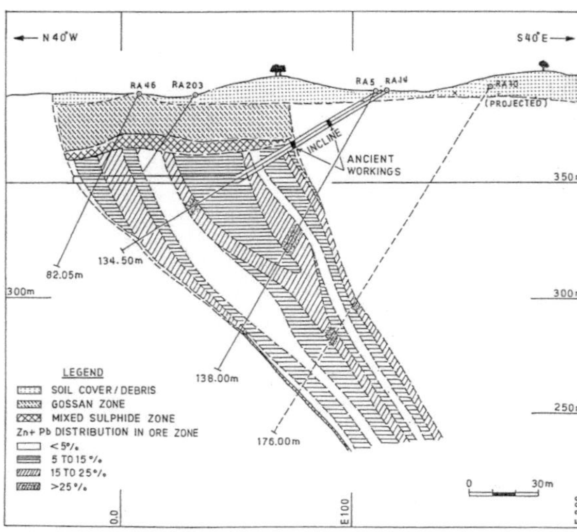

Fig. 3.42 Transverse section of the Rampura-Agucha orebody (after HZL)

contains a reserve of 59 Mt @ 1.98% Pb, 13.53% Zn and 50 g/t of Ag. The orebody, exposed on the surface and tapering at depth (Fig. 3.42), is hosted by graphite mica sillimanite schist close to the contact of BGC II of Heron (1953). The ore and the host rocks have been metamorphosed to upper amphibolite (650 °C/6 kb; Deb and Sehgal 1997) to granulite facies (780 °C/6.2 kb; Mishra and Bernhardt 2009). The ore is stratabound and represented by coarse grained sphalerite, galena, pyrite, and pyrrhotite. Pb-isotope data of galena gives a model age of 1804 Ma (Deb et al. 1989).

The *Rajpura-Dariba* group of deposits occurs in a 19 km long belt in the Rajsamand district in five clusters at Dariba, Rajpura, Sindeswar Khurd, Sindeswar Kalan, and Bamnia Kalan (Fig. 3.43). The multiple ore lenses and bands in these areas are hosted by recrystallized siliceous dolostone and graphite mica schist and are found to occur from shallow to great depths. The cumulative reserve is about 261 Mt @ 3–4% Pb–Zn (45 Mt @ 2.4% Pb, 6.5% Zn, 100–150 g/t Ag). The orebodies are both stratabound and stratiform and remobilized into veins at places. The ores and their host rocks have been isofacially metamorphosed at middle amphibolites facies (Deb and Bhattacharya 1980; Mishra et al. 2006). At Dariba at shallower levels a clear metal zonation existed with Cu-rich zone in the footwall, Pb–Z zone in the middle and Fe-rich zone in the hanging wall. The ores are commonly represented by sphalerite, galena, pyrite, pyrrhotite, tetrahedrite-tennantite, arsenopyrite (Fig. 3.44) and a host of rare sulpho-salts in the remobilized sections (cf. Deb and Pal 2004a). Pb-isotope data of galena from these ores give a model age of around 1800 Ma (cf. Deb et al. 1989). U–Th–Pb chemical dating of zoned monazite grains from the host rocks of Rajpura–Dariba (RD) and Rampura–Agucha (RA) deposits (Hazarika et al. 2013) indicate that while RD deposit experienced a single tectono-thermal event at ca. 1.62 Ga, the RA deposit has undergone two metamorphic events at ca. 1.62 and 1.0 Ga.

The *Zawar* group of deposits in the Udaipur district occurs as multiple ore lenses in a 16 km long carbonate ore zone with ore concentrations at Balaria, Mochia Magra, Zawarmala, and Baroi Magra (Fig. 3.45). The total ore reserve is about 50 Mt @ 2.0% Pb, 4.7% Zn, 40 g/t Ag. The orebodies are stratabound (cf. Fig. 1.9c) and hosted by middle greenschist facies dolostone and have simple mineralogy of sphalerite, galena, and pyrite. Pb isotope data gives an age of 1710 Ma for this deposit (cf. Deb et al. 1989; vide also Bhattacharya 2004; Sarkar and Banerjee 2004).

The *Kayar* Zn–Pb deposit is located roughly 11 km northeast of Ajmer, Rajasthan. Geologically, the Kayar deposit is hosted by intensely folded metasediments (quartz mica schist) of the Precambrian Delhi Supergroup comprising arenaceous, argillaceous, and calcareous sediments. Amphibolites and pegmatites have intruded the host rocks; amphibolites generally occur as sills. The

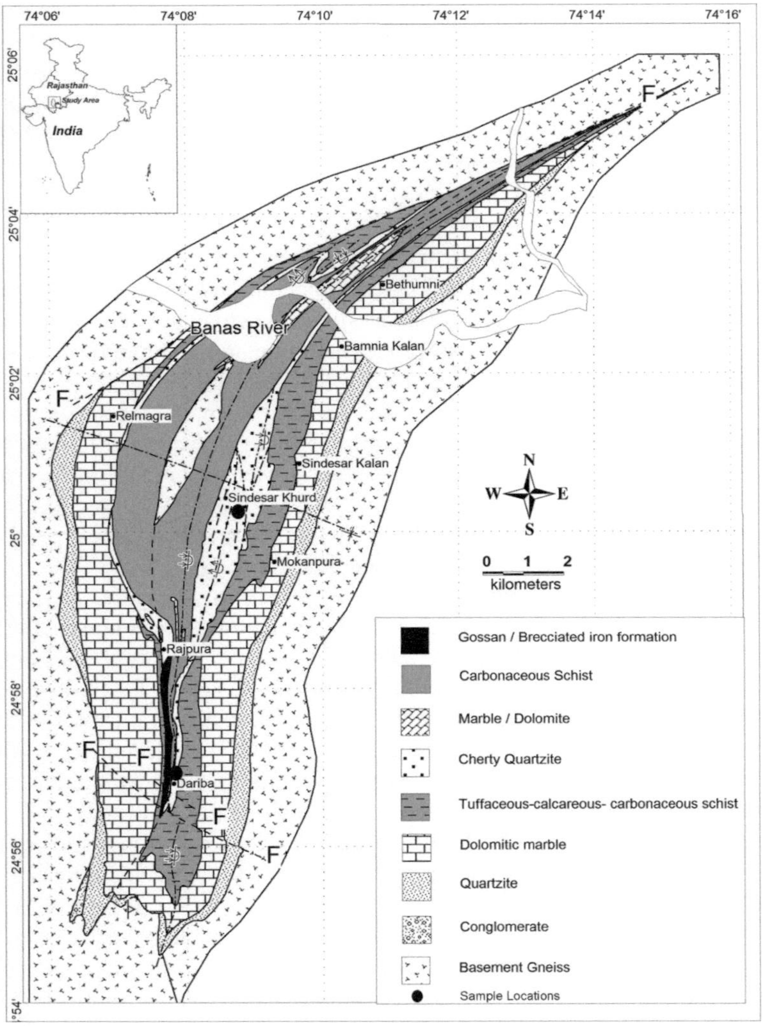

Fig. 3.43 Geological map of Dariba-Rajpura-Bethumni belt, Rajsamand district, Rajasthan (after GSI 1995)

ore body is usually flanked by pegmatite in the hanging wall. General trend of lithologies is NNE–SSW with 65–70 dip in upper section which gradually reduces to almost horizontal at depth of +300 m from surface suggestive of a broad secondary synformal structure with a shallow northeasterly plunging hinge. Metamorphism reached P/T conditions of the amphibolite facies. The Kayar deposit is estimated to contain 7.99 Mt ore @ 11.9% Zn, 1.6% Pb, and 36 g/t Ag.

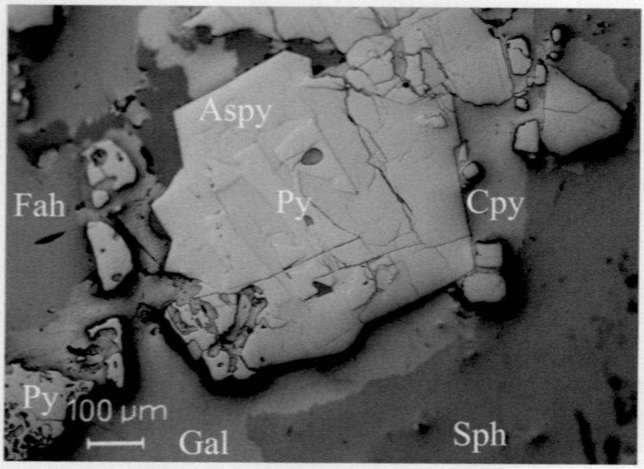

Fig. 3.44 Photomicrograph (under reflected light) of a polished section of the Rajpura-Dariba ore showing arsenopyrite (*Aspy*)-pyrite (*Py*) porphyroblast in a groundmass of galena (*Gal*), sphalerite (*Sph*), chalcopyrite (*Cpy*) and fahlore (*Fah*)

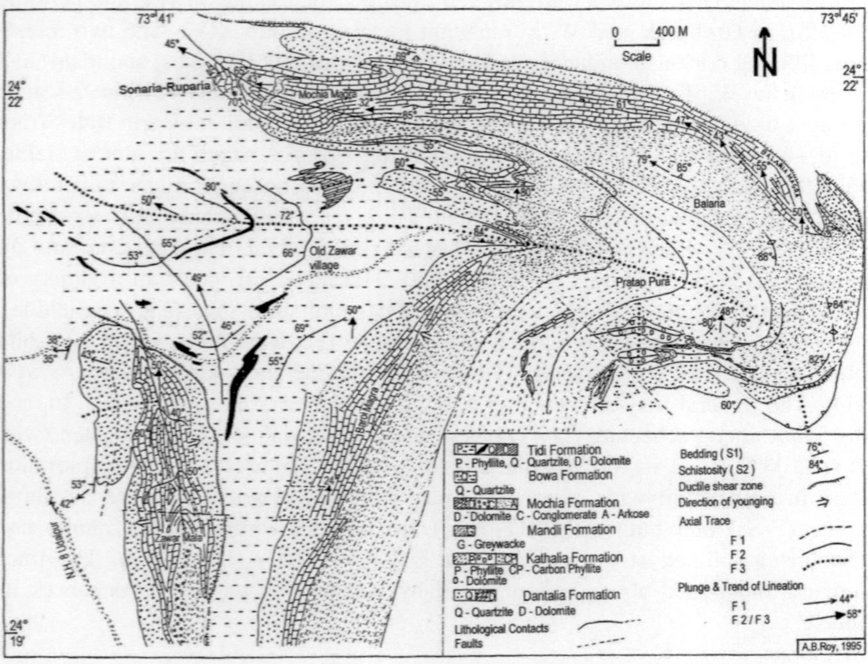

Fig. 3.45 Geological map of Zawar area showing the dolomitic ore band extending from Balaria, through Mochia Magra, Baroi Magra and Zawar Mala (after Roy 1995)

3.9 Tin–Tungsten Ores

Tin is a metal used from historical times. Its alloys are equally important: they include bronze (Cu and Sn, used since 3000 BC), pewter (historically, Sn with some lead, presently Sn with Cu and Sb), phosphor-bronze etc. It gives shine and prevents corrosion and therefore, used extensively as coating on steel, tin cans being the most common product. Various tin compounds (e.g., stannous chloride) have wide ranging commercial use, including in glass making.

Tungsten metal is used mainly for the manufacture of tungsten carbide, ferro-tungsten, high grade steel, and tungsten filament for lamps and other applications. Tungsten carbide is next to diamond in hardness and thus finds use in many types of equipment for cutting, drilling, etc.

3.9.1 *Geochemistry and Mineralogy*

Tungsten deposits are often found closely associated with tin deposits although their chemical characteristics are different, placing them quite apart in the periodic table (Sn in Group IV and W, a transition metal in Group XIV). The two metals have different crustal abundances as well: 2.3 ppm for Sn (49th most abundant) and 1.2 ppm for W. Tin has two common oxidation states (+4 and +2) but the most common oxidation state for W is +6, though it can exhibit all oxidation states from −2 to +6. Tungsten has five naturally occurring isotopes which are almost stable, while Sn has the largest number of 10 stable isotopes. Consequently the mineralogy of tin is more varied than that of tungsten. Tin is recovered mainly from cassiterite (78.8 wt% Sn), which is a tetragonal dioxide of the metal. The next common tin mineral is a sulfide, stannite (27.8 wt% Sn). The metal can occur in a variety of deposits in the form of complex oxides, sulfides and sulfo-salts (e.g., canfieldite). The most common tungsten mineral is wolframite (Fe, Mn) WO_4, which is a solid solution of Fe (Ferberite) and Mn (Huebnerite) end members, each having 76 wt% WO_3. The mineral may also contain a few trace elements like Nb, Ta, Sc, In, and Mo. The mineral scheelite, $CaWO_4$, is also a commercial source of tungsten (with 80 wt% WO_3) and is strongly fluorescent, a character useful for its exploration when in association with other leucocratic minerals. Two other minerals, tungstenite (WS_2) and tungstite ($WO_3{\cdot}H_2O$) are far less common. Wolframite and scheelite are mined at an ore grade of 0.2–0.4% WO_3 (Dill 2010). The trace elements mentioned above accompanied by Au and Ag in the ore enhances its value.

3.9.2 Types of Ore Deposits

Granites are the most common host rocks for both tin and tungsten mineralization, particularly when they are produced by anatexis in the lower crust and then emplaced in mid-crustal levels after undergoing enough fractionation to develop S-type characteristics (vide Chap. 2). The granitic skarns in calcareous sediments in association with both I and S type granites are also hosts to these metals. Proximal skarn deposits of Sn-(W) are known in Cornwall, England and in Erzgebirge in Germany where W is mostly a by-product obtained during the extraction of Sn. Such deposits are also found in Russia, Tasmania, and Alaska. Proximal W skarns and other contact metasomatic deposits may be reduced or oxidized depending on host rock compositions, which may be carbonaceous or hematitic (Newberry and Swanson 1986). Stratiform Sn-W deposits are found associated with volcanic rocks, with W considered to have concentrated by exhalative process in Austria and South Korea. Thrust-bound Au-W hydrothermal veins and stockwork occur in the Bohemian massif (Dill 2010). W is considered to have transported as chloride complexes as polytungstate ions in hydrothermal fluids (Wood and Samson 2000). Concentration of Sn and W occur in the beach placers of Indonesia, Malaysia, Thailand and Myanmar, close to volcanic arcs of Cenozoic and Mesozoic age. Alluvial and elluvial placers of the metals are known in Brazil, Bolivia, DR Congo, and Nigeria.

3.9.3 Sn-W Mineralization in India

The pegmatites in the Bastar-Malkangiri belt are well known and extend from near Dantewara-Bacheli in Chhattisgarh in NNW to Koraput in Odisha in ESE, covering an area of >1000 km^2. In this region several tin-bearing pegmatites also containing Nb, Ta, Li, Be, F, Zr, Hf, and REE (cf. Pal et al. 2007) are spread over a cumulative stretch of about 15 km. The major locations are Tongpal-Govindpal-Mundval area of south Bastar region and Mundaguda-Maithili area of Koraput district. Bengpal-Sukma high-grade metamorphic suites have been affected by major acid igneous plutonic activity around 2.6, 2.2, 1.5, and 0.7 Ga. The granitic phases are predominant before 2 Ga while pegmatitic activity dominated around 1.5 Ga. The early phase of pegmatites are barren but the later phases with pneumatolytic activity are mineralized (Babu 1994). About 16,000 tons of tin metal is reported from this belt (Sarkar and Gupta 2012) from primary ores. Besides, colluvial tin ores of economic interest have been known from this zone. Cassiterite in these ores is associated with columbite–tantalite or with lepidolite, beryl, fluorite and fluor-apatite at different places.

Bihar mica belt, now in Jharkhand, is known for the occurrence of a large number of quartz-feldspar pegmatites containing white mica (muscovite) that is commercial in quantity and quality. Such pegmatites contain beryl, columbite-tantalite, lepidolite, and cassiterite that have been mined locally. More than 200 tons of columbite-tantalite (Ta_2O_5 10–50%, and Nb_2O_5 40–65%) and several hundred tons of beryl have been mined by the Atomic Minerals Division (Directorate) during several past decades. These pegmatites are believed to have formed from granitic melts generated by fractional melting of Chotanagpur Gneissic Complex at 960 ± 50 Ma ago.

The Tosham deposit, situated 120 km NNW of Delhi in Bhiwani district of Haryana is known for the unique polymetallic mineralization of Cu–(Au)–Ag–Sn–Bi–In (Seetharam 1986; Murao et al. 2008). The disseminated, crude stockwork and vein mineralization is hosted by greisenized metasedimentary rocks of Delhi Supergroup, intruded by a porphyritic granite stock and by later rhyolitic effusive which forms the dome-shaped hill (Fig. 3.46a). Ar–Ar plateau age of the granite is 818.3 ± 3.6 Ma, of the rhyolite is 793 ± 18 Ma and K–Ar age of hydrothermal phlogopite associated with Sn–Cu veins (Fig. 3.46b) is 704 ± 36 Ma suggesting a rough time gap of ~100 My between the emplacement of the host rocks and the hydrothermal mineralization (Murao et al. 2000). Roquesite and stannite coexist with chalcopyrite while sphalerite in these ores is In-bearing. The resource estimated by MECL (Kinney 1986) is 18 Mt tin ore grading 0.15%. The crustal abundance of indium, which is in high demand in the electronic industry the world

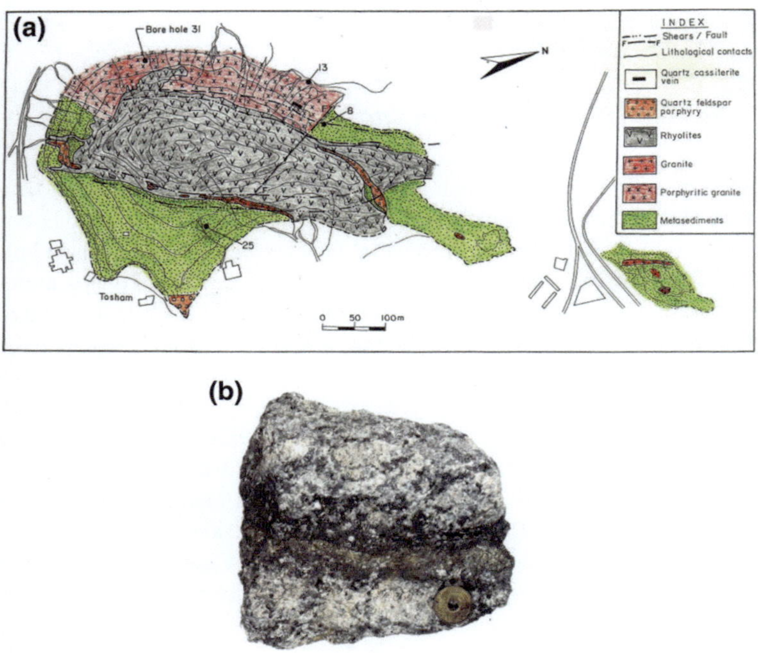

Fig. 3.46 a Geological map of Tosham prospect (after Murao et al. 2008); **b** Chalcopyrite-cassiterite vein in porphyritic granite at Tosham (Photo: S. Murao)

over, is ca. 0.1 ppm. The In-content of tin-polymetallic veins can reach up to 3000 ppm (Murao et al. 2008). Thus Tosham has the potentiality of becoming the only deposit of this kind in the country.

Tungsten deposits in India are located mainly in three different regions: (i) West-central Rajasthan; (ii) Nagpur district of Maharastra and (iii) West-central West Bengal (Dekate 1979). A few other smaller occurrences have no economic significance.

The Precambrian terrain of west-central Rajasthan hosts well known tungsten deposits at Degana in Nagaur district, Sewariya in Pali district and Balda in Sirohi district – all related to Neoproterozoic acid magmatism (S-type granites) at ~740 Ma (cf. Verma and Pandian 1983; Bhattacharjee et al. 1993; Krishnamurthi and Prabhakar 2000; Fareeduddin and Bhattacharjee 2000 among others). In fact, Bhattacharjee et al. (1993) delineate a 500 km long Sn-W metallogenic belt, starting from Tosham in Bhiwani district of Haryana to Balda in Rajasthan (vide Fig. 3.40).

The Degana mineralization is associated with a polyphase intrusion within phyllites of Delhi Supergroup, comprising a main body of medium grained S-type granite that has been intruded by aplitic granite and granite porphyry. It is considered as a high level intrusion correlated with the 735 Ma-old Jalor granite (Chattopadhyay et al. 1982). Wolframite occurs (i) in greisen-bordered veins in granitoids and phyllites; (ii) stockwork masses in brecciated phyllite and (iii) as elluvial placers surrounding the Rewat hill. Fluorite, topaz, and apatite are common in the wall rocks and common sulfides are associated with the tungsten ores. The average grade of the deposit is 0.25–0.54% WO_3 in veins and about 0.4% WO_3 in the elluvial gravel bed.

At Sewariya, a polyphase granitoid batholith outcrops for more than 100 km with a narrow width of 4–6 km, intrusive into Banded Gneissic Complex to the west and Barr metasediments to the east. It comprises a main biotite-rich S-type orogenic granitoid body and a cross-cutting leucocratic granitoid body. W–Sn–Li mineralization, represented by wolframite, sicklerite, triphylite, and cassiterite, occurs in associated quartz veins restricted to ductile shear zones and rarely in pegmatitic veins. Wall rock alteration includes greisenization, sericitization, kaolinization, tourmalinization, and K-metasomatism. Five prospects have been identified at Pipaliya, Motiya, Kotariya, Richmaliya, and Bijathal. The mineralized zones here range from 200 to 600 m in length and 20 to 50 m in width. WO_3-content of the ore is between 0.18 and 1.0%.

W-mineralization at Balda and Deva-ka-Bera are associated with the Balda granite, a leucocratic, well-foliated and discontinuous body that has intruded the argillaceous and calcareous metasedimentary rocks of the Delhi Supergroup and has given rise to extensive pneumatolytic alteration along the contact. Mineralization occurs in quartz and greisenized pegmatite veins located mainly along deformed and silicified shear zones within the Balda granite and adjoining metasediments. The mineral association in the veins is represented by wolframite + quartz + tourmaline + fluorite + topaz + sulfide minerals. These veins are between 170–600 m in length and 1–5 m in width, with a grade of 0.2–2.2% WO_3.

A WSW to SW-trending, 16 km long belt of tungsten mineralization, represented by wolframite and scheelite, closely associated at places with minor tin, niobium and tantalum, occurs in the Nagpur district of Maharastra, between Agargaon to Chapegarhi in the NE and Khobna to Kuhi in the SW, within Precambrin Sakoli Group of rocks (Bhoskar and Chande 1979) (Fig. 3.47). Granitoids and pegmatites are emplaced within the pelitic schists of Sakoli Group which shows zones of intense greisenization and tourmalinization. The mineralized zone at Agargaon is lenticular in shape, traceable over a length 1.4 km and over 10–115 m width, and is represented by numerous small quartz veins along a shear zone within intricately folded chlorite-muscovite-tourmaline schists. Near Khobna, the Sakoli metasediments host two NW–SE-trending parallel zones of W-bearing greisens occurring as small pods, lenses, veins and bands and extending over a length of 1 km each with width varying from 10 to 100 m. The deposit is likely to contain a reserve of 2.46 Mt of tungsten ore, equivalent to nearly 7000 tons of W, up to 60 m depth (Jain and Mohan 1988). At Kuhi, two parallel zones of tungsten mineralization have been delineated. In one of the zones, the mineralization extends over a strike length of 1.65 km with width between 10 and 85 m and a vertical continuity of 170 m (Jain and Mohan 1984). The mineralization in the belt is ascribed to hypothermal polyascendant solutions in the temperature range of 400–270 °C (Dekate 1979), which have been derived from granitic bodies at depth.

Chendapathar (22° 45′: 86° 45′) and Porapahar (22° 57′ 30″: 86° 49′) in Bankura district, West Bengal contained tungsten mineralization which was mined with occasional breaks during the last century. It is considered to be hypothermal to mesothermal, formed in the temperature range of 365–190 °C with a monoascendant character. Wall rock alteration here has given rise to the assemblage of muscovite-sillimanite-kyanite (Dekate 1979). More studies are desirable.

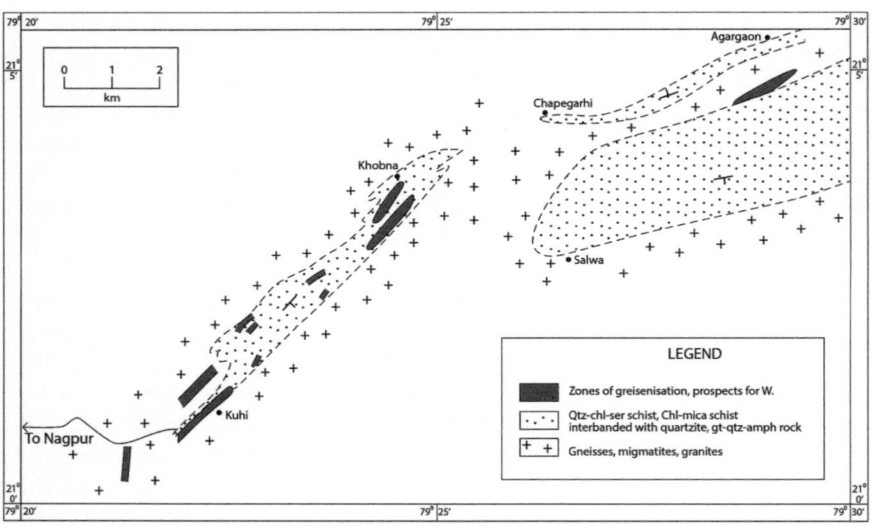

Fig. 3.47 Geological map of Agargaon-Khobna-Kuhi tungsten belt, Nagpur district, Maharashtra (after Jain and Mohan 1984)

3.10 Rare Earth (REE) and Other Associated Metals Deposits

The rare earth elements (REEs), or chemically speaking lanthanides, comprise 15 elements bearing the atomic numbers 57–71. Of these the first six elements, Lanthanum (La), Cerium (Ce), Praseodymium (Pr), Neodymium (Nd), Promethium (Pm) and Samarium (Sm) and Europium (Eu) are called light rare earth elements (LREE). To this subgroup Yttrium (Y) (At. No. 39) is commonly added because of similarities of properties. The heavy REEs (HREEs) comprise the rest, i.e., Gadolinium (Gd), Terbium (Tb), Dysprosium (Dy), Holmium (Ho), Erbium (Er), Thulium (Tm), Ytterbium (Yb), and Lutetium (Lu). Because these elements resembled the oxides of alkaline earths (Ca, Ba, etc.) and were not commonly found as minerals, these were called rare earths. Today they are important tools in petro-mineralogical studies. Besides, the REEs are high-tech metals and have a large number of uses in modern equipments such as, smart phones, LCD screens, Computer hard discs, CFL, MRI, Positron emission tomography, fiber optics, lasers, permanent magnets, hybrid cars, atomic reactors and defense hardware, petroleum refining, etc. During upper mantle partial melting, due to lanthanide contraction, LREEs get concentrated into the melt forming crustal rocks, while the HREEs are left behind in the mantle minerals.

More than 200 minerals contain REE in tangible proportions, of which bastnaesite, monazite, and xenotime are the three most significant REE minerals. Their compositions respectively are $(CeF)CO_3$, $(CeY)PO_4$, and $(CeLaTh)PO_4$. They contain most other REEs in varying proportions. Less important minerals are allanite, eudialyte, and loparite. REE minerals generally contain Th and U in tangible proportions in their atomic structures.

3.10.1 Types of REE Deposits

REE mineralization is known from

1. Carbonatites and the allied rocks, such as syenite and nepheline syenite
2. Beach placers
3. Pegmatites and hydrothermal quartz fluorite veins
4. Hematite granite breccias (IOCG) as in Olympic Dam, Australia
5. Skarn deposits
6. Ion absorbant clays
7. Residual enrichment blankets

Production and trade in REE is highly imbalanced at this moment. China is overwhelmingly the largest producer of REEs, producing 1,30,000 tons or 97% of the world production in 2010. India produced 2700 tons. Estimated reserves of REEs are 110 Mt, ~50% of which occur within Chinese territory. Presently the world electronic industry is in a state of crisis because of monopoly REE business by China. However, some new large discoveries have recently been made, such as in a carbonatite complex in Helmand province, Afganistan. We discuss the important deposits in some details so that this information may be useful in exploration in new sites.

As of today, the first group of deposits are most important, because the world's largest repository of REE, the Bayan Obo (Baiyenabo) in Inner Mongolia, China, though genetically is still considered very complex, has a great resemblance to this group. Its mineral/metal reserves are as follows: 1500 Mt of hematite and magnetite, 100 Mt of LREEs in monazite and bastnaesite (6% LREEs; bastnaesite > monazite) and 1 Mt ores with 0.13% Nb_2O_3. The rock association consists of Mesoproterozoic slates, sandstones, and quartzite. The mineralization is hosted by a Mg-rich carbonate rock, which could be of sedimentary (diagenetic) or igneous origin, i.e., a carbonatite rock (Fig. 3.48). The fluids in ore and gangue minerals apparently have affinity towards carbonatite and alkali complexes (Smith and Handerson 2000). According to Wu (2008) the deposit shows features, shared by both magmatic (carbonatite) and magmatic hydrothermal (IOCG) type.

From 1965 to 1995, the mineralization at Mountain Pass, California, USA was the world's largest producer of REE. Soon thereafter its operations were suspended and may not have been resumed as yet. Castor (2008) estimated a reserve of >20 Mt ores with 8–9% REEOs (cut-off grade 5%). Mineralization occurs in an area of 10 × 2.5 km^2, where metamorphic rocks are intruded by ultra-potassic granites, syenites and shonkinites.

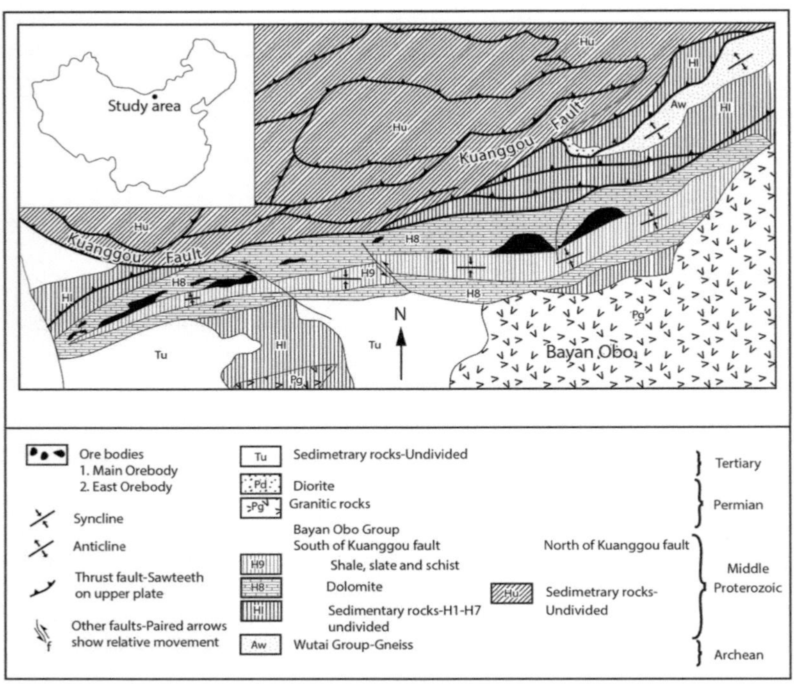

Fig. 3.48 Geological map of the Bayan Obo area (after Chao et al. 1995)

Beach placer deposits of India and Australia, containing several economic and subeconomic minerals, are well known by now. Monazite is the principal REE-bearing mineral. Ilmenite and rutile account for the Ti resource in these placers. These Indian deposits are discussed in moderate details in the following section.

The Olympic Dam deposit of South Australia is an example of IOCG-type deposit, also in the context of REE mineralization. The IOCG model has already been discussed vide Chap. 2, in a different context though. It can only be mentioned here that the deposit reportedly contains 3.4 Mt ores containing 0.17% La, and 5.4 Mt ores containing 0.27% Ce (Laznicka 1993), besides large reserve of Au, Cu, U, and Ag.

"Blanket" type deposits are a rather new find. It is best represented by Chinese deposits (Jiangxi, China). There it occurs on the top part of Mesozoic granites having a mixed REE composition (Y, Sm, Eu, Gd) forming part of a sapropelitic profile. The primary hosts of REE are zircon, apatite, and allanite. REE-fluorocarbonates are found along grain boundaries and internal fractures and fissures. These developments are assigned to a post-solidus (magmatic-) hydrothermal stage of formation. The fairly soluble fluorocarbonates were subjected to supergene leaching and consequent enrichment (Ishihara et al. 2008). These secondary REE ores are presently exploited at Jianxi, China.

Mary Kathlene skarn deposit was worked for uranium in Australia. It contains REE in allanite.

In earlier days, magmatic magnetite-apatite deposits were processed also for REE extraction. However, those were the days when REE-bearing mineral deposits were worked for academic studies rather than for industrial use.

3.10.2 REE Mineralization in India

India's REE resources are mainly exogenic, occurring as beach placers. About 75% of India's placer deposits occur in the western (southern tip) and eastern coast. The best deposits are located at Chavara in Kerala and Manavalakurichi in Tamil Nadu. Another deposit is located at Chatrapur on the Odisha beach (vide Fig. 6.31). Monazite is the main container of REE (about 60%) in India. Other associated minerals of economic importance are ilmenite and rutile as well as zircon and garnet. The average REE composition of Indian monazite is presented in Table 3.8.

The REE content of monazite, as given above, is expected to produce a reserve of $8 \times 0.6 = 4.8$ Mt of REEO. Australia has a comparable stock of monazite. But the Indian monazite is poorer in La and Y and richer in Ce. REE-bearing carbonatite deposits at Samalpatti and Pakkanadu area, Dharampuri district, Tamil Nadu are interesting but not exploitable.

Table 3.8 Average REE composition (wt%) of Indian beach monazite (Sarkar et al. 1995)

La_2O_3	22	Gd_2O_3	1.0
CeO_2	47	Tb_4O_7	0.1
Pr_2O_3	6	Dy_2O_3	0.2
Nd_2O_3	18	Ho_2O_3	0.1
Sm_2O_3	3	Y_2O_3	0.5
Eu_2O_3	0.01		

In the Singhbhum Cu–U belt xenotime and allanite occurs at several places, the best concentration of xenotime being found at Kanyaluka.

Broadly, three types of REE mineralizations with other associated metals are reported from the Purulia district of West Bengal.

(i) La, Ce, Nb mineralization along with carbonatite rock along the "northern shear zone".
ii) Apatite-magnetite-REE (La, Ce) mineralization along North Purulia shear zone.
iii) Relatively rarer metals – Li, Cs, Rb are found to occur along a tectonic lineament in the Chotanagpur Gneissic Complex

All of them are subeconomic at the present state of knowledge.

In Rajasthan-Gujarat area REE-bearing pegmatites with other associated metals occur at several places. The dominant associated metals are Be, Nb, Ta. The beryl-bearing pegmatites occur along the entire length of the Aravalli region running for about 500 km. The Rajasthan mica belt is bigger than the Bihar and Nellore mica belts. Rajasthan has produced several thousand tons of high-grade beryl ore, analyzing between 10 and 14% BeO. Nb–Ta occurrences are concentrated in the central part of the belt, e.g., at Malpura, Ajmer, and Bhilwara areas.

Occurrence of Li minerals (lepidolite and spodumene) at Rajgarh and Potlan are interesting. Recently swarms of lepidolite-bearing pegmatites have been noticed in Sirohi district in the south (Maithani and Nagar 1999). Other minerals, particularly beryl, columbite-tantalite are found at places in Sabarkantha district of Gujarat.

Interestingly, in Rajasthan and Gujarat there are two types of pegmatites of two different ages. The older ones intrude into the basement gneisses and the younger ones only in the Delhi and Aravalli sequences. The younger pegmatites containing muscovite, beryl, REE and other associated minerals, are absent in the older ones. Zoned pegmatites are generally REE-bearing and also contain other metals such as, Be, Nb, and Ta.

3.11 Aluminum Ores

Aluminum is in great use in today's life beginning with the foil that wraps the candies you like through the many utensils in your family kitchen to the aeroplane you may have traveled in. They are all made up entirely or largely of aluminum.

3.11.1 Chemical and Physical Properties of Aluminum

Aluminum is the third most abundant element in the Earth's crust and the most abundant metal. Occurring in different rock-forming minerals, there is an inexhaustible reserve of aluminum in the earth's crust. But unfortunately most of these, in the present state of knowledge, are not amenable to an economically viable process of Al-extraction, except bauxite.

Aluminum is a moderately soft silvery white light metal (2.7 g/cc). Melting point is 658.7 °C. It is a good conductor of heat and electricity and resistant to corrosion. For making aluminum extremely corrosion resistant, it is given a heavy coat of aluminum oxide by making it an anode in an electrolytic cell. There is an industrial name for it, i.e., anodized aluminum, which is much in demand.

3.11.2 Geochemistry and Mineralogy of Aluminum Ores

Today bauxite is the main ore of aluminum. Bauxite deposits are masses of lateritic matter in which aluminum dominates over iron. According to Valeton (1981) it should not contain <45% Al_2O_3 and not more than 20% Fe_2O_3 and combined SiO_2 in the range of 3–5%. The principal Al-mineral in bauxite is gibbsite (α-$Al_2O_3{\cdot}3H_2O$). The abundant Al-phase is boehmite (γ-$Al_2O_3{\cdot}H_2O$) though gibbsite generally is $\gg$ boehmite and diaspore (β-$Al_2O_3{\cdot}H_2O$). Industry prefers gibbsite, because it is amenable to leaching by hot NaOH solutions (Bayer process). Kaolinitic clay may be present in small proportions, which is the source of reactive silica, unwanted in industry. Iron is present as oxides and hydroxides. TiO_2 is present up to a few percent in different oxidic phases.

3.11.3 Distribution Through Space and Time

Five major bauxite provinces are delineated throughout the world

1. Indian province (East Coast, plateau regions of eastern and central India, Western Ghats, a number of other places)
2. South American province (Guinea and Central Brazilian shield)
3. West African province (Guinea, Ghana, and Mali)
4. South east Asian province (Vietnam, Laos, and Cambodia)
5. Pre-coastal province of North and South-West Australia

Bauxites of Archean age (3100 Ma) are reported from Swaziland, South Africa. The first peak of bauxitization correlates with the Lower Carboniferous. A much larger one, however, coincided with the late Tertiary to the present.

3.11.4 Typology of Bauxite

Bauxite is commonly divided into two types: lateritic bauxite and karst bauxite. Lateritic bauxite occurs in the superficial parts of exposed Al-silicate rocks, such as, granites, granite gneisses, basalts, alkaline rocks, arkose, etc., and represents about 85% of the world reserve of ~35 Gt (Brimhall and Lewis 1992). There may be long distance transport of the bauxidized material, in which case the deposit is called allochthonous. Karst bauxite develops by reworking of clay-bearing carbonate rocks.

Tardy (1997) proposed a threefold classification of lateritic bauxite as follows:

1. Orthobauxites
2. Metabauxites
3. Cryptobauxites

Orthobauxites, as the name suggests, are typical bauxites. The nature of the weathering profile in metabauxites is virtually the same as that of orthobauxites, excepting that it has a high Al/Fe ratio. Cryptobauxites, according to this classification, are those bauxites which are covered by a clay-rich blanket. This is rarer still.

3.11.5 Bauxite Formation in a Weathering Profile

Looking up from the protolith (primary or original rock) at the bottom of a weathering profile containing lateritic bauxite, we see the following features:

The protolith changes to a partially altered zone, i.e., saprolites, in which the original minerals (relics) and rock textures may coexist along with the new-formed minerals, such as kaolinite, halloysite, and Fe-hydroxides. Halloysite and goethite are more common in the lower part of this zone, while kaolinite and halloysite are more common in the upper part. The saprolite zone is overlain by the zone of bauxitization. The most important chemical reactions involved in the process are the following:

$$2KAlSi_3O_8 + 2H^+{}_{aq} + 7H_2O \rightarrow Al_2Si_2O_5(OH)_4 + 2K^+{}_{aq} + 4H_4SiO_{4aq}$$

[Plagioclase feldspar, and even muscovite mica may be involved in such a reaction]

$$\underset{\text{(Kaolinite)}}{Al_2Si_2O_5(OH)_4} + 5H_2O \rightarrow \underset{\text{(Gibbsite)}}{Al_2O_3.3H_2O} + 2H_4SiO_{4aq}$$

The deposition of gibbsite and the transport of SiO_2 to solution (H_4SiO_4) are controlled by a pH of 4.5–9 of the system. A supportive drainage system is very important, to the extent that it may help skip the intermediate stage of kaolinitization and the feldspar may be directly converted to gibbsite. Bauxite develops above the groundwater table, or in the zone of its annual fluctuations. Bauxite is commonly overlain by a duricrust which is an indurated concretionary rock containing more Fe-oxides than alumina and more hematite than gibbsite.

3.11.6 Why Bauxites Occur Where They Do?

Like all mineral deposits, bauxite deposits do not occur everywhere. What are the factors that favor their occurrence at a place and absence in another?

1. Favorable climate: tropical monsoon climate with a mean temperature of 23–26 °C and annual rainfall of 1200–4000 mm have been found favorable.
2. Suitable drainage: In addition to 1, drainage through the rock must be good to take out the unwanted reaction products.
3. Geomorphology: Most laterite-bauxite deposits form as blankets covering the ancient planation surfaces dissected by valleys.
4. Forest cover: A tropical to Savannah forest cover generally help in decomposing the minerals in the rocks, increasing the penetration of ground water into the rocks and mediating the pH of the ground water.
5. For bauxite formation it is not important if a rock is igneous, metamorphic, or sedimentary. What is important is that it must not be poor in aluminum as in ultramafic rocks or quartzite or sandstone.
6. Tectonic conditions: Most lateritic bauxite deposits formed on tectonically quiet continental platforms, whereas Karst bauxite deposits formed mainly in orogenic belt, preceding the orogenesis.

3.11.7 India's Resources of Al Ores or Bauxites

With an estimated reserve of >3 billion tons of bauxitic ores, India is indeed rich in aluminum resource, being one of the five principal Al-ore containing countries of the world.

The principal bauxite provinces of India are (Fig. 3.49):

1. Eastern Ghats, extending from Odisha through Andhra Pradesh up to northern Tamil Nadu
2. Plateaux bordering Jharkhand and Madhya Pradesh
3. The Amarkantak plateau, particularly the Maikala range of Madhya Pradesh
4. Western Ghats

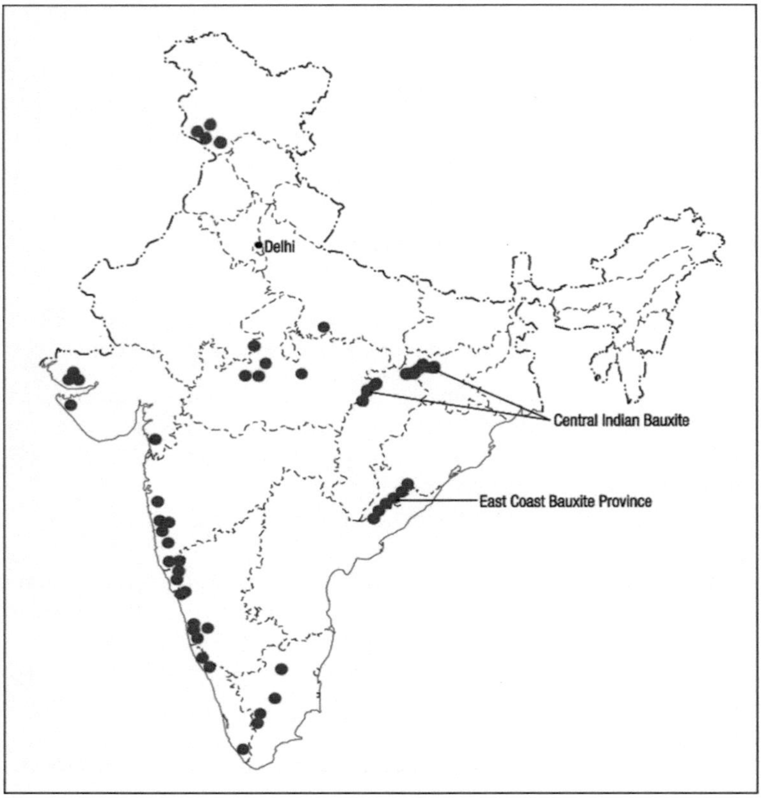

Fig. 3.49 Location of different bauxite deposits on a political map of India

2 and 3 could be grouped together into the Central Indian bauxite belt and there are a few smaller deposits found at a number of places such as, in Riasi district of Jammu & Kashmir, Coastal Gujarat, etc.

East Coast Bauxite

Many of the East coast bauxite deposits came to be confirmed only in the seventies of the last century, although laterite cappings were reported from a number of places about hundred years earlier (Krishnan 1935). Eastern Ghat bauxite or East Coast bauxite as some prefer to call them, occur along the coastal region of Odisha and Andhra Pradesh over an area of $400 \times 30\ km^2$ (approax). Odisha has a larger share of $\sim$1600 Mt, which constitutes $\sim$53% of the country's total reserve. Andhra Pradesh contains 658 Mt or 22% of the national reserve.

All the major bauxite deposits along the East Coast are "high level bauxites", being located at elevations >900 m above MSL. However, some Monadnocks along the coastal planes are at places capped by laterites. These are blanket-type

Fig. 3.50 ~20 m thick lateritic bauxite cap on Galikonda hill, Andhra Pradesh (Photo: M Deb)

deposits with large aerial extent (~0.5 km), appreciable thickness (<1.0 to max. ~50 m thick) and usually good profile differentiation with little or no overburden (Deb and Joshi 1984). The well-known aluminous laterite occurrences in this East Coast Bauxite Province are at Anantagiri area in Vishakhapatnam district of Andhra Pradesh where bauxite cappings at altitudes of 1090–1445 m are found at Galikonda (Fig. 3.50), Raktakonda, Katuki, and Chittamgondi (Raman 1976). In Odisha, aluminous laterite deposits occur at Pottangi, Panchpatmalli, and Baphlimalli hills in the Koraput district, around Kashipur in Kalahandi district and on the Gandhamardan plateau of Bolangir district.

The protoliths of bauxite in the East Coast are khondalites (>90%) and charnockites. The ores are generally spongy and holocrystalline and commonly characterized by the presence of "pseudofolia", i.e., a false foliation. A generalized weathering profile in the East Coast bauxite (GSI 1994) is presented in Table 3.9.

Bauxite Deposit in Central India

The central Indian bauxite belt trend ENE–WSW, from Ranchi and Palamau districts of Jharkhand to Balaghat district of Madhya Pradesh, passing through Surguja district of Chhattisgarh and Sadol, Bilaspur and Mandla districts of Madhya Pradesh. The bauxitized mesas along this belt include Samirapat and Mainpat

Table 3.9 A generalized weathering profile in East Coast bauxite

Alteration zones	Thickness (m)
Fe-lateritic duricrust[a]	1–5
Bauxite with lateritic intercalations	2–54
Partially lateritised protolith/lithomarge	3–25
Kaolinitized protoliths[b]	5–8
Unaltered protoliths (khondalite, charnockite)	

[a]Duricrust: A blanket of surficial rocks formed by the cementation of loose rock matter by silica, carbonate or iron oxide/hydroxide
[b]Protolith: The first-formed rock from which the rock under reference formed

(Chhattisgarh) and Amarkantak, Phutkapahar (Madhya Pradesh). There are two more clusters of bauxite deposits, i.e., in Jabalpur–Sidhi and Rewa–Banda–Satna sectors.

Amarkantak deposits occur as lenses and pockets within the lateritic cappings on the dissected plateaus, which are quite high (1050–1100 m) and occupying the eastern part of the Maikala range. Best developed large deposits are at the Hazaridadar section of the Amarkantak area. A typical weathering profile at the Amarkantak area starts with basalt (Deccan trap) at the base, gradually giving way upward to altered basalt, underclay, ferruginous laterite, aluminous laterite, bauxite, ferruginous laterite and ends up with a top soil. It is presented in the following Table 3.10 and the geochemical changes involved are shown in Fig. 3.51.

Mineralogically, the Amarkantak bauxite consists predominantly of gibbsite, with some goethite and hematite. Kaolinite and cryptocrystalline silica do not exceed 5% together. The total reserve of Amarkantak plateau deposits in Sahdol district is about 20 Mt bauxite (with 45–61% Al_2O_3; 0.85–11% SiO_2; >3.5 to <8% TiO_2).

Supkhar and Kotapahar areas in Balaghat (MP) are moderately rich in bauxite ores. Supkhar contains a deposit of 13.5 Mt, Kotapahar contains another 4.2 Mt (40–56% Al_2O_3, 0.85–11% SiO_2). Deccan basalt is at the base of the weathering profile.

Bauxite deposits at Katni, Jabalpur district, MP have been mined for quite some time. The ore zone is 2–12 m thick. There is a good reserve of refractory type of ores (>52 to <56% Al_2O_3, 2–11.7% SiO_2, 4.6–7.7% TiO_2). Gibbsite is the principal bauxite mineral in Katni ores. The Katni bauxite is generally believed to have formed from Vindhyan limestone and shales. Unexpectedly, it is a low-level bauxite.

Table 3.10 A generalized lateritic profile of the Amarkantak area (Jagannatha Rao and Krishnamurthy 1981)

Rock types exposed	Brief description of the weathering zones	Depth (m) from the surface
Top soil	Lateritic soil (red, loamy, ferruginous)	0–1
Ferruginous laterite	Pisolitic and nodular ferruginous laterite	1–3
Bauxite	White to cream colored hard bauxite	3–5
Aluminous laterite	Buff colored non-pisolitic material With brown and white patches	5–6
Ferruginous laterite	Reddish brown to brick red, cavernous Ferruginous laterite	6–17
Goethite	Dark brown, compact material	17–18
Underclay	Pink brown, purple clayey material, gritty bands	18–19
Altered basalt Basalt	Dark brownish gray, gritty, porous altered Volcanic rock	19–23

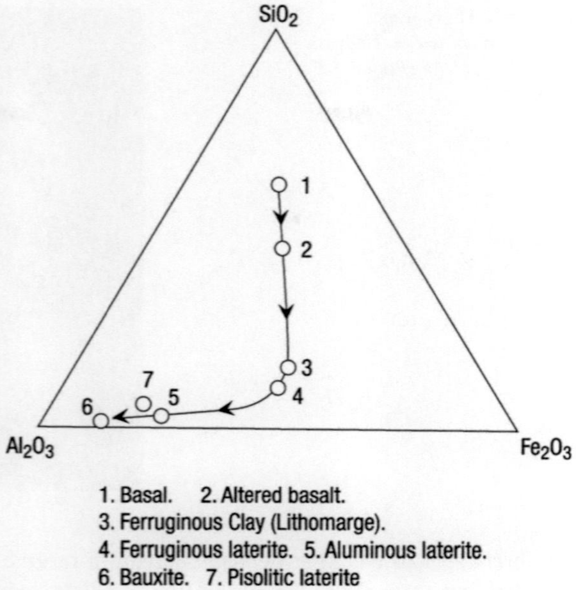

Fig. 3.51 Evolution of basaltic protolith into bauxite at Amarkantak, defined by SiO_2–Al_2O_3–Fe_2O_3 system

Bauxite in laterite cappings occur at Jamirapat and Mainpat areas of Chattisgarh. Deccan traps lie at the base of the weathering profile at both the areas. GSI proved a reserve of 36.6 Mt (52% Al_2O_3) and a reserve of 33 Mt at Mainpat.

Bauxite Deposits in Eastern India

The bauxite deposits in eastern India may be considered as the eastern extension of the bauxite-bearing regions of Madhya Pradesh, Chattisgarh, and Jharkhand. They comprise chains of bauxitised mesas that include those of Bagru hills (Lohardaga district), Richigunda (Ranchi district) and Shrendag and Netarhat (Gumla district) in Jharkhand. The country rocks are represented by the Chotanagpur Granite Gneiss Complex.

Bagru deposits comprising the lease-holds at Bagru, Bhusar, Hisri and Hisri New have a mining history predating Indian independence. They are located between 23° 28′ 20″N and 84° 35′ 05″–84° 36′ 40″E. The above lease-hold areas, collectively called Bagru Pat is located at 1025–1055 m above MSL. The principal bauxite phase in the ores (Fig. 3.52) is gibbsite with goethite as the subordinate phase and Ti- and Fe–Ti-oxides as minor/trace phases. In the clay zone the principal phase is kaolinite with quartz, illite and muscovite as minor phases and tourmaline, zircon, rutile, kyanite, and sillimanite as trace phases. The mineral composition of the clay zone material and the field relationship of the clay zone with the overlying bauxite and underlying gneissic rocks suggest that clay and the overlying bauxite ore developed by supergene alteration of Chotanagpur Granite Gneiss. The weathering profile at Bagru is shown in Fig. 3.53.

Fig. 3.52 High grade pisolitic bauxite ore, Bagru Hills, Lohardaga (Photo: S.C. Sarkar)

Shrendag and Netarhat deposits, having a reserve of $\sim$19 Mt of bauxitic ores are also important. Here bauxite occurs as bands, lenses and impersistent sheets or blankets. The generalized weathering profile of the area, as discerned from escarpment and drill holes, is as follows:

Top soil and murrum
Lateritic bauxite (pisolitic)
Bauxite
Aluminous laterite
Lithomarge
Altered basement gneiss

An uncommon feature of the bauxite ores of the region is the development of boehmite, rather than gibbsite in them. Boehmite commonly forms by the dehydration of gibbsite.

The bauxite deposits of Khamar Pat and Mandua Pat in Ranchi district are close to the town of Lohardaga. They have high-grade bauxites with Al_2O_3 content varying between 49.7 and 60%. Isolated pockets and bands of high-grade bauxite also occur in Pokhra Pat and Dudhia Pahar areas on the Ranchi plateau. High level laterites with thickness varying from 9 to 61 m, containing irregular patches of bauxite, also occur in the Kharagpur hills in the Monghyr district of Bihar at Khapra, Maira-Maruk, and Maira areas.

West Coast Bauxite

West coast bauxite refers to the bauxite deposits occurring in the pre-coastal areas of Maharastra and Karnataka. About 100 Mt of bauxite has been estimated by the GSI (1994) along this region. Important bauxite deposits of Maharastra are located

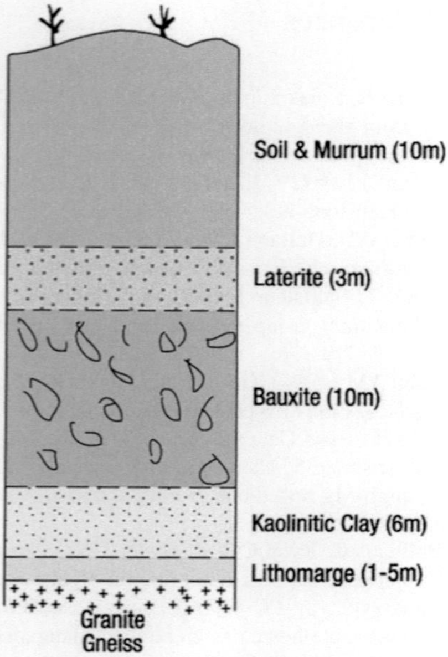

Fig. 3.53 Weathering profile showing the location of the bauxite zone at Bagru

at Kolhapur, Kolaba, Ratnagiri, and Satara districts on plateau tops. Deccan basalts are the ultimate protoliths of these bauxite ores.

More than 50 small pocket deposits, the aggregate reserves not exceeding 20 Mt have been located along the Konkan coast in Raigarh, Ratnagiri, and Sindhudurg districts. They remain more or less unexploited. Average of 50 samples of bauxite from this zone analysed Al_2O_3: 51.65%, SiO_2: 2.42%, Fe_2O_3: 6.37%, TiO_2: 2.92%. Sindhudurg ores (overlying the Dharwar rocks) are still better in grade.

Bauxite deposits occur in Karnataka as well, most important of which are located in Belgaum district, particularly in the Boknur-Navage plateau and the Paduvari plateau in south Karnataka. The weathering profile at Boknur-Navage plateau is as follows (Krishna Rao et al. 1989):

Bauxite-bearing lateritic zone ($\sim$10 m thick)
Lithomargic clay zone ($\sim$2 m thick)
Altered basalt zone
Unaltered basalt

Generalized chemical composition of Belgaum bauxite is: SiO_2 2.68%, Al_2O_3 57.93%, Fe_2O_3 5.04%, TiO_2 5.81%, Gibbsite is the principal bauxite mineral.

At Paduvari the deposit is richer in Fe, though categorized as economic. Bauxite development is not great in Kerala ($\sim$13.2 Mt; GSI 1994). These are reported from Trivandrum, Quilon, Allepey, and Cannanore districts. Ores are both authigenic and allogenic (transported and re-deposited as sediments). There is even clay–bauxite mixture.

References

Anand R, Balakrishnan S, Mezger K (2005) Geochemical and geochronological studies of titanites from the granitoid rocks of the eastern Dharwar craton, South India. Geophys Res Abstr, European Geosci Union 7:1565

Anantha Iyer GV, Vasudev VN (1985) Copper metallogeny in the Jogimardi volcanic, Chitradurga greenstone belt. J Geol Soc Ind 26:580–598

Anbar AD, Holland HD (1992) The photochemistry of manganese and the origin of banded iron formation. Geochim Cosmochim Acta 56:751–765

Auge T, Salpateur I, Bailly L, Mukherjee MM, Patra RN (2002) Magmatic and hydrothermal platinum group minerals and base metal sulfides in Baula complex, India. Can Mineral 40: 277–309

Babu TM (1994) Tin in India. Geol Soc Ind, Sr Mineral Resour India 7:217 pp

Bache JJ (1987) World gold deposits: a geological classification. North Oxford Academic, London & Orleans Cedex 179 pp

Balakrishnan S, Hanson GN, Rajamani V (1990) Pb and Nd isotope constraints on the origin of high-Mg and tholeiite amphibolites, Kolar schist belt, Southern India. Contrib Mineral Petrol 107:272–292

Ballhaus C, Ryan CG, Mernach TP, Green DH (1994) The partitioning of Fe, Ni, Cu, Pt and Au between sulfide, metal and fluid phases: a pilot study. Geochim Cosmochim Acta 58:811–826

Bandopadhyay PC (1996) Facies associations and depositional environment of the Proterozoic carbonate-hosted microbanded manganese oxide ore deposit, Penganga Group, Godavari rift basin, India. J Sed Res 66:197–208

Bandopadhyay BK, Roy A, Shukla RS, Guha Sarkar TK (1992) SEDEX-type copper and zinc deposits in the Proterozoic Sakoli Group, Nagpur and Bhandara districts, Central India. In: Sarkar SC (ed) Metallogeny related to tectonics of the Proterozoic mobile belts. Oxford and IBH Publishing Co. (P) Ltd., pp 53–101

Banerji AK (1962) Cross folding, migmatisation and ore localization along parts of the Singhbhum shear zone, south of Tatanagar, Bihar, India. Econ Geol 57:50–71

Basu A, Maitra M, Roy PK (1997) Petrology of mafic-ultramafic complex of Sukinda valley, Orissa, India. Ind Minerals 50:271–290

Beus AA (1976) Geochemistry of the lithosphere. MIR Publishers, Moscow 366 pp

Bhattacharjee J, Fareeduddin Jain SS (1993) Tectonic setting, petrochemistry and tungsten metallogeny of the Sewariya granite in the South Delhi fold belt, Rajasthan. J Geol Soc Ind 42:3–16

Bhattacharya HN (2004) Analysis of sedimentary succession hosting the Paleoproterozoic Zawar zinc-lead sulphide deposit, Rajasthan, India. In: Deb M, Goodfellow WD (eds) Sediment-hosted lead-zinc sulphide deposits. Narosa Publishing House, Delhi, pp 350–361

Bhoskar KG, Chande VD (1979) Tungsten mineralisation in the Sakoli Group of rocks in Nagpur and Bhandara district, Maharashtra. Mineralisation associated with acid magmatism. In: Proceedings of IGCP-26 workshop, Nagpur, vol 13, pp 134–141 (Geol Surv Ind)

Boyle RW (1979) The geochemistry of gold and its deposits. Geol Surv Can Bull 280:584 pp

Brimhall GH Jr, Lewis CJ (1992) Bauxite and laterite ores. Encyclopedia of Earth System Science, vol 1. Academic Press, London, pp 321–335

British Geological Survey (BGS) (2015) World mineral production 2009–2013. Nottingham, pp 79

Castor SB (2008) Rare earth deposits of North America. Resour Geol 58:337–347

Chakraborty KL, Chakraborty TL (1976) Source and fixation of nickel in the nickeliferous limonite profile of Sukinda valley, Orissa. J Geol Soc Ind 17(2):186–193

Chakraborty KL, Chakraborty TL (1984) Geological features and the origin of chromite deposits of Sukinda valley, Orissa, India. Mineral Deposita 19:256–265

Chao ECT, Tatsumoto M, Mackee EH (1995) Caledonian subduction, repeated activation and multiple episodes of mineralization of the Bayen Obo REE-Fe-Nb ore deposit, Inner Mongolia, China. Global Tectonics Metallogeny 5(1–2):37–39

Chatterjee A (1964) Geology, mineralogy and genesis of iron ores of some deposits of the Bailadila Range, Bastar district, M.P. Q J Geol Min Met Soc Ind 36:12–57

Chattopadhyay S, Saha AK (1998) Lower Proterozoic gold-sulfide mineralization in Pular-Parsori area, Maharastra, with a probable acid magmatic linkage (Abstr vol). UNESCO IUGS-IGCP 368, Bhubaneswar, India

Chattopadhyay B, Mukhopadhyay AK, Singhal RK (1982) Post-Erinpura acid magmatism in Sirohi, Rajasthan and its bearing on tungsten mineralization. In: Symposium metallogeny of the Precambrian, IGCP project 91, Bangalore, pp 115–132

Chernyshev IV, Safonov Yu G (1988) The source and age of mineralization. In: Chukhrov FV et al (eds) Kolar Goldfield, India. Moscow, Nauka, pp 187–195 (in Russian)

Chinnaswamy SS, Mishra B (2013) Greenstone metamorphism, hydrothermal alteration and gold mineralization in the genetic context of the granodiorite-hosted gold deposit at Jonnagiri, Eastern Dharwar craton, India. Econ Geol 108:1015–1036

Cox DP, Singer DA (1986) Mineral deposit models. U S Geol Surv Bull 1693:379 pp

Crocket JH (2002) Platinum group element geochemistry of mafic and ultramafic rocks. In: Cabri LJ (ed) Geology, geochemistry, mineralogy and mineral beneficiation of platinum group elements, vol 54. Montreal, Quebec, pp 177–210 (Can Inst Mining Metall and Petroleum)

Crookshank H (1963) Geology of South Bastar and Jeypore from Boiladila range to Eastern Ghats. Geol Surv Ind Mem 87:149 pp

Dasgupta S, Banerjee H, Fukuoka M, Bhattacharya PK, Roy S (1990) Petrogenesis of metamorphosed manganese deposits and the nature of the precursor sediments. Ore Geol Rev 5:359–384

Dasgupta S, Roy S, Fukuoka M (1992) Depositional models for manganese oxide and carbonate deposits of the Precambrian Saussar Group, India. Econ Geol 87:1412–1418

Deb M (2014) Precambrian geodynamics and metallogeny of the Indian shield. Ore Geol Rev 57:1–28

Deb M, Bhattacharya AK (1980) Geological setting and conditions of metamorphism of Rajpura–Dariba polymetallic ore deposit, Rajasthan, India. In: Proceeding of 5th quadrennial IAGOD symposium, Schweizerbartse Verlag, pp 679–697

Deb M, Joshi A (1984) Petrological studies on two East Coast bauxite deposits of India and implications on their genesis. Sed Geol 39:121–139

Deb M, Sehgal U (1997) Petrology, geothermobarometry and C-O-H-S fluid compositions in the environs of Rampura-Agucha Zn-(Pb) ore deposit, Bhilwara District, Rajasthan. Proc Ind Acad Sci Earth Planet Sci 106:343–356

Deb M, Pal T (2004a) Geology and genesis of base metal sulphide deposits in the Dariba-Rajpura-Bethumni belt, Rajasthan, India in the light of basin evolution. In: Deb M, Goodfellow WD (eds) Sediment-hosted lead-zinc sulphide deposits. Narosa Publication House, Delhi, pp 304–327

Deb M, Thorpe RI (2004b) Geochronological constraints in the Precambrian geology of Rajasthan and their metallogenic implications. In: Deb M, Goodfellow WD (eds) Sediment-hosted lead-zinc sulphide deposits. Narosa Publishing House, Delhi, pp 246–263

Deb M, Thorpe RI, Cumming GL, Wagner PA (1989) Age, source and stratigraphic implications of lead isotope data for conformable sediment-hosted base metal deposits in the Proterozoic Aravalli-Delhi orogenic belt, northwestern India. Precambrian Res 43:1–22

Deb M, Chattopadhyay A, Deol S (2008) Conditions and timing of gold mineralization in Gadag schist belt, constrained by structural and fluid inclusion data. In: Pre-workshop international field workshop gold metallogeny in India, Delhi University & NGRI, Hyderabad, pp 90–96

Dekate YG (1979) Tungsten ore deposits of India and their genesis. Mineralisation associated with acid magmatism. In: Proceedings of IGCP-26 workshop, Nagpur, vol 13, pp 25–30 (Geol Surv India)

Deol S, Deb M, Chattopadhyay A (2010) Geology of the Bhukia-Jagpura prospect, Banswara district, SE Rajasthan, India and a model for gold mineralization. In: Deb M, Goldfarb RJ (eds) Gold metallogeny: India and beyond. Narosa Publishing House, New Delhi, pp 234–255

Deol S, Chattopadhyay A, Deb M (2014) Deformation and metamorphism of gold-sulphide lodesin the Bhukia-Jagpura gold prospect, Rajasthan: implications for ore genesis. J Earth Syst Sci 123(1):1–13

Dill HG (2010) The "chessboard" classification scheme of mineral deposits: mineralogy and geology from aluminium to zirconium. Earth Sci Rev 100:1–420

Dunn JA (1937) The mineral deposits of eastern Singhbhum and surrounding areas. Geol Surv Ind Mem 69(1)

Emmons WH (1937) Gold deposits of the world. McGraw Hills Book Co, New York 562 pp

Emsbo P (2000) Gold in Sedex deposits: a new look. Rev in Econ Geol 13:427–431

Fareeduddin BJ (2000) Tectonics and metallogeny along the Phulad Lineament Zone, Rajasthan. In: Deb M (ed) Crustal evolution and metallogeny in the northwestern Indian shield. Narosa Publishing House, New Delhi, pp 395–416

Fermor LL (1909) The manganese ore deposits of India. Geol Surv Ind Mem 37:1–4

Fleet ME, Tronnes RG, Stone WE (1996) Partitioning of platinum group elements (Os, Ir, Ru, Pt, Pd) and gold between sulfide liquid and basalt melt. Geochim Cosmochim Acta 60:2397–2412

French B (1973) Mineral assemblages in diagenetic and low grade metamorphic iron formations. Econ Geol 68:1063–1074

Gandhi SM (2000) Ancient mining and metallurgy in Rajasthan. In: Deb M (ed) Crustal evolution and metallogeny in northwestern Indian shield. Narosa Publishing House, New Delhi, pp 29–48

Garrels RM, Christ CL (1965) Solutions, minerals and equilibria. Harper and Row, 450 pp

Genkin AD, Safonov Yu G, Bronikhin VA, Krishna Rao B, Vasudev VN (1988) Mineralogy and geochemistry of the Kolar Goldfield. In: Chukhrov FV et al (eds) Kolar Goldfield, India. Moscow, Nauka, pp 94–156 (in Russian)

Geological Survey of India (1994) Detailed information on bauxites in India. GSI, New Delhi

Geological Survey of India (1995) Geological map of Dariba-Rajpura-Bethumni belt, Rajsamand district, Rajasthan, GSI, Calcutta, 1:50,000

Geological Survey of India (GSI) (1998) Geological map of India, Calcutta, 1:2,000,000

Glasby GP (1988) Manganese deposition through geological time: dominance of the post-Eocene deep-sea environment. Ore Geol Rev 4:135–144

Golani PR (2008) Evidence of regional scale mantle flux in the Paleoproterozoic gold- and sulfide-bearing carbonate rocks in southeastern Rajasthan, western India. In: Pre-workshop, international field workshop 'Gold Metallogeny in India', pp 117–118

Golani PR, Pandit MK (1999) Evidence of epithermal activity and gold mineralization in Newania carbonatites, Udaipur district, Rajasthan. J Geol Soc Ind 54(3):251–257

Gross GA (1973) The depositional environment of the principal types of Precambrian iron formations. In: Genesis of Precambrian iron and manganese deposits, vol 9. UNESCO, Earth Sciences, Paris, pp 15–21

Grover AK, Verma RG (1993) Gold mineralization in the Precambrian (Bhukia area) of southeastern Rajasthan—a new discovery. J Geol Soc Ind 42:181–188

Groves DI, Goldfarb RJ, Robert F, Hart CJR (2003) Gold deposits in metamorphic belts: overview of current understanding, outstanding problems, future research, and exploration significance. Econ Geol 98:1–29

Gupta A (1975) A study of geology and ore mineralization at Lawa area, district Singhbhum, Bihar. Unpublished PhD thesis, Jadavpur University, Calcutta, India, 245 pp

Gupta A (2010) Gold mineralization in the eastern segment of Indian Precambrian shield: a review. In: Deb M, Goldfarb RJ (eds) Gold metallogeny: India and beyond. Narosa Publishing House, New Delhi, pp 256–280

Gutzmar J, Beukes NJ (1998) The manganese formations of the Neoproterozoic Penganga Group, India—revision of an enigma. Econ Geol 93:1091–1102

Gutzmar J, Beukes NJ (2000) The manganese formations of the Neoproterozoic Penganga Group, India—revision of an enigma—a reply (to Supriya Roy's discussion). Econ Geol 95:239–240

Haldar SK (2004) Grade and tonnage relationships in sediment-hosted lead-zinc sulphide deposits of Rajasthan, India. In: Deb M, Goodfellow WD (eds) Sediment-hosted lead-zinc sulphide deposits. Narosa Publishing House, Delhi, pp 264–272

Hamilton JV, Hodgson CJ (1986) Mineralization and structure of the Kolar Goldfield, India. In: Macdonald AJ (ed) Proceedings of Gold'86, Toronto, pp 270–283

Hazarika P, Upadhyay D, Mishra B (2013) Contrasting geochronological evolution of the Rajpura-Dariba and Rampura-Agucha metamorphosed Zn–Pb deposit, Aravalli-Delhi Belt, India. J Asian Earth Scs 73:429–439

Heron AM (1953) The geology of central Rajputana. Mem Geol Surv Ind 78:389 pp

Hitzman MW (2002) Iron oxide-Cu-Au deposits: what, where, when and why? In: Porter TM (ed) Hydrothermal iron oxide copper-gold and related deposits: a global perspective, vol 1. PGC Publishing, Adelaide, pp 9–25

Hitzman MW, Reynolds NA, Sangster DF, Allen CR, Carman CE (2003) Classification, genesis and exploration guides for non-sulfide zinc deposits. Econ Geol 98:685–714

Indian Bureau of Mines (2005) IBM Minerals Yearbook, Nagpur

Indian Bureau of Mines (2012) Statistical profile of minerals 2010–2011. Nagpur, pp 127

Institute MI (2011) Gold. Retrieved from http://www.eoearth.org/view/article/153052

Ishihara S, Hua R, Hosino M, Murakami H (2008) REE abundance and REE minerals in granitic rocks in the Nanling Range, Jiangxi Province, Southern China and the generation of the REE-rich weathered crust deposits. Resour Geol 58:355–372

Jafri SH, Khan N, Ahmed SM, Saxena R (1983) Geology and geochemistry of Nuggihalli schist belt, Dharwar craton, Karnataka, India. Mem Geol Soc Ind 4:110–120 (Naqvi SM, Rogers JJW (eds) Precambrian of South India)

Jagannatha Rao J, Krishnamurthy CV (1981) Some observations on the mineralogy and geochemistry of Hazaridadar and Raktidadar plateaus, Amarkantak area, M.P., India. In: Lateritisation processes. Oxford and IBH Publishing Co, New Delhi, pp 89–108

Jain SS, Mohan M (1984) Prospects of tungsten mineralization in Kuhi-Chapegarhi area, Nagpur district, Maharashtra. Rec Geol Surv Ind 113(6):141–149

Jain SS, Mohan M (1988) Tungsten mineralization in Khobna area, Nagpur district, Maharashtra. Rec Geol Surv Ind 117(3–8):29–34

Jain SC, Nair KKK, Yedekar DB (1995) Geology of Son-Narmada-Tapti lineament zone in Central India. Geol Surv Ind (Nagpur) 10:1–154

Jones HC (1934) The iron ore deposits of Bihar and Orissa. Geol Surv Ind Mem 63(2):167–302

Kasting JF (1993) Earth's early atmosphere. Science 259:920–926

Kerrich R (1993) Perspectives on genetic models for lode gold deposits. Mineral Deposita 28:362–365

Khan MW, Bhattacharya TK (1993) A reappraisal of the stratigraphy of Bailadila Group, Bacheli, Bastar district, M.P. J Geol Soc Ind 42(6):549–562

Kinney GL (1986) The mineral industry of India. In: Bureau of mines/minerals YB area reports: International, vol 3, pp 395–411

Klein C (1973) Changes in mineral assemblages with metamorphism of some banded Precambrian iron formations. Econ Geol 68:1075–1088

Knight J, Lowe J, Joy S, Cameron J, Merrillees J, Nag S, Shah N, Dua G, Jhala K (2002) The Khetri copper belt, Rajasthan: iron oxide-copper-gold terrane in the Proterozoic of NW India. In: Porter TM (ed) Hydrothermal iron oxide-copper-gold and related deposits, a global perspective, vol 2. Australian Mineral Foundation, Adelaide, pp 321–341

Kolb J, Rogers A, Meyer F, Vennemann T (2004) Development of fluid conduits in the auriferous shear zones of the Hutti gold mine, India: evidence of spatially and temporally heterogeneous fluid flow. Tectonophys 378:65–84

Krienitz MS, Trumbull RB, Hellmann A, Kolb J, Meyer FM, Wiedenbeck M (2008) Hydrothermal gold mineralization at the Hirabudini gold mine, India: constraints on fluid evolution and fluid sources from boron isotopic compositions of tourmaline. Mineral Deposita 43:421–434

Krishna Rao B, Srinivasan R, Ramchandra BL (1982) Mode of occurrence and origin of manganese ores of Shimoga district, Karnataka. J Geol Soc Ind 23:226–235

Krishna Rao B, Satish PN, Sethumadav MS (1989) Syngenetic and epigenetic features and genesis of the bauxite-bearing laterite of Boknur-Nagve plateau, Belgaon district, Karnataka. J Geol Soc Ind 34:46–60

Krishnamurthi R, Prabhakar T (2000) The nature, fluid characteristics and genesis of tungsten mineralization in Rajasthan: a brief review. In: Deb M (ed) Crustal evolution and metallogeny in the northwestern Indian shield. Narosa Publishing, House, New Delhi, pp 417–429

Krishnan MS (1935) Lateritisation of khondalite. Rec Geol Surv Ind 68(4):392–394

Krogstad EJ, Balakrishnan S, Mukhopadhyaya DK, Rajamani V, Hanson GN (1989) Plate tectonics 2.5 billion years ago: evidence at Kolar, South India. Science 243:1137–1340

Laznicka P (1993) Precambrian empirical metallogeny. In: Precambrian lithologic associations and metallic ores, Pt b. Elsevier, Amsterdam, pp 997–1622

Lindgren W (1933) Mineral deposits, McGraw Hill, New York, p 930

Maithani PB, Nagar RK (1999) Rare metals and rare earth pegmatites of western India. Expl Res Atom Minerals 12:101–131

Manikyamba C, Naqvi SM, Rammohan M, Gnaneshwar Rao T (2004) Gold minerlization and alteration of Penakacherla schist belt, India: constraints on Archean subduction and fluid processes. Ore Geol Rev 24:199–227

Mishra B, Panigrahi MK (1999) Fluid evolution in the Kolar gold field: evidence from fluid inclusion studies. Mineral Deposita 34:173–181

Mishra B, Bernhardt H-J (2009) Metamorphism, graphite crystallinity, and sulfide anatexis of the Rampura–Agucha massive sulfide deposit, northwestern India. Mineral Deposita 44:183–204

Mishra B, Upadhyay D, Bernhardt H-J (2006) Metamorphism of the host and associated rocks at the Rajpura-Dariba massive sulfide deposit, Northwestern India. J Asian Earth Scs 26:21–37

Mondal SK, Baidya TK, Gururaja Rao KN, Glascock MD (2001) PGE and Ag mineralization in a breccias zone of the Precambrian Nuasahi ultramafic-mafic complex, Orissa, India. Can Mineral 39:979–996

Mondal SK, Ripley EM, Li Chusi, Frei R (2006) The genesis of Archean chromitites from Nuasahi and Sukinda massifs in the Singhbhum craton, India. Precambrian Res 148:45–66

Mukherjee R, Mondal SK, Rosing MT, Frei R (2010) Compositional variations in the Mesoarchean chromites of the Nuggihalli schist belt, Western Dharwar Craton (India): potential parental melts and implications for tectonic setting. Contrib Mineral Petrol 160:865–885

Mukherjee R, Mondal SK, Frei R, Rosing MT, Waight TE, Zhong H, Ravindra Kumar GR (2012) The 3.1 Ga Nuggihalli chromite deposits, Western Dharwar craton (India): geochemical and isotopic constraints on mantle sources, crustal evolution and implications for supercontinent formation and ore mineralization. Lithos 155:392–409

Mukhopadhyay J, Chaudhuri AK, Chanda SK (1997) Deepwater manganese deposits in mid- to late Proterozoic Penganga Group of the Pranhita-Godavari valley. In: Nicholson K et al (eds) Manganese mineralization, geochemistry and mineralogy of terrestrial and marine deposits, vol 119. Geological Society London Special Publications, pp 105–115

Mulay VV (2001) Manganese deposits of central India—a review. Geol Surv Ind 64:341–352

Murao S, Deb M, Tagaki T, Seki Y, Pringle M, Naito K (2000) Geochemical and geochronological constraints for tin-polymetallic mineralization in Tosham area, Haryana, India. In: Deb M (ed) Crustal evolution and metallogeny in the northwestern Indian shield. Narosa Publishing House, New Delhi, pp 430–442

Murao S, Deb M, Furuno M (2008) Mineralogical evolution of indium in high grade tin-polymetallic hydrothermal vein. A comparative study from Tosham, Haryana state, India and Goka, Neigi district, Japan. Ore Geol Rev 33(3–4):490–504. doi:10.1016/j.oregeorev.2007.02.004

Murthy PSN, Chatterjee AK (1995) The origin of the iron ore deposits of Donimalai area of Sandur schist belt, Karnataka state, India. J Geol Soc Ind 45:19–31

Naganna C (1977) Manganese ore deposits of North Canara. In: Proceeding of symposium on geological exploring mining mineral processing and metallurgical ferrous and ferroalloy minerals, Bangalore University, pp 110–113

Nair KK, Jain SC, Yedekar DB (1995) Stratigraphy, structure and geochemistry of the Mahakoshal greenstone belt. In: Sinha-Roy S, Gupta KR (eds) Continental crust of northwestern and central India. Geol Soc Ind Mem 31:403–433

Narayanaswami S (1962) Manganese ores. Bull Geol Surv Ind A Econ Geol 1–114

Narayanswami S, Ziauddin M, Ramachandra AV (1960) Structural control and localization of the gold-bearing lodes, Kolar gold field, India. Econ Geol 55:1429–1459

Nayak VK (1980) The manganese ores of Kajlidongri, district Jhabua, Madhya Pradesh, India. In: Varentsov IM, Grasselly G (eds) Geology and geochemistry of manganese, vol 2. Schweizerbart, Sttutgart, pp 265–277

Newberry RJ, Swanson SE (1986) Scheelite skarn granitoids: an evaluation of the roles of igneous source and process. Ore Geol Rev 1:57–81

Nijagunappa R, Damodaran KT, Somasekar B, Naganna C (1986) Genesis of the chromite deposits of Karnataka craton, Southern India. In: Carter B et al (eds) Metallogeny of basic and ultrabasic rocks. Theophrastus Publications SA, Athens, pp 190–221

Omana PK, Santosh M (1996) Laterite profile geochemistry in outlining supergene gold deposits: a case study from Nilambur, Kerala. J Geol Soc Ind 48(2):139–150

O'Neil HSC, Dingwell DB, Borisov A, Spattel B, Palme H (1995) Experimental petrochemistry of some high siderophile elements at high temperatures and some implications for core formation and mantle's early history. Chem Geol 120:255–273

Pal N, Mishra B (2003) Alteration geochemistry and fluid inclusion characteristics of the greenstone hosted gold deposits of Hutti, Eastern Dharwar craton, India. Mineral Deposita 37:722–736

Pal DC, Mishra B, Bernhardt H-J (2007) Mineralogy and geochemistry of pegmatite-hosted Sn-, Ta-Nb-, and Zr-Hf-bearing minerals from the southeastern part of the Bastar-Makangiri pegmatite belt, central India. Ore Geol Rev 30(1):30–55

Pal DC, Barton MD, Sarangi AK (2009) Deciphering a multistage history affecting U-Cu (–Fe) mineralization in the Singhbhum Shear Zone, eastern India, using pyrite textures and compositions in the Turamdih U–Cu (–Fe) deposit. Mineral Deposita 44(1):61–80

Patel SK, Jena BK, Das KC (2005) Manganiferous formations of Bonai-Kendujhar belt and a proposal for revision of its lithostratigraphy. In: Proceeding of seminar in mineral and energy resources of Eastern and Northeastern India, MGMI, Kolkata, pp 163–179

Pathan AM, Govindaiah S (1986) Vanadiferous titanomagnetite and associated sulphide occurring in a layered complex from Masanikere and Tagdur-Nuggihalli, Karnataka. Q J Geol Min Met Soc Ind 58(4):252–259

Poulsen KH, Robert F, Dubé B (2000) Geological classification of Canadian gold deposits. Geol Surv Canad Bull 540:106 pp

Radhakrishna BP (1996) Mineral resources of Karnataka. Geol. Soc. India, Bangalore, 471 pp

Rajamani V (1996) Gold mineralization in the Kolar schist belt (Abstr vol). In: National workshop exploration and exploitation of gold occurrences of India, Hyderabad, December 2–4, pp 255–257

Ram Mohan M, Sarma DS (2010) Geochemistry and evolution of the Hutti-Maski greenstone belt: link between Archean crustal evolution and gold mineralization. In: Deb M, Goldfarb RJ (eds) Gold metallogeny: India and beyond. Narosa Publishing House, New Delhi, pp 191–205

Ramakrishnan M (1990) Crustal development in southern Bastar, Central Indian craton. Geol Surv Ind 28:44–66

Raman PK (1976) Bauxite deposits of Ananthagiri, Vizag district, Andhra Pradesh. J Geol Soc Ind 17:236–244

Riyaz Ulla MS, Pathan A, Maaskant P (1996) The characterization of metamorphic facies of metabasalts of Hutti greenstone belt. J Geol Soc Ind 47:547–554

Rogers AJ, Kolb J, Meyer FM, Vennemann T (2013) Two stages of gold mineralization at Hutti mine, India. Mineral Deposita 48:99–114
Roy S (1966) Syngenetic manganese formations of India. Jadavpur University Press, Calcutta, p 219
Roy S (1981) Manganese deposits. Academic Press, London
Roy AB (1995) Geometry and evolution of superposed folding in Zawar lead-zinc mineralized belt, Rajasthan. Proc Ind Acad Scs (Earth Planet Sci) 104:349–371
Roy S (1997) Genetic diversity of manganese deposition in the terrestrial geological record. In: Nicholson K, Hein JR, Buhn B, Dasgupta S (eds) Manganese mineralization: geochemistry and mineralogy of terrestrial and marine deposits (Geol Soc London Spec Publ No 119), pp 5–27
Roy S (2006) Sedimentary manganese metallogenesis in response to the evolution of the Earth system. Earth Sci Rev 77:273–305
Safonov Yu G, Krishna Rao B, Vasudev VN (1988) The structure of the Kolar goldfield. In: Chukhrov FV et al (eds) Kolar Goldfield, India. Nauka, Moscow, pp 761–781
Saha AK (1994) Crustal evolution of Singhbhum-North Orissa, Eastern India. Geol Soc Ind Mem 27:341 pp
Santosh M, Nadear S, Javoy M (1995) Stable isotopic evidence for the involvement of mantle-derived fluids in Wynad gold mineralization, South India. J Geol 103:718–728
Sarkar SC (1984) Geology and ore mineralization of the Singhbhum copper-uranium belt. Jadavpur University Press, Eastern India 263 pp
Sarkar SC (1988) Genesis and evolution of the ore deposits in the early Precambrian greenstone belts and adjacent high grade metamorphic terrains of peninsular India—a study in similarity and contrast. Precambrian Res 39:107–130
Sarkar SC (2000) Orogenic gold mineralization: the Indian scenario. In: 31st International Geological Congress, Rio de Janeiro, Brazil, pp 11–7
Sarkar SC (2010) Gold mineralisation in India: an introduction. In: Deb M, Goldfarb RJ (eds) Gold metallogeny in India and beyond. Narosa Publ House, New Delhi, pp 95–122
Sarkar SC, Deb M (1974) Metamorphism of the sulphides of the Singhbhum copper belt, India—the evidence from the ore fabric. Econ Geol 68:1282–1293
Sarkar SC, Dasgupta S (1980) Geological setting and transformation of sulfide deposits in the northern part of the Khetri copper belt, Rajasthan, India: an outline. Mineral Deposita 15:117–137
Sarkar SC, Banerji S (2004) Carbonate-hosted lead-zinc of Zawar, Rajasthan in the context of world scenario. In: Deb M, Goodfellow WD (eds) Sediment-hosted lead-zinc sulphide deposits. Narosa Publishing House, Delhi, pp 328–349
Sarkar SC, Gupta A (2012) Crustal evolution and Metallogeny in India. Cambridge Univ Press, Delhi, pp 840
Sarkar SC, Dwivedy KK, Das AK (1995) Rare earth deposits in India—an outline of their type, distribution, mineralogy-geochemistry and genesis. Global Tectonics Metallogeny 5(1&2): 53–61
Sarkar SC, Kabiraj S, Bhattacharyya S, Pal AB (1996) Nature, origin and evolution of the granitoid-hosted early Proterozoic copper-molybdenum mineralization at Malanjkhand, Central India. Mineral Deposita 31:419–431
Sarma DS, McNaughton N, Fletcher IR, Groves DI, Mohan MR, Balaram V (2008) Timing of gold mineralization in the Hutti gold deposit, Dharwar craton, south India. Econ Geol 103:1715–1727
Scott SD, Barnes HL (1972) Sphalerite-wurtzite equilibria and stoichiometry. Geochim Cosmochim Acta 36:1275–1295
Seetharam R (1986) Argentiferous roquesite (CuInS2) from the Tosham tin prospect, Bhiwani district, Haryana. J Geol Soc Ind 28:21–28
Sillitoe RH (1991) Intrusion-related gold deposits. In: Foster RP (ed) Gold metallogeny and exploration, Blackie & Sons, Glasgow, pp 165–209

Singer DA (1993) Development of grade and tonnage models for different deposit types. In: Kirkham RV, Sinclair WD, Thorpe RI, Duke JM (eds) Mineral deposit modeling, GAC Spec Pap 40, pp 21–30

Singh RN, Sharma RR, Kishku SR, Soney Kurien P, Lahiri TC (2005) Gold mineralization in Singhbhum greenstone-granite terrain, Jharkhand. In: Proceeding of seminar in mineral and energy resources of eastern and northeastern India, Kolkata, pp 243–252

Siva Siddaiah N, Rajamani V (1989) The geologic setting, mineralogy, geochemistry and genesis of gold deposits of the Archean Kolar schist belt, India. Econ Geol 84:2155–2172

Smith MP, Handerson (2000) Preliminary fluid inclusion constraints on fluid evolution in the Bayen Obo Fe-REE-Nb deposit, Inner Mongolia. Econ Geol 95:1371–1388

Som SK, Joshi R (2002) Chemical weathering of serpentinite and Ni enrichment in Fe oxide at Sukinda area, Jajpur district, Orissa, India. Econ Geol 97(1):165–172

Srikantappa C (2001) Composition and source of deep crustal fluids and their role in the evolution of charnokitic granulites of Southern India. Geol Surv Ind 55:15–30

Stein HJ, Hanna JL, Zimmerman A, Markey RJ, Sarkar SC, Pal AB (2004) A 2.5 Ga porphyry Cu-Mo-Au deposit at Malanjkhand, Central India: implications for late Archean continental assembly. Precambrian Res 134:189–226

Tardy Y (1997) Petrology of the laterites and tropical soils. Oxford and IBH Publishing Co, New Delhi 408 p

Tripathi S (2008) Geological characterization of the carbonate-hosted polymetallic prospect at Imalia, Mahakoshal Belt, Central India. Unpublished PhD thesis, University of Delhi, 315 pp

Valeton I (1981) Bauxites on peneplained metamorphic and magmatic rocks, on detrital sediments and on karst topography, their similarities and contrasts of genesis. Lateritisation processes. Oxford and IBH Publishing Co, New Delhi, pp 15–23

Vasudev VN, Chadwick B, Nutman AP, Hegde GV (2000) Rapid development of the late Archean Hutti schist belt, northern Karnataka: implications of new field data and SHRIMP U/Pb zircon ages. J Geol Soc Ind 55:29–540

Venkatesh AS (2001) Geochemical signatures and auriferous implications in Sonakhan greenstone belt. Geol Surv Ind 55:219–228

Verma OP, Pandian SM (1983) Conceptual model for exploration of tungsten deposits in India. Their integrated exploitation and economics. Proc Ind Nat Sci Acad 50(5):523–540

Verma RG, Golani PR (1999) Investigation for gold in Bhukia-Jagpura area, Banswara district, Rajasthan. Rec Geol Surv Ind 130(7):23–24

Wedepohl KH (1971) Geochemistry. Holt Rinehart and Winston, New York 231 pp

Wilkinson JJ (2003) On diagenesis, dolomitisation and mineralisation in the Irish Zn-Pb orefield. Mineral Deposita 38:968–983

Wood SA, Samson IM (2000) The hydrothermal geochemistry of tungsten in granitoid environments: relative solubilities of ferberite and scheelite as a function of T, P, pH and m_{NaCl}. Econ Geol 95:143–182

Wu C (2008) Byan Obo controversy: carbonatites versus iron oxide-Cu-Au = (REE-U). Res Geol 58:348–354

Zhou MF, Robinson PT (1997) Origin and tectonic environment of podiform chromite deposits. Econ Geol 92:259–262

Ziauddin M, Narayanswami S (1974) Gold resources of India. Bull Geol Surv Ind, Sr A (Econ Geol), 38(I):186 pp

Chapter 4
Nonmetals, Industrial Minerals and Gemstones

Abstract Nonmetallic and industrial minerals are the essential raw materials for a number of industries. Gemstones have been in demand throughout much of human history. Nonmetallic and industrial minerals are divided into the following groups: refractory minerals, fertilizer minerals, minerals used in cement industry, chemical industry, electrical industry, glass and ceramic industry, abrasive industry, and those used as fillers and pigments, and as building stones. Refractory minerals are those that can, besides withstanding high temperatures (up to ~1500 °C), endure thermal and mechanical shocks. They are mostly used in the internal lining of metallurgical furnaces. They are categorized into three sub-types: acid, neutral, and basic, depending on their relationship with various kinds of slags and furnace linings. The acidic variety includes quartz, fire clay, ball clay, kyanite, sillimanite; the neutral variety includes chromite, graphite and asbestos, while the basic variety comprises magnesite, dolomite, and bauxite. India has got a reasonably good reserve of refractory minerals, being the largest repository of alumino-silicate minerals. The principal fertilizer mineral is a phosphate (phosphorite, apatite), now used mostly for the manufacture of superphosphate using H_2SO_4. Bulk of phosphate mineral deposits are sedimentary (-diagenetic) in origin. A number of phosphate deposits are in Rajasthan, followed by Madhya Pradesh, Uttarakhand, and West Bengal. However, India imports phosphate ore from West Asia. K-bearing fertilizer minerals occur in large quantities in Nagaur-Ganganagar basin, Rajasthan. Portland cement requires limestone, clay, and some 5% gypsum for its manufacture. India has a satisfactory reserve of these raw materials. Plaster of Paris is made with gypsum. Minerals needed in chemical industry are native sulfur, Fe-sulfides, barite, and fluorite. In fact, >80% of barite mined is used in making "heavy mud" for the oil drilling industry while fluorite's main use is as flux in metallurgical processes. Minerals used for thermo-electrical insulators comprise muscovite mica, asbestos, steatite, talc, vermiculite, and pyrophyllite. Li is extractable from lepidolite. Micas are available aplenty from the mica-pegmatites of Bihar, Rajasthan, and Andhra Pradesh. Glass and ceramic materials are in great demand for making quite a few things necessary for our everyday life, as well as in industries. We have necessary reserves of quartz, feldspar, and clays for this industry. Industrial abrasives vary in hardness and may be natural or manufactured. Hard abrasives are more

M. Deb and S.C. Sarkar, *Minerals and Allied Natural Resources and their Sustainable Development*, Springer Geology,
DOI 10.1007/978-981-10-4564-6_4

manufactured these days. Natural hard mineral matter for abrasives includes diamond (industrial variety or bort) and corundum. Natural soft abrasive materials are many. Building stones comprise granite gneisses, granites, marble, limestone, sandstone and slate. Charnockites and khondalites also make good building stones. Quality of a building stone is determined by its beauty and durability. India's annual business in this trade exceeds INR 10 billion. The most precious gemstones found in India are diamond, followed by ruby, sapphire and emerald. Diamonds are obtained from kimberlites and lamproites and related coarse sediments in central and south India. Ruby, sapphires, and emerald are found in the Kashmir Himalaya, Rajasthan, and Karnataka. Precious quality zircon (*gomed*) is found in the beach sands of peninsular India. The bulk of India's need of gemstones is, however, satisfied by other countries including South Africa, Sri Lanka, and Myanmar.

Keywords Industrial minerals · Refractory minerals · Phosphorite · Portland cement · Plaster of Paris · Mineral insulators · Heavy mud · Ceramic material · Abrasives · Gem minerals · Beach sands

4.1 Introduction

Nonmetallic and industrial minerals are essential raw materials for a variety of industries. Building stones and gemstones have been in great demand throughout much of human history. In fact, the first earth materials used by man must have been hard rocks, shaped to act as hunting tools and mineral pigments to color their bodies. In present industrialized developed nations, the use of nonmetallic and industrial rocks exceed that of metallic minerals. The situation in India is fast becoming so, as infrastructure development is going on at a frantic pace. In the following pages, these commodities are discussed at length, particularly with reference to the Indian situation, under three sub-heads: Nonmetals, Building stones and Gemstones.

4.2 Nonmetals

Stone implements, clays, and pigments were the earliest earth resources used by man. In the present day, production of certain nonmetallic minerals, industrial rocks and building stones, in particular the mineral aggregates, constitute the majority of natural resources used by modern societies in terms of their total output and value. These minerals provide the material support for several industries such as infrastructure, chemical, ceramic, fertilizer, refractories, etc. Unfortunately, necessary R&D studies have not been carried out on these important economic minerals in

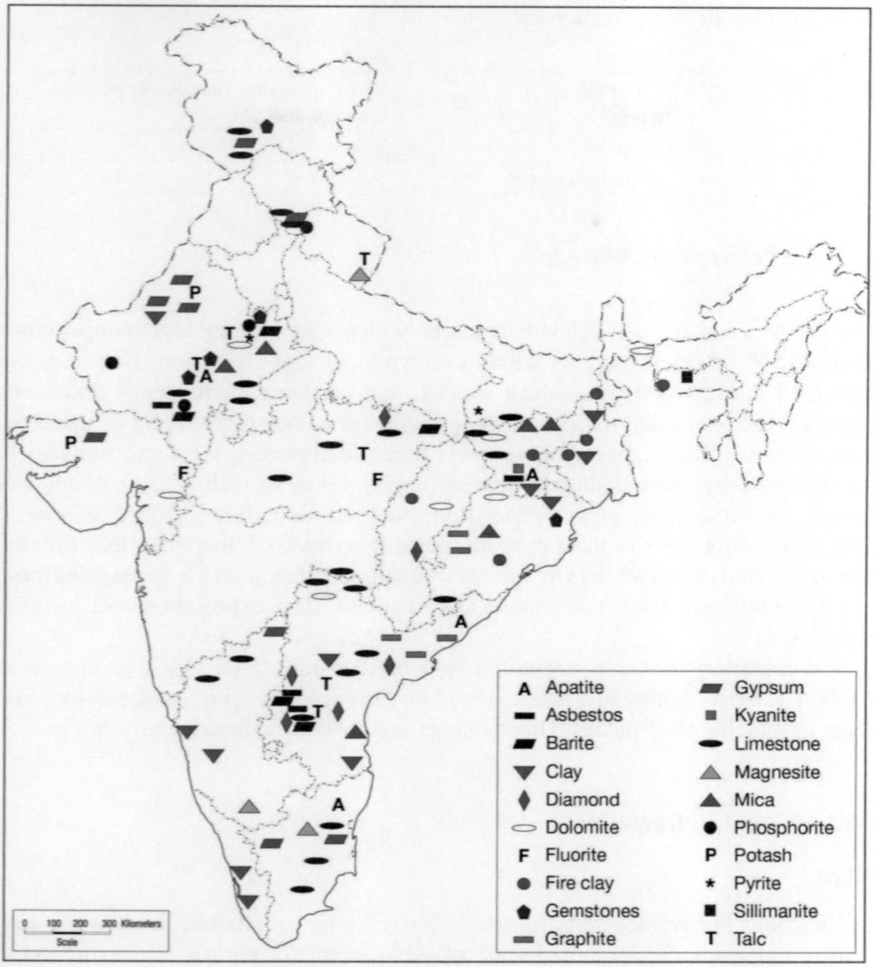

Fig. 4.1 Distribution of important nonmetallic mineral deposits, diamond and other gemstones in India

contrast to their metallic counterparts. As a result, adequate scientific literature on them as well as relevant reference textbooks in the field are wanting.

India is home to some of the largest deposits of these industrial minerals (cf. Deb 1980). The distribution of important nonmetallic deposits, diamonds, and gemstones is shown in a map of India (Fig. 4.1). These commodities are discussed in this chapter as raw materials for different industries. When a nonmetallic commodity finds use in more than one industry, the mineral is discussed at length in this chapter under that industry for which it is a main raw material.

Table 4.1 Classification of refractories

Basic	Neutral	Acidic
Magnesite	Chromite	Silica: quartzite, vein quartz etc. fire clay, ball clay
Dolomite	Graphite	Kyanite
Bauxite	Asbestos (minor)	Sillimanite

4.2.1 Refractory Minerals

Refractory minerals are defined as those which can sustain high temperatures around 1500 °C or even higher without showing any signs of fusion. They can also withstand thermal and mechanical shocks, and are resistant to many chemicals. They are used for many purposes, the principal ones being the lining of metallurgical furnaces. Their other uses comprise lining incinerators, kilns and furnaces in ceramic industry, glass industry, cement-making kilns, as well as coke ovens and boilers used in electric power generation. Automobile industry, which is now in spurt, uses refractories in making spark plugs. It is expected, therefore, that with the rise of demand of refractories in various industries, research on the geology of these raw materials and their use should also increase. The expectation still remains unfulfilled.

The refractory minerals are commonly divided into three types, as shown in Table 4.1: basic, neutral and acidic, based on their reaction with furnace linings and slags in metallurgical plants. These linings are covered with refractory bricks.

4.2.1.1 Acid Refractories

Silica

Acidic Silica for refractory purposes is derived from quartzites, sandstones, vein quartz, and sands. Such quartzites are of metamorphic origin, sandstones and sand of sedimentary derivation and vein quartz from hydrothermal source. These above-mentioned mineral deposits occur in almost all the geological formations, starting from the Precambrian to the Recent.

Because of their heat resistant property, these refractories are widely used in the arches and crowns of furnaces. Open hearth furnaces, acid Bessemer converters and electric furnaces utilize silica bricks. The raw material should contain at least 96–98% SiO_2. Magnesia and alkalis must be as low as possible but iron oxide in limited quantities helps in acting as a binder. Silica bricks should have a specific gravity of 2.3–2.4 and can withstand temperatures up to 1650 °C under load.

Indian distribution

Quartzites:

Refractory-grade quartzites occur commonly in the geological formations belonging to the Dharwar, Aravalli, Kurnool, Cuddapah, and Vindhyan supergroups. State-wise, the major exploitable occurrences are as follows:

Bihar: In the Kharagpur hills, fine-grained, massive quartzite occurs, composed predominantly of quartz, with garnet, magnetite, biotite, and muscovite as minor phases. Silica content varies between 97.5 and 98%.

Jharkhand: Refractory grade quartzites occur in both Singhbhum districts, between Rakha Mines and Kendadih in the east, and near Chandil and Chaibasa in the west. Gangpur group of rocks in western Jharkhand and Odisha also have large quantites of such quartzites.

Karnataka: Quartzites consisting almost entirely of pure quartz occur in the Dharwar schist belts of Karnataka. Near Bhadravati alone, in the Shimoga schist belt, about 2 Mt of quartzites have been estimated. Refractory-grade quartzites also occur near Bangalore, Krishnarajasagar, Holalkere, Holenarsipur and in the neighborhood of Mysore city.

Andhra Pradesh: Ferruginous silica beds are found in the Chaiyar Group of rocks in Cuddapah and Kurnool formations occurring on the Oopalpad Plateau and in the vicinity of Yadakee. They have constrained use.

Sands from Jabalpur area in Madhya Pradesh are also used for refractory lining of furnaces.

Fire clay

Fire clays are refractory clays of sedimentary origin comprising structurally disordered kaolinite, with some mica or illite and some quartz. They are characterized by very low alkali content and small amounts of other constituents, such as, compounds of Fe, Ti, Ca, Mg, Li, and free silica. They are commonly very plastic but may also be nonplastic in nature. Interestingly, plasticity of fire clays increase when very finely ground. The SiO_2 content ranges between 45 and 55% and Al_2O_3 content varies between 30 and 45%. Bulk density is 1.98 and the porosity of normal fire clay bricks is 21.7. Refractoriness and plasticity are the two main properties needed for manufacture of refractory bricks. Plastic fire clays resemble ball clays, and are used as bond clays for the manufacture of saggers, glass pots, crucibles, mortars, and refractory cement. Fire clays that are nonplastic usually contain a high proportion of silica and are low in clays and aluminous minerals. The refractoriness of fire clays varies between 1400 and 1600 °C. Fire clays bricks are classified by ASTM as: Low heat duty (1515–1615 °C; with pyrometric cone equivalent (PCE = 19–28); Intermediate heat duty (up to 1650 °C; with PCE = 30); High heat duty (up to 1700 °C; with PCE = 32) and Super heat duty (beyond 1775 °C; PCE = 35). They also find use in pottery industry and in vitrified clay pipes.

Indian distribution

Common occurrence of fire clays are found in the coal measures of the Gondwana coal fields. Jharkhand is the largest repository of fire clays from the coal fields of Karanpura and Jharia. West Bengal comes next in importance, producing fire clays from the Raniganj coal field (cf. Fig. 6.14). Fire clays are generally restricted to Barakar Formation. Rajhara and Daltongunj coalfields are also important suppliers of plastic fireclay. Chhattisgarh also contributes to fire clay production, particularly from Korba coal field. Odisha also hosts good quality fire clay from Sambalpur and Denkanal districts. Fire clay from Belpahar is extensively used in steel plants. In some other states such as Karnataka and Tamil Nadu refractory kaolin, occurring in Precambrian formations, are regarded as alternatives of fire clays or refractory clays. Good refractory fire clay also occurs in Nizamabad, Asifabad, East Godavari, and Cuddapah districts of Andhra Pradesh. In Madhya Pradesh, highly plastic and refractory clays are known from Jabalpur, Betul, Bilashpur, Drug, Hoshangabad, and Satna districts. In Rajasthan, fire clays are distributed in Jaisalmer, Sawai Madhopur, and Bikaner districts. The average chemical composition of fire clays from Ranigunj and Belpahar is presented in Table 4.2.

Ball clays

Ball clays are grayish-white to creamish in color, very fine-grained, sometimes carbonaceous and are moderate to highly plastic in nature. They have high bonding capacity and tensile strength. The ball clays are derived from lumps or ball-shaped forms in which they occur in sediments. Mineralogically, they comprise kaolinite of poorly ordered crystal structure with varying amounts of quartz, feldspar, mica or illite. The chemical composition of ball clay is almost similar to kaolin or China clay, except that it is richer in silica and poor in alumina. Ball clays take shape with difficulty when used in casting. They are dewatered easily by filter pressing. Ball clays produce a vitreous substance on firing, but at a much lower temperature compared to kaolin. These clays find extensive use in bonding furnace sands and refractory materials. They are also used in pottery and other ceramic industries, as fillers in plastic and rubber, in insecticides and fungicides.

Ball clays in India occur mainly in states of Maharashtra, Rajasthan, Gujarat, Kerala, and Tripura. Several occurrences of ball clays are located near Mumbai and surrounding areas in Maharashtra and near Barmer in Rajasthan.

Table 4.2 Chemical composition of fire clays (after Gokhale and Rao 1973)

Constituents	Ranigunj	Belpahar (%)
SiO_2	50–60%	50
Al_2O_3	27–38%	31–33
Fe-oxides	1.2%	3.5
CaO	1.0%	0.42–0.61
MgO	0.30%	0.22–0.38
TiO_2	Traces	1.38
Alkalis	1.36–2.0%	0.34–0.44
Loss on ignition	9–14%	10.11

Alumino-silicate refractories

Kyanite and sillimanite belong to this class of refractories. They have similar chemical composition but different crystal structures and hence, different physical behavior. They can withstand temperatures around 1650 °C. The desirable chemical composition of aluminous refractories is: $Al_2O_3 \geq 56\%$, $SiO_2 = 42\%$, $Fe_2O_3 < 1\%$, $TiO_2 < 1.3\%$, $CaO + MgO$ not $> 0.1\%$.

Kyanite (Al_2SiO_5)

Kyanite is characterized by its white to bluish color, bladed to needle-like form, good cleavage, and varying hardness in different cleavage directions. It usually occurs in Al-rich crystalline schistose rocks formed under moderately high P-T conditions and not uncommonly associated with minerals, such as corundum, staurolite, and andalusite. The transparent crystals imbued with color tinges are used as semiprecious gemstones.

Kyanite in India is classified into three grades: Grade I, Al_2O_3 62–64%; Grade II, Al_2O_3 58–62%; Grade III, Al_2O_3 54–58%. Refractory properties of Indian kyanite are: PCE 35 or above; 1785 °C refractoriness under load and porosity 24–25%.

Indian distribution

India has the largest reserve of kyanite in the world. The state-wise distribution of this important refractory mineral is as follows:

Jharkhand: A number of quartz-kyanite bodies occur all along the northern boundary of the Singhbhum shear zone, west of Jamshedpur. Massive pegmatoid kyanite deposits occur at Lapsa Buru where a number of such veins with a total reserve of more than 2 million tons of kyanite are of great economic importance. The Lapsa Buru hill forms the central high portion of a long ridge, about 8 km in length, and rises about 300 m above the plains, where there were kyanite-producing quarries of Tata Iron and Steel Company (TISCO). The country rocks are garnetiferous mica schists with occasional staurolite within which there are two large patches of muscovite-biotite-chlorite schists, each with a prominent lense of cherty quartzite. The kyanite pegmatite bodies are located both inside the cherty quartzite and also in the enclosing schists (Fig. 4.2a) (Banerji 1981). The pegmatoid kyanite masses produce unusual fabric at places (Fig. 4.2b). Along the extension of the Lapsa Buru hill, toward east and southeast in Seraikela near Kharswan, massive kyanite rocks occur in association with aluminous mica schists and kyanite-quartz rocks. Kyanite has been quarried in recent past from the Dhalbhumgarh–Baharagora region. In and around Sini, in West Singhbhum district, coarse bladed kyanite occurs within mica schistsas segregation veins at several places.

Based on detailed study, Dunn (1929) suggested that kyanite veins in Singhbhum are produced by metamorphic segregation from the enveloping country rocks which represented a variety of aluminous (bauxitic) tuff or clay that was derived from weathering of pre-existing basaltic rocks. Later, Dunn and Dey (1942)

(a)

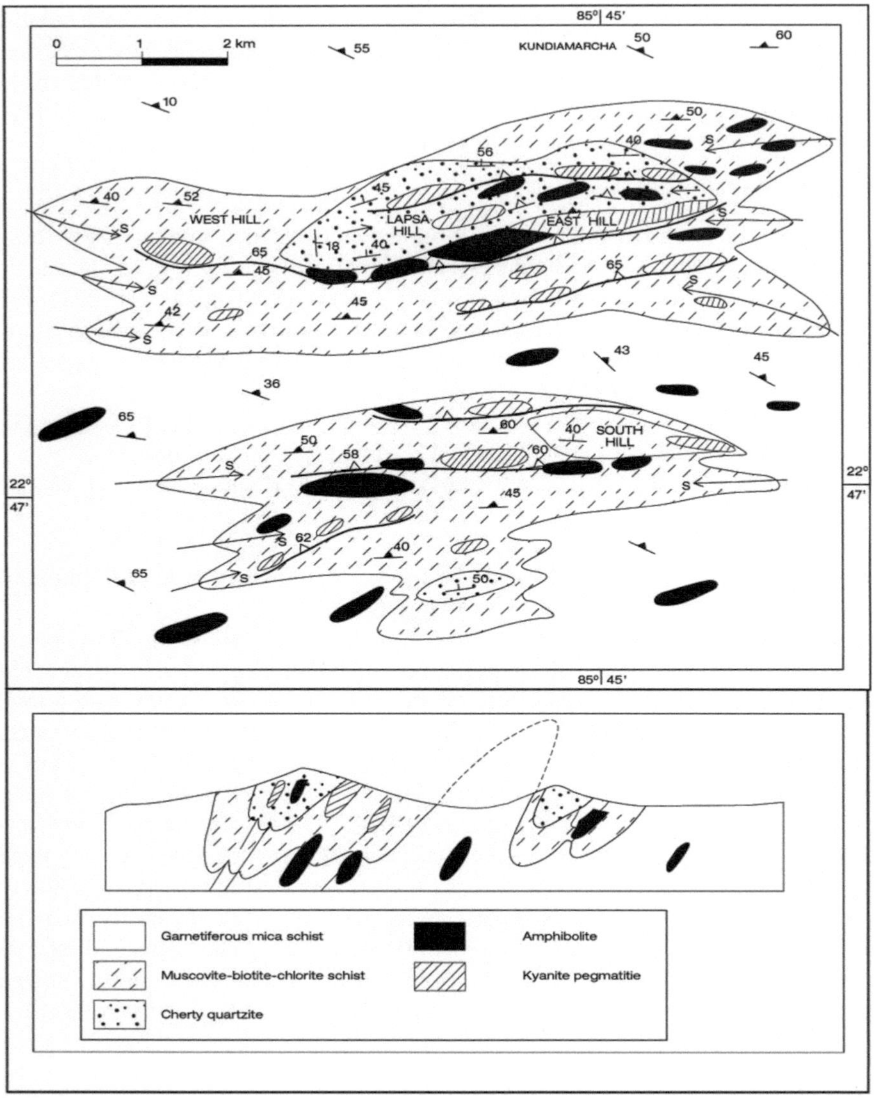

Fig. 4.2 **a** Geological map and a N–S section of Lapsa buru kyanite deposit, Singhbhum, Jharkhand (after Banerji 1981). **b** Elephant's egg texture in pegmatoid kyanite in Lapsaburu (Photo: Courtesy, Dr. A. Gupta)

Fig. 4.2 (continued)

inferred that hydrothermal emanations from granitic rocks within the shear zone aided the process of segregation. More recently, Banerji (1981) commented on the genesis of Lapsa Buru kyanite. According to him the kyanite appears to have segregated during a late phase of shearing when temperature around 450–500 °C and pressure around 5–6 kbar prevailed along the shear zone. Presence of tourmaline, topaz, and dumortierite in the kyanite veins indicate that boron and fluorine may have aided the segregation process. However, none of the existing views on the concentration of such a huge quantity of alumina locally, seems adequate at the moment.

Andhra Pradesh: Kyanite-quartz rocks occur in Nellore district of Andhra Pradesh. The deposits are located in mica schists of Nellore mica-belt, northwest of Saidapuram. Kyanite is found in pockets and lenticular bands extending over lengths up to 150 m. Kyanite also occurs intercalated with quartz schists and quartzites. In Khammam district also, there are several exploitable deposits of kyanite and garnet. The productive area, about 15 km^2 in extent, is underlain by garnetiferous kyanite mica rocks.

Rajasthan: Pockets and lensoids of quartz and kyanite, up to 1 m in size, occur randomly along the contact of pegmatites and quartz veins which have intruded biotite-garnet-kyanite schists in Dungarpur district. All these highly metamorphosed rock formations overlie the Banded Gneissic Complex (BGC). Contribution of pegmatite formation to the segregation of kyanite is a possibility.

Maharashtra: Large deposits of kyanite are also mined in Bhandara district.

Sillimanite (Al_2SiO_5)

The mineral, with gray to light brown or pale green color and vitreous luster, occurs in compact aggregates of radiating needles and other fibrous aggregates in high-grade metamorphic rocks. It is identified easily by its acicular needle-shaped crystal habit. Sillimanite is a product of high-grade metamorphism of aluminous rocks often occurring directly at the contact with igneous rocks. The mineral also occurs in crystalline schists in association with metamorphic minerals such as, andalusite, corundum, cordierite, and spinel. Indian sillimanite can be used as a refractory without calcinations since it exhibits very little expansion.

Indian distribution

Sillimanite deposits in India are widely distributed, particularly in the Precambrian crystalline complexes in Meghalaya, Rajasthan, Madhya Pradesh, Jharkhand, Odisha, Kerala, Andhra Pradesh, and Tamil Nadu.

Meghalaya: This state is endowed with nearly 70% of the total reserves of sillimanite in India. The largest reserve of sillimanite in India is located in the unique deposit at Sonapahar, within Shillong Meghalaya Gneissic Complex (SMGC) which has a U–Pb zircon age of ca. 1600 Ma (Yin et al. 2010). The mineral occurs in 21 scattered deposits in an area of 78 km^2 in the vicinity of Sonapahar, Nongpur, and Nongbain villages in the Nongatoin area of Khasi Hills in Meghalaya. The deposits are associated with alumina-rich rocks, such as cordierite-biotite-quartz-microcline gneiss and sillimanite-quartz schist metamorphosed to upper amphibolites to granulite facies and enveloped in granite. The total estimated reserves are at 250,000 tons (up to a depth of 7 m). The economic variety of sillimanite occurs as huge boulders of massive ore. The deposit is worked out by the Assam Sillimanite Ltd.

According to Golani (1989), the massive sillimanite-corundum-bearing schists occur as lenticles and stringers varying in thickness from <1 m to 4–5 m. The schists occur as a horizon which can be traced for 25 km in the Sonapahar area. The massive sillimanite-corundum deposits are stratabound and considered to be metamorphosed Precambrian mature weathering product, confined to the quartz-sillimanite schists which represent a less mature regolith. The major-element chemistry of the massive sillimanite-corundum and quartz-sillimanite schist does not correspond to that of any common sedimentary or igneous rock. The author suggests that these unusual rocks were formed by isochemical metamorphism of Precambrian bauxitic paleosols.

Madhya Pradesh: Pipra in Sidhi district is known to contain economic deposits of sillimanite. It is popularly known as Rewa sillimanite. Both boulder and vein deposits are reported.

Maharashtra: A 5 km stretch is known to contain corundum, kyanite and sillimanite-bearing deposits in Bhandara district.

Bihar-Jharkhand: Sillimanite is reported from Gaya, Ranchi, Munger, and Singhbhum districts.

Sillimanite in small quantities is also reported from Tannilai mine in Tiruchirappalli district, Tamil Nadu and Madar mine, Udaipur district, Rajasthan.

India also possesses a large reserve of sillimanite (5–6%) associated with the beach sands of Kerala (Kozhikode, Palakkad, Ernakulam and Kottayam districts), Tamil Nadu and Odisha. The mineral is extracted while separating monazite, zircon, rutile, etc. from the sands.

4.2.1.2 Neutral Refractories

Chromite ($FeCr_2O_4$)

High-grade chromite ore is used in the manufacture of chrome bricks, chrome-magnesite bricks, and allied refractory products. The refractory-grade chromite ore should contain >40% Cr_2O_3 and have Al_2O_3 content between 12 and 30%. They should have low Fe_2O_3 (<15%) and SiO_2 (<5%) as well. High percentage of iron is detrimental as the bricks expand on heating. In industrial practice, an excess of MgO is added to the chrome ore in order to neutralize all the excess silica. Chromite bricks have PCE above 35 and refractoriness under load is up to 1430 °C.

Vide the earlier chapter (Chap. 3) for further discussion on chromite and its distribution in India.

Graphite (C)

Graphite, theoretically a mineral of carbon, occurs in two forms: natural graphite which includes (a) crystalline and (b) amorphous varieties. A third variety is known as artificial or manufactured graphite. The inherent qualities of graphite, for which it is so much in demand in the manufacturing industries, are its high lubricity, ability to withstand high temperature (up to 3000 °C in the absence of O_2), low co-efficient of thermal expansion, good electrical and thermal conductivity, and resistance to reaction with ordinary chemical reagents. It is also generally not wetted by metals. That is why flaky graphite is used in the manufacture of crucibles for melting metals. Graphite is a common foundry facing material and is also used in the manufacture of lead pencil, batteries, lubricants and brushes. It is an essential material in atomic reactors. The commercial graphite is graded mainly on its carbon content. The common impurities in graphite are silicates and sulfides. In industry, flaky graphite is preferred to fine-grained varieties. Even if lower in grade, the flaky varieties are easier to concentrate. For some special uses, C > 99.9% may be required. This is done by chemical processing of concentrates. Graphite can also be manufactured from petroleum, coke, or anthracite in electric furnaces.

Graphite can develop by several geological processes, such as (i) regional metamorphism and contact metamorphism; (ii) crystallization in igneous rocks, such as in basalts and nepheline syenites, and (iii) through hydrothermal solutions from deep-seated magma, such as vein graphite in pegmatites. Hydrothermal graphite formation has been conjectured as follows (Frost 1979):

$$2C_{(Solid)} + 2H_2O \rightarrow CH_4 + CO_2$$
$$CH_4 + CO_2 \rightarrow 2C_{(graphite,ppt)} + 2H_2O$$

Hydrothermal C–O–H forms at high T/P and in the presence of water, dissolving solid carbon or pre-existing graphite from the country rocks.

Majority of graphite deposits form by metamorphism (both contact and regional) of sedimentary carbonaceous matter, such as those present in black shales. Graphite is also found in iron meteorites. The best known graphite deposits in the world are located in Sri Lanka and Madagascar.

Indian distribution

Bulk of graphite production in India comes from the states of Odisha, Jharkhand, Karnataka, and Andhra Pradesh.

Odisha: The main concentration of graphite is reported from khondalitic rocks (garnetiferous quartz sillimanite rocks), garnetiferous quartzites, calciphyres and graphite schists of the Eastern Ghats. The deposits occur in the form of veins, lenses, and pockets within the host rocks. Two varieties are found: flaky and amorphous. The rich deposits of graphite occur in Sambalpur, Bolangir, Dhenkanol, Phulbani, Koraput, and Kalahandi districts of Odisha. Good-quality graphite from Odisha has fixed carbon content in the range of 55–60%.

Jharkhand: Khondalites, calc-granulites and quartzites, schists and gneisses, pegmatites and limestones in Daltonganj, and Palamau districts host good graphite deposits. Rich graphite concentrations are located in the Sokra, Khandih, and Rajhara areas. The carbon content of graphite ore in these occurrences is around 50%.

West Bengal: In the district of Purulia (erstwhile Manbhum district), adjacent to Jharkhand, good grade graphite deposit was reported (Deb 1949) from Precambrian mica schists and quartzites traversed by innumerable pegmatite veins near Anara, located between Adra and Purulia. Graphite deposits are concentrated as veins near the crushed contact of pegmatites with the country rocks or as lenticular pockets in the altered mica schists. The deposits contained 55–60% graphitic carbon.

Andhra Pradesh: Graphite deposits are found in khondalites of East Godavari district of the Eastern Ghat belt. Graphite is present in the form of irregular lenses and pockets and rarely found as veins of varying thickness. Well-known deposits are found around Peddanakonda in Bhadrachalam taluk. Graphite occurs in the form of irregular veins and as highly disseminated material in Vishakhapatnam district around Marupalli. Graphite occurrences are also known in Krishna and west Godavari districts of the state.

Karnataka: Fine-grained graphite is reported from Kolar schist belt in Bangarpet taluk and flaky variety is found in crystalline schists near Mavinhalli and Tonavalli areas of Mysore district. Graphite is also found in several areas of Chitradurga schist belt and in the Bababudan hills. Their mineability is not very high.

Asbestos

The heat resistant character of asbestos (melting point varying between 1200 and 1500 °C, depending on the type of asbestos) along with its infusibility makes it commercially important in various industries. Some of its uses are thus in woven brake linings, clutch facings, gaskets, safety clothing in fire services, chemical filters, and heat-resistant conveyor belts. More information on asbestos is presented in a later section.

A quality refractory is produced by mixing 85% of magnesite ($MgCO_3$) with 15% of asbestos powder and fusing the mixture.

4.2.1.3 Basic Refractories

Magnesite ($MgCO_3$)

This carbonate of magnesium is found to occur in different ultrabasic igneous rocks and in dolomitic limestones. In the former, magnesite forms by alteration of Mg-rich silicates. Its main use is in refractories. Dead burnt magnesite, which is calcined at high temperature between 1400 and 1500 °C and converts to crystalline periclase (MgO), is used in the manufacture of bricks for furnace linings. Magnesite is also used as an ore of metallic Mg, the lightest structural metal. Magnesite also finds use in chemicals (some of therapeutic use, such as, milk of magnesia) and fertilizer industries, in special cement manufacture, in paper, as well as in the preparation of pulp and rubber. At present, a large part of the production of Mg comes from brines and seawater.

Magnesite is commonly derived from the alteration of Mg-rich metamorphic and igneous rocks (serpentinites and peridotites) through the action of waters containing carbonic acid and occurs in the form of veins and irregular masses. Such magnesites are compact, cryptocrystalline and often contain opaline silica. Mainly three varieties of magnesite are recognized: (i) sparry crystalline, (ii) cryptocrystalline, and (iii) amorphous.

Magnesite deposits form by the following processes (Pohl 2011):

(1) by desilicification and carbonatization of Mg-rich igneous rocks, such as, serpentinites or peridotites, by CO_2-rich fluids contributed by diagenetic, magmatic or mantle degassing processes:

$$\underset{\text{(Forsterite)}}{Mg_2SiO_4 + 2CO_2} \rightarrow \underset{\text{(Magnesite)}}{MgCO_3} + SiO_{2aq}$$

The product is massive or vein-like.

(2) by metasomatic replacement of marine dolomite and limestone by exchanging Mg for Ca. The responsible fluid is believed to form by evaporation and/or diagenesis. The product is generally coarse grained and stratabound.

Indian distribution

The major repositories of magnesite in the country are the states of Tamil Nadu, Karnataka, Uttarakhand, and Jammu & Kashmir.

Tamil Nadu: Magnesite, occurring in dunites of Chalk Hills in Salem district, popularly known as "Salem magnesite" in India, is found as irregular veins in ultrabasic intrusive masses over an area of 11 km^2. The magnesite is of very good quality (95–97% $MgCO_3$), white to grayish in color, compact and massive in general. Estimated reserves are more than 100,000 tons up to a depth of 15 m from the surface.

Karnataka: Magnesite in this state occurs as an alteration product of ultrabasic rocks. It forms network of veins of various shapes and sizes, in the serpentinized rocks. Magnesite of massive amorphous variety occurs particularly in Hassan, Mysore, and Coorg districts.

Uttarakhand: Magnesite deposits are hosted mainly by dolomites and also at places, associated with talc in Almora and Someshwar districts of the state. The mineral occurs in the form of veins, stringers and as massive crystalline deposits. Valdiya (1968) reported stromatolite-bearing dolomitic rocks with lentiform deposits of coarsely crystalline magnesite in the Gangolighat Formation of the calc-zone of Pithoragarh district in Kumaon Himalaya. These magnesite deposits are very extensive, originating from the Kali river valley toward east and continuing to the Alaknanda valley in the northwest, stretching over a distance of 130 km. The average chemical composition of Pithoragarh magnesite, according to this author, is: MgO 42.2%, CaO 1.6%, SiO_2 + insolubles 5.1%, R_2O_3 3.1%, LOI 47.3%. He suggests that Mg-rich basin waters reacted with algal bioherms and associated limestone beds and gave rise to discordant magnesite bodies. Coarse granularity developed later by recrystallization during regional metamorphism.

Jammu & Kashmir: Magnesite mineralization occurs near Katra town in Udhampur district of J & K, associated with Proterozoic Sirban limestone. Pockets, lenses, and bands of magnesite of varying sizes are found at Panthal, Adhkanwari, and Pie-nala. The major mineralized body comprises sparry magnesite with coarse grained magnesite veins within them and is considered to be of replacement origin. National Mineral Development Corporation (NMDC) estimated a reserve of 2.5 million tons of refractory-grade magnesite ore with 40.5–45.7% MgO, 0.1–5.8% CaO and 50.3% LOI (GSI 2005).

Dolomite [$CaMg(CO_3)_2$]

Dolomite and calcite are major constituents of carbonate rocks. Dolostone, mostly referred to as dolomite in literature, is considered as a principal raw material in the iron and steel industries. It is also used as a refractory material, for which calcined products are preferred. Theoretically, dolomite should have an equal proportion of $CaCO_3$ and $MgCO_3$, a condition which is hardly obeyed. It should be low in SiO_2, Fe_2O_3, and Al_2O_3 (together $<3\%$). Open hearth furnaces use dolomite for basic lining and Bessemer converters use dead-burnt material for the same purpose. Dolomite is also used in the manufacture of high magnesia lime, basic magnesium carbonate, Epsom salts, and for the manufacture of elemental magnesium. Dolomite is used also in chemical industry, in manufacture of paper, leather, glass, etc. and as building material, as terrazzo stucco and also as crushed stones.

Dolomite is usually a part of a sedimentary carbonate facies and occurs in all geological ages but the economically important deposits are mostly confined to the Proterozoic and Palaeozoic eras. Usually the dolomites are associated with limestones and sometimes they occur as irregular beds, lenticles, and pockets and rarely do they occur as hydrothermal vein deposits. Gypsum, anhydrite and alkali salts are often found associated with dolomite in saline evaporative basins. Dolomite also results from partial or complete dolomitization of the marine calcium carbonate including marine shells and organic debris.

Indian distribution

Dolomite (dolostone) and dolomitic marble occur extensively in several states in different stratigraphic horizons.

West Bengal: Large deposits of dolomite occur in Buxa Duars area in the northwest of Jalpaiguri district, spread over an area of about 13 km^2. There are several varieties of dolomite in this region out of which two are important: one is massive, compact, and light gray in color, and the other is dark brecciated type, possessing distinct bedding planes. Chemically both are pure dolomites. A part of this dolomite band enters into Bhutan. This deposit has been discontinuously mined for several decades now. Recent studies have shown that dolomite mining increases the pH of ground water in tea plantations, adversely affecting them.

Rajasthan: This state has a large resource of dolomite, mostly occurring in the Delhi Supergroup of rocks. High-grade dolomite deposits are found in Kesarpura, Hatondi, and Akhri areas in Ajmer district. They are usually high magnesium crystalline dolomites, low in silica, alumina and iron-oxides.

Karnataka: The dolomite deposits which form part of the Dharwar Supergroup, occur chiefly in Dharwar–Shimoga and Gadag–Chitradurga schist belts. Dolomite deposits are also found in the lower Kaladgi Group of Cuddapah age. Large reserves of dolomite also occur in Belgaum district.

Jharkhand: Dolomite occurs as bands or as thin lenticular patches associated with calc-silicate rocks and magnetite-tremolite schist near Daltonganj in Palamau district. Some outcrops of crystalline magnesium limestones are also found in the Iron ore group of rocks in Singhbhum district.

Odisha: Enormous deposits of dolomitic marble belonging to the Gangpur Group of rocks occur in the Birmitrapur and Purnapani areas of Sundergarh district of the state. Tatas exploited them for a long time for their steel plant at Jamshedpur.

Olivine $(Mg, Fe)_2SiO_4$

Olivine, particularly its forsteritic variety is in demand in metallurgy today. It is required for the manufacture of special refractories, but is mainly ($\sim 75\%$) used as a slag conditioner, similar to dolomite in pig iron production. Replacing dolomite it reduces coal consumption, thereby reducing CO_2 emission. Olivine makes good foundry sand (Pohl 2011).

Dunite is the common source for industrial olivine. It occurs at several places in India, particularly in south India.

4.2.2 Minerals for Fertilizer Industry

Plant growth and high crop yield require three principal elements: nitrogen, phosphorous, and potassium. Natural nitrates are not commonly used now as they have been widely replaced by nitrogenous fertilizers made from atmospheric nitrogen. Phosphatic fertilizers, earlier produced from bones of dead animals, are now produced from phosphate rocks (phosphorites) using sulfuric acid and soluble super-phosphate in successive stages. Potassium fertilizers are at places extracted from evaporate deposits. Other minerals used as sulfate fertilizers include gypsum and pyrite.

Rock phosphates (phosphorites)

Phosphorus is present in minor to trace amounts in most rocks. However, only in phosphorites the values can be as high as 40% P_2O_5. With appropriate process of beneficiation, ores as low as 4% P_2O_5 can be utilized. More than 180 mineral species are known to contain 1% or more of P_2O_5. However, the mineral apatite, which is a phosphate of calcium, with fluorine and chlorine [$Ca_5(F, OH, Cl)(PO_4)_3$], is the main source of phosphorus in the Earth's crust. P may be variously replaced by Mn, Sr, Na, Mg, U or REE. The cryptocrystalline variety of apatite is called collophane or graneolite. The major part of phosphorite production is utilized in the manufacture of super-phosphates used for increasing soil fertility. A small part of the phosphate rock is used for the recovery of elemental phosphorus, for the manufacture of phosphate chemicals used in different industries. Elemental phosphorous is used in match industry as well as in the manufacture of incendiary bombs and fireworks. Trace elements in apatite, such as, U, Th, F, V, REE are recoverable as by-products in some cases.

Phosphate deposits can form in different geological situations, two or three of which are important. Marine phosphorites, which constitute the principal resources of this material, accounting for roughly 80% of world's phosphate rock production. Major accumulations of this resource appear to have taken place in a marine environment, particularly where upwelling of cool phosphate-saturated sea waters moved across near-shore continental margins (Fig. 4.3). Here the phosphate precipitated by complex microbiological processes, producing phosphatic mud, nodules and crust in units a few cm to tens of meters thick. However, there is a recent view that phosphates of wide, shallow (epicontinental) seas, as of Northern Africa and Arabia cannot be suitably explained by upwelling. According to Glenn et al. (1994) fluvial introduction of phosphorous explains the situation better. Igneous rocks represented by alkaline complexes, including carbonatites, are another type of source for these mineral resources. A third source is the apatite-rich iron ores (Kiruna type, a variant of IOCG?) or apatite-magnetite-rich rocks (vide Chap. 3 for more discussion). Lastly, phosphate concentrations or Guano deposits form in some oceanic islands due to bird droppings.

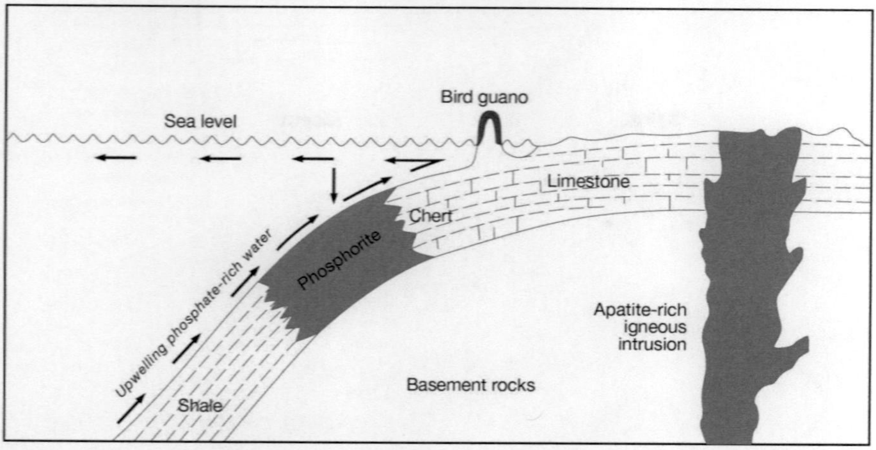

Fig. 4.3 Sketch depicting various modes of formation of phosphate concentrations: bedded phosphorites, apatite-magnetite vein and guano deposits (after Craig et al. 2011)

Indian distribution

Rajasthan: Phosphorite deposits occurring in the rocks of the Aravalli Supergroup in the vicinity of Udaipur (Banerjee 1971) and in Jhabua district of Madhya Pradesh are the most important phosphate deposits in India. There are two zones of phosphorite in the vicinity of Udaipur city: Jhamarkotra, Maton, Kanpur, and Kharbaria-ka-Gurha deposits occur close to the base of the Aravalli stratigraphy to the southeast of Udaipur; Bargaon, Nimachmata, Sisarma, and Dakankotra, occur to the west of Udaipur. The largest rock-phosphate deposit in India is represented by the Jhamarkotra phosphorite deposit, located about 26 km southeast of Udaipur. This deposit, like other deposits in the Aravalli Supergroup of the Udaipur area, is associated with dolomitic limestone beds which have been metamorphosed in the greenschist facies and are silicified and brecciated at places. They are also commonly stromatolitic. In Jhamarkotra, the phosphorite-bearing unit has thickness varying from 1 to 25 m and forms a broad arcuate belt that extends for more than 16 km in length (Fig. 4.4). The phosphorite horizon, belonging to the Maton Formation overlies a very thick orthoquartzite and cherty sequence. The phosphorite occurrences at Jhamarkotra comprise three different types: (i) columnar stromatolitic (algal) phosphorite (Fig. 4.5a, b); (ii) laminar algal phosphorite; (iii) reworked, silicified and brecciated phosphorite. The P_2O_5 content of Jhamarkotra phosphorites varies between 12 and +30% with an estimated reserve of 36 million tons of the former and 17 million tons of the latter grade (Banerjee 1986). The Maton deposit lies about 15 km southeast of Udaipur. The phosphate horizon (Fig. 4.6), traceable for 3.5 km with an average thickness of 20 m, follows the S-surface of dolomitic marble in the north, jaspery quartzite in central and brecciated quartzite and sandy phyllite in the southern parts. The estimated reserves are 9.2 million tons with average grade of 26% P_2O_5. The other deposits in the two

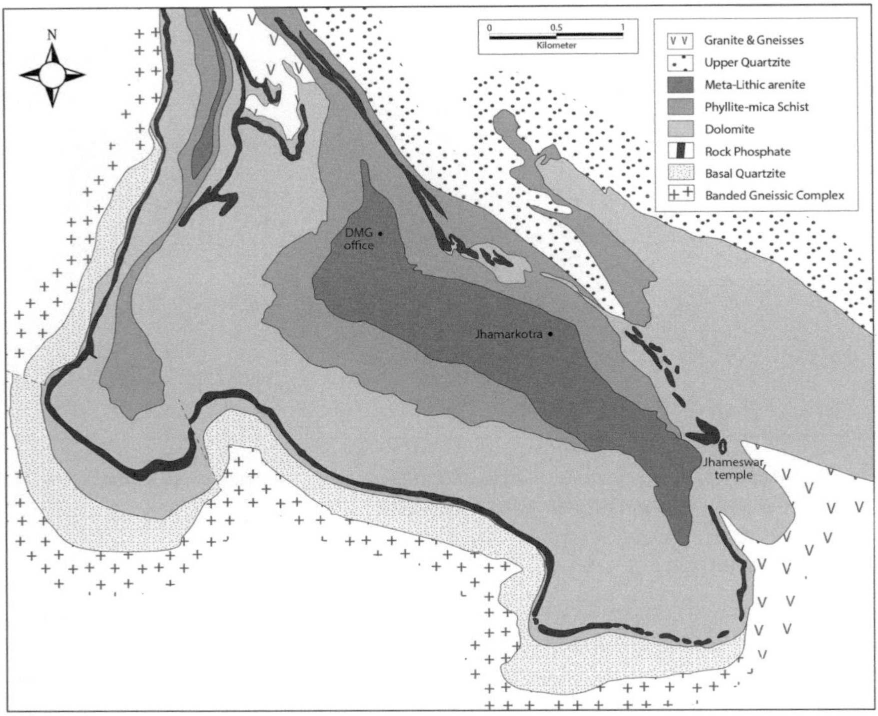

Fig. 4.4 Geological map of Jhamarkotra phosphorite deposit (after Roy et al. 1980)

groups do not have any sizeable reserves except the one at Kanpur which has 4 million tons with 10–12% P_2O_5.

The Siriska and Kushalgarh formations of the lower Ajabgarh Group of Delhi Supergroup have minor occurrences of phosphorites near Achraul, Baraud, Siriska, and Rajgarh (Dayal et al. 1981). A bed of phosphorite, 2.5–3.9 m thick, and containing 12.5–31.3% P_2O_5 occurs at Achraul, near Siriska. In the Rajgarh-Ramgarh area phosphorite occurs interbanded with calcareous quartzites but the associated stromatolites do not contain any phosphatic material. P_2O_5 content varies up to 15%. Economically significant sedimentary phosphate deposits have been found in Birmania area of Jaisalmer district. The deposits are intimately associated with a sequence of dolomite, chert, carbonaceous shale, and sandstone, correlated with the Marwar Supergroup, time-equivalents of the Vindhyans. The phosphorite deposit here, 1–9 m thick, extends as a single horizon for 6.8 km. It has 10–15% P_2O_5, and has an estimated reserve of 4.90 million tons (Banerjee 1986).

Madhya Pradesh: Phosphorite-bearing dolomitic marble and chert bands, with associated stromatolites, occur in phyllite and quartzite country rocks of Aravalli Supergroup in the Jhabua district, adjoining the state of Rajasthan. The ore zone forms the eastern limb of a regionally overturned asymmetric synform and is a

Fig. 4.5 **a** A view of the Jhamarkotra phosphorite mine. **b** Highly deformed algal stromatolites in Jhamarkotra mine, Rajasthan (Photos: M. Deb)

persistent horizon, 12 km long, from Amlamal in the north to Ramphapur in the south. The phosphorite grade and reserves are variable: 5.67 million tons of 29% P_2O_5 and 2.04 million tons of 14.5% P_2O_5 (Banerjee 1986). Phosphorite associated with ferruginous shale, quartzite, and quartz breccia of Gangau Formation overlies dolomites and quartzites of Bajna Formation lying above the Bundelkhand granite in the western part of the Bijawar basin. The phosphorite bodies are small in size and scattered. The estimated reserve of Hirapur deposit is 1.44 million tons with an overall grade of about 20% P_2O_5 (Banerjee 1986).

Uttarakhand: A few sedimentary basins contain phosphagenic sequence in the Himalayan region. Most well known is the Krol basin in the Mussoorie hills of the lesser Himalaya where the Precambrian–Cambrian transition recorded in the Krol-Tal sequence is characterized by low-grade phosphorites in close association with pyrite mineralization at Maldeota and Durmala (Fig. 4.7a). The chert-phosphorite member is in the lowermost Tal overlying the upper Krol

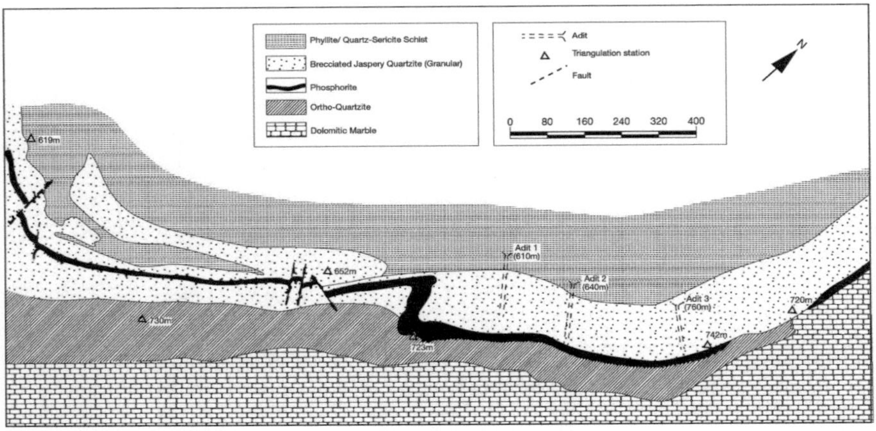

Fig. 4.6 Geological map of Maton phosphorite deposit (after GSI)

(E) carbonates and siltstones at the bottom and lower Tal shale-siltstone sequence on top (Jiang et al. 2002). The phosphorites have many modes of occurrence: fine grained granular, nodular and laminar or stromatolitic forms closely associated with carbonaceous black shale and chert (Fig. 4.7b). Most of the phosphate samples show low but varying eU_3O_8 contents, between 0.005 and 0.03% while some show as high as 0.11% eU_3O_8. The total rock phosphate reserve was estimated at 18 million tons with P_2O_5 content ranging from 16.49 to 30%, and averaging about 20% (Banerjee 1986).

Andhra Pradesh: Phosphorite in the Cuddapah basin of Andhra Pradesh (Dwivedy 1995) (cf. Fig. 4.8) was first reported from Chelima area of Kurnool district. Phosphorite beds are confined to Cumbum, Tadpatri and Srisailam Formations. They are generally of low grade, analyzing about 9% P_2O_5, with occasional higher values of as much as 35% P_2O_5. The phosphorites are pelletal, fragmental or nodular types with stromatolitic phosphorite noted near Peddasettypalle. In parts of the Kurnool Formation, the Owk shale contains phosphorite bands of non-stromatolitic origin. These phosphorites show higher REE abundance and some U_3O_8 content.

Apatite

The apatite-magnetite mineralization of Singhbhum region of Jharkhand and Purulia district of West Bengal are minor sources of fertilizer raw material in eastern India. The former occurs as lensoid bodies in different host rocks along and close to the footwall of the Singhbhum Shear Zone (vide Fig. 3.34). Three zones are recognized with economic potentiality (Sarkar 1984):

(i) Khejurdari-Dhantuppa-Gohala
(ii) Badia-Pathargora-Samaidih

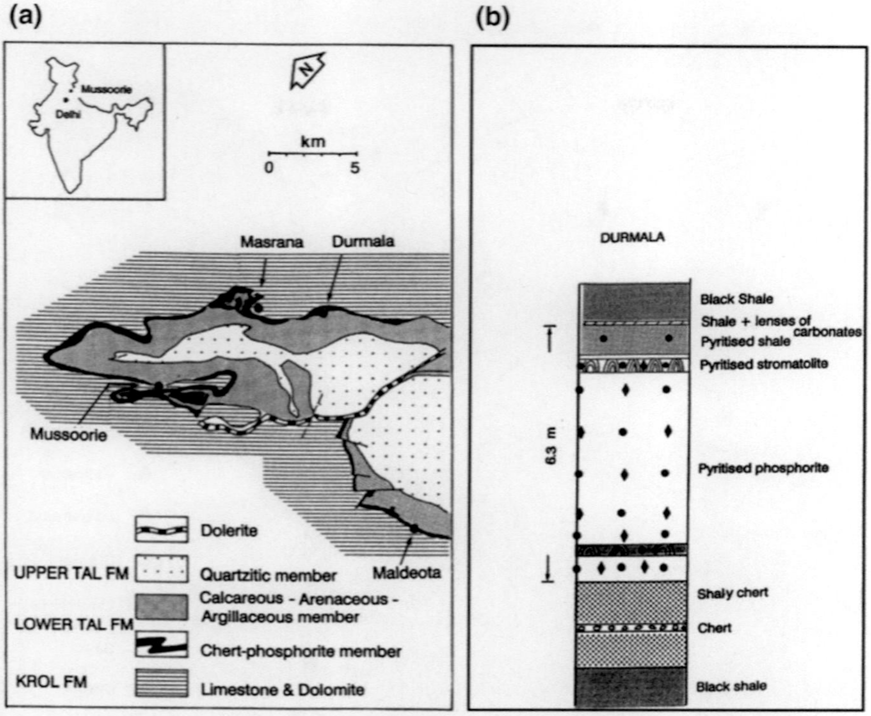

Fig. 4.7 **a** Generalized geological map of Krol basin in Mussoorie hills of Lesser Himalaya. **b** Section showing phosphorite-chert-black shale association at Durmala

(iii) Patharchakri-Kudada-Dudra

In the first, the host rocks are biotite schists, chlorite-biotite schists or soda granites. An uncommon assemblage of quartz-kyanite-magnetite-apatite is noted at Gohala. East of Badia and Mosabani (the well known copper mines in Singhbhum copper belt), the biotite schists contain pockets and lenses of apatite with varying content of magnetite. The mineralized zone is on the hanging wall side of Soda Granite/feldspathic schists at Rangamatia, east of the Pathargora mine shaft. At Dhobani, Tamajhuri, Kasidih and Samaidih the mineralization is confined within Dhanjori metabasites. However, at Barjudi, the mineralization is found to occur in mica schists of Chaibasa Formation. In the third locality, south of Tatanagar, the mineralization is hosted by quartz-chlorite schists ± Soda Granite/feldspathic schists (Sarkar and Gupta 2012).

The conspicuous features of the apatite-magnetite mineralization, briefly described above, include (i) shear zone control of the mineralization; (ii) occurrence mostly in Dhanjori basic volcanic or their derivatives and (iii) intense hydrothermal

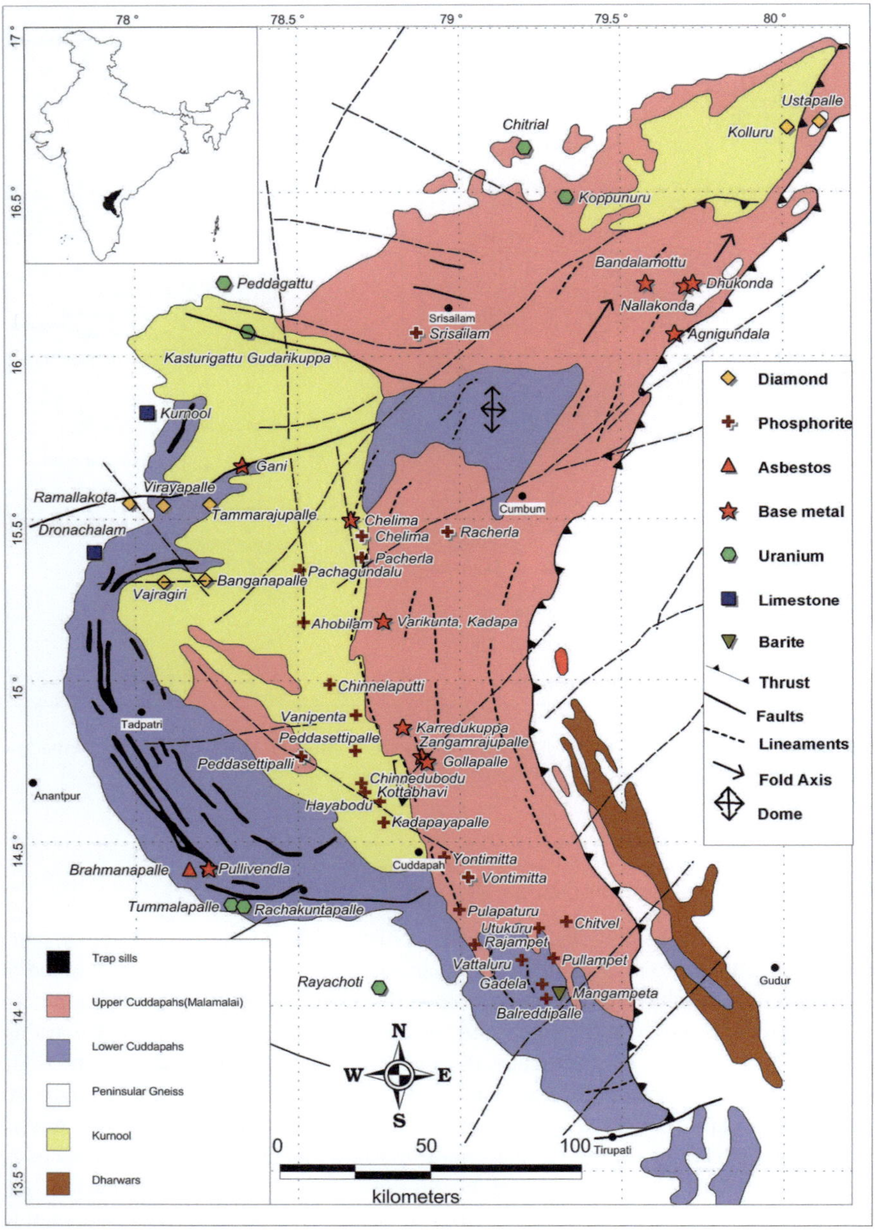

Fig. 4.8 Geological map of Cuddapah basin showing the locations of different mineral deposits (after Dwivedy 1995)

alteration in and around the ore lenses. Based on these and other geochemical features, it may be surmised (Sarkar 1984; Sarkar and Gupta 2012) that hydrothermal solutions were released from clastic rocks in the sequence during regional metamorphism and during their passage through the pre-existing/penecontemporaneous shear zones they could have leached the Fe, P and F from the volcanics and deposited the phosphatic mineralization in suitable locales. It could have been metasomatically redistributed locally during Soda Granite emplacement.

Similar phosphatic mineralization has been recorded from the northern shear zone, north of the Dalma lavas in Singhbhum, near Purulia town in West Bengal. The mineralization at Beldih, Mednitanr, Kutni and Churingora are associated with carbonatites within metamorphosed volcani-sedimentary rocks. A reserve of 4.56 Mt of ore with average P_2O_5-content of 13 wt% has been estimated at the first location. Apatite mineralization in the Chotanagpur Granite Gneiss is also reported from Panrkidih in north Purulia district of West Bengal (Sarkar and Gupta 2012).

In Kashipatnam, 60 km northwest of Vishakapatnam in Andhra Pradesh, steeply dipping veins of apatite-magnetite-vermiculite rocks are known to be exploited for their phosphate content in the Eastern Ghat belt (Narasayya and Sriramdas 1974). The orbodies are tabular and lenticular, filling fractures in Precambrian migmatitic rocks. Their thickness may be more than 2.5 m with lengths between 30 and 50 m. Depth persistence is noted till 35 m. 33% average P_2O_5 content in these veins has been reported.

Several carbonatite bodies lie along a fault zone in Tamil Nadu. Small Sevathur carbonatite in this zone, 200 km SW of Chennai contains 14.2% P_2O_5. The size suggests that the resource is likely to be limited. The Newania carbonatite complex, east of Udaipur, Rajasthan is estimated to contain one million ton of 6–10% P_2O_5.

Potash

K is an essential element in agricultural fertilizers and is also important in human and animal metabolism. Sylvite (KCl) with 63% K_2O, Nitre or Chile salt petre (KNO_3) with 47% K_2O and Glaserite ($Na_2SO_4 \cdot 3K_2SO_4$) with 43% K_2O are the major sources of potassium. About 5% of potash production is consumed by the chemical industry for manufacture of dyes, drugs, detergents, soaps, glass and analytical reagents. The major world production of potash comes from marine evaporate deposits (Germany, France, etc.). Other significant deposits include non-marine Sabkha deposits of Salt Petre (Chile), sea brines of the Dead Sea (Israel and Jordan) and lake brines (California, Utah in USA, Tunisia, Africa). Glauconitic greensands have been used in England, France, Belgium, USA and Australia as slow-acting potash manure in horticulture.

Evaporation in a closed body of sea water under arid conditions lead to syn-sedimentary deposition of sodium, potassium, calcium, and sometimes, borate,

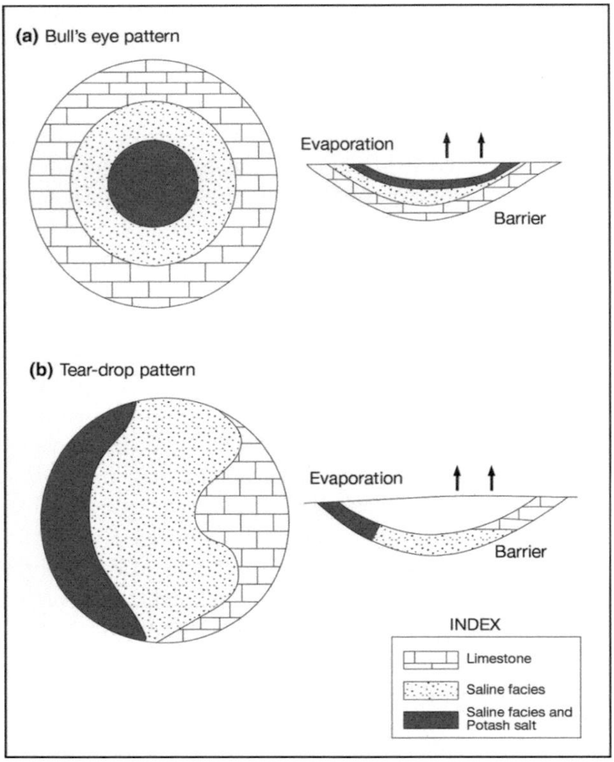

Fig. 4.9 "Bull's eye" and "tear-drop" shaped disposition of evaporate deposits (after Zharkov 1984)

nitrate and other salts, constituting a marine evaporite deposit. A typical evaporite sequence starts with calcite and dolomite at the base, followed upward by gypsum or anhydrite, rock salt or halite, and potassium and magnesium salts at the top when sea water concentration reaches almost 100 times the original volume. The distribution of evaporitic rocks in a depositional basin may produce different patterns, such as "bull's eye" or "tear drop" (Fig. 4.9) depending on the symmetry of the basin (Zharkov 1984). The non-marine evaporites include both coastal and continental Sabkha deposits as well as inland lake deposits. The Sabkhas are salt-encrusted supratidal surface or coastal flats bordering lagoons or inner shelf region, such as those found in the Little Rann of Kuchchh. Inland lakes with salt deposits are common in western Rajasthan such as, at Didwana (Sambhar Lake).

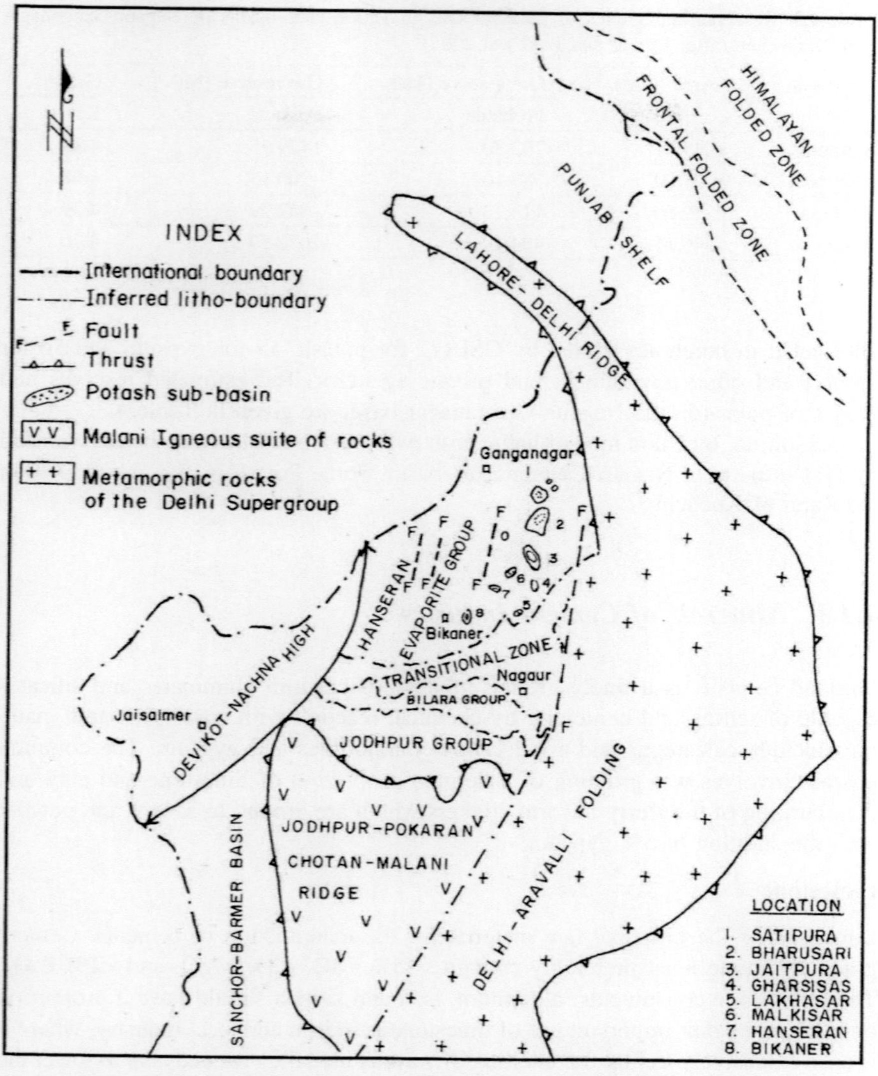

Fig. 4.10 Geological setting of Hanseran Evaporite Group in NW India (after Kumar and Bakliwal 2005)

Potash-bearing marine evaporatic sediments are recorded from northern and northwestern part of Rajasthan and the adjoining states of Haryana and Punjab (Fig. 4.10) (Kumar and Bakliwal 2005). The host sequence is a late Neoproterozoic unmetamorphosed marine sediments of the Nagaur-Ganganagar basin, now represented by the Marwar Supergroup. North of Nagaur, halite-bearing thick evaporites of Hanseran Evaporite Group (HEG) of Eocambrian (Vendian) age have been

Table 4.3 Reserves and grades of potassic salts in various sub-basins of Nagaur-Ganganagar basin, Rajasthan (after Kumar and Bakliwal 2005)

Sub-basin	Area (km^2)	Ore reserve (Mt)	Ore reserve (Mt)	Grade
		Probable	Possible	K%
Satipura	245.61	202.30	1429.95	4.80
Bharusari	71.00	88.16	300.15	4.68
Lakhasar	29.50	113.73	342.29	4.39
Total	346.11	404.19	2072.39	4.70

intersected in boreholes drilled by GSI (72 for potash, 45 for gypsum and 50 for lignite) and other government and private agencies. The estimated reserves and grades of potash in the Nagaur–Ganganagar basin are given in Table 4.3.

In addition, 6 trillion tons of halite with over 80% NaCl has also been estimated by GSI in parts of Nagaur-Ganganagar basin. Some Potash is also reported from the Rann of Kuchchh.

4.2.3 Minerals of Cement Industry

Portland cement is a finely ground mixture of calcium aluminates and silicates capable of setting and hardening by chemical reaction with water. The raw materials include calcareous and argillaceous commodities and gypsum. The common process involves wet grinding of a definite proportion of limestone and clay and then burning of the slurry to form clinkers which are ground to a very fine powder with the addition of 5% gypsum.

Limestone

Limestone is the principal raw material for the manufacture of cements. Cement grade limestone must preferably contain >45% CaO, <3% MgO, and <1% P_2O_5. The other three constituents: aluminum, iron and silicon should have a proportion of 1:2:5. The other important use of limestone is in iron and steel industry, where it is used extensively as flux for smelting of various metallic ores and also to lower the temperature of melting in blast furnaces. It is an important raw material for a number of chemical industries, such as manufacture of calcium carbide and bleaching powder by use of lime containing 95% CaO. Lime is produced by calcination of limestone or dolomite. It is used as an aggregate in concrete, and as a road material. Finely crushed limestone is used as a soil conditioner, for whitening and whitewashing and as a filler. Limestone is also used in lithography, as a dimension stone for decorating exterior facings.

Limestones are sedimentary rocks, deposited in shallow or deep water marine environment. Accumulations of calcareous materials take place at first as calcareous silts containing dead marine plants and invertebrate animal shells, in time being progressively converted into limestones which are composed of calcite crystals, with varying percentage of magnesium carbonate and mechanically admixed impurities. Limestones may form at the sea bottom by the accumulation and lithification of particles of calcareous materials, originally secreted in the sea water by living marine organisms. Usually oolitic limestones are formed in littoral zones due to the coagulation of colloidal solutions of calcium carbonate around minute sand grains. At first the oolites are composed of aragonite which is later converted into calcite. However, calcite is relatively unstable in the weathering environment due to its high solubility in acidic waters.

Indian distribution

Calcareous rocks occur in all the principal geological formations of India, right from the Precambrian to the Recent. The most important economic concentrations are found (Fig. 4.1) in the Vindhyan sequences of Bihar, Uttar Pradesh, Madhya Pradesh and Rajasthan, The deposits in *Bihar* are mostly in the Rohtas Formation of the Vindhyan Supergroup in Sahabad district. In *Uttar Pradesh*, extensive deposits of limestone are found in Mirzapur district, while in *Uttarakhand* they occur in Dehradun district. The former are in Vindhyan sequence while the latter are in Krol rocks. How much of these deposits are of limestone or of dolostone remains doubtful. In *Madhya Pradesh*, large deposits occur in the Semri Group rocks of the Jabalpur district. In *Rajasthan*, dolomitic limestone in the Raialo Group occurs around Alwar, Nagaur, Kishengarh, and Udaipur. Tertiary limestones are distributed near Jaisalmer. Large deposits of limestone also occur in the Saurashtra region of *Gujarat*. In *Odisha*, limestone is found in the Birmitrapur Formation of Gangpur Group in the vicinity of Sundergarh. Reserves estimated are in the range of 250 Mt. Important occurrences are also known from Sambalpur and Koraput districts. Large workable limestone deposits occur in the Cuddapah basin of *Andhra Pradesh*. Massive and flaggy limestone of Narji Formation in the Kurnool Group has 4 billion tons of reserve of the cement grade variety. Vempalle limestone in Anantapur district is also exploited extensively. Additionally, in the Palnadu area of Guntur district, large reserves of cement grade limestone have been located.

Gypsum ($CaSO_4{\cdot}2H_2O$) and **Anhydrite** ($CaSO_4$):

Gypsum in its pure form is colorless to white. The mineral is also found in a fibrous variety called satin spar, which is characterized by silky luster; alabaster is the fine-grained massive variety whereas a variety that yields broad colorless and transparent cleavage folia is named selenite. Gypsum when calcined at 200 °C loses about 75% of its water content. This product, finely ground, is principally used for production of Plaster of Paris which finds extensive use in building industry and in medical aid. Satin spar and alabaster are cut and polished for various ornamental purposes. Gypsum also serves as a soil conditioner and finds use in the fertilizer industry, as mentioned earlier. Gypsum for the fertilizer industry should have a minimum of 87% $CaSO_4{\cdot}2H_2O$.

Gypsum is a common mineral in sedimentary rocks, often occurring as thick beds. It is frequently found interstratified with limestones and shales and usually occurs as a layer underlying beds of rock salt, having been deposited there as one of the first minerals to have crystallized on evaporation of salt waters. Thin layer of gypsum is thus found on the surface of many desiccating lakes in arid regions. It is also found as lenticular bodies or scattered crystals in clays and shales, found in volcanic regions, especially where limestones have been acted upon by sulfur vapors.

Indian distribution

Rajasthan: This state is the largest repository of gypsum in the country sharing almost 95% of the country's reserve. The deposits are confined to the Tertiary rock formations of Jaisalmer, Jodhpur, and Bikaner regions, with potential deposits of desert gypsum in Barmer district also. Deb (1952) described the occurrence of natural salts and also estimated a reserve of 100,000 tons of selenite gypsum from Lunkaransar Tal, north of Bikaner.

Tamil Nadu: This state contains the largest resource of gypsum in south India. Usually the deposits are found in highly fossiliferous rocks of Uttatur and Tiruchirappalli Formations of Cretaceous age. They are intimately mixed up with the black cotton soil and estuarine clays of Pleistocene period or even the recent sediments. Coimbatore, Ramanathapuram, and Tiruchirappalli districts stand out in this context.

Gujarat: Rich deposits of gypsum have been reported from several areas of Bhavnagar, Jamnagar, Junagarh, and Kuchchh districts. The richest deposits are found to occur in Rann associated with Gaj Formation.

Gypsum mineralizations are reported from different places in Karnataka, Himachal Pradesh, and Jammu & Kashmir too.

4.2.4 Minerals of Chemical Industry

Sulfur (S)

In nature sulfur occurs in the following modes: native sulfur, sulfides of base metals, and sulfates of calcium, magnesium and rarely potassium. The first two are the principal sources of sulfur. A moderate amount of sulfur is recovered from gases from roasters treating sulfide minerals.

Native sulfur of commercial interest is virtually absent in India. Sulfur emanations are reported from the bore-holes sunk by GSI in Puga valley in Ladakh region of Jammu and Kashmir, which is associated with a small amount of arsenic. Sulfur was recovered as a by-product (in the form of H_2SO_4) from sulfurous fumes of copper smelter of HCL at Moubhandar in Jharkhand and is presently recovered in the Khetri copper smelter in Rajasthan.

Fig. 4.11 Field photograph of lower and upper Kaimur sandstone at Amjhore with carbonaceous Bijaigarh shale in between. (Photo: M. Deb); *Inset* on *upper right corner*: Litholog of the pyritiferous zone in Bijaigarh shale (after Banerjee et al. 2006)

Pyrite (FeS_2) and **Pyrrhotite** ($Fe_{1-x}S$):

Both pyrite and pyrrhotite are used as source of sulfur for industrial purposes.

Indian distribution

Bihar: Sedimentary pyrite deposit occurs at Amjhore, about 20 km south of Dehri-on-Son in the Vindhyan hills of Rohtas district. Carbonaceous Bijaigarhshales, up to 70 m thick, (Guha 1971; Banerjee et al. 2006) and sandwiched between the lower and upper sandstones of the Kaimur Group (Fig. 4.11) of the Vindhyan sequence host the pyritic lenses and beds. The pyrite bed has an average thickness of 1.1 m and divides the carbonaceous unit into two parts, the lower shale (average thickness 32 m) and the upper shale (average thickness 15 m). In details, the lower shale has fine intercalations of sandstones and siltstones, is less carbonaceous and shows gradational contact with the lower Kaimur sandstone whereas the upper shale contains only a few siltstone beds, is more carbonaceous and has abundant pyrite disseminations. These features suggest a deeper environment of deposition of the upper Bijaigarh shale. The pyrite bed was mined by Pyrites, Phosphate and Chemicals Ltd. (PPCL), a Government of India Undertaking, till 2003 through a number of adits on the hill face.

Rajasthan: Pyrite-pyrrhotite deposit, occurring as several concordant, stratiform sulfide bodies, and co-folded with the host amphibolites, belong to the Ajabgarh Group rocks of Delhi Supergroup at Saladipura, Sikar district, Rajasthan. It does not form an obvious part of the Khetri copper belt, but occurs about 15 km to the

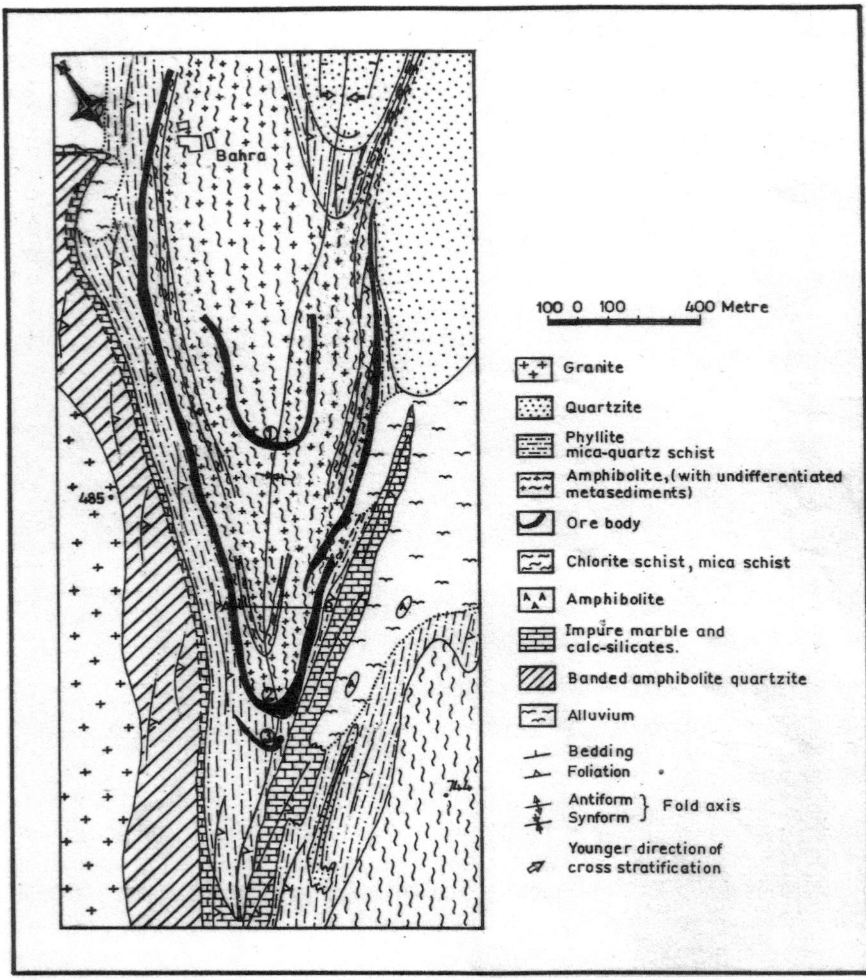

Fig. 4.12 Geological map of Saladipura pyrrhotite deposit, Rajasthan (after Sarkar et al. 1980)

east. The largest of the orebodies is about 7 km in strike length (Fig. 4.12), has a reserve of 115 million tons of ore with an average sulfur content of 22.5%. The ores are considered to be of sedimentary-diagenetic in origin and then underwent the same thermo-tectonic modifications along with the country rocks at peak temperatures around 600 °C and 5.5 kb pressure (Sarkar et al. 1980). The Pb–Pb model age of Saladipura ore is ~1700 Ma (Deb et al. 1989)

Himachal Pradesh: Several lenticular bands of pyrite, averaging 35–45% sulfur, occur east of Taradevi, near Chota Shimla within Simla slates.

Karnataka: The copper sulfide deposit in mafic volcanic rocks at Ingaldhal, Chitradurga schist belt, has two types of mineralizations: stratiform and vein types

(Anantha Iyer and Vasudev 1985). The stratiform ore in the footwall of the mineralized zone occur as a banded sulfide-chert body, 3–20 m thick and extending for about 7 km along strike. The ore consisted principally of pyrite, pyrrhotite and magnetite with local concentrations of chalcopyrite and sphalerite. Mining on the deposit is abandoned.

Rock salt (NaCl)

Rock salt is not only an essential component of human and animal diet, but also is an important raw material for the chemical industry. Most halite ($\sim$50%) is used for the production of chlorine, soda ash (Na_2CO_3), caustic soda (NaOH) that are basic chemicals for glass, paper, PVC and aluminum metal manufacture.

The only known rock salt deposit occurs in the Mandi district of Kulu Himalaya associated with Shali Formation. Much of the rock salt consumed in India is imported from Pakistan where it occurs in the Salt Range of West Punjab with a well-known deposit at Khewra. The salt lakes of Rajasthan are also considered as important source of salt production in the country. The principal salt lakes are at Samber, Didwana, Pachbadra, and Lunkaransar.

Barite ($BaSO_4$)

Barite when pure is white, opaque to transparent. It is very heavy (specific gravity: 4.3–4.6), which distinguishes it from most other nonmetallic minerals that otherwise may resemble it. The mineral is used in paint and varnish industry, as an extender and filler in paper, linoleum and rubber. It is also used extensively as an ingredient in the "heavy drilling muds" in oil industry. In fact, the major part of barite mined is utilized in this industry.

Barite deposits are chiefly of three different types: (a) veins replacing limestones and dolomites; (b) residual deposits in argillaceous formations derived from the weathering of barium-bearing rocks; (c) bedded deposits of barite in volcano-sedimentary successions. Barite occurs in Pb and Zn sulfide ores along with fahlore, Mn oxide, fluorite, carbonate, and quartz. In some deposits, some of these metals may even be by-products of barite mining.

Indian distribution

Kurnool, Cuddapah, and Anantpur districts of Andhra Pradesh are the most important repositories of barite in India. Smaller deposits of barite are found in Rajasthan, Jharkhand, Odisha, Madhya Pradesh, and Karnataka. Most of these deposits are of the vein type. Rare bedded deposits of barite are much rarer. They are recorded from the Archean Sargur succession at Ghattihosahalli in Karnataka and Proterozoic volcano-sedimentary successions in Cuddapah district of Andhra Pradesh and Udaipur district of Rajasthan.

Andhra Pradesh: At Mangampeta in Cuddapah district occurs the largest barite deposit in the world, with a resource of 75 million tons, which accounts for about 28% of world reserves and 98% of that in India. Lensoid bodies of barite occur interlayered with tuffs, carbonaceous shales and dolomitic carbonates of the Pullampet Formation ($\sim$1300 Ma old) of the Cumbum sequence of the Nallamalai

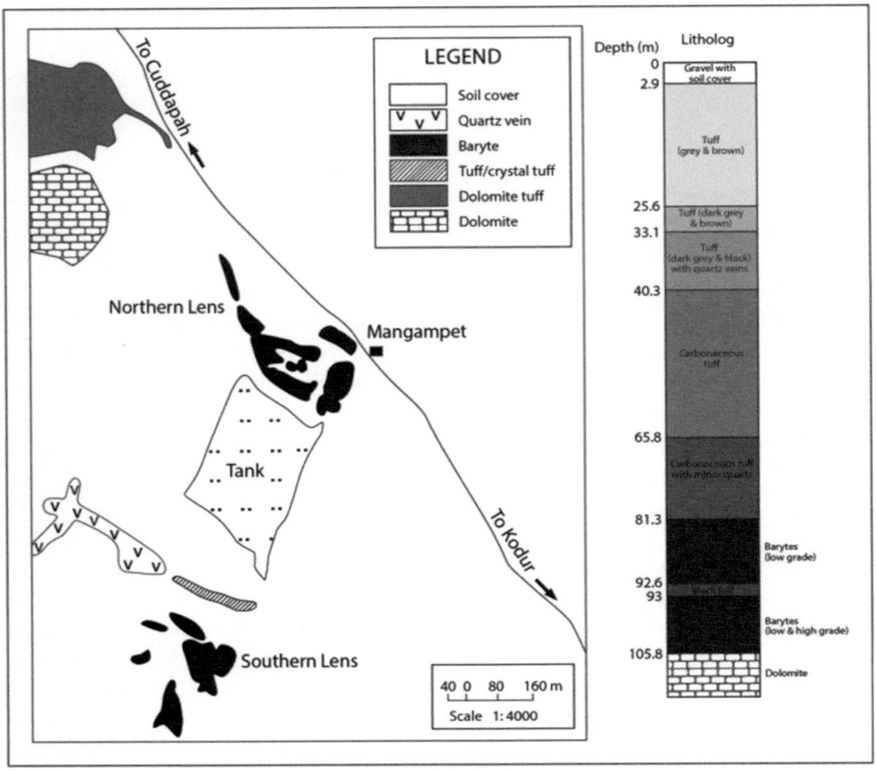

Fig. 4.13 Geological map of Mangampet barite mine and litholog of the northern lens (after Deb and Pal 2015)

Group of Mesoproterozoic age (Neelakantam 1987). The deposit comprises two separate lenses (Fig. 4.13): the northern lense, at least 1.2 km long and 900 m wide with 21 m average thickness, covering an area of 3/4th km^2 and having the major portion of the reserves, is being presently mined; the southern lens is 300 m in strike length with a width of 70 m and steep dips toward ENE. The barite is of four types: granular, lapilli/rossette, vein and replacement (Neelakantam 1995). The barite on the average analyzes 75% $BaSO_4$. Field and petrographic evidence suggest that the barite deposit is of volcanogenic origin whereas sulfur isotope data of the barite and associated pyrite indicate that they were derived from sea water sulfate by bacteriogenic reduction (Karunakaran 1976; Neelakantam 1995). Total Carbon values in the tuff samples range between 0.01 and 5.49 wt% and $\delta^{13}C_{org}$ values, range between −28.0 and −32.4‰ and $\delta^{34}S_{SO4}$ between +43.4 and +44.1‰. These values indicate biogenic source for carbon and a modified seawater for the barite sulfur respectively (Deb and Bheemalingeswara 2008). The layering in the barite ore is somewhat irregular to wavy, which suggests that it may have been of algal origin. Deb and Pal (2015) envisage (Fig. 4.14) mixing of seawater

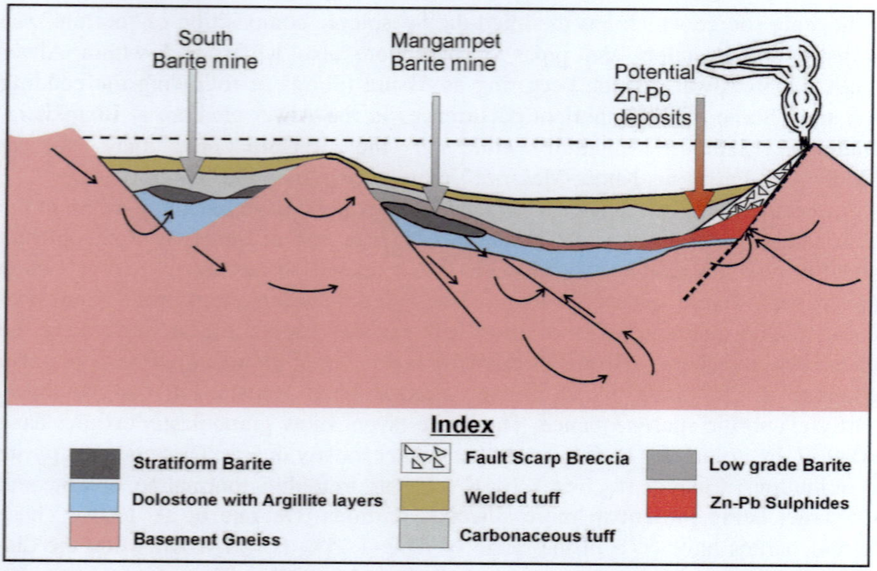

Fig. 4.14 Model for the origin of Mangampet barite deposit (after Deb and Pal 2015)

with hydrothermal exhalative solutions, carrying barium, in a restricted basin with prolific biologic activity, receiving argillaceous sediments from continental sources and pyroclastic matter from explosive volcanism. The restricted nature of the basin resulted in a gradual decrease of oxygen content in the bottom waters and low recycling rates for the organic matter. Change from normal marine to euxinic conditions helped the formation of syn-diagenetic pyrite. Barite veins are also reported from the neighborhood of Mittamidapalle, Uppalapalle, and Rajupalem, all in Cuddapah district. Most of the veins appear to be related to the Cuddapah traps and appear to have formed by replacement in the Vempalle limestone.

Kurnool district also contains high grade deposits of barite at Ippatla, Midipenta, Nadipalle, Kottapalle, and Balapalapalle. In Kottapalle area the barite veins are about a meter wide and extend for 2.5 km. They are fissure-filling type veins in dolomitic country rocks. Barite deposits occur at Nerijumapalle, Mutsokota, and Chandana areas of Anantapur and also near Khammam.

Rajasthan: Five major "en echelon" lenses of barite occur interbedded with mafic metavolcanics and some recrystallized carbonates in a ~1.6 km linear zone within a quartzite band trending NW–SE, about 3 km northwest of Jagat near Udaipur. The band is surrounded on all sides by Banded Gneissic Complex rocks close to their western contact with the basal Aravalli metasediments (Deb et al. 1991). These barites have exceptionally high $^{87}Sr/^{86}Sr$ ratios (0.72284–0.72382) and $\delta^{34}S$ between 17.11 and 21.24‰ (x = 18.75‰). Thus these barites bear the signature of Proterozoic sea water whose sulfur isotope composition is nearer to present day oceans. The Sr isotope values, however, point to a local highly radiogenic source which has modified the Sr isotopic composition of "normal" sea water. Veins, veinlets and pipes of barite are also widespread within Alwar quartzites in Alwar district, occurring as fissure fillings or following the bedding foliation. Some of the important occurrences in the Alwar area are at Bhankhera, Bhurasid, Sainpuri, Jharauli while in the Rajgarh area they are at Madhogarh-Ferozpur, Khora-Makrora, Jamroli, etc. (Jain 1987).

Karnataka: The occurrence of bedded barite deposit of sedimentary origin in the Chitradurga district, west of Ghattihosahalli, was first recorded by Radhakrishna and Srinivasaiah (1974). The deposit is considered to belong to Sargur Group ($\geq$3.0 Ga). Barite occurs as thin (<1 m), discontinuous stratiform bands and lenses within quartzite, in association with metamorphosed limestone close to the underlying metabasic and meta-ultramafic rocks. Graphitic metasediments are also common in this group, which is considered to have deposited in shallow water, partly evaporitic shelf sequence. The barite layers show granoblastic texture, have 60–90% by volume of barite and the remainder mostly quartz. Disseminated pyrite is ubiquitous; Cr-mica (fuchsite), Ba-K feldspar, tremolite, tourmaline, epidote and Cr-garnet being present in more siliceous portions (Devraju et al. 1999). These Sargur barites have $\delta^{34}S$ in the range of 4.02–7.45‰ (x = 5.40‰) while the disseminated pyrites have $\delta^{34}S$ between −3.62 and −5.87‰. These barites are also characterized by low $^{87}Sr/^{86}Sr$ ratios ($\sim$0.7018 $\pm$ 7). The isotopic data most probably reflect Archean coeval sea water composition while their mantle-like values indicate buffering of the sea water by mantle-derived basaltic crust (Deb et al. 1991).

Fluorspar (CaF_2)

This mineral has a wide industrial use, such as in open hearth steel plants, in the manufacture of aluminum, enamel, glass and ceramic products, refrigerants, insulating foam and in production of hydrofluoric acid. Metallurgical grade requirement is CaF_2 >60% with good granularity and low contents of SiO_2, S and Pb. For chemical grade, CaF_2 should be >97% and very low in S and P.

Flourspar usually occurs as fissure filling and replacement veins in association with calcite and quartz in a variety of rocks, such as, limestone and dolomite, nepheline syenite, carbonatite, and various volcanic rocks.

Most economically important fluorspar deposit in India occurs in the Ambadongar carbonatite complex in the Panchmahal district of Gujarat. Fluorite-quartz mineralization occurs as veins, cavity fillings and replacement bodies in the feldspathized sandstone (Nimar sandstone) and at the contact of sandstone with the carbonatite. Fluorite is found in various colors but blue and yellow colors predominate. Subramaniam and Parimoo (1963) described it as an example of mineralization related to Deccan volcanism. Magmatic fluids emanating from the basaltic magma supplied the fluorine to the calcareous Bagh sediments (Bagh limestone)to produce the fluorspar mineralization. This is more or less supported by Palmer and Williams-Jones (1996). However, Sukheswala and Udas (1967) regarded the fluorspar mineralization to be genetically related to the carbonatite occurrence recorded by them.

Fluorspar is found in a number of locations in Madhya Pradesh amongst which the deposits at Khairagarh, Nandgaon, Chandidongri, and Malhan (Deb 1980) are better known. Important deposits of fluorspar are also located in south Rajasthan at Mando-ki-pal, Chowkri Chapoli, Jhalara and Asind in Dungarpur district. The deposits occur in sheared granites within Aravalli rocks.

Borax and Borates

Na borate, known as borax, is the main salt of boric acid ($Na_2B_4O_7{\cdot}10H_2O$). This is used to produce other commercially important borates, such as, monosodium metaborate ($NaBO_2{\cdot}4H_2O$), ammonium pentaborate [$(NH_4)_2{\cdot}5B_2O_3{\cdot}5H_2O$]. Important borate minerals are kernite, colemanite, ulexite, priceite and sassolite.

Borax is easily soluble in water producing a mildly alkaline, antiseptic solution. It has low melting point and good fluxing properties for brazing, welding, soldering of metals. It is an excellent cleansing agent and is widely used at homes for washing, cleansing and laundering constituent of soaps. It also has medicinal use, as a disinfectant, mouth wash, tooth powder, cosmetics, deodorants, etc., making it one of the most useful salts.

Borax is generally a constituent of evaporate sequence.

4.2.5 Minerals for Electrical Industry

Asbestos is essentially a commercial name for a group of silicate minerals characterized by fibrous habit. These minerals have wide variety of compositions and thermal resistivity. They can be easily split into fibers of extremely fine filaments capable of being felted or spun together. These minerals are used in making heat-resistant roof-sheets, pipes and in spinning heat-resistant fabrics. Principal producers of asbestos in the world are Russia, China, Brazil, Kazakhstan, and Canada.

Asbestos is divided into two distinct groups, the serpentine group and the amphibole group. The serpentine group includes the elastic and the silky chrysotile variety, and accounts for 95% of world's asbestos production. The latter comprises short and brittle-fibred orthorhombic amphiboles, anthophyllite and amosite, and the monoclinic varieties, tremolite, actinolite, crocidolite (a fibrous form of the mineral riebeckite). Chrysotile is generally of short cross-fiber. The fibers are flexible, less brittle than amphibole asbestos with a fair to high tensile strength. They are thus more suited to spinning into yarn and woven into cloths. The commercial utilization of asbestos depends to a large extent on its chemical composition, length and strength of its fibers. Chrysotile is the more valuable asbestos. When low in iron, it is a very good electrical insulator. Tremolite asbestos is more resistant to chemicals and is thus better suited for chemical filters. Asbestos is always associated with serpentine which is commonly a product of hydrothermal alteration of ultrabasic igneous rocks such as, peridotite and dunite or a product of metamorphism of dolomitic limestone.

Indian distribution

High-quality chrysotile asbestos in India is found in the state of Andhra Pradesh (Fig. 4.8) while the amphibole asbestos is distributed in places of Jharkhand, Rajasthan, Tamil Nadu, and Karnataka.

Andhra Pradesh: The most important occurrences are in the districts of Kurnool and Cuddapah. In the Pulivendla taluk of Cuddapah district, between Brahmanapalle and Gopalanntula, asbestos formed over a distance of 15 km by hydrothermal activity of doleritic intrusions on Vempalle limestone. The asbestos is cross-fibred and chrysotile in variety with thickness of veins and seams varying from <1 mm to 15 cm (Anantha Iyer et al. 1987). Reserves are estimated at 0.25 million tons (Dwivedy 1995). Veins of chrysotile asbestos occur also in Rajupalam area of Cuddapah district. In Kurnool district asbestos occurs in Dhone taluk in Vempalle limestones associated with trap rocks.

Jharkhand: Amphibole asbestos is reported extensively from Singhbhum district of southern Jharkhand where a total of about 99 occurrences are known. Crysotile asbestos is however rare. Amphibole asbestos occurs in actinolite-tremolite talc chlorite rocks, northeast of Chaibasa and also around Manpur. Asbestos occurs as veins, which die out at shallow depths. The length of the veins ranges from only 2.5 cm to 21 m and the width from a cm to 0.9 m. Some amphibole asbestos occurs in the dunite and peridotite rocks, exposed in the chromite quarries west of Chaibasa. Asbestos mineralization also exists in the Saraikela area. The probable reserve in the more important 32 localities is estimated at 0.4 million tons within a depth of 6–30 m (Deb 1980).

Rajasthan: Chrysotile variety of asbestos is obtained from six localities of Udaipur district and two in Ajmer district. All other occurrences that are found in Udaipur, Dungarpur, Bhilwara, Ajmer, Jodhpur and Pali districts are of amphibole variety only.

Micas

Since the dawn of the electrical age in the late nineteenth century, mica has been an important raw material for the electrical industry. In India, *abhra* or mica finds mention from the ancient times and is still an essential ingredient in *Ayurveda*. Mica sheets and powder, either in its natural state or after tinting, has been extensively used for decorative and ornamental purposes. Mica is recovered from its associated minerals solely by hand picking. Crude mica is usually in the form of rough books of varied shapes and sizes. These books are first cobbed to remove dirt and gangue minerals and then rifted or split into suitable thickness, after which major flaws are removed by dressing and trimming. The mica scrap produced during dressing is used for the manufacture of micanite, insular mica bricks used in steel plants and allied products. A small amount is used for manufacture of mica powder used for rubber and tyre industries.

Among the micas, three members: muscovite, phlogopite and vermiculite (vide Appendix A for compositions) are economically useful. Muscovite is the most important mica. Lepidolite mica is a source of lithium. Good quality phlogopite is also used. Workable muscovite deposits are confined to three belts in Bihar and Jharkhand, Andhra Pradesh and Rajasthan (Murthy 1964). The first one, known commonly as the *Bihar mica belt* is the most important. It extends for about 160 km with an average width of 16 km from Gaya district in Jharkhand on the west through Hazaribagh and Monghyr districts to Bhagalpur district to the east, covering an area of over 3900 km^2. The mica belt is constituted by a series of metasediments: fibrolite-muscovite-biotite schists, calc silicate granulites and gneisses, hornblende schists, micaceous quartzites, and conglomerates. These rocks have been metamorphosed under amphibolite facies conditions. Granitic plutons have intruded these metasediments. The long, narrow, lensoid mica pegmatites are localized mainly in the mica schists and micaceous quartzites and rarely in the hornblende schists, and show sharp contact with the country rocks. They have been emplaced along favorable structures, such as, noses and limbs of folds, foliations, etc. The largest of the pegmatite bodies are 500 m long and 30 m wide. The maximum vertical depth of pegmatite mines is about 200 m. Zoned or composite pegmatites have petrographically distinct mappable units which may have been emplaced at different times and can also occur individually as discrete homogeneous pegmatites. The muscovite-quartz (±sillimanite) pegmatites seem to be the earliest while the lepidolite-beryl-cleavlandite pegmatite is the latest, occurring always as a replacement unit. In a few zoned pegmatites, the central quartz body is separated from the wall rocks by mica-rich pay shoots. Economically, clear ruby muscovite is the most common with books as much as half a ton at places. Green muscovite is common in the northeastern part of the belt. Mineralogically, the essential minerals are quartz, microcline-perthite, plagioclase and muscovite. Accessory minerals include biotite, tourmaline, apatite and garnet. Beryl, garnet, arsenopyrite, magnetite, ilmenite, columbite-tantalite and pitchblende, zircon, allanite, monazite have been reported from different pegmatite bodies.

The *mica belt* in *Nellore* district, Andhra Pradesh is sickle shaped and extends for about 96 km with an average width of 16 km in a NNW–SSE direction. The country rocks include biotite- and hornblende schists, interbanded with minor quartzites, feldspathic schists, etc. The rocks are metamorphosed under amphibolite facies conditions. The mica pegmatites are most frequently found in biotite schists and are commonly simple, though zoned pegmatites with a central quartz core have also been recorded at a few localities. They occur as discrete or connected lenses, often arranged in en echelon pattern. The length of the pegmatites is variable: 10–550 m or more with an average thickness of 3–5 m. Some pegmatites, e.g. at Saidapuram, have been mined to a depth of 160 m. The essential minerals are quartz, albite, microcline-perthite and muscovite. Muscovite occurs in shades of green or in between green and light ruby, known as "honey mica". Green potash feldspars are only found in Nellore pegmatites. Biotite and tourmaline are common accessory minerals. Less common are apatite, beryl, samarskite, columbite-tantalite and allanite.

The *Rajasthan mica belt* extends for about 320 km from Jaipur district to Udaipur district, with a NNE–SSW trend through Ajmer, Bhilwara, Tonk and Pali districts with an average width of 96 km and covering a total area of 3000 km^2. The pegmatites are confined to Precambrian metasedimentary rocks of Aravalli and Delhi supergroups, metamorphosed to amphibolite facies and intruded by granites of different ages. They are generally intrusive into steeply dipping muscovite-biotite schists, micaceous quartzites and less commonly into amphibolites, dolomites and calc silicate rocks. Tabular, branching and lensoid bodies are common. The larger bodies could be 500 m long and 100 m in width. Some of them are now exploited in open cast mines (cf. Fig. 2.11). The essential minerals are quartz, microcline, microcline-perthite, plagioclase, and muscovite with subordinate amounts of biotite and tourmaline. Muscovite generally occurs in shades of pink, but black and green spotted varieties are also common. Accessory minerals are garnet, apatite and beryl. These minerals along with tourmaline are known to attain very large size in some of these pegmatites. Cassiterite, spodumene, lepidolite, samarskite, columbite-tantalite, zircon and some radioactive minerals have been also reported from some of the pegmatites, among other minerals.

Origin of pegmatites in the mica belts of India is generally considered in terms of the role of fluids, rich in volatiles (H_2O, F, B, etc.) and K, Na, Li, Al, Si and the contribution made by the mica schist country rock. The pegmatites occurring within mica schists are richer in mica books compared to those occurring within other rock types, such as, granites, hornblende schists, etc. While it is reasonable to assume that the mica schists have contributed in part toward the contribution of mica, the other important accessory minerals such as, beryl, columbite-tantalite, samarskite, pitchblende, lepidolite, etc. were evidently derived from other sources. The formation of mica pegmatite is favored in a geological setting where the country rocks have been metamorphosed under amphibolite facies conditions at a level higher than where granitization takes place.

Steatite, talc and pyrophyllite

Steatite is a massive, impure variety of talc. When heated to a temperature above 1000 °C for 6 h it converts to a hard material with minimum shrinkage. Cut into blocks it makes a good electrical insulator.

Steatite occurs in dolomitic marbles or ultrabasic igneous rocks. High grade deposits are found in Andhra Pradesh, particularly in Anantapur, Cuddapah and Kurnool districts. Important deposits also exist in Rajasthan (Jaipur, Bhilwara and Udaipur districts), Madhya Pradesh (Beraghat near Jabbalpur) and Uttarakhand (Almora district).

Pyrophyllite [$Al_2Si_4O_{10}(OH)_2$], white to pale green in color, is commonly grouped with talc [$Mg_3Si_4O_{10}(OH)_2$] because of similarity of physical and optical properties. With a hardness of 1–2 it is used in crayons, as filler and extender for paints and insecticides. Finds use as refractory. Good pyrophyllite deposits occur in Udaipur district of Rajasthan at the contact of the Banded Gneissic Complex with the sedimentary cover of the Aravalli Supergroup rocks. These deposits are considered to be a part of a metamorphosed paleosol sequence (Pandit et al. 2008)

Vermiculite

Composition in a simple form is $(OH)_2(Mg, Fe)_3(Si, Al, Fe)_4O_{10}{\cdot}4H_2O$. Micaceous structure, hardness 1.5. When heated, vermiculite exfoliates by expanding at right angles to the cleavage into worm like pieces, leading to an increase of bulk volume by 8–12 times. Vermiculite is used for home insulation, in cold storage, refrigeration and in various types of high temperature and low temperature industrial equipment. Plasters with vermiculite are fireproof and crack resistant. Expanded vermiculite is used in horticulture for retaining moisture in soil. Also used as a lubricant and filler.

Vermiculite is formed by hydrothermal alteration of ultrabasic igneous rocks such as, pyroxenites.

4.2.6 Minerals Used in Glass and Ceramic Industry

Porcelain and China wares are produced from a mixture of silica and alumino-silicates, the raw materials commonly being kaolin, feldspar, and quartz. In addition, a few other commodities, such as, limestone, gypsum, talc, and steatite find use in glass and ceramic industry. Finely ground mixture of the three commodities is fired in a kiln in sagger containers so that the furnace flames do not affect the wares. For high class porcelain the firing takes place for several hours at 1400 °C. Molds of Plaster of Paris, made from gypsum are used for shaping the wares. Talc is utilized in the manufacture of white wares, electrical porcelain and high tension insulators.

Silica (Quartz)

Sources of this very useful material are glass sands, orthoquartzites, highly siliceous sandstones, vein quartz and quartzo-feldspathic pegmatites. For glass making, granularity and chemical composition are important. Silica content should be very high (between 92.5 and 97.5%) with limited alumina (< 0.6%). Chromium and iron oxides are undesirable because they make the glass colored. For optical glass their content should be very low (between 0.005 and 0.008%) but for dark glass, such as those used in bottles, it could be as high as 0.7%. The silica source after grinding to the required size is generally beneficiated to get rid of the impurities such as, clays, mica, iron and titanium oxides etc.

The glass sands are the disintegrated products of siliceous sandstones, but may also be obtained from rivers sands or beach sands. The Naini sands of Uttar Pradesh are found extensively in Lohgarh and Bargarh areas and represent weathering product of upper Vindhyan Kaimur sandstones. These deposits extend into Banda district of U.P and Rewa district of M.P. Good quality glass sands are also found in Bundi district of Rajasthan at Borodhia. Occurrence of glass sands are also known in Churulia area, Burdwan district of West Bengal, being weathering product of Lower Barakar sandstones. Similarly the Rajmahal area of Bihar has glass sand deposits from the Damuda sandstones. Milky white quartz veins and reefs are widespread in the state of Karnataka and are used for glass manufacture at several places including the Kolar district.

Feldspar

Feldspar for ceramic industry should not have more than 68–72% silica and not more than 0.10% Fe_2O_3. The potash feldspars, orthoclase and microcline are economically more important than the other varieties. They are found in pegmatites and various acid igneous rocks. Microcline is used mainly in glass and ceramic products while Soda feldspars are used for glazing purposes. Feldspars are found in the three mica pegmatite belts in Bihar, Rajasthan and Nellore described earlier. Well known occurrences are in Kodarma area and in the region of Mihijam and Jamtara in the border of Bihar and West Bengal. Other producing areas are in Salem and Tiruchirapalli district of Tamil Nadu and Hassan district of Karnataka.

Clays

Kaolin (or China clay) and Ball clay are mainly used in porcelain and ceramic industries. China clay also finds use as filler in paper and rubber industries. Kaolin and Ball clay are represented mainly by the kaolinite group of minerals and are formed by the decomposition of feldspathic rocks such as, granite gneiss, granites and pegmatites. The plasticity of clays depends on viscosity and bonding strength. The casting properties of clays, important for various porcelain wares, are controlled by viscosity, texture, grain size and soluble salts in clays. Good deposits of China clay are found widely in India: in Bhagalpur and Monghyr districts of Bihar, Ranchi and Singhbhum districts of Jharkhand; Mayurbhanj district of Odisha; Bankura and Birbhum districts of West Bengal; Banda district of UP.; Gwalior, Katni and Chindwara districts in MP; Barmer, Bikaner, Pali and Ajmer districts of Rajasthan; Sabarkantha district of Gujarat; Chanda and Ratnagiri district of Maharashtra and Udhampur in Jammu and Kashmir; Arcot and Salem districts of Tamil Nadu; Adilabad, Anantapur, Nellore and Guntur districts of Andhra Pradesh and Shimoga and Hassan districts of Karnataka. There are also some deposits of China clay in Meghalaya and Tripura in northeast India. Special mention may be made of the Pathuriaghata hills' China clay on Damuda arkosic sandstone in West Bengal that was exploited to make the first porcelain factory in India, the Bengal Potteries Ltd in Calcutta, more than 100 years ago. A young nationalist entrepreneur, Satyasunder Deb, persevered to receive process training in Japan to pioneer this industry in India. There are also several other deposits in the nearby Rajmahal hills and the well-known Karanjia mine near Chaibasa which catered to this growing industry in eastern India.

4.2.7 Mineral Fillers

Filler is a substantially inert material not used as a reactive or essential component, the primary function of which is to modify the properties or processing of the composition of a commodity or fabricated article. Specifically the function of mineral fillers is to modify optical, physical or chemical properties or fill voids and modify working characteristics.

Some important fillers and their applications are:

Asbestos—Floor tiles, plastics
Carbon—Rubber, printing ink, paint
Clay (Kaolin)—Paper, rubber, linoleum, paint, textile
Clay (Bentonite, Fuller's earth)—Insecticides
Micas—Roofing, paint, rubber, wall paper, plastics etc.
Carbonates—Fertilizers, asphalt, oil paint, rubber, linoleum, paper.
Talc—Paint, rubber, roofing, insecticides, paper, textile etc.
Crystalline silica—Cement and plasters, asphalt, paint, rubber etc.
Diatomite—Paint, plastic, pesticide, paper, rubber etc.
Lime—Asphalt
Cement—Asphalt
Calcium sulfate—Fertilizer, paint, paper, cloth, insecticide

4.2.8 Mineral Pigments

These are minerals that give color, opacity or body to paint, plaster, stucco, mortar, cement, linoleum, rubber or similar material. They are classified as:

1. Natural mineral pigments
   (a) Iron oxides, e.g., hematite, limonite
   (b) Clayey material such as, ochre, sienna, umber and colored shale. Common composition of ochre is Fe_2O_3 17–60%, SiO_2 35–50%, Al_2O_3 10–40%, LOI 10–12%; Sienna is Fe_2O_3 25–75%, SiO_2 10–35%, Al_2O_3 10–20%, MnO_2 small amount, LOI 10–12%; Umber is Fe_2O_3 37–60%, SiO_2 16–35%, Al_2O_3 3–13%, MnO_2 11–23%, LOI 10–15%.
   (c) Graphite
2. Synthetic pigments: various colors including lithophane, Zn white, Ti white, ultramarine, Prussian blue etc.

4.2.9 Abrasives

These are natural minerals or synthetic materials that are used to grind, polish, abrade or clean solid materials through rubbing action or by impact, such as in sand blasting. The important physical properties of such abrasives are hardness, brittleness, grain shape and size, character of fracture or cleavage. When abrasives are bonded, such as on grinding wheels, additional characteristics include stability under high heat and bonding character of grain surface. Obviously, economic factors of cost and availability are also important.

Abrasives are of two types: (1) Natural abrasives and (2) Manufactured or synthetic.

The natural abrasives are classified according to their hardness:

(a) High hardness (>7): Diamond 10, Corundum 9, Emery 7–9, Garnet 6.5–7.5
(b) Intermediate hardness (5.5–7): Silica abrasives—chalcedony, chert, quartz, quartzite, sandstone, silica sand.
(c) Low hardness (<5.5): Apatite, Dolomite, Sillimanite, Iron oxide, Talc, Pumice, Chalk, Clay, Diatomite, Siliceous shale, Tripoli.

Manufactured abrasives include several carbides, such as, boron carbide, silicon carbide, tantalum carbide, tungsten carbide; zirconium silicate, tin oxide, chromium oxide, magnesia, rouge, ferric oxide, fused alumina, various bonded and coated abrasives.

Industrial diamonds are one of the most important and essential abrasives in modern industries. Diamond drilling is required for exploration of all minerals, oil and gas, for technical studies related to foundation of dams and large buildings, etc. Other uses include diamond tipped tools, diamond toothed saw, diamond wheels, diamond dust bonded with resins, etc. There are three main types of industrial diamonds:

(i) Bort: These are off-colored, flawed or broken diamonds unsuitable for use as gem.
(ii) Carbonado or black diamond: Very hard and tough aggregate of very small diamond crystals.
(iii) Ballas: Very hard and tough mass of diamond crystals radiating from a common center.

Corundum is naturally crystallized oxide of aluminum, which is next to diamond in hardness. Corundum occurs in peridotites, nepheline syenites and feldspathic rocks; also in granulites and khondalites. Emery is a natural mixture of corundum, iron oxide and other impurities and has been used as an abrasive for centuries. Emery usually occurs in crystalline limestone and schists and also in peridotites and diorites. Emery papers of varying grades (coarseness) are widely used for polishing ores and other solid materials.

Garnets are a common abrasive. They vary widely in color, hardness, toughness and fracture. Color does not have any effect on abrasive quality. Garnets are usually crushed in different sizes and pasted in coated papers. The raw material for the high hardness abrasives, other than diamond, is available in the states of Andhra Pradesh and/or Rajasthan in India.

4.3 Industrial Rocks

4.3.1 Building Stones

Stones have always found use in building industry in a variety of ways. In recent years, spurt in construction, including building, structural works, road and pavement making, has created a huge demand for minerals and rock-based aggregates of inorganic origin. Usually all types of stones which are hard, tough, and can withstand weathering and abrasion, that is, have high durability, are given preference in this industry. The workability of building materials depend on their hardness. Colors and directional properties of building stones are also considered to be the other important criteria. All types of rocks, in particular, massive and compact sandstones and limestones, dolomites, marble, slates, granite-gneisses, crystalline schists, khondalites, charnockite, and compact laterites, are used extensively as roofing and building material. These rocks are used directly as substitutes for bricks with some amount of trimming of the surface and sometimes by processing them in measured dimensions. Such building stones which are cut and dressed after quarrying are called "dimension stones". The dimension stones are also used for other structural purposes such as construction of bridge pillars, abutments, fences, retaining walls, monuments, paving stones, switch-boards, etc.

The industry of building and dimension stones in India today reportedly does a business of more than INR 10 billion per year.

Indian building stones

The gneisses, granites, charnockites, slates, crystalline limestones, marbles, and sandstones and metamorphosed basic rocks are considered as excellent building materials in India, as elsewhere.

Jointed granite gneisses and granites are easier to quarry. Such rocks are abundant in Peninsular India. They also occur in different localities of extra-Peninsular India. The banded gneiss rocks and Erinpura granites of Rajasthan, the Bundelkhand gneisses of Uttar Pradesh, and similar granite-gneissic rocks of Jharkhand, Madhya Pradesh, Andhra Pradesh, Tamil Nadu, and Karnataka have provided excellent building materials for the construction of temples, palaces, monuments and tomb stones, etc. throughout history in almost all the states in India. The pink, white, and grayish white Precambrian granites of Rajasthan are used in royal palaces in Bikaner, Jodhpur, Mewar and other regions. Granite-gneisses of different Archaean complexes, such as those occurring in Chitradurga and other adjoining districts of Karnataka, are used as size stones, slabs, pillars and pedestals and are exported to foreign countries to serve as kerb-stones and tomb stones. The oft quoted "black granite" is in fact a metabasite.

The charnockites of Tamil Nadu, Karnataka and Andhra Pradesh are considered as one of the most durable stones in the world. The tomb of Job Charnock, long credited with the foundation of the city of Calcutta (now Kolkata), is made of charnockite from the St. Mary's hill in the outskirts of Chennai. The temples and monuments of Mahabalipuram, south of Chennai, have been carved out of solid monoliths of charnockitic rocks. These represent great feats of artistry and craftsmanship in ancient India.

Khondalites of Odisha and Andhra Pradesh are not as durable as the granites but are still used extensively as building stones in these two states. In Andhra Pradesh, most of the buildings are made of khondalites and compact laterites. In Konark and Puri temples of Odisha, most of the stones are either khondalites or hard and compact laterites. The statues and figures are invariably carved out of khondalites, particularly the gigantic wheels of the chariot in the Sun temple of Konarak. Laterites of Western Ghats are used as building stones in different parts of Maharashtra, Karnataka and Kerala. Laterite can be cut and shaped very easily into required sizes before it hardens considerably on exposure.

Crystalline limestones and marbles of Rajasthan, particularly the Makrana marble, are being used for many centuries as building and ornamental stones. The great Taj Mahal of Agra and Victoria Memorial of Kolkata are built partly with Makrana marble (Fig. 4.15a). Marbles of Raialo Group of Rajasthan are extensively quarried in Raialo, Alwar, and Jaipur. Marbles are exploited in Rajnagar, Kankroli and Nathdwara in Mewar. The white marble of Betul and multicolored marbles of Chindwara, Nagpur and Narsinghpur are also valuable building stones. Motipura marbles from Baroda district, Gujarat are serpentinous marbles, mottled with pink and rose striations, which have preferred use in the construction of temples. In Koraput district and Gangpur region of Odisha, there are several varieties of crystalline limestones which are used as building materials. Vindhyan limestones of lower Bhander Formation, occurring near Gwalior, contain spherulitic structures, in which the semi-circular shells display different colors and make the rock more attractive.

The limestone and sandstone deposits of Vindhyan Supergroup are quarried at many places of their occurrence, particularly along the Son valley in Bihar and Uttar Pradesh and also in Rewa and Jabalpur in Madhya Pradesh. Extensive and large scale quarrying of Vindhyan limestone is carried out near Kota in Rajasthan. These stones provide the most common greenish colored flooring material and sold as "Kota stone". The Khatu sandstone from Jodhpur district, Rajasthan, makes very good flagstones, particularly suitable for fine carvings. They are also considered good for fabricating perforated and ventilating windows and screens, usually found in palaces and big buildings. Vindhyan sandstones of Bhander Group of uppermost Vindhyan age serve as excellent building stones due to their regular bedded nature, uniform grain-size, soothing colors, high durability and easy workability. The stones are cream and light gray in color with crimson and pinkish tints. The famous Sanchi Stupa and stupas of Sarnath and Barhut, which represent some of the great

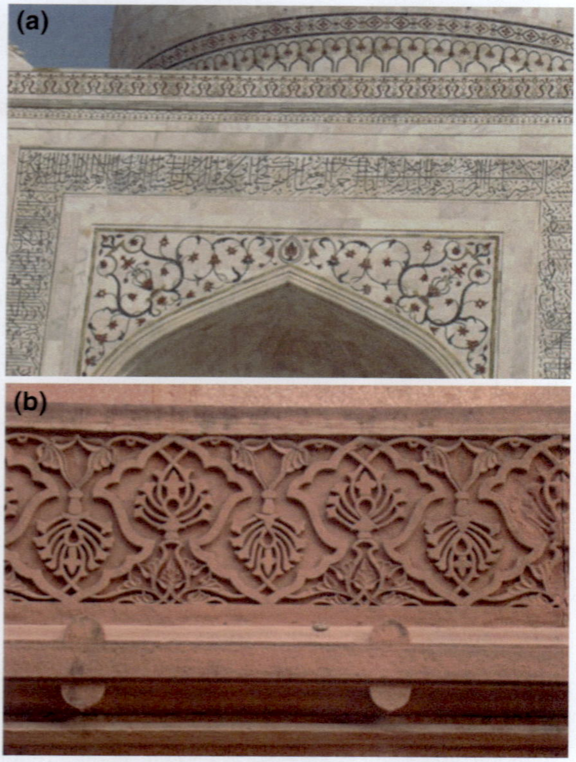

Fig. 4.15 **a** Exquisite inlay work with semiprecious stones (e.g. cornelian, malachite, jade. Lapis lazulli etc.) on Makrana marble making up the façade of Taj mahal at Agra (Photo: M. Deb). **b** Beautiful carvings on red Vindhyan sandstone in Agra fort (Photos: M. Deb)

Buddhist heritages in India, are built of Vindhyan sandstones. The famous royal township of Fatehpur Sikri and the Agra fort, built by Emperor Akbar, are entirely made of red Vindhyan sandstones (Fig. 4.15b). The landmarks of old Delhi, Red Fort and Jama Mashjid and the Rashtrapati Bhawan, designed by Lyutens on Raisinha Hill, New Delhi are also made of red sandstones of the Bhander Group. A major part of these sandstones are quarried in Rajasthan, particularly in Bundi, Kota, Dholpur, Jaipur and Bharatpur and also in Mirzapur district of UP.

The Aravalli (Jharol) slates of Rajasthan, which can be easily cleaved, are used as roofing material. The slates occur profusely in the vicinity of Ajmer and Jharol. The Alwar quartzites of Moundla and the micaceous gritty rocks of Ajmer and Nasirabad produce thick and durable building slabs and blocks. The serpentinized ultrabasic rocks along the Rakhabdev lineament, from near Dungarpur in the south and Gogunda in the north in southern Rajasthan, are quarried extensively for flooring material. They take excellent polish, have imposing structural fabric and sold commercially as "green marble".

The Cretaceous Deccan Traps. In Maharashtra produce compact, hard, and durable building materials. In the vicinity of Mumbai, however, the light buff and cream colored trachytes are very much in use locally and they are preferred to the dark colored basalts. The 'Gateway of India' in Mumbai is entirely made of this rock. The trachytes occur extensively in Salsette Island and are also quarried in Malad and Kharodi in the neighborhood of Mumbai.

4.3.2 Crushed Stones and Aggregates

When the large rock pieces are broken into smaller pieces for subsequent compaction, they are called "crushed stones" and aggregates. They are derived out of any of igneous, sedimentary and metamorphic rocks as well as glacial, alluvial, beach and marine gravels. Sand and grit, extracted out of river beds, river terraces, alluvial cones and eskers are also used extensively as constructional material. The fine aggregates, up to 5 mm in size, are used in producing concrete by mixing with cement and lime. The recent spurt in infrastructure development in India has put an enormous pressure on river beds across the country where large scale sand mining has become a cause of serious ecological and environmental concern.

4.4 Gemstones

4.4.1 Diamond

Diamond, the hardest substance known, is also reputed for its dazzling brilliance. When properly faceted, light falling on the stone undergoes total internal reflection giving it this uncommon property. Major part of diamond recovered from its host rocks is of industrial variety. Only a minor part is of gemstone quality. Primary sources of diamonds are kimberlite pipes and vents, lamproite or peridotite dykes and hypabyssal dykes of lamprophyres. Secondary source of diamond is in conglomerate beds, alluvial gravels and sand. The kimberlites are compact, dark-colored potassic ultrabasic rocks, rich in magnesium. Mineralogically, they contain phenocrysts of olivine, enstatite-bronzite, chrome-diopside, phlogopite, and pyrope garnet with minor amounts of ilmenite and perovskite. Lamproites are similar to kimberlites, rich in potassium and magnesium, but with a dyke-like form. Their varied mineralogy includes phenocrysts of phlogopite, leucite, olivine, garnet, perovskite and other accessory minerals. Lamprophyres have phenocrysts of mafic minerals and may range from metaperidotite to mica peridotite. Diamonds are formed under very high pressures. Their growth is not expected to be realized in the normal crust of the earth. It is therefore believed that diamonds originate in the upper mantle or in the root zone of thickened continental crust and kimberlite pipe act as vehicles to bring them to the crust as inclusions (vide Sarkar and Gupta 2012 and references therein).

Indian distribution

Lamproites and kimberlites in India are reported from Bundelkhand craton and Dharwar craton in the northern and southern cratonic blocks of the Indian shield. Lamprophyres are known from the Gondwana rocks of eastern India only (Fig. 4.16). Out of the nine locations in India where lamproites and kimberlites are reported (Scott Smith 2007), only Majhgawan and Hinota in Bundelkhand craton and Wajrakarur and Lattavaram in Dharwar craton are diamondiferous. This author (Scott Smith, op.cit.) also suggests that Majhgawan and Hinota rocks are olivine lamproite lapilli tuffs and not kimberlite as suggested by other workers earlier. While Lattavaram is represented by kimberlite alone, Wajrakarur has two phases,

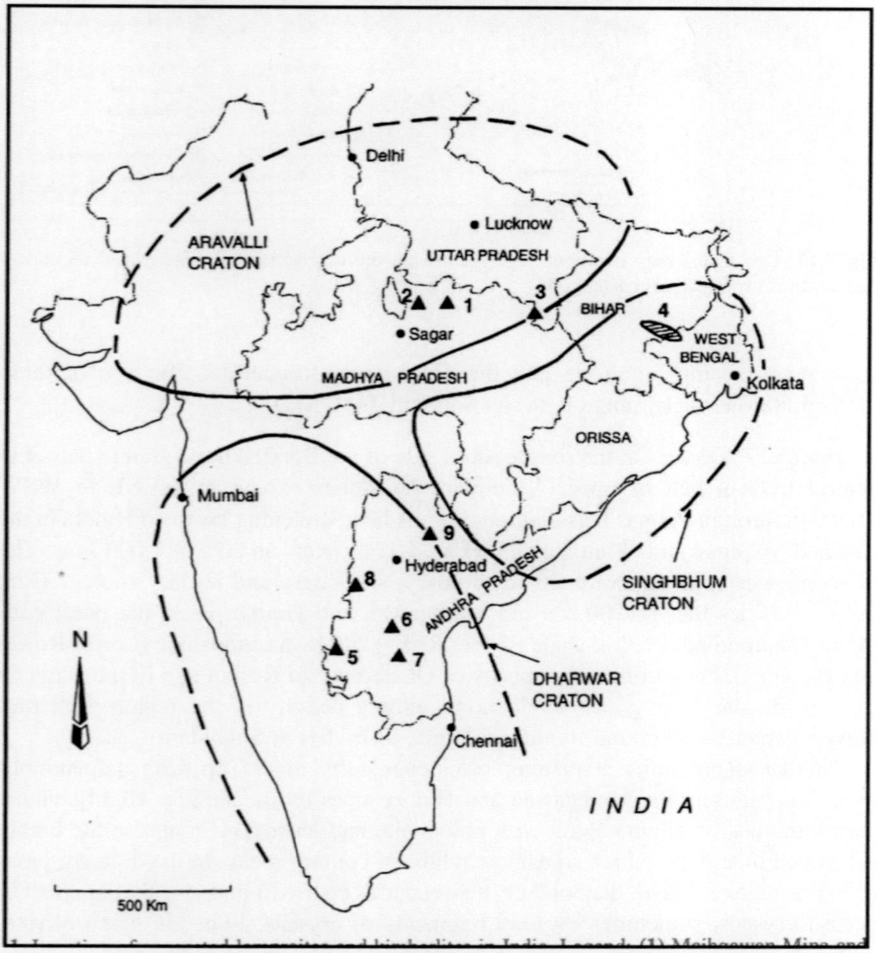

Fig. 4.16 Map showing the locations of lamproite and kimberlite in India (after Scott Smith 2007). (1) Majhgawan and Hinota, (2) Angor, (3) Jungel, (4) Intrusives in Gondwana coalfields, (5) Wajrakarur & Lattavaram, (6) Chelima, (7) Zangamrajupalle, (8) Maddur, (9) Warangal

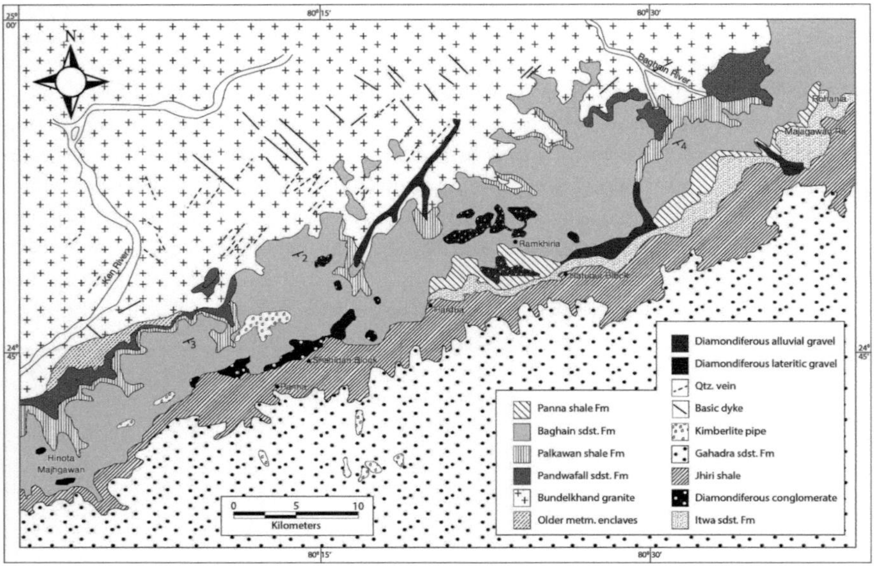

Fig. 4.17 Geological map of Panna region showing the disposition of primary and secondary diamond occurrences (after Rau 2007)

the younger being lamproite and the older being kimberlite. The age of these diamondiferous rocks range between 840 and 1630 Ma.

Madhya Pradesh: On the southeastern side of the Bundelkhand granite massif in central India a belt of upper Vindhyan sandstones extend in a ENE to WSW direction through Panna. This diamondiferous belt, stretching between Hinota in the east and Majhgawan R.S. in the west (Fig. 4.17), covers an area of 1,000 km^2. The diamond workings are confined to primary, secondary and tertiary sources (Rau 2007). Besides the $\sim$1100 Ma old Majhgawan and Hinota pipes, the basal conglomerate interbeds of Jhiri shale (Upper Rewa) and Itwan sandstone (Lower Rewa) and the alluvial and colluvial deposits of Quaternary to Recent age in the Baghain plateau are the main zones of diamond mining activity in the region. The best known center for diamond mining in Panna district is at Shahidan.

Circular depressions containing calcareous tuffs mixed up with serpentinous materials represent the Majhgawan and Hinota pipes on the surface. Highly altered pseudomorphs of olivine along with phlogopite and leucoxene comprise the highly brecciated pipe rock, which contain xenoliths of country rocks. In the different pipes of Panna diamond field, diamond crystals occur as coarse to fine, mostly imperfect to perfect crystals, sometimes even as fragments of crystals. Euhedral green olivine, diopside, garnet, and spinel occur as primary inclusions in the diamonds.

As part of a regional exploration program, Rio Tinto had discovered in 2004 a cluster of eight diamondiferous pipes at Bunder in Madhya Pradesh. The giant mining company estimated 53.7 million tons of kimberlite with diamond content of 34.2 million carats. However, for environmental clearance the company was asked to explore the possibility of underground mining to avoid loss of high quality forest area and wait till the Ken-Betwa River linking project was finalized. Besides, the National Tiger Conservation Authority was also concerned that the project "has the potential to disrupt tiger dispersal around Panna landscape". It is probable that as a reaction, Rio Tinto decided to close this Rs. 2200-crore diamond mine project, which would have created considerable employment of local people and fetched Rs. 208 crore toward royalty and taxes to the government of Madhya Pradesh once excavation began.

Andhra Pradesh: In Andhra Pradesh, several districts, namely Cuddapah, Anantapur, Kurnool, Krishna, and Godavari are famous for diamond production. In all these areas, loose diamond crystals are occasionally picked up from the alluvium. Sometimes diamonds are recovered also from the conglomerates and sandstones of Banganapalli stage of the Kurnool Group, underlying the Palnad limestones (Sarkar and Gupta 2012). The region from where many valuable diamond crystals have been recovered is Wajrakarur in Anantpur district.

4.4.2 Other Common Gemstones

Gemstones usually occur in alluvial gravels due to their highly resistant and chemically inert character and also sometimes in mineral veins of igneous origin. Sapphire and pyrope may occur in some of the diamondiferous kimberlites. Usually potash-rich or soda-lithium-rich pegmatites are the host rocks of many beautiful gemstones, such as, topaz, sapphire, ruby and zircon. Basic and andesitic lava flows, and granite intrusives also host gemstones at places. Some contact-metamorphic limestones may contain lapis lazuli and ruby. Opals are deposited from volcanic waters while amethyst develops in vein deposits. Turquoise is a gemstone of supergene origin in regions of cuprous mineralization. Distribution of these gemstones in the Indian subcontinent is shown in Fig. 4.18.

Ruby, Sapphire and Emerald

Both ruby and sapphire are the gem varieties of crystallized alumina called corundum. Sapphire of different colors occurs in nature but the red and pink varieties are usually called ruby. In Kashmir Himalaya, in and around the remote Nanga Parbat, sapphire deposits are known to occur. The formations comprise granites and other igneous intrusions particularly pegmatites penetrating the crystalline metamorphic schists. Intimately intermixed with the sapphire deposits are aquamarine, rubicelle or green tourmaline.

Al_2O_3 is the main constituent of ruby and sapphire. It, however, forms spinel instead of corundum in the presence of magnesia. Spinel and corundum of gem quality are found in several places in Karnataka, particularly in Kadmane and

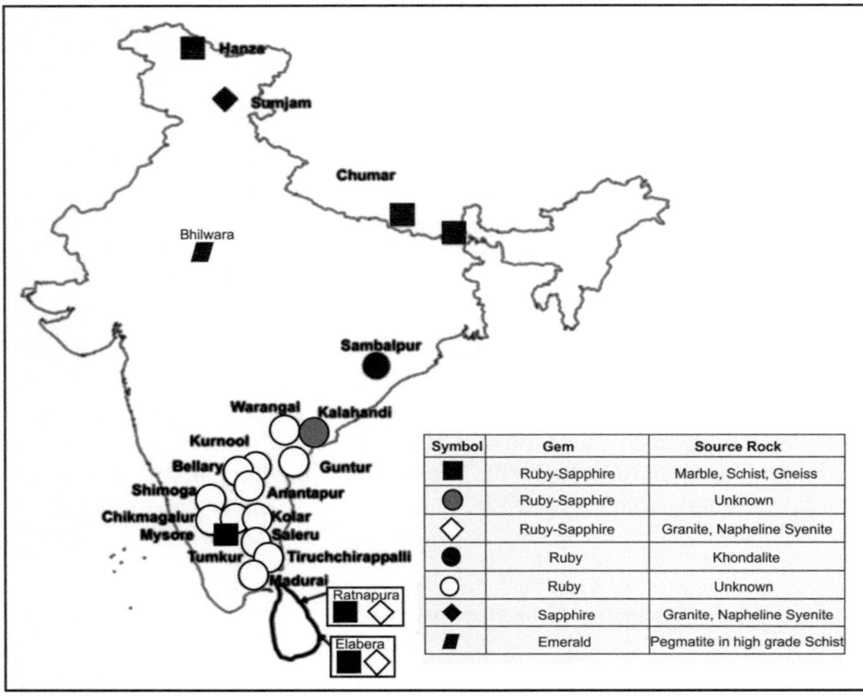

Fig. 4.18 Distribution of gemstones (other than diamond) in India and in some adjoining countries

Kelkoppa areas. They have deep colors and are harder than the Burmese rubies. Translucent dark-colored rubies are found at Adihalli near Bageshpura and in Hardur district. Karnataka also produces rock crystals, opal, garnets, aquamarine, and some emeralds.

In Tamil Nadu transparent varieties of corundum, i.e. ruby and sapphire varieties, such as, blood red, oriental sapphire, oriental emerald along with chrysoberyl, zircon, topaz, etc. have been occasionally found around Kangeyam and Padiyur in Periyar district and in the Manavadi area, Karur district (Singh et al. 2001). Ruby occurs in the paleo-gravels/pebbles in Manavadi area.

Emerald and aquamarine represent the colored and transparent varieties of beryl. The former has deep green color while the latter is characterized by light bluish green color. Because of its attractive deep green color and rarity of occurrence, emerald is the most expensive gemstone next to diamond only. Presence of chromium renders the green color to the mineral. In India, the famous emerald deposits are found in Mewar region of Rajasthan (Fig. 4.19). The area is Kaliguman where emerald mines are present near the village of Amet in the neighborhood of the old fortress of Kumbhalgarh. The precious mineral occurs within bands of biotite schists, very much like garnets and andalusite. Emeralds are also found at Banas in

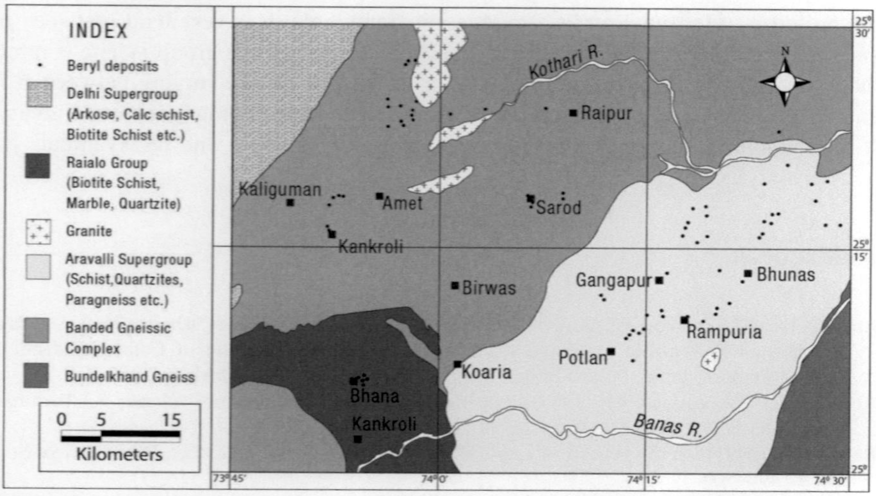

Fig. 4.19 Geologic setting of beryl (emerald) occurrences in Rajasthan (after GSI)

Ajmer district, Rajasthan. The best quality of emerald is, however, found in Myanmar (Burma) and Sri Lanka.

Zircon ($ZrSiO_4$)

Granites and gneisses have zircon as a common accessory mineral. Indian gemologists call it *gomed*. Zircon crystallizes in tetragonal system and is prismatic in habit with pyramidal terminations. Hardness is high around 7.5. Zircon comes next to diamond in brilliance and internal glow. Zircon is transparent to translucent and its color is commonly yellowish brown. It usually contains monazite as small inclusions which render it radioactive. Also a little U may replace Zr. Much of the colorless or transparent zircons are used as gemstones. It is colorless when unmodified but turns yellowish or light brown when metamicted due to radiation from radioactive inclusions. Interestingly, when such grains are heated, they become transparent again through structural restoration.

Zircon occurs in the beach sands of southwest India as placer deposits. In India, gem variety of zircon is found in small quantities in nepheline syenite rock, in the neighborhood of Kangayam, in Coimbatore district, and also in the pegmatite veins of Kadavur associated with charnockite complexes, in Tiruchirapalli district of Tamil Nadu. The Seitur graphite mines of Ramnad taluk of Tamil Nadu produce considerable quantity of gem variety of zircon. Gem variety is also found in Kerala, near Travancore, in pegmatite veins.

Jade and Nephrite

The mineral includes two varieties: jadeite and nephrite. The latter is the more common form of jade. This mineral, usually leaf green to grass green in color due to the presence of ferrous iron, is a monoclinic amphibole. It is very hard (between 6 and 6.5) and compact, with a splintery fracture. The pyroxene group jadeite is rarer than nephrite, and is very tough and compact with hardness varying between 6.5 and 7. It is also splintery in character. The color is greenish white to emerald green. It is available in ample quantity in the Darjeeling Himalayas and in Myanmar.

References

Anantha Iyer GV, Ramana PK, Kutty TR, Subbanna GN (1987) Chrysotile asbestos mineral along stylolites in Proterozoic carbonate sediments of Vempalle formation of Cuddapah district, Andhra Pradesh, India. Purana basins Peninsular India Geol Soc Ind Mem 6:459–469

Anantha Iyer GV, Vasudev VN (1985) Copper metallogeny in the Jogimardi volcanic, Chitradurga greenstone belt. J Geol Soc India 26:580–598

Banerji AK (1981) On the genesis of Lapsa Buru kyanite, Singhbhum district, Bihar. J Geol Soc Ind 22:496–501

Banerjee DM (1971) Precambrian stromatolitic phosphorites of Udaipur, Rajasthan, India. Geol Soc Am Bull 82:2319–2330

Banerjee DM (1986) Proterozoic and Cambrian phosphorites—regional review: Indian sub-continent. In: Cook PJ, Shergold JH (eds) Phosphate deposits of the world, vol 1. Proterozoic and Cambrian Phosphorites. Cambridge Univ Press, UK, pp 70–90

Banerjee S, Dutta S, Paikaray S, Mann U (2006) Stratigraphy, sedimentology and bulk organic geochemistry of black shales from the Proterozoic Vindhyan supergroup (central India). J Earth Syst Sci 115(1):37–48

Craig JR, Vaughan DJ, Skinner BJ (2011) Earth resources and environment. Prentice Hall, 528 pp

Dayal B, Jhamwal RS, Negi RS, Sinha AP, Chaurasia PK, Hore MK (1981) Phosphorite occurrences in Delhi supergroup, northeastern Rajasthan. In: Proceeding of 4th international seminar phosphorites, Udaipur, 1984, pp 129–133 (Geol Surv Ind Spec Publ 17)

Deb M, Thorpe RI, Cumming GL, Wagner PA (1989) Age, source and stratigraphic implications of lead isotope data for conformable sediment-hosted base metal deposits in the proterozoic Aravalli-Delhi orogenic belt, NW India. Precambrian Res 43:1–22

Deb M, Hoefs J, Baumann A (1991) Isotopic composition of two precambrian stratiform barite deposits from the Indian shield. Geochimica Cosmochimica Acta 55:303–308

Deb M, Bheemalingeswara K (2008) Carbonaceous tuff-stratiform barite association in Mangampet deposit, Andhra Pradesh, India: implications on depositional environment and origin. In: 33rd international geological congress, Oslo, Abstr No MRD13407L, Symp MRD-13. Ore deposits associated with black shales: from their origin to their environmental impacts

Deb M, Pal T (2015) Mineral potential of Proterozoic intra-cratonic basins in India. In: Mazumder R, Eriksson PG (eds) Precambrian basins of India: stratigraphic and tectonic context, pp 309–325 (Geol Soc London Mem 43)

Deb S (1980) Industrial minerals and rocks of India. Allied Publ Pvt Ltd, New Delhi, p 603

Deb S (1952) Occurrence of natural salts and selenite-gypsum in the neighbourhood of Lunkaransar Tal, North Bikaner, Western Rajputana. Bull Nat Inst Sci Ind 1:94–99

Deb S (1949) Graphite deposit of Anara, Manbhum district. Sci Cult 15:162–163

Devraju TC, Raith MM, Spiering B (1999) Mineralogy of archean barite deposit of Ghattihosahalli, Karnataka, India. Can Min 37:603–617

Dunn JA (1929) The aluminous refractory materials: kyanite, sillimanite and corundum in northern India. Mem Geol Surv Ind 52(2):145–274

Dunn JA, Dey AK (1942) The geology and petrology of eastern Singhbhum and surrounding areas. Mem Geol Surv Ind 69(2):443

Dwivedy KK (1995) Economic aspects of Cuddapah basin with special reference to uranium—an overview. Proceedings Annual convention on Geological Society India, Tirupati'95, pp 116–138

Frost BR (1979) Mineral equilibria involving mixed volatiles in a C-O-H fluid phase, the stabilities of graphite and siderite. Am J Sci 279:1033–1059

Glenn CR, Föllmi KB, Riggs SR et al (1994) Phosphorous and phosphorites: sedimentology and environments of formation. Eclogie Geol Helv 87:747–788

Gokhale KVGK, Rao TC (1973) Ore deposits of India: their distribution and processing. Thompson Press (India) Ltd, Delhi 223 pp

Golani PR (1989) Sillimanite-corundum deposits of Sonapahar, Meghalaya, India: a metamorphosed Precambrian palaeosol. Precambrian Res 43:175–189

Guha J (1971) Sulphur isotope studies of the pyrite deposits of Amjhor, Sahabad district, Bihar. India. Econ Geol 66(2):320–330

Jain VK (1987) Mining and geology of barytes deposit of district Alwar, Rajasthan. J Mines, Metals Fuels, July: 328-330

Jiang G, Christie-Blick N, Kaufman AJ, Banerjee DM, Rai V (2002) Sequence stratigraphy of the Neoproterozoic Infrakrol formation and Krol group, lesser Himalaya, India. J Sed Res 72:524–542

Karunakaran C (1976) Sulphur isotopic composition of barites and pyrites from Mangampeta, Cuddapah district Andhra Pradesh. J Geol Soc Ind 17(2):181–185

Kumar V, Bakliwal PC (2005) Potash in India. Geol Surv Ind Misc Publ 65:131 pp

Murthy MVN (1964) Mica fields of India. International Geological Congress, 22nd session, New Delhi, p 23

Narasayya BL, Sriramdas A (1974) Apatite from apatite-magnetite deposits of Kashipatnam, A. P. Proc Ind Acad Sci 80(A) V: 207–210

Neelakantam S (1987) Mangampeta barite deposits, Cuddapah district, Andhra Pradesh Purana basins of Peninsular India. Mem Geol Soc Ind 6:429–458

Neelakantam S (1995) Volcanogenic bedded barite, Mangampeta. Proceedings Annual convention Geological Society India, Tirupati'95, pp 174–185

Pandit MK, de Wall H, Chauhan NK (2008) Paleosol at the Archean-Proterozoic contact in NW India revisited: evidence for oxidizing conditions during paleoweathering? J Earth Sys Sci 117:201–209

Pohl WL (2011) Economic geology, principles and practice. Metals, minerals, coal and hydrocarbons—introduction to formation and sustainable exploitation of Mineral Deposits. Wiley, Sussex, UK, p 663

Radhakrishna BP, Srinivasaiah C (1974) Bedded barytes from the precambrian of Karnataka. J Geol Soc Ind 15(3):314–317

Rau TK (2007) Panna diamond belt, Madhya Pradesh—a critical review. J Geol Soc Ind 69:513–521

Roy AB, Nagori DK, Golani PR, Dhakar SP, Choudhuri R (1980) Structural geometry of the rock phosphate bearing aravalli rocks around Jhamarkotra mine area, Udaipur district Rajasthan. Ind J Earth Sci 7(2):191–202

Sarkar SC (1984) Geology and ore mineralisation of the Singhbhum copper-uranium belt. Jadavpur Univ Press, Calcutta, India 263 pp

Sarkar SC, Bhattacharya PK, Mukherjee AD (1980) Evolution of the sulphide ores of Saladipura, Rajasthan, India. Econ Geol 75:1152–1167

Sarkar SC, Gupta A (2012) Crustal evolution and Metallogeny in India. Cambridge Univ Press, Delhi 840 pp

Scott Smith BH (2007) Lamproites and kimberlites in India. J Geol Soc Ind 69:443–465

Singh H, James M, Vasudevan D (2001) Gem occurrences in Tamil Nadu. Geol Surv Ind. Spec Publ No. 58:665–670

Subramaniam AP, Parimoo ML (1963) Fluorspar mineralization related to deccan basalt volcanism of Ambadongar, Baroda district, India. Nature 198:563–564

Sukheswala RN, Udas GR (1967) Fluorspar mineralization related to carbonatite-alkalic complex at Amba Dungar, Gujarat state. Curr Sci 36:4–6
Valdiya KS (1968) Origin of the magnesite deposits of southern Pithoragarh, Kumaon Himalaya, India. Econ Geol 63:924–934
Yin A, Dubey CS, Webb AAG, Grove M, Gehrels GE, Burgess WP (2010) Geologic correlation of the Himalayan orogen and Indian craton (part 1): structural geology, U-Pb Zircon geochronology and tectonic evolution of the Shillong plateau and its neighboring regions in NE India. Bull Geol Soc Am 122(3/4):336–359
Zharkov MA (1984) Paleozoic salt-bearing formation of the world. Springer, Heidelberg

Chapter 5
Geodynamic Context of Metallogeny

Abstract Geodynamics is the dynamics of the Earth. Metallogeny is the genesis of metals, rather the metal-bearing mineral deposits. So, 'Geodynamic context of metallogeny' relates metallogeny to geodynamics. We generally refer to geodynamics in terms of plates and plumes, that ultimately control magmatism, sedimentation, and metamorphism and hence ore genesis in the crust-mantle system. The solid earth is made up of physically distinct layers (from surface downward) of lithosphere, asthenosphere, mesosphere, outer core, and an inner core. The lithosphere is presently divided in to seven large and a few small pieces ('plates'). They underwent making, breaking, and remaking several times during much of the Earth's history, giving rise to what we call supercontinents and supercycles today. Formation of mineral deposits, which are in fact products of petrogenesis of sorts, controlled by tectonics, therefore varied in time and space. Tectonic setting of mineralization are today accepted to be as follows: continental hot spot produce Cr, Ni–Pt–Cu deposits, diamonds, certain stratiform base metal deposits, and even granite-related mineralization of Sn. Constructive plate boundaries host Fe-, Mn-oxides along the rifted ridge, and also Alpine-type chromite deposits. Passive continental margins localize deposition of evaporites, phosphates, BIFs, carbonate-hosted Pb–Zn deposits. Subduction zones produce porphyry type Cu–Mo–(Au) deposits and may be also Cu $\pm$ Au skarn deposits. Continental collision zones contain podiform chromite, VMS, U and porphyry Cu deposits. Greenstone belts (ancient volcano-plutonic arcs?) are known for being the major localizer of orogenic gold, komatiite-related Cu–Ni deposits and also Cu–Zn mineralizations. Intracratonic settings are credited with IOCG mineralization. Tectonically stable regions with plateau-like relief and located in tropical and sub-tropical regions are best suited for lateritization.

Keywords Geodynamics · Lithospheric plates · Plumes · Continental hot spots · Subduction · Greenstone belts · Supercontinents · Supercycles · Metallogenic provinces · Collision-type metallogeny · SEDEX-type mineralization · IOCG mineralization

M. Deb and S.C. Sarkar, *Minerals and Allied Natural Resources and their Sustainable Development*, Springer Geology,
DOI 10.1007/978-981-10-4564-6_5

5.1 Introduction

Dynamics is a branch of mechanics. It is defined as the mathematical and physical study of the behavior of bodies under action of forces, which produce changes in motion in them (Uvarov et al. 1977). Geodynamics, therefore, should mean the dynamics of the Earth, the solid Earth as we understand it. The nature of geological materials being what they are, geological systems subjected to changed geodynamic situations not only change their size, shape, density and may be position in space, but also their petrochemical characters due to change in the P-T regime. These dynamic movements may be the causes of large-scale aqueous fluid distribution, resulting in deposition of mineral matter. New-formed magmas may also be sources of ore melts or sources of hydrothermal solutions in particular geodynamic regimes. Therefore, understanding of the regional metallogeny requires as clear a view as possible of the geodynamic history of a region. This fundamental aspect of Earth Science not only deals with structures and their development in the lithosphere, but also controls magmatism, sedimentation, and metamorphism and since mineral deposit formation is closely associated with the last three processes an understanding of mineral deposit formation needs proper appreciation of their geodynamic settings in terms of geochronological constraints. As we shall see in the following pages, we need to be concerned about both the 'plates' and the 'plumes' (mantle plume) in this context. However, it needs to be mentioned at this stage that the present discussion on geodynamic settings of mineral deposits does not include the geophysical components of the solid earth and neither involves any physico-chemical and mathematical modeling to understand mantle convection and plate tectonics, formation of mobile belts and intra-cratonic basins, etc.

Earth scientists were obliged to think of geodynamics way back in the eighteenth to early nineteenth century when they were interested to explain orogenic belts. Forerunners amongst them believed that geosynclinal accumulation was initiated by an ophiolite sequence, consisting of oceanic ultramafic and mafic igneous rocks succeeded by deep marine siliceous and argillaceous sediments. Other contributions (Stille 1936; Kay 1951; Aubouin 1965 amongst others) followed. But it was Turneare (1955) first, who comprehended formation of mineral deposits in the tectono-lithological context, while Russian geologist Bilibin (1955) turned his attention to metallogenic provinces and epochs. Other scientists who followed this trend and deserve a mention are Pereira and Dixon (1965), Stanton (1972), and Meyer (1988). The concept of plate tectonics made its appearance in the 70s, but was not yet in full bloom or all-embracing. There was yet no discussion of metallogeny in the context of plate tectonics. But soon contributions worth mentioning relating to this aspect were made by Rona (1980), Mitchel and Garson (1981), Sawkins (1984, 1990), together with a couple of individual articles, positing the same. We, in this chapter, subsume these concepts and yet present a critical discussion of our own, involving the world picture in general and the Indian scenario in particular. A somewhat similar review was made earlier by Sarkar (1985).

In the last decade of the past century, the concept of supercontinent and/or supercontinent cycles made its first appearance (Rogers 1996). Barley and Groves (1992) utilized this concept to explain the space–time distribution of mineral deposits in the context of Meyer's compilations (Meyer 1981, 1988). Not everything about supercontinent making and breaking and their relationship with metallogeny is beyond dispute today. But the concept, as of today, best explains the geodynamic basis of space–time distribution of mineral deposits, at least since the Late Archean times.

Another concept relevant to the present discussion, as mentioned above, is mantle plume activity. Obviously it escaped our attention for a long time. We shall briefly discuss this topic in a subsequent section. Before that we define a few things for the sake of convenience of the following discussion. These are related to the basic structure of the Earth.

5.2 Continental and Oceanic Lithosphere, Asthenosphere, Upper Mantle, Lower Mantle and the Core of the Earth

From the surface downwards, the Earth is made up of physically distinct layers: the lithosphere (comprising 'cool', hard rocks), asthenosphere (weak, 'soft' rocks), mesosphere (dense, rigid rocks), outer core (liquid, Fe and Ni), and an inner core (solid, Fe and Ni) (Fig. 5.1).

Modern oceanic lithosphere has ~6 km-thick basaltic crust and a ~30 to ~50 km-thick lherzolitic mantle lithosphere, near ocean ridges. The oceanic crust is generally <200 My old. The mantle lithosphere thickens to a maximum of ~70–100 km as it progressively loses heat with increasing distance from the spreading axis by accretion of the underlying hot, slowly flowing asthenosphere. Compared to the underlying asthenosphere, oceanic lithosphere, away from the ridges, is comparatively cool, rigid, and negatively buoyant, i.e., heavy. The continental crust is ~35 km thick on the average and may have components that are giga years old. The subcontinental lithospheric mantle (SCLM) is thicker, particularly underneath Archean cratons. As we shall see in the following pages, we are concerned with the lithosphere possibly more than with the 'other spheres' of the solid earth.

The lithosphere is divided into seven large and a number of small pieces ('plates'). These plates together constitute a laterally mobile upper boundary layer of the 3-dimensional convection cells in the asthenospheric mantle. The boundary between the top of the lower mantle and bottom of the upper mantle (D') is at 670 km. The mantle-core boundary, generally referred to as D'', is located at a depth of 2900 km. D' and D'' are mineralogical phase transition boundaries and are defined seismologically (Fig. 5.1). The upper and the lower mantles probably convect independently, albeit with episodic overturns. Heat is transferred from the

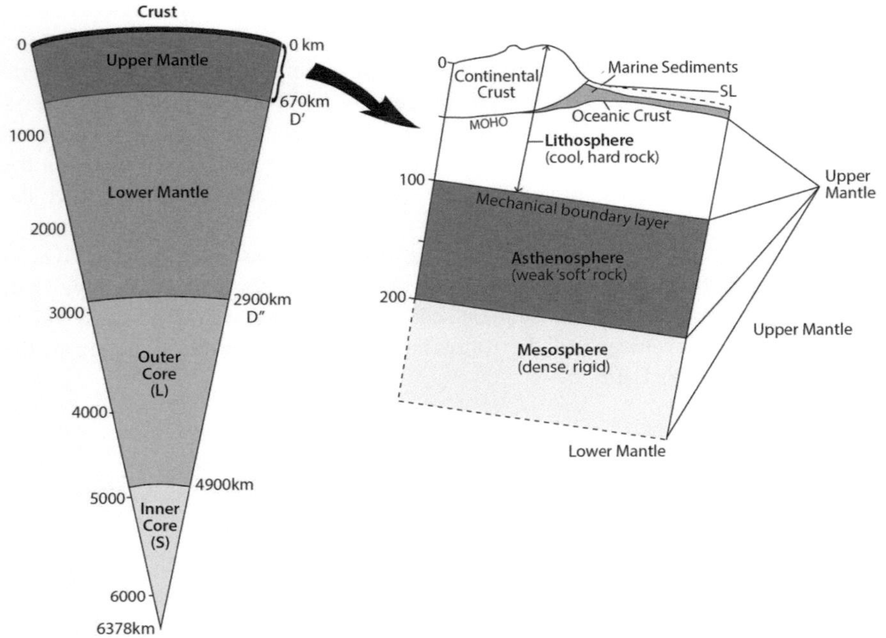

Fig. 5.1 A radial section of the earth showing its internal structure and the different constituent spheres. *Note* the *D′* and *D″* phase transition boundaries, defined seismologically

core and the mantle toward the surface, not only by convection, but also physically by hot mantle plumes that rise from the core-mantle boundary.

Core is the innermost spherical mass of the Earth, consisting of a liquid outer shell and a solid inner mass. It is thought to be composed of metallic iron and nickel (Fig. 5.1). The liquid outer core is dynamic and greatly influences the overlying mantle.

5.3 An Outline of Plate Tectonics

The lithospheric plates comprising Earth's outermost shell, mentioned above, are 100–150 km in thickness and are perpetually in motion with respect to one another and the Earth's axis of rotation. The rate of plate motion varies from 1 to 15 cm per year. Thus they migrate across the surface of the Earth, interacting at divergent, convergent and fault boundaries, producing volcanism, earthquakes and are responsible for mountain building and rearrangement of the positions of continents and oceans on a time scale of 10s to 100s of million years. A plate is accreted along the divergent boundary and destroyed along the convergent boundary (Fig. 5.2). The rates of production of new lithosphere at spreading centres are balanced by

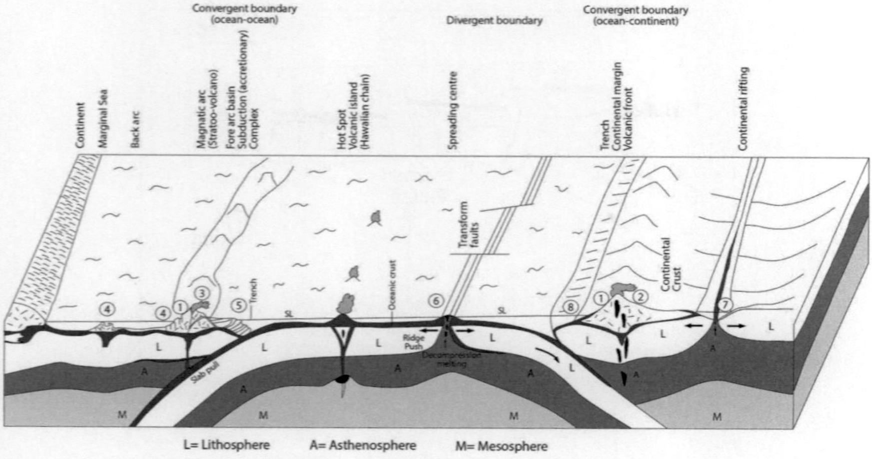

Fig. 5.2 Sketch showing divergent and convergent plate boundaries. Hot spot igneous activity represents intra-plate volcanism located spatially between the two tectonic environments. The *circled numbers* denote mineral deposit types: *1* Cu (±Au) porphyry; *2* Mo ± Sn–W greisens; *3* Epithermal Au ± Cu etc.; *4* VMS deposits; *5* Orogenic Au; *6* Cypruss-type VMS (later transported); *7* Red Sea type metalliferous sediments; *8* MVT deposits

consumption at the subduction zones. This allows the size of the Earth to remain constant, without growth or shrinkage. The principal cause of the plate movement is the negative buoyancy of the subducting block. The transform boundary (fault) is a plate boundary fault, parallel to the direction of relative plate motion.

A divergent plate boundary may be initiated in the continental domain or in the intra-oceanic domain. A divergent plate boundary is created when oceanic plates separate due to far-field extensional forces. A ridge is created along this zone through the input of mafic magmas, produced by decompressional melting of asthenospheric mantle. The thickness of the lithosphere increases to 70–100 km from zero at the beginning. In the continental domain, divergence is initiated by the development of a three-armed or 'trilate' intra-continental rift system in which two arms develop into a ocean basin, while the third turns inactive after a stage of development, the trilate having developed superjacent to an underlying mantle plume (Morgan 1972; Burke and Wilson 1976). The rift sequence may have important volcanic members in it, in addition to sediments. Predominantly, alkali granite provinces are associated with hot spots related to the rifts. The failed rift is called 'aulacogen', following the Russian terminology. The new-formed ocean gradually develops its mid-zone ridge of mafic-ultramafic rocks and the two borders move away from each other. At an early stage the ridge may not appear, as is the case in the Red Sea (Fig. 5.3). As the Earth surface is conservative, before long the plate boundary opposite to the constructive one will sink below (subducted) and resorbed in the upper mantle, or otherwise conserved, as is conspicuous in some seismic tomograms.

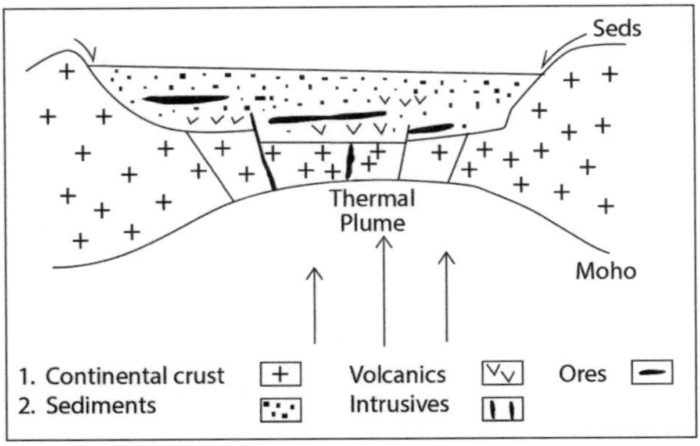

Fig. 5.3 Cartoon showing basic characteristics of a hot spot rift basin like the Rea Sea

The convergent plate margin/boundary has the following tectonic elements: (i) a deep marine trench, seaward of the fore arc; (ii) a subduction-cum-accretion complex; (iii) a fore-arc basin between the arc axis and the subduction-accretion complex; (iv) a magmatic arc; (v) continental back arc. Magmatic arcs are in cases divisible into extensional, neutral and compressional arcs (Fig. 5.2). The trench is the steep-sided elongate depression on the ocean floor, locating the zone of subduction. The volcanic-plutonic arc is parallel with the trench and is made up of volcano-plutonic rocks (±their derivatives). It generally starts with the emplacement of basic volcanic rocks, followed by emplacement of andesite, diorite, granodiorite (±granite). These are still sites of active arcs (cf. 'Ring of Fire' along the Pacific boundary). The arcs are commonly split into horst-graben structures and the grabens may be partly filled up by locally derived sediments. It is important to note that arc magmatism varies along and across the strike (and this is significant for metallogeny, as we shall see later). For example, tholeiitic magmatism is dominant before the fore arc basin and the arc axis; calc-alkaline magmatism occurs mainly in the central region of the arc, while the late alkaline igneous rocks are distributed between the arc axis and the back arc region.

There is another type of arc system. This develops when a plate of oceanic lithosphere subducts below another one of the continental type, the resulting volcano-plutonic arc is emplaced upon the continental crust at its periphery (Fig. 5.2) and not as typical island arcs, where both the converging plates are of oceanic lithosphere, such as, in Japanese islands. The western margins of South and Central Americas are typical examples of this situation. These are called 'Continental' or 'Cordilleran' arcs.

A domain in the Plate tectonics parlance is the 'Passive Continental Margin'. It is defined as the major morphological feature of ocean margins of relatively 'quiet type', as along the Atlantic (and Indian) ocean margins. The geological environments may include continental talus fans or epeiric seas and extend through the coastal plain and continental shelf to deep marine environments of the continental rise.

Another tectonic environment related to Plate tectonics is 'collisional' setting, a conspicuous example of which is the Himalaya, but in reality, there are many other examples of collisional setting. In this setting two continental blocks jostle against each other, following or accompanying a subduction stage.

The 'transform' plate boundaries are in fact conservative boundaries accommodating motion from the divergent to the convergent plate boundaries (Fig. 5.2). Transform faults that separate continental lithospheric blocks are called transcurrent faults.

5.4 Mantle Plumes

Mantle plumes constitute another source of dynamism in the crust-mantle system and ultimately contribute to mineral deposit formation either individually or in collaboration with plate tectonics. Their importance, however, has been fairly adequately emphasized, amongst others, by Pirajno (2000, 2004).

What is a mantle plume after all? It is an anomalously hot jet of mantle material, ejected from thermal boundary layers, probably the mantle-core boundary (MCB/D'') at 2900 km. This material advects through the mantle because of buoyancy. Time taken from the beginning to the end is 10–50 my. The plume-head varies from 500 to 1000 km in diameter and the tail, ~100 km. The plume loses some heat as it advects. At the top of the upper mantle, the plume-head is ~1500 °C hot and the tail is ~1700 °C. Being hotter, the plumes conductively heat the ambient mantle. On meeting the normal lithosphere at ~150 km, the plume-head flattens to ~1000–2000 km, undergoing decompressional melting (White 1992) (Fig. 5.4).

Plume activity, particularly of superplumes, is episodic with maxima at ~3.8, 3.4, 3.0, 2.7, 2.4, 1.9 and 1.7 Ga together with one at ~250 Ma and another superplume in Cretaceous. While the plume erupts proximal to a spreading centre, it may capture the ridge, as has happened in case of the Iceland plume in the Mid-Atlantic Ridge.

Mantle plumes are divided into three broad types:

(i) Long-lived hot spots with low magma flux. They generate ocean islands or hot spot tracks on continents.
(ii) Short-lived plumes that generate flood basalt provinces. They include plumes that erupt through oceanic lithosphere and form oceanic plateaus or continental flood basalts, such as that in Siberia in Russia and the Deccan in India.
(iii) Superswells or mantle upwellings. These have diameters as large as ~40,000 km and spawn hotspots.

Pirajno (2004) provided an elaborate discussion on the relationship of mineral deposits and mantle plumes. He divided mineral deposits in this context in two end members:

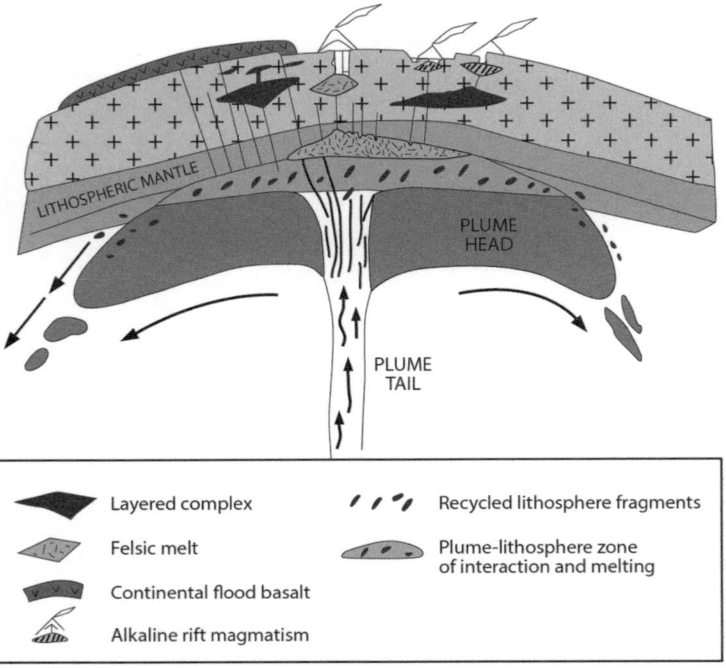

Fig. 5.4 Sketch of a mushroom-shaped mantle plume advecting underneath sub-continental lithospheric mantle. Note the felsic melts generated in the SCLM and in the crust, the mafic-ultramafic layered complexes and the continental flood basalt (after Pirajno 2000)

1. Magma-associated ore deposits
2. Mineral deposits formed by giant hydrothermal systems related to anorogenic magmatic systems.

5.4.1 Mantle Plume-Related Magma-Associated Principal Ore Deposits

1. Layered mafic-ultramafic complexes and associated Cr, PGE, V, Fe, Ti (Cu, Ni) as in Bushveld (South Africa), Great Dyke (Zimbabwe), Stillwater, Duluth complexes (USA) are well-known examples of this class. Abbott and Isley (2002) suggested that layered intrusions formed by deep mantle plume events contain high levels of PGE and Cr, because these plumes are associated with high degree of melting in the mantle and therefore, have higher concentration of compatible elements such as, Pt and Cr.

2. Flood basalts and sill complexes of Ni–Cu–(PGE) are represented by Norilsk-Talnakh deposits, Siberia (Russia), Ni–Cu ores associated with Archean komatiite rocks at Kambalda, Western Australia. Archean greenstone komatiite-tholeiite sequences are suggested to be products of mantle plumes (because of high heat requirement for melting and differentiation) and moreover these sequences may be fragments of plume-related oceanic plateaus (Tomlinson et al. 1998; Storey et al. 1991). Some other workers, however, drawing parallels from TTG origin in Precambrian, prefer subduction tectonics for the origin of komatiites. The latter are mostly those who believe Ni–Fe–Cu-bearing komatiites were extruded in a rifted continental setting or during island arc volcanism (Naldrett 2010).
3. Anorogenic granites containing Sn, Nb, Ta, W, U, Th, F, Be, Zn, Cu as in Josh plateau, Nigeria, Bushveld Granite (Africa), Rapakivi Granites (Scandinavia) and Proterozoic deposits from Rondonia, Brazil. Mineralization occurs as veins, dissemination and greisens.
4. Massif-type anorthosite containing Fe–Ti–V ores and troctolite-anorthosite containing Cu–Ni–Co in Voisey Bay, Labrador-Grenville Province, Canada. A giant deposit containing Fe–Ti–V is the 1.5 Ga Suwalski anorthosite massif in Poland (Morgan et al. 2000). Tectonic environment is rift setting. Windley (1995) argued that the supercontinent breakup was initiated as a result of extension above a mantle plume-head with the production of lower crustal melts that included anorthosite.
5. Anorogenic granitoids associated with Olympic Dam style mineralizations, also called Iron oxide–Cu–Au–REE-type hydrothermal systems. Included in the anorogenic magmatism are kimberlites, lamproites, and carbonatites. These rocks have isotopic signatures similar to ocean island basalt (Nelson et al. 1988) and as such, are considered as part of plume magmatism (Bell 2001).

5.4.2 Plume-Related Hydrothermal Ore Deposits

In addition to direct production of magmas, thermal anomalies related to mantle plumes constitute powerful heat sources in the crust. These may be, at least in cases, responsible for the creation of crustal-scale hydrothermal circulation systems and production of a number of ore deposits. The majority of intracontinental SEDEX deposits formed between Mid- and Early Proterozoic ($\sim$1.9–1.0 Ga), which interestingly coincided with plume events at 1.9, 1.6, 1.3–1.2, and 1.1 Ga. Red Sea brine pools may be present day example of the setting.

Iron Formations: we have in an earlier section discussed in some details about the BIFs. They are generally believed to have formed from hydrothermal effluents produced by submarine LIPs (Oceanic plateaus) developed through plume-related mafic volcanism (Barley et al. 1997; Isley and Abbot 1999).

5.5 Archean Geodynamics as Read from Greenstone Belts

Greenstone belts, consisting mostly of mafic-ultramafic (komatiitic) rocks and sediments (dominantly volcanogenic), variously deformed and metamorphosed and intersected by granodiorite-granite intrusive, are common in many Precambrian (Late Archean – Early Proterozoic) terranes. They are the repositories of a fair part of gold reserves of the world and not a negligible part of Ni–Cu sulfide deposits. Kolar gold deposit of India (already discussed) and the Kambalda Ni–Cu–PGE-bearing deposit in Western Australia are two well-known examples of the above two types respectively, amongst many others. Their tectonic setting(s) is still not beyond controversy. But many Earth scientists are of the view that they are ancient volcano-plutonic arc systems.

Neoarchean greenstone-granitoid belts/terranes show both similarities to and differences with Proterozoic and even Phanerozoic Cordilleran type orogenic belts (Burke et al. 1976; Windley 1981, 1995; Kusky and Kidd 1992; Polat et al. 1999). Komatiites are ubiquitous in Archean greenstone terranes but are rare in Proterozoic and Phanerozoic counterparts. Komatiite liquids get extracted from melts produced in anomalously hot mantle plumes and thus, we may infer higher mantle temperatures in the Archean. Assuming smaller plates and a reasonably longer global ridge system in the Archean, ridge subduction would have been more frequent, accounting for higher heat supply to the convergent margin, explaining production of more TTGs (Hargraves 1986; Polat and Kerrich 2004).

Neoarchean greenstone belts are now commonly thought to be Cordilleran-type collages of oceanic arc and plateau terranes in which orogenesis was induced by plateaus jamming against arcs.

5.6 Major Type of Mineral Deposits and Their Geodynamic (Plate Tectonic) Setting

5.6.1 Mineral Deposits Related to Intracratonic Magmatism

A number of mineral deposit types related to anomalous mantle magmatism in intra-cratonic setting occur above thick subcontinental lithospheric mantle (SCLM), probably during initial extension related to the failed break up of supercontinents. These include chromite, PGE and Ti-V-bearing magnetite deposits in layered igneous intrusions, diamond in alkaline pipes and Fe-oxide Cu–Au (IOCG) deposits related to alkaline to A-type granitic intrusions (Fig. 5.5; vide Groves and Bierlein 2007).

5.6.1.1 Deposits Related to Mantle Basic Magmatism

The most well-known deposits of this type include PGE, chromite and vanadiferous and titanium-bearing magnetite deposits in large layered mafic to ultramafic

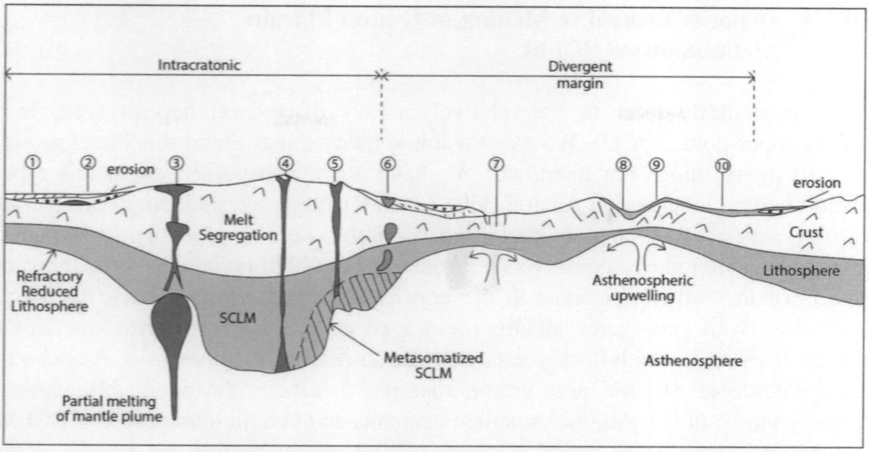

Fig. 5.5 Location of major mineral deposit types (shown as *circled numbers*) in an intracratonic setting, passing on through a passive margin to a divergent margin setting. Note the role of (generally Archean) sub-continental lithospheric mantle (SCLM) in generating at least four types of mineral deposits. *1* Banded iron formation + Mn deposits; *2* beach sands and placer deposits; *3* bushveld-type Ni–Cu–PGE deposits; *4* diamonds in kimberlite; *5* lamproite diamonds; *6* IOCG (Olympic Dam type); *7* Zambian Cu–Co? SEDEX? BHT? *8* Ni–Cu sulfides; *9* VMS; *10* Paleozoic SEDEX? Kupferschiefer (after Groves and Bierlein 2007)

intrusions. Chromite and PGE accumulate at lower levels and V-Ti-bearing magnetite at the upper. Giant deposits of the type (Bushveld Complex, Stillwater Complex for example) occur toward the centre of Archean cratons, whereas the smaller ones tend to lie close to the craton margin. It has been postulated that thick buoyant Archean SCLM is required for producing the giant deposits (Groves et al. 2005a). The basic magmas form in the hotter parts of the mantle possibly aided by plume impaction beneath the Archean SCLM. The parent magmas must be highly enriched in ore elements, particularly PGE.

5.6.1.2 Deposits Related to Deep Alkaline Magmatism

The classic deposits of this type are diamonds. Primary origin of diamond has been discussed in an earlier section (Chap. 4). Diamonds are found in Precambrian (mainly Archean) cratons where the relatively low-T and high-P conditions near the base of the underlying thick SCLM favored the growth of diamonds in metasomatized (by likely subduction-related potassic carbonatite fluids) garnet harzburgite or eclogite (Gurney et al. 2005, 2010). Diamonds were brought to the surface by rapid uprise of alkaline kimberlite magmas throughout the craton or lamproite magmas toward its margins (Jaques et al. 1984). Diamond deposits have formed since inception of widespread thick SCLM in the Neoarchean.

5.6.1.3 Deposits Related to Melting of Craton Margin Metasomatized SCLM

These particularly refer to the relatively newly discovered deposit type, Iron oxide-copper-gold (IOCG). We by now know many things about this broad group, but still many things are unknown. We have already discussed about this type, briefly though, in an earlier section (vide Chap. 2). Here, we shall emphasize on its tectonic aspects. Like many other aspects of this deposit type, its geodynamic setting also is not above controversy. Groves et al. (2010) reviewing some relevant aspects of these deposits came to the conclusion that their spatial and temporal association with anorogenic alkaline or A-type granitoids were consistent with a genetic relationship with deeply sourced, mantle-derived magmas from Archean to Paleoproterozoic SCLMs near craton margins or other lithospheric boundaries. A likely source of Cu, Au, incompatible elements and volatile-enrichment is SCLM that was metasomatized during previous subduction on continental margins. This model is extended by Haywood (2008), when he suggests that the sites of IOCG deposit formation were reactivated back arc settings in which domains of metasomatized SCLM below the back arc with anomalous volatile contents and ore metal contents, were the source of local partial melts, the ultimate source of IOCG mineralization.

5.6.2 Mineral Deposits Related to Intracontinental Rifting

5.6.2.1 A Deposit on Rifted Cratonic Margins

Magmatic Ni–Cu ± PGE sulfide deposits in mafic-ultramafic intrusions are the type examples of this geodynamic setting (Fig. 5.5). They show a classic temporal distribution related to rifting of Archean or younger SCLM in the Paleoproterozoic to Mesozoic times. This is probably related to mantle plume events and following supercontinent amalgamation, but preceding intense rift volcanism or ocean formation. The Ni–Cu sulfide deposits form adjacent to and commonly outboard of Precambrian lithospheric margins.

Other deposits formed near continental margins as a result of mantle plume impingement on rifted Archean SCLM include the Fe–Ti oxide deposits related to anorthosite magmatism (Groves and Bierlein 2007).

5.6.2.2 Deposits in Intracratonic Rift Settings

A number of sediment-hosted mineral deposits occur in sedimentary basins that developed in intracratonic rift settings, although connected in cases to oceanic basins. Important members of this group are the giant stratabound Cu–Co deposits of the Zambian Cu belt, the giant Proterozoic SEDEX Pb–Zn deposits of Northern

Australia and the giant Broken Hill type (BHT) Pb–Zn–Ag deposits of Australia and South Africa (Fig. 5.5). Many of these deposits are thermo-tectonically modified to varying extent, the Broken Hill deposit having almost touched granulite facies of metamorphism. Thus, their geodynamic settings are not always free from debates. We will return to this issue later.

5.6.3 Mineral Deposits Related to Divergent Margin Tectonics

Mineral deposits formed in divergent margins during supercontinent break-up include a variety of essentially sediment-hosted syngenetic and epigenetic deposits in passive continental margins and deposits associated with oceanic spreading.

5.6.3.1 Syngenetic Deposits of Passive Continental Margins

Large, syngenetic sediment-hosted deposits that form on passive continental margins are beach sands (ilmenite, rutile, zircon/South Africa, Australia) and more importantly, Fe–Mn ores (Fig. 5.5). Banded iron formation (BIF) and Minette-type ironstones as well as manganese ores originally deposited through sedimentation-diagenesis, ultimately get enriched to ores by subsequent hydrothermal or supergene alterations. The largest BIF deposits formed during the period, ca. 2.65–1.85 Ga, when several phenomena such as large supply of iron from sea floor hydrothermal systems, probably related to mantle plumes (Isley and Abott 1999), and large continental shelves for deposition of these chemical sediments and at least partially anoxic oceans were available at the same time. Mn deposits were less abundant in the Archean. They also were deposited on passive continental margins, as did the evaporates (Miocene/Red Sea), phosphorites (Miocene/Florida; Recent/West Africa) and, carbonate-hosted Pb–Zn deposits (Mississippi valley; Southern Alps). These are suitable locales for hydrocarbon accumulation too.

5.6.3.2 Deposits Formed During Ocean Spreading

Mineralization related to this tectonic setting has been divided into two principal types.

1. Mineralization in ocean ridges and ocean basins
2. Mineralization associated with ophiolite sequences

Ocean ridges are internally rifted parallel to their lengths (Fig. 5.2), as a result of tension to which the domain is continually subjected. These rifts contain sediments, particularly chemical precipitates of Fe- and Mn-oxides. These are hydrothermal deposits. But those overlying the sediments away from the ridge crest are believed to be hydrogenous, or authigenic in origin.

Hydrogenous deposits have Fe/Mn ratios of ~ 1, high trace metal (Ni, Co, Cu) contents, high Th and low U/Th ratios (Bonatti et al. 1976a, b). These deposits occur as crusts and more as nodules. They have been reported more from summits and flanks of sea mounts, from oceanic plateaux and at places of deep oceanic basins. The seamount nodules are rich in Co (Cu + Ni + Co $\simeq$ 2%). With improved technology and the situation becoming more demanding, these may be exploited one day. Some near-analogues of such deposits are reported from land occurrences.

Another variety of ores that develop in this tectonic domain has much similarity with the ore deposits in the Red Sea basin (Figs. 5.2 and 5.3). Here also the brine entering the mafic (-ultramafic) fractured rocks in the ridge-crest region reacted with the wall rocks, leaching ore-forming metals and sulfur and being warmed up by the ambient geothermal energy moved upward by convection and precipitated as ores at higher levels where the situation became appropriate. These hydrothermal fluid vents give rise to what is called submarine *Black smokers* (cf. Sect. 2.2.2.2).

Mineral deposits that formed at ancient spreading centres (or in primitive back arcs) are rare due to their poor preservation. However, such deposits do occur in slices of ophiolites obducted on to continental crust due to ocean closure, which is in convergent margin setting in their final tectonic position. Ore deposits in ophiolitic associations are divisible into two, depending on their composition. These would be:

(a) Alpine-type chromite (±Pt) deposits
(b) VMS deposits containing sulfides of Fe, Cu, and Zn (±Ag, Au)

Some examples of the first type include Semail complex of Oman and numerous podiform chromite deposits in the Urals, Albania, Turkey, Philippines, New Caledonia, and Cuba. The Cyprus sulfide deposit (Fig. 5.2) is commonly referred to as the typical case of the type (b). But most ophiolitic sulfide deposits are pyrite-rich and poor in sulfides of other metals. Deformation ± recrystallization is a common feature of ophiolite-associated ores. Precambrian examples of both the types are rare (Kerrich et al., 2005; Groves and Bierlein 2007).

5.6.4 Mineral Deposits Related to Convergent Margin Tectonics

A much larger variety and number of mineral deposits occur in these geodynamic settings because of the development of a large number of magma types that are the primary sources of ore materials and/or secondary hydrothermal fluids produced during involvement in different geodynamic regimes during inversion. Island arcs (including back arcs) are important in this respect.

5.6.4.1 An Arc-Related Magmatic-Hydrothermal and Hydrothermal Mineral Deposits

The most outstanding example of ore mineralization in this broad group of mineral deposits is the porphyry Cu–Au(–Mo) deposits (Seedorf et al. 2005; Sillitoe 2010). Basic characteristics and genesis of these deposits are well established and already discussed in an earlier section (vide Chap. 2). Porphyry deposits are generally of Phanerozoic (commonly Cenozoic and Mesozoic) age. They are extremely rare in Proterozoic and Archean. A commonly held belief is related to their preservation-potential. Since these deposits occur at rather shallow depths in environments of tectonic uplift, often followed by collision, older deposits must have been eroded away. Hence they are rare in older times. This opinion is now contested. Many Precambrian supracrustal formations including arc-successions are preserved around the world. Such situations should be competent to retain porphyry Cu deposits, if there was originally any (Seedorf et al. 2005). Richards and Mumin (2013) have an elegant, yet simple explanation in this context. They propose that subduction-related magmas would have been S-poor in the Precambrian – a situation that inhibited the formation of S-rich porphyry-type Cu deposits in the Precambrian, except in local, somewhat unusual situations.

Porphyry Cu systems develop mainly in the magmatic arc setting (Fig. 5.2) that is subjected to regional stress regimes that vary from extensional through oblique to contractional. Strongly extensional settings, represented by the basalt-rhyolite bimodal volcanism, however, are not characterized by well-developed porphyry Cu systems (Sillitoe 1999; Tosdal and Richards 2001). However, there is an observable relationship between broadly contractional settings suggested by crustal thickening, surface uplift and rapid exhumation and intense porphyry Cu mineralization (Sillitoe 2010, references therein). It is believed that crustal compression aids development of large middle-upper crustal magma chambers, capable of efficient fractionation, magmatic fluid generation and release, particularly at times of rapid uplift and erosional unroofing (Sillitoe 1999).

Porphyry deposits in more primitive intra-oceanic arcs (e.g., SW Pacific) tend to be more gold-rich, compared to those in more continental settings (e.g., North American Cordillera), which may be enriched in Mo, or even in Sn or W in rare cases. Epithermal high sulfidation Cu–Au–Ag deposits represent upper parts of porphyry systems.

Intra-arc fault-systems which have been active before or during magmatism and porphyry mineralization are apparently important localizers of mineralization. Some researchers have emphasized the importance of intersections between continental scale transverse fault zone and arc-parallel structures for porphyry Cu formation (Richards et al. 2001).

The Sn–Ag occurs in a belt, sub-parallel with the Cu belts but toward the continent (in the back arc region) and related to a different group of granitic rocks, i.e., peraluminous, ilmenite series and locally S-type magmas (Fig. 5.2). The Bolivian Sn–Ag belt that occupies a unique behind-arc position in the Andean cordillera is an interesting example.

5.6.4.2 Back Arc-Related Submarine Mineral Deposits

These are VMS deposits of Cu–Zn–Pb (±Au–Ag–Ba) that are represented in many mining camps from the Neoarchean to the present. More than 300 sites of hydrothermal activity and sea floor mineralization are known today on the ocean floor. About 100 of these are sites of high temperature venting and polymetallic mineral deposition (Hannington et al. 2005). Barrie and Hannington (1999) classified VMS deposits into 5 types depending mainly on the composition of the associated rocks (up to 3 km distance in the footwall and 5 km along strike). The principal metal-content of the VMS deposits are broadly reflective of the bulk composition of the associated rocks. In Table 2.4 and Sect. 2.2.2.1, we have presented and discussed the litho-types of these deposits, bulk lithology and principal ore metals in terms of their geodynamic setting (Fig. 5.2).

5.6.4.3 Mineral Deposits in Foreland Basins

Economic mineral deposits in foreland basins can be broadly divided into two groups: (i) Placers (including paleoplacer gold); (ii) Sediment-hosted hydrothermal deposits. The former are concentrated within the active orogen or form along the mountain front where the detrital gold gets continuously accumulated. Most of the giant placer gold deposits were deposited in Mesozoic to Recent convergent margins along the Pacific Rim. Paleoplacer gold deposits of Neoarchean Witwatersrand deposits were long debated on whether they were paleoplacer, or hydrothermal in origin. Detailed zircon studies by Kosticin and Krapez (2004) clearly demonstrate that the host Central Rand Group was deposited in a retro-arc foreland basin.

Although a few MVT deposits have been arguably related to extension tectonics, the majority of them are believed to have formed in foreland basins, related to convergent margins (Leach et al. 2005).

The restriction of large MVT deposits to the Phanerozoic, almost certainly relates to the emergence of coralline limestone reefs from the Devonian onwards. This fact qualified the carbonate rocks to be more porous and permeable and hence more suitable for ore deposition.

Paleoproterozoic unconformity-related uranium mineralization, famously known from Canada, Australia, Africa, and many other countries, are sited in foreland basins. Their origin is linked to supercontinent assembly at ca. 1.8–1.7 Ga (of Columbia).

5.6.5 Mineralization at the Continent–Continent Collision Regions

"Collisional tectonics" is the story of two continental plates confronting each other with the consequent results. But it does not take place all of a sudden and indeed takes place in several steps. The process starts by the movement of a subducting plate below an overriding plate. The two plates may do it by colliding with an active continental margin arc on the overriding plate, forming a continent–continent collision (as in the Himalaya) or more commonly with an island arc on the overriding plate. The remnant of an oceanic basin between the two may be represented by an ophiolitic suture zone. Examples may be quite a few. We refer here only to the Indus-Tsangpo suture along the northern boundary of the modern-day Himalaya. Mitchell and Garson (1981) were generous in assigning up to seven tectonic elements in such a zone: pre-collision remnant and residual ocean basins, syn-collision suture zone, the "hinterland" margin of the overriding plate, syn- to post-collision foreland thrust belts, foreland basins, hinterland basins and the intermontane troughs of the foreland and hinterland (Fig. 5.6).

Remnants of residual ocean deposits, barring a few phosphate and carbonaceous deposits, are negligible. The syn-collisional ophiolites along the suture zone contain only a few sub-economic chromite deposits in some places such as, Oman, Cuba, Turkey, New Guinea. Margins of the overriding plate are usually not rich in ore deposits. Syn- to post-collision thrust belts often contain mineral deposits that are generally inherited, except some Sn, W, U-bearing deposits which might have formed from hydrothermal fluids derived from crustal anatexis and generation of S-type granites. The best example of this is provided by the Hercynian Sn granites of Southwest England and Sn–W deposits related to the Hercynian granites of Portugal. The Rossing U deposit of Namibia may also be mentioned. Some Himalayan ore deposits occurring in the Lesser Himalaya are essentially inherited (Sarkar et al. 2000). The hinterland and foreland basins may be obliterated or only partially preserved during collision, along with the mineral deposits, if originally present.

This is about what we knew about ore mineralization in the collision–collision geodynamic setting. Until recently the India-Asia collision zone has been thought to be too sterile compared to its magnitude. But discoveries in recent years of several large porphyry Cu–Mo ± Au deposits in the Miocene Gangdese belt of southern Tibet ($\sim$89°E–93°E) (Qulong porphyry Cu–Mo and Jiama porphyry Cu–Mo–Au deposits), emplaced in the post-subduction and collisional settings, have changed our perception to a fair extent. There are Paleocene-Eocene granitic intrusions and co-eval volcanic rocks in the western part of the belt, which are water-poor (<4 wt %) and weakly oxidized (ΔFMQ < 1.0). Porphyry Cu mineralization in this sector is too weak compared to that in the eastern Gangdese. In contrast, the eastern magmatic rocks have higher fO_2 values (ΔFMQ = 0.8–2.9) and more water content. Two giant porphyry Cu deposits, Qulong porphyry Cu–Mo deposit containing 1420 Mt ores (0.5% Cu, 0.03% Mo) and the Jiama porphyry Cu–Mo–Au deposit containing 1054 Mt ores (0.44 wt% Cu), besides some smaller ones are located in this sector. This is now viewed as a result of post-subduction or collision-related magmatism.

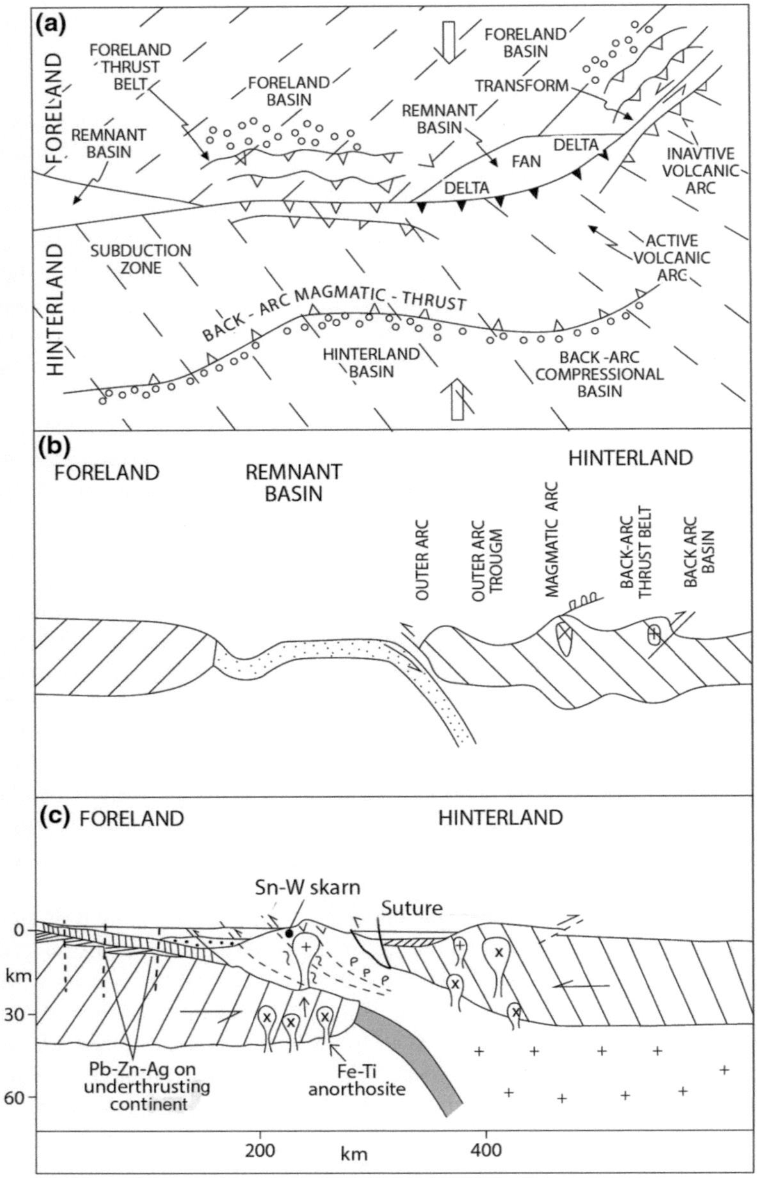

Fig. 5.6 Cartoon of a foreland-hinterland zone (**a** in plan; **b** and **c** in section) in a collisional setting (after Mitchell and Garson 1981); **b** and **c** depict stages of collision with some locations of mineral deposits in such setting shown in **c**

We have already briefly discussed how mineralization takes place along the volcanic arc, following the MASH model of Hildreth and Moorbath (1988). It was an extensional regime. The contractional collision stage may be accompanied by crustal thickening, delamination of SCLM (sub-crustal lithospheric mantle), or detachment of a major slab from the subducting underplate). The detachment is likely to trigger upwelling of asthenospheric melt, which interacting in the subduction zone modified the Tibetan lithosphere, causing partial melting and subsequently producing voluminous quantities of ore-bearing fluids. There is also the suggestion that residual sulfides from the earlier Gangdese arc magmatism might have been remobilized during the Miocene phase of asthenosphere-triggered magmatism (Fig. 5.7) (Richards 2009; Wang et al. 2014). This is a new addition to our knowledge of metallogeny in collisional setting.

5.6.6 *Some Controversies About the Geodynamic Status of Some Sediment-Hosted Ore Deposits*

After an excellent discussion on the geodynamic setting of many ore deposits, Groves and Bierlein (2007) direct our attention to some controversies that are still making the rounds with respect to the geodynamic settings of ore deposits hosted by sediments. It is their conclusion that the sediment-hosted stratiform to stratabound deposits having Pb, Zn, and Cu in variable proportions, such as those of the Zambian copper belt (Africa), Kupferschiefer deposits (Europe) and the so-called SEDEX deposits, are difficult to classify in terms of geodynamic setting at the time of their formation. Although there is a general agreement that they formed in intracratonic rift basins during crustal extension, there is appreciable controversy regarding their broader scale tectonic setting and the driving force for hydrothermal fluid flow needed for ultimate ore formation.

The hosting sequence of some important sediment-hosted stratabound copper deposits, such as those of Zambian Cu belt, almost certainly formed in the intracratonic rift setting (Tembo et al. 1999), although a passive margin setting has also been suggested (Binda 1995). Kerrich et al. (2005) on the other hand related their occurrence to the initial dispersal of Rodinia at ca. 800 Ma. There is a crucial debate about whether the economic deposits formed mainly during diagenesis (e.g., Shelly et al. 2005) or during the inversion of the intracratonic basins during later orogeny (McGowan et al. 2006)

SEDEX deposits, as we have discussed earlier, are stratiform to stratabound massive Zn–Pb $\pm$ Cu deposits that formed during sedimentation or early diagenesis from basinal brines. They illustrate well the problems of categorization of the geodynamic setting of some mineral deposits (Groves and Bierlein 2007). We take Proterozoic example from Australia (Large et al. 2005). These giant deposits pertain to Mt. Isa, McArthur River and Century deposits and formed at ca. 1650–1600 Ma, following shortly after the final assembly of Proto-Australia at ca. 1780–1700 Ma,

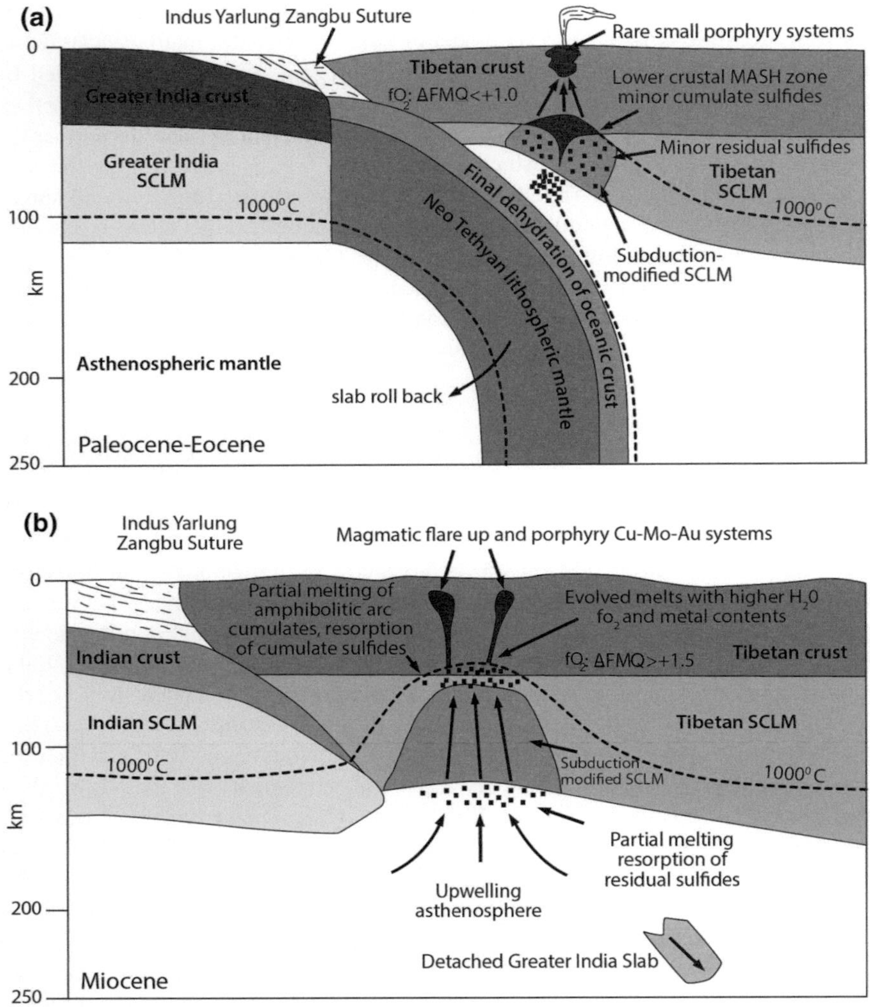

Fig. 5.7 Sketch of post-subduction or collision-related magmatism which has given rise to several large Porphyry Cu–Mo ± Au deposits in the Miocene Gangdese belt of southern Tibet. (See text for more details) (after Wang et al. 2014)

as part of the supercontinent Columbia, which formed via collision between Laurentia, Baltica, and Siberia at 1850–1700 Ma. Groves et al. (2005a) argued that this was the first supercontinent assembly of major crustal pieces, providing a foundation for the extensive sedimentary rift basins in which the SEDEX deposits could form in response to episodic extension related to tectonism at the margins of the cratonic blocks. Reactivation of deep extensional faults, close to the suture along which the old cratonic pieces were assembled, appears critical for the generation of SEDEX deposits in the rift basins. Paleozoic SEDEX deposits (e.g., Sullivan, Red Dog) occur in rifted continental margins, and therefore, could also be included in the category of deposits related to divergent margins.

Broken Hill-type deposits show similarities to SEDEX deposits though they are enriched in Pb and much enriched in Ag compared to most others. But in contrast to reduced dolomitic siltstones and shales by which SEDEX deposits are commonly hosted, Broken Hill deposit is hosted by highly metamorphosed quartzo-feldspathic sediments. Another example of uncertainty about the geodynamic model of a mineral deposit or belt stemming from the uncertainty of its origin is well illustrated by the Mt. Isa deposit, Australia. Until some years ago it was believed to be syngenetic. Now, it is mostly believed to have a secondary origin like the orogenic gold (cf. Perkins 1984).

Finally, regarding the relative importance of plates and plumes in the geodynamic evolution of the Earth there is not much to dispute about. Both were effective but their difference varied with time. According to the arguments of Groves et al. (2005a, b) and Kerrich et al. (2005) amongst others, Mesoproterozoic represents the transition from mantle plume-dominated plate tectonics to modern style plate tectonics.

A synthesis of geodynamic setting, lithological association(s), and mineral deposit(s) is presented in Table 5.1.

5.7 Supercontinents and Superevent Cycles

The concept of the supercontinent cycle made its first appearance in late 1980s, based on the belief that the continental masses assemble and disperse cyclically on a time scale that range between 200 and 500 My (Hoffman 1988; Rogers 1996; Rogers and Santosh 2004; Santosh et al. 2009). To be somewhat precise, a supercontinent is defined as a quasi-rigid or rigid assembly of most of the Earth's continental landmasses at a specified time. The word "most" is somewhat vague. Numerically it should be $\geq 75\%$ of preserved crustal area from a given age (cf. Meert 2012). The supercontinent cycle involves rifting and breakup of the supercontinent, followed by a stage of reassembly when the dispersed continents come together and collide to form a supercontinent once again. The idea of supercontinent cycle has grown from strength to strength with time. The first conjectural supercontinent was *Pangea*, during Permian and Triassic times that broke up $\sim$180 Ma ago to produce the present-day configuration of continents and oceans. Looking backward in time, we could speculate on some more, less-constrained supercontinents such as, *Rodinia* at 1.1–1.0 (assembly) to 0.8–0.7 Ga (break-up), *Nuna/Columbia* at 1.9–1.8 Ga (assembly) to 1.4–1.2 Ga (break-up) (Evans 2013) and *Kenorland* at $\sim$2.7–2.5 Ga (assembly) to 2.45–2.1 Ga (breakup) (cf, Barley et al. 2005). There were a few more suggestions also. But the majority opinion favored the above few only.

Table 5.1 Tectonic settings, lithological associations and mineral deposits

	Tectonic settings	Lithological associations	Mineral deposits
1.	Continental hotspots including rifts and aulacogens	Mafic-ultramafic rocks, carbonatites, clastic and chemical sediments (dominantly shales and dolostones) Granite, granite gneiss	Cr and Ni–Pt–Cu deposits (Great Dyke, Zimbabwe), diamonds. Variously transformed sedimentary-diagenetic deposits, like the Sullivan, B. C., Mt. Isa, Broken Hill, and McArthur deposits (Australia), Kupferschiefer (Central Europe) and the Cu–Co deposits of Zambia-Zaire etc. Sn-mineralization, Josh plateau (Nigeria), Rondonia (Brazil)
2.	Constructive plate boundaries (a) Ocean ridges, sea mounts, ocean basins (b) Obduction zones	Mid-oceanic ridge basalts; precipitates of Fe and Mn oxides Ophiolitic associations of ultramafic and mafic rocks	Fe-, Mn oxides along the rifted ridge and also crusts of Fe and nodules of Mn (containing Cu, Co, Ni $\simeq$ 2 wt%) on summits and sea mount flanks Sulfides of Fe, Zn $\pm$ Cu usually at places on the ridge Alpine type chromite ($\pm$ Pt); cyprus type Fe–Cu–Zn ($\pm$Au $\pm$ Ag) deposits
3.	Passive continental margins	Clastic and chemical (mainly carbonate) sediments	Evaporites, phosphorites, BIFs and Minette type Fe ores; clastic- and carbonate-hosted Pb–Zn deposits; placer deposits; hydrocarbon accumulations
4.	Subduction zones	Dominantly basic volcanics, arc and continent provided sediments and intermediate to felsic calc-alkaline rocks	Porphyry Cu–Mo $\pm$ Au deposits, skarn Cu + Au and epithermal Au mineralizations in Cordillera type belts and behind-arc Sn–Ag mineralization in the peraluminous and Ilmentite series S-type granitoids
5.	Continental collision zones	Remnants of island arcs on the overriding plate and that of an ocean between the two continents by an ophiolite zone. There are usually reactivated and regenerated country rocks from the marginal parts of the two continents	Podiform chromite deposits; VMS deposits; Sn and W, U deposits in felsic rocks. U mineralization in molasse deposits as in the Siwaliks of India and Pakistan. Most spectacular porphyry Cu deposits have been recently discovered in the

(continued)

Table 5.1 (continued)

	Tectonic settings	Lithological associations	Mineral deposits
			Miocene Gangdes belt of Southern Tibet
6.	Greenstone belts (ancient volcano-plutonic arc systems)	Mafic-ultramafic (komatiitic) rocks and TTG-dominant granitoids with subordinate proportions of clastic and chemical sediments	Au, Cu, Ni and Algoma type Fe and Cu–Zn mineralization
7.	Intracratonic settings, which are suggested to have been anorogenic at formation	Highly metasomatized (Na_2O, K_2O) rocks of various compositions	Iron oxide- Cu sulfide-gold deposits ± REE (IOCG type ores)
8.	Tectonically stable, peneplaned or nearly so, rock masses in tropical-sub-tropical regions	Weathered profiles in Al-bearing rocks	Lateritic deposits (including bauxite, Ni-laterite, U)

Many authors (Pirajno 2000; Condie 2004 amongst others) believe that the breakup of a supercontinent is triggered by enhanced mantle plume activity at specific time intervals while catastrophic mantle plumes in between increase

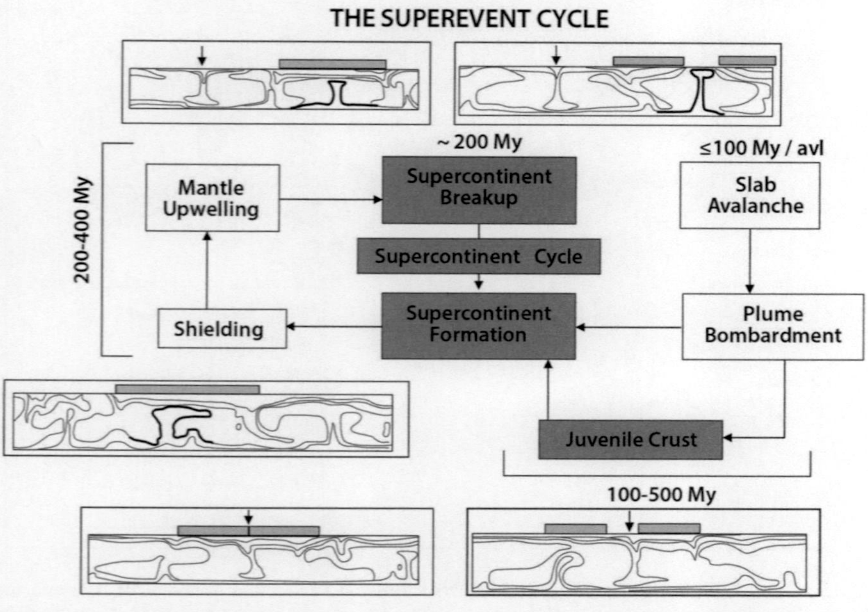

Fig. 5.8 Interplay of supercontinent cycle, catastrophic mantle avalanche and mantle plumes, and juvenile crustal growth produce a super event cycle (after Condie 2004; Pirajno 2000)

juvenile crustal growth and bring about supercontinent assembly (Fig. 5.8). Condie (2004) further postulated the "superevent" cycles at 2.7, 1.9, and 1.2 Ga in which the "graveyards" of subducted oceanic lithosphere, stored at the 670 km D' boundary, catastrophically avalanched to the core-mantle boundary (i.e., the D'' layer). This material, molten, produces the plume which moves rapidly upward, ultimately causes plume bombardment under the lithosphere, which directly or indirectly is responsible for crustal growth. Sill-sediment complexes of the Mesoproterozoic Sullivan Pb–Zn deposit in British Columbia, Canada and Neoproterozoic basal sequences associated with the Central African Cu province are evidence of mantle plumes that caused dispersal of the supercontinents Columbia and Rodinia respectively. Evidence of lithospheric delamination has been provided by tomographic imaging of the mantle under Tibet, India and the adjacent Indian Ocean (Van der Voo et al. 1999). These studies reveal several zones of relatively high *P*-wave velocities at various depths (Fig. 5.9) which have been interpreted as the remnants of delaminated SCLM that avalanched when Greater India continued to converge northward with Asia after ~45 Ma.

Murphy and Nance (1992) recognized two principal styles of supercontinent aggregation, which they called *internal* and *external*. The internal aggregation refers to continent–continent collision, as in the case of Alpine-Himalayan,

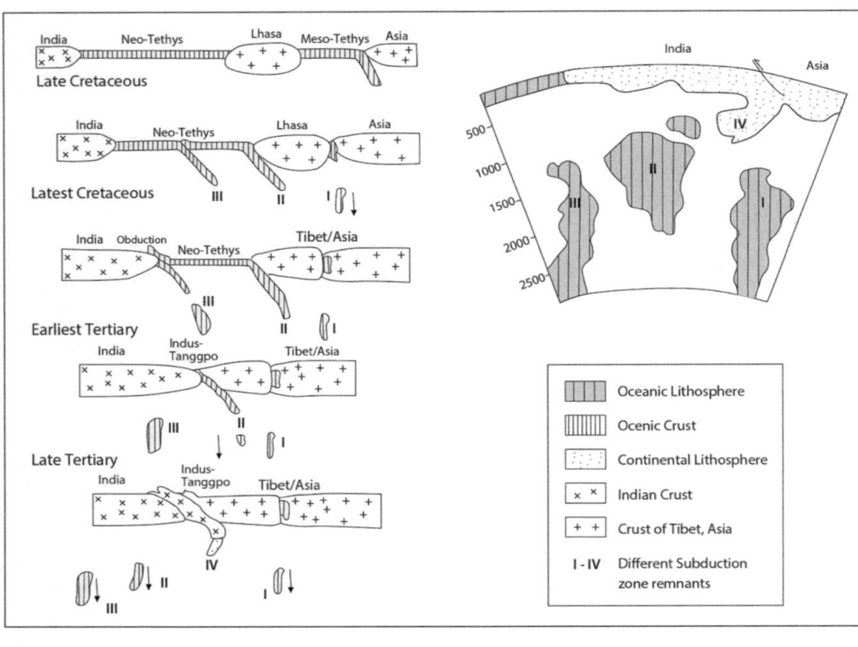

Fig. 5.9 Sketch showing the probable Mesozoic Tethyan subduction zones (I, II, III) and the Tertary lithospheric delamination (IV) during northward convergence of Greater India with Asia. The right hand side of the diagram depicts the results of tomographic imaging of parts of the mantle in the region (after Van der Voo et al. 1999)

Appalachian-Grenville orogenic belts. External aggregation refers to Cordilleran type tectonics where, as we have already discussed, allochthonous tectonostratigraphic terranes are transpressively accreted to a continental margin. Neoarchean magmatic-accretionary events in Canada, South Africa, India, Western Australia, probably relate to an early external supercontinent aggregation, with which was apparently associated the development of orogenic gold provinces of these regions.

Supercontinents may assemble in two configurations: introversion and extroversion. Introversion comprises break-up, opening, and then closing of the internal oceans and finally, reassembly. In extroversion, following supercontinent dispersal, margins of continental fragments rotate and collide during reassembly. The processes may combine, however.

5.8 Metallogenic Provinces in a Supercontinent Cycle Framework

The making and breaking of supercontinents have come to be believed as the most important geological events in Earth's history. Its relationship with mineralization, if any, cannot therefore be ignored. Most mineral deposit types form in particular types of intracratonic, and divergent and convergent margin geodynamic settings, and the temporal distribution of those deposits should correspond with the supercontinent cycles mentioned above (Barley and Groves 1992; Groves and Bierlein 2007). The accompanying diagram (Fig. 5.10) shows the temporal distribution of different deposit types vis-a-vis their broad geodynamic setting and their broad compatibility with the supercontinent cycles.

Archean crust is resistant to reworking during later geological events due to its thick refractory continental lithospheric mantle (CLM). This characteristic accounts for the preservation of outstanding reserves of gold ores in Neoarchean greenstone terranes, VMS (back arc) camps of the Superior province and komatiite-associated Ni-sulfide deposits. The abundance of VMS deposits in the Superior province in contrast to sparseness of the same in some other Neoarchean terranes, such as, in India, South Africa, and Western Australia appears somewhat anomalous. But we do not always have all the answers for all the questions readily!

During the formation of a supercontinent, the interface of two accreting blocks will be characterized by subduction, marginal basin development and collision. This will accompany ore mineralization, as expected from discussions above. Its breaking will take place by rifting followed by mineralization. Repetitive development of arc, marginal basin and plume magmatism and subduction in the convergent margin will convert the *tectogen* into a *metallogen*. It has been suggested that the long-term enrichment of the convergent margin lithosphere will ultimately culminate in a bonanza metallogenic event (Barley et al. 1998).

Few deposits appear to form during periods of supercontinent stability. Sedimentary rock-hosted Cu and Pb deposits will form mainly during the beginning

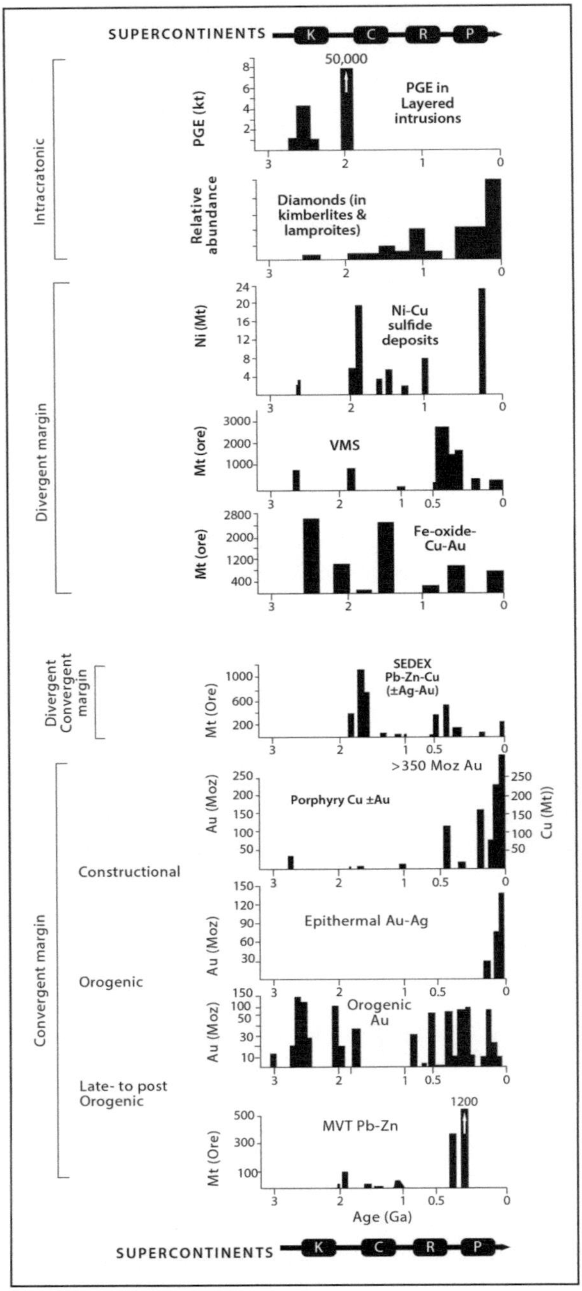

Fig. 5.10 Temporal distribution of different deposit types in terms of their geodynamic settings viewed against the supercontinent cycles (*K* Kenorland; *C* Columbia; *R* Rodinia; *P* Pangea) (after Groves and Bierlein 2007)

of supercontinent fragmentation, whereas deposits related to convergent tectonics (porphyry Cu, VMS, orogenic Au) develop predominantly during periods of subduction and supercontinent aggregation. The relationship between supercontinent cycles, juvenile crustal growth and superplume/plume events and metallogeny is best exemplified by the formation of orogenic gold deposits (vide Goldfarb et al. 2001; Groves et al. 2005a) (Fig. 5.11). Gold deposits formed during the older two Precambrian supercontinents, Kenorland and Columbia, are preserved in cratonic blocks. However, the gold deposits formed during the growth of Rodinia have been lost from geologic records due to erosion of Meso-Neoproterozoic blocks to their deep crustal levels which are below gold-favorable zones (Goldfarb et al. 2010). Although orogenic Au deposits are poorly represented during most of Earth's Middle Ages (i.e., from 1.7 to 0.9 Ga; Cawood and Hawkesworth 2014), significant Au accumulation is recorded in this time period in IOCG deposits formed in extensional environments in supra-subduction and post-collision settings and temporally linked to A-type granites and kindred intrusions. Noting this different kind of Au endowment in the Mid- to Neoproterozoic time span, Richards and Mumin (2013) identify the IOCG deposits as the Precambrian equivalent of Phanerozoic porphyry-type deposits and relate the change in mineralization style to increasing sea water sulfate in the subducted oceanic lithosphere as well as to decreasing lithospheric geothermal gradients during these times, resulting in a change from sulfur-poor to sulfur-rich arc magmas.

The change in style of base metal-bearing VMS deposits from the Archean Abitibi style to the Phanerozoic Kuroko and Cyprus types may be due to pertinent geological reasons, such as the change in the style of subduction, change in the arc-magma composition etc. The VHMS-gap in the Earth's middle ages (1.7–0.75 Ga) (Cawood and Hawkesworth 2014) is also difficult to explain given the well-developed convergent plate boundaries along the margins of Laurentia and Baltica which could have acted as fertile upper mantle source regions for such mineralization. The scarcity of porphyry Cu and epithermal Au deposits in rocks older than 200 My is believed to be due to their high erosion potential. It is further noteworthy that the preservation potential is high in 'external' (Cordilleran type) than in 'internal' type mountain belts (Barley and Groves 1992).

It is worth noting that well-dated mineral deposits can provide an important constraint on the veracity of supercontinent reconstruction because they should lie adjacent to proposed sutures, or on the margins of external oceans in such reconstructions. Kerrich et al. (2005) provide excellent examples of SEDEX Pb–Zn, Fe oxide–Cu–Au and anorthoste-associated Fe–Ti–V deposits in a reconstruction of Columbia (Fig. 5.12). In an opposite way, accurate supercontinent reconstruction can suggest the potential for as yet undiscovered ore deposits in the extension of metallogenic provinces, later dispersed during supercontinent break-up and hence may serve as a regional ore guide (Groves and Bierlein 2007).

In summary, metallogenic provinces and epochs are defined by episodic development of metal deposits in space and time respectively (Meyer 1981). They represent the products of the interplay of plume and plate tectonics associated with juvenile crustal growth within the larger framework of supercontinent cycles

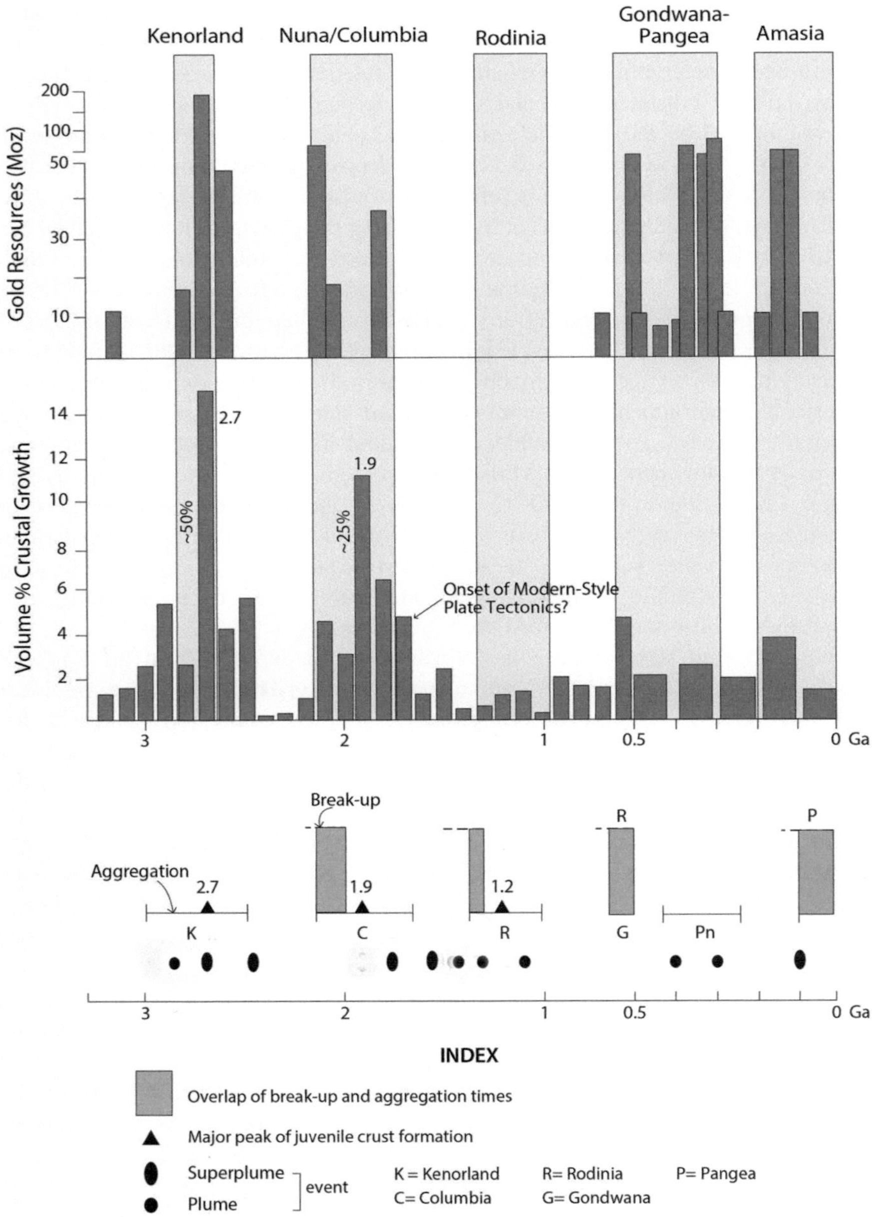
Kenorland
Nuna/Columbia
Rodinia
Gondwana-Pangea
Amasia
Gold Resources (Moz)
200
100
50
30
10
Volume % Crustal Growth
14
12
10
8
6
4
2
2.7
1.9
~50%
~25%
Onset of Modern-Style Plate Tectonics?
3
2
1
0.5
0 Ga
Break-up
Aggregation
2.7
1.9
1.2
K
C
R
G
Pn
R
P
INDEX
Overlap of break-up and aggregation times
Major peak of juvenile crust formation
Superplume
Plume
event
K = Kenorland
C= Columbia
R= Rodinia
G= Gondwana
P= Pangea

◀**Fig. 5.11** Correlation of gold resources in orogenic gold deposits with peak periods of crustal growth (based on data from Condie 2000) and the formation of Kenorland, Columbia, Gondwana/Pangea and perhaps a new supercontinent, Amasia (after Goldfarb et al. 2010). The overlap of break-up and aggregation times of supercontinent cycles and the major peaks of juvenile crust formation are taken from Condie (1998). Mantle plume/superplume events are based on Abott and Isley (2002)

(Barley and Groves 1992). Metallogeny is also connected to the evolution of the subcontinental lithospheric mantle with time and the types of orogenic belts that developed at periods of earth history. Finally, the preservation potential plays an important role in the life cycle of a metallogen. This complex interplay of various factors that are responsible for the secular changes in metallogeny, as discussed above, is shown schematically in Fig. 5.13. This last aspect, though raised years ago by Veizer et al. (1989) as birth and death of ore deposits, understood best through concepts of demography, has not received much attention until recently. Kaur and Chaudhuri (2014) argue that the three phases of supercontinent cycle – subduction, collision, and break up – produce different volumes of magmas and also have different preservation potential of rocks formed during these phases. The preservation potential of rocks, and by analogy, of mineral deposits, is high during

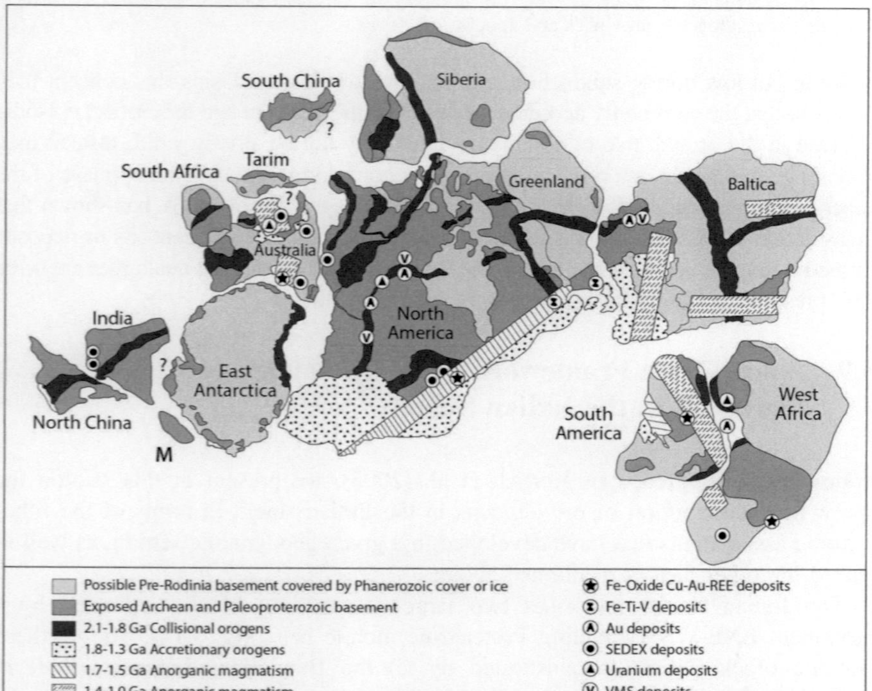

Fig. 5.12 Configuration of Paleo- to Mesoproterozoic supercontinent Columbia (of Zhao et al. 2004) depicting the belts of anorogenic magmatism and the location of various mineral deposit types (after Kerrich et al. 2010)

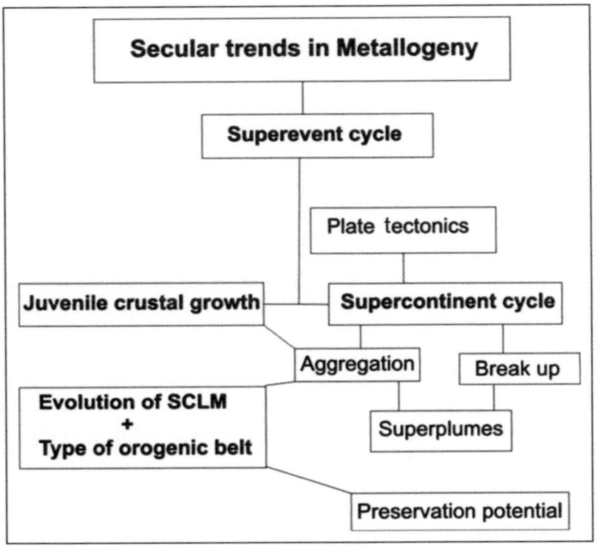

Fig. 5.13 Schematic diagram showing the interplay of various factors responsible for secular changes in metallogenic provinces and epochs

collision but low during subduction and periods of break up. Using this concept they point out that the commonly accepted premise that the peaks in age data reflect episodic increase in the growth rate of continental crust may not be strictly valid. Instead they may reflect the preservation potential of rock record during the collisional phase of the supercontinental cycle. However, discussions in Chap. 2 in this book has shown that many products of subduction, such as, the sub-volcanic porphyry deposits or deposits formed during rifting such as those in the Red Sea and their ancient analogues are often well preserved making this concept somewhat contentious.

5.9 Space–Time Framework of the Metallogenic Provinces in the Indian Shield

Following the approach of Kerrich et al. (2005), we present in this section the space–time distribution of ore deposits in the Indian shield in terms of the litho-tectonic associations that have developed in a given geodynamic setting, as well as due to the other factors mentioned above.

The Indian shield comprises two large-scale crustal blocks separated by a prominent ENE-WSW-trending Proterozoic mobile belt (Fig. 5.14). The northern cratonic block (NCB) is constituted by (i) the Bundelkhand massif (BM) in north-central India and (ii) the Aravalli-Delhi orogenic belt (ADOB) enclosing two tectonic blocks of the Banded Gneissic Complex (BGC) basement in north-west India, along with (iii) the outlying segment of Shillong Plateau in the north-east and (iv) the Himalayan Proterozoic belts in the north. The larger southern cratonic block

(SCB) comprises the Singhbhum-North Odisha or Eastern Indian craton (SC), Bastar or Central Indian craton (BC), Dharwar or South Indian craton (DC, comprising the western, WDC and eastern, EDC segments) as well as the Southern Granulite Terrain(SGT), and the Eastern Ghat mobile belt (EGMB). The Central Indian Tectonic Zone (CITZ) of Proterozoic age separates the two cratonic blocks and forms a composite zone comprising the Mahakoshal belt in the north followed southwards by the Betul belt, Sausar and Sakoli belts, and the Kotri belt. The CITZ is ca. 100 km wide and is marked by an aborted rift and intense shearing that got reactivated from time to time since the Proterozoic (Acharyya and Roy 2000) and now channels two important west-flowing rivers, Narmada and Tapti. Interestingly,

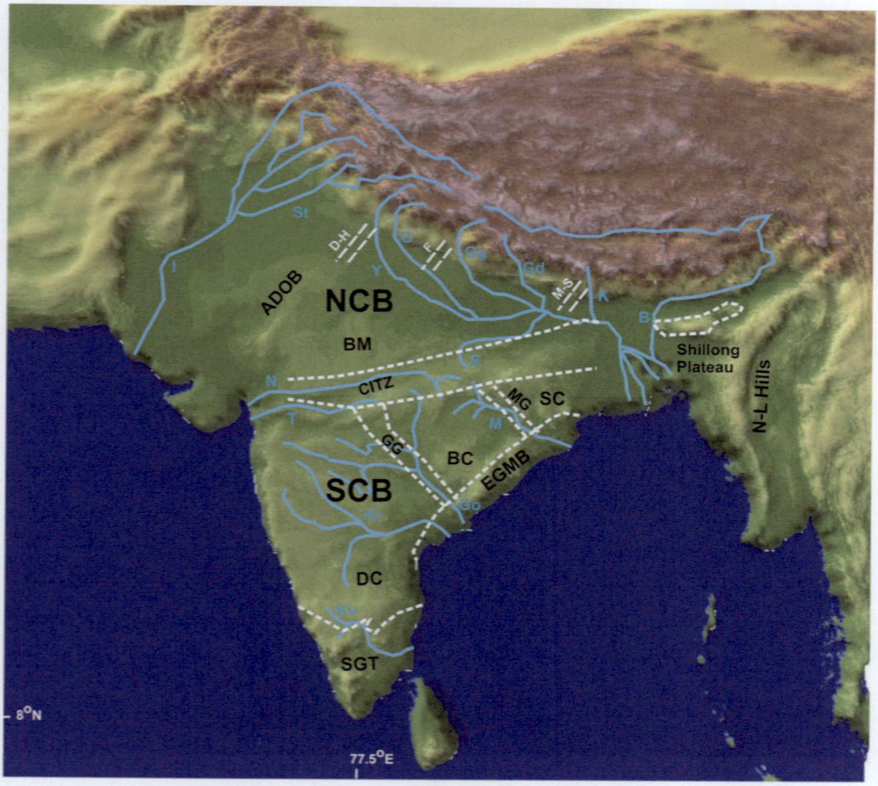

Fig. 5.14 Crustal architecture of India, showing the two cratonic blocks, various cratons, rifts, subsurface ridges, and major rivers. Abbreviations used: *ADOB* Aravalli–Delhi Orogenic Belt; *BC* Bastar craton; *BM* Bundelkhand Massif; *CITZ* Central Indian Tectonic Zone; *D–H* Delhi–Hardwar ridge; *DC* Dharwar craton; *EGMB* Eastern GhatMobile Belt; *F* Faizabad ridge; *GG* Godavari Graben; *M–S* Mungher–Saharsa ridge; *MG* Mahanadi Graben; *N-L Hills* Naga-Lusai hills; *NCB* Northern cratonic block; *SCB* Southern cratonic block; *SC* Singhbhum craton; *SGT* Southern Granulite Terrain. Rivers: *B* Brahmaputra; *G* Ganga; *Gd* Gandak; *Gg* Ghagra; *Go* Godavari; *I* Indus; *K* Kosi; *Kr* Krishna; *Kv* Kaveri; *M* Mahanadi; *N* Narmada; *S* Son; *St* Satluj; *T* Tapti; *Y* Yamuna (after Deb 2014)

the mighty Brahmaputra River in Assam flows along the extension of the same trend in the northeast (Fig. 5.14), making it a fundamental structure between the two cratonic blocks of the Indian subcontinent.

There are various lithotectonic associations in the geodynamic domains of the Indian shield, e.g., the cratonic nuclei, Archean-Proterozoic gneisses and granulites, Archean-Proterozoic granites and greenstones, Proterozoic mobile belts, intra-cratonic basins and anorogenic granitoid massifs. Within them are concentrated vast resources of iron, manganese, and aluminum and notable resources of chromium, copper, lead and zinc, and titanium, a metal now largely concentrated in coastal placer deposits as ilmenite along with REE-containing monazite. The latter hosts substantial amounts of uranium and thorium too. There are also significant primary reserves of uranium and limited reserves of gold and tungsten (Fig. 5.15), as well as rare concentrations of tin and platinoid metals. Out of these, the Al and REE (including Th and Ti) resources, although derived from Precambrian rocks, are the products of metallogenic processes in the Mesozoic and Cenozoic eras (Deb 2014).

In terms of distribution of metallic deposits in the two cratonic blocks, the southern cratonic block (SCB) is distinctly richer in terms of resources of iron, manganese, chromium, copper, gold, and uranium. Rare occurrences of PGE mineralization is also recorded from this block. The northern cratonic block (NCB) is only richer in lead–zinc resources.

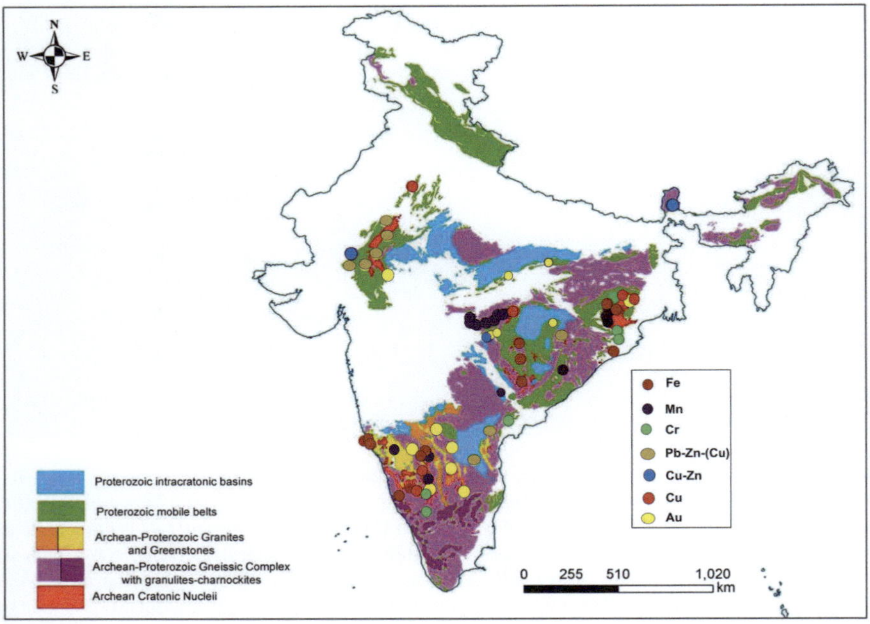

Fig. 5.15 Distribution of major metallic ore deposits in India, in terms of the geodynamic domains, mentioned in the text (after Deb 2014)

5.9.1 Archean Metallogeny

The earliest mineralization in the Indian shield (Fig. 5.16) is recorded from the northern Odisha-Singhbhum craton within Iron Ore Group (IOG III) sequence in the Tomka-Daitari basin (cf. Fig. 3.6). The BIF unit here is considered to be early Paleoarchean, since it is overlain by a dacitic lava with a U–Pb zircon age of ~3.5 Ga (Mukhopadhyay et al. 2008). Thus, the IOG III mineralization can be compared to the Algoma-type Isua Iron Formation of Greenland, believed to have been deposited at about 3.8–3.7 Ga (Rosing et al. 1996). Metallogeny in the SC is best understood in the backdrop of its geological evolution, as summed up recently by Nelson et al. (2014): (i) Between 3530 and 3300 Ma tonalities were emplaced with volcanics, clastic and carbonates and BIF (cycle 1); (ii) Between 3325 and 3300 Ma burial, deformation, and uplift transformed the central part of the Singhbhum basement to tonalitic gneisses, with the cycle 1 sedimentary rocks incorporated into the gneisses as enclaves but preserved within synforms around the basement margins; regional metamorphism and granodiorite intrusions took place until ca. 3285 Ma; (iii) BIF and clastic sedimentary rocks (cycle 2) were deposited around the margins of the craton into the older (cycle 1) sedimentary rocks and adjacent gneissic basement until ca. 3.1 Ga; (iv) A further episode of granite intrusion at 3090 Ma was followed by uplift and erosion of the central part of the

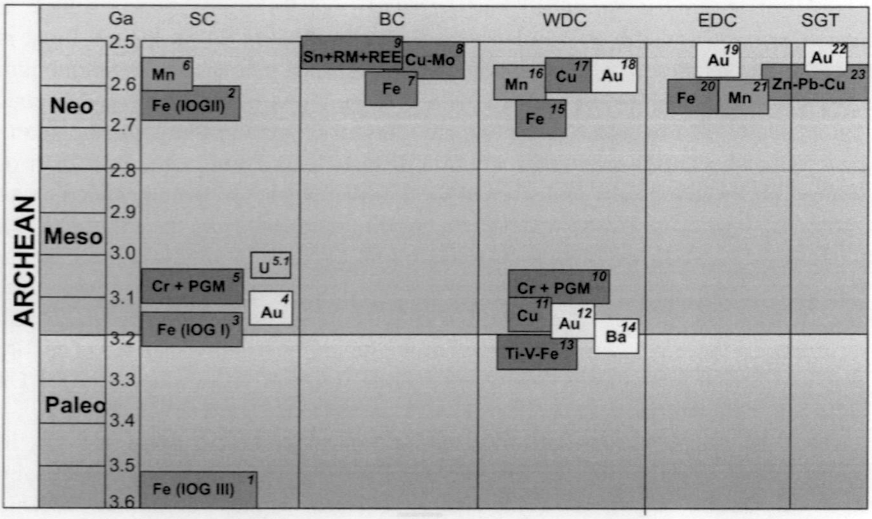

Fig. 5.16 Space-time distribution of Archean ore deposits of India. Numbers in boxes refer to: *1* Tomka-Daitari BIF; *2* Jamda-Koira valley BIF; *3* Gorumahisani-Sulaipat-Badampahar BIF; *4* Kunderkocha; *5* Boula-Nuasahi; *5.1* Mahagiri uraninite placer; *6* Joda-Jamda-Koira; *7* Dalli-Rajhara-Rowghat-Bailadila, *8* Malanjkhand, *9* Bastar-Malkangiri; *10* Byrapur-Nuggihalli belt; *11* Kalyadi, Alladihalli; *12* Kempinkote; *13* Shimoga-Nuggihalli; *14* Ghattihosahalli; *15* Kudremukh, Bababudan; *16* Tumkur; *17* Ingaldhal; *18* Gadag, Ajjanahalli; *19.* Kolar-Hutti-Ramagiri; *20* Bellary-Hospet; *21* Sandur Schist belt; *22* Wynad Gold Field; *23* Mamandur (after Deb 2014)

craton prior to 2806 Ma; (v) Volcano-sedimentary and BIF rocks were deposited during a third sedimentary cycle at 2.8 Ga. Metallogeny in the Singhbhum craton took place in the early stage of shield formation (Fig. 5.16). Here, Algoma-type BIF (IOG I) in Gorumahisani-Sulaipat-Badampahar region of Mayurbhanj district of Odisha hosts large Fe-concentration in a possible greenstone belt environment, in which the shallow marine syn-sedimentary volcanism points toward basinal rifting. This BIF-hosted mineralization took place possibly between 3.3 and 3.1 Ga, making it slightly older than the large chromite (±PGM) mineralization in the Sukinda and Baula-Nuasahi intrusive complexes in the southern edge of the craton. Chromitite seams occur as bands and lenses within serpentinized dunite in the former and within a dunite-peridotite unit of a gabbro-anorthosite intrusive in the latter. There are three chromitite lodes in the Boula mine, the PGM mineralization occurring in the brecciated and hydrothermally altered 'Ganga-Shankar' lode, intruded by the younger 3.1 Ga old Bangur gabbro. No unanimity of opinion exists in the classification of the deformed chromitite orebodies of Sukinda-Nuasahi zone as Bushveld-type (Chakraborty and Chakraborty 1984) or Alpine type (Banerjee 1972; Page et al. 1985), since the features noted are intermediate between the two end members and may form a distinct clan (Haldar et al. 2005). The chromite orebodies in this zone are considered to have formed from mantle-derived boninitic magma in a supra-subduction zone setting and later overprinted by hydrothermal activity (Mondal et al. 2006). Os-isotope systematics in the Cr-PGM mineralization indicates extraction of the initial melt from the SCLM under SC around 3.7 Ga (Mondal et al. 2007). The small but irregularly rich lode-gold mineralization at Kunderkocha is probably of similar antiquity and based on its geologic setting, is considered to be hosted by a relict, deformed greenstone belt within the Singhbhum Granite batholith. Mineralization is confined to gray quartz veins in carbonaceous phyllite and Cr-bearing mica schist that are transected by barren milky white quartz veins. Four locations were mined within a distance of 1.5 km in the early part of 20th century, where a lode was exposed for 250 m and parts of which yielded on an average 28.9 g/t of Au over a width of 90 cm (Ziauddin and Narayanaswami 1974). Uraninite paleoplacers, dated ca. 3.02 Ga based on their youngest detrital zircon population, occur within Mahagiri quartzites which developed unconformably over Mesoarchean Singhbhum granite (3.44–3.1 Ga) (Mukhopadhyay et al. 2016). According to these workers, the detrital uraninite points to the existence of highly felsic and K-rich granodiorite-granite-monzonite suites of rocks older than 3.1 Ga in the Singhbhum craton.

The WDC displays features of the earliest stages of shield formation too, in which large expanse of Peninsular Gneiss (3.3–3.0 Ga), represented by gneiss-migmatites of TTG composition, enclose highly deformed oceanic mafic-ultramafic complexes, supracrustal volcano-sedimentary sequences and chemical sediments of the Sargur Group (3.3–3.1 Ga) as mega-xenoliths. This whole basement sequence is unconformably overlain by the Dharwar supracrustal rocks (2.9–2.6 Ga) found in the younger schist belts of Bababudan, Shimoga, and Chitradurga, constituted by conglomerates, orthoquartzites, mafic and felsic volcanic, BIF and greywackes (Radhakrishna and Naqvi 1986). The Sargur sequence is considered to have formed by magmatism in a subduction setting, related to partial melting of an oceanic slab (Martin 1986). Komatiites in the Sargur Group possibly formed in oceanic plateaus from plumes with elevated Cr–Ni values. The low content of incompatible elements in these komatiites are indicative of a depleted mantle during ca. 3.2 Ga which in turn implies an earlier episode of mantle differentiation, greenstone volcanism, and continental growth probably during ca. 3.6–3.3 Ga, which is substantiated by Nd and Pb isotope data of gneisses and komatiites in WDC (Jayananda et al. 2008). Existence of older crust in the region is also indicated by detrital zircon ages of ~3.58 Ga in the Dharwar supracrustal rocks (Nutman et al. 1992). The later mafic-felsic igneous activity bears signatures of subduction processes. The relatively small deposits of Cr, Ti–V, Cu, and Au in the Mesoarchean Sargur belt (Nuggihalli belt in particular) (Fig. 5.16) can be related clearly to the partial melting processes in such mafic-ultramafic oceanic environment and related magmatism. Recent studies (Mukherjee et al. 2010) have established that the primary chromite composition from Nuggihalli deposits also point to a parental melt of high-Mg komatiitic composition, formed above a supra-subduction zone setting from compositionally similar boninitic magma. The lensoid chromite bodies underwent multiple deformation and compositional modification during low-T hydrothermal processes during serpentinization. The isotopic data of the bedded Sargur barites of Ghattihosahalli (vide Sect. 4.1.4) have been suggested (Deb et al. 1991) to reflect Archean coeval seawater composition, with the mantle like values indicating buffering of the sea water by mantle-derived basaltic crust.

In the Bababudan hill ranges of WDC gigantic concentrations of Algoma-type BIF (~10 Gt reserve at Kudremukh) took place around ca. 2.7 Ga through warm submarine exhalations during volcanic quiescence in the greenstone belts. Almost synchronously (~2.6 Ga), Superior-type BIF got deposited by syn-diagenetic process in the Bellary-Hospet region of the Sandur schist belt of the EDC and in the Jamda-Koira basin of SC as IOG II, in close association with Mn oxide ores (Fig. 5.16). Though good age constraints are not available, the Bailadila Group of BC is considered stratigraphically equivalent to the younger IOG of SC, and thus, presumably the Dalli-Rajhara-Rowghat-Bailadila iron ores of BC also deposited in the same late Archean time slot. The BIF formed by upwelling of reduced bottom waters dissolving Fe^{2+} and Si and precipitating them in the oxygenated waters of the continental shelves (cf. Klein and Beukes 1993). Some of the largest BIF-Fe ores were deposited worldwide in the time slot of 2.7–2.45 Ga (Barley et al. 1998), just as in the Indian shield. Along with the Mn-oxide ore association in Tumkur area of WDC, Sandur belt of EDC and Joda area in SC, these BIF mineralizations indicate prevalence of sufficient oxygen level in the late Archean atmosphere above the proto-Indian shield. In fact, based on a pyrophyllite-bearing Archean paleosol near Keonjhar in Singhbhum craton, which formed between 3.29 and 3.02 Ga, Mukhopadhyay et al. (2014) suggest that atmospheric oxygen at these earlier times was sustained at a relatively narrow concentration level between 10^{-4} and 10^{-3} PAL which is above those that could have been produced by photo-dissociation of atmospheric water. These authors propose that this early transient atmospheric oxygenation was due to the existence of microorganisms capable of oxygenic photosynthesis. The timing and process of Fe-ore formation from the BIF in the different iron ore fields, however, remains a controversial issue and it has been suggested (Sarkar and Gupta 2005) that both hypogene and supergene processes, at places in combination, may have brought about the ore concentration.

Gold mineralization took place extensively in the southern part of the Indian shield in the late Archean times (Fig. 5.16). The timing of such gold concentration in India conforms to one of the age clusters (2.8–2.55 Ga) of orogenic gold deposit worldwide (Goldfarb et al. 2010) (cf. Fig. 2.54). A hotter mantle in the Archean was responsible for the abundance of short-lived mantle plumes that produced the greenstone terranes in a thin lithosphere along the edge of older continental blocks. The underlying buoyant SCLM, a residual product of massive plume-induced melting in the upper mantle, consequently formed extensive granitoid terrains which were critical for long-term stabilization of Au-rich mid-crustal zones (Groves et al. 2005b). The geological lay-out of EDC, on the eastern side of an Archean proto-continental shield, matches well with this model. Chadwick et al. (2000) proposed that WDC was the foreland to an accretionary arc to the east, represented by a late Archean batholith (their Dharwar batholiths). However, how the penecontemporaneous greenstone belts fit into the accretionary arc remains to be understood (Sarkar 2010). EDC has at least two world class ore systems in the now-inactive Kolar deposit, which has produced about 800 t of the yellow metal in last 100 years of its operation, and the Hutti group of deposits (Hutti, Uti, Hira Buddini) with a total reserve of ≥ 120 t of Au (Mishra and Pal 2008). The age of first stage gold mineralization at Hutti deposit was deciphered to be 2547 ± 10 Ma from hydrothermal monazite from the wall rocks (Sarma et al. 2008), followed by the second stage 10 Ma later (Rogers et al. 2013), post-dates greenstone formation by about 40–50 My but overlaps with regional magmatism (Yallagatti radiogenic granite: 2532 ± 3 Ma; Anand et al. 2005). A similar situation probably exists in Kolar. The age of mineralization is about 2550 Ma (Au-sulfide-telluride sample from 108th level of the Champion lode, using Pb–Pb method and a Proterozoic model; Ralph Thorpe GSC 2008 Open File Report) which overlaps with the age of granodiorite near the ore zone (2551 ± 2.5 Ma) but is 140–180 Ma later than greenstone volcanism (cf. Ram Mohan and Sarma 2010). Recently, Rogers et al. (2013) have proposed a two-stage model of ore genesis at Hutti, based on stable isotope studies. The first stage is represented by a typical orogenic gold mineralization (grade < 5 g/t) in a compressional setting with a low salinity metamorphic fluid source. The second stage produced high grade (up to 1 kg/t) auriferous zones in a transcurrent tectonic setting through a mixture of magmatic fluid and isotopically light formation fluids contemporaneous with the highly radiogenic Yelagatti granite intrusion. This inference is supported by bimodal distribution of $\delta^{11}B$ of tourmaline in Hira Buddini mine (Krienitz et al. 2008) which requires an origin from at least two isotopically distinct fluid sources that subsequently mixed. The lighter fluid (+1‰) had the B released from metamorphic devolatilization of greenstone lithologies whereas the ^{11}B-rich fluid is attributed to degassing of I-type granite that intruded the greenstone sequence, providing the heat and fluids to the hydrothermal system. Though such detailed studies are not available in Kolar, one may recall the study of Mishra and Panigrahi (1999) in which they invoked gold deposition by mixing of two fluids—a low T ($\leq$200 °C)-low salinity (< 7 wt% NaCl equiv.) fluid with a high T ($\geq$400 °C)-high salinity ($\geq$50 wt% NaCl equiv.) fluid. The latter may surely be derived from the contemporaneous granite mentioned above. Given the favorable ages of rocks, lithologies, structures, and metamorphic setting in the EDC, it is surprising that Hutti remains presently the only producing mine in the country. The suggestion that middle to upper greenschist facies metamorphism is most conducive for orogenic gold endowment (Groves et al. 2003) is not quite convincing. This may be generally true but not inviolable, as the two biggest gold endowments in EDC are in higher grade amphibolites facies rocks, with the Kolar deposit located close to the border of granulite facies terrain. Rather, the answer seems to lie (Deb 2014), in line with the above discussion, in the multistage history of gold metallogeny where a superimposition of a magmatic process on the metamorphic fluid focusing along a structural conduit (cf. Fig. 3.29) can increase the economic potentiality of an individual deposit manifold, like in Hutti and Kolar.

The main gold resources in the WDC are the Gadag gold field in the north and the BIF-hosted Ajjanahalli deposit, which is structurally controlled (Kolb et al. 2004) in the central part of Chitradurga schist belt. Monazite and xenotime have been dated at 2522 $\pm$ 6 Ma in the Gadag ore and 2520 $\pm$ 9 Ma in the Ajjanahalli ore whereas the associated felsic volcanic and intrusive granites in Gadag schist belt are 2588 $\pm$ 10 Ma and $\geq$2555 $\pm$ 6 Ma, respectively (Sarma et al. 2011), suggesting thereby that the volcanics and granites are at least 65 and 33 My older, respectively, than the mineralization. The trend of the primary lode-gold mineralization in the Wynad gold field in northern Kerala – northwestern Tamil Nadu in SGT appears to be an extension of the arcuate Chitradurga schist belt, across the transition zone to the south. However, the mineralization in magnetite-quartzites is apparently affected by the E-W trending Moyar-Bhabani shear zone. The ore-bearing fluids in this field have been ascribed to a mantle-derived source (Santosh et al. 1995). While no age constraints are available for the Wynad gold, Nair et al. (2005) correlate primary Au-sulfide mineralization in Attapady valley with a Paleoproterozoic (2.5–2.0) tectonic event with remobilization having taken place during the Pan-African ($\sim$500 Ma) event. The only Zn–Pb–Cu prospect in granulite terrain of Tamil Nadu is at Mamandur. The pelitic granulites in the ore zone have been dated at 2.6 Ga (Sarkar 1988). The deposit is sub-economic.

Porphyry copper deposits are commonly located in the apical zones of calc-alkaline plutons intruded at shallow levels and therefore, such deposits have poor preservation potential. This is particularly true for EDC where calc-alkaline granitoids of Neoarchean age abound but none have shown the potentiality of hosting porphyry type Cu (-Mo) deposits. The giant Malanjkhand deposit in BC (Fig. 5.16) is an exception, having been saved from normal erosion by normal faulting involving the deposit (cf. Sarkar and Gupta 2012). The host Malanjkhand batholith is a composite one with compositions ranging from TTG to granite with the majority falling in the trondjhemite-granite field. The ore and the host rock have shown the same age of around 2490 Ma by different geochronologic (U–Pb zircon SHRIMP, Re–Os chronometry) techniques. The composition of the host rock, close temporal relationship between the ore and the host rock, wall rock alteration, low grade and high tonnage led workers to identify the mineralization as porphyry type (Sikka and Nehru 2002; Stein et al. 2004), though with some dissimilarities (Sarkar et al. 1996). The mineralization, however, lacks stock work morphology and comprises mega-quartz reefs. This has prompted other workers to classify Malanjkhand as a granite-related vein-type copper sulfide deposit (Panigrahi and Mookherjee 1997; Rai and Venkatesh 1990). The typical stockwork that occurs at the apical zone might have been eroded. In reviewing the Neoarchean Cu deposits, another enigmatic fact comes into focus. Though there are so many greenstone belts with Algoma type BIF of similar age in the DC, there is hardly any VMS Cu deposit, except a small stratiform, pyritic Cu–Zn deposit at Ingaldhal associated with ~2.6 Ga old volcanic rocks in the Chitradurga schist belt.

The Bastar-Malkangiri pegmatite belt of southern part of BC is the main host of primary Sn mineralization along with concentration of rare metals (common RM minerals being beryl, columbite-tantalite, and lepidolite) and REE minerals (commonly xenotime, monazite, and allanite) (Fig. 5.16) in the country. This is a narrow WSW-ESE trending belt within Archean Bengpal Group comprising metasediments and metabasic rocks (Sarkar and Gupta 2012). The regional metamorphic event here is dated at 2530 ± 89 Ma (Ramesh Babu 1999) and hence, the pegmatitic mineralization can be considered to be of latest Archean age.

5.9.2 *Proterozoic Metallogeny*

The space–time distribution of Proterozoic metallogeny in India is presented in Fig. 5.17. The oldest volcano-sedimentary basin in SC, the Dhanjori basin, developed on the Singhbhum Granite basement (SGB) at ~2.8 Ga. However, mineralization in the Dhanjori basin took place in its uppermost part only around 1.9 Ga, which got affected during the development of the Singhbhum shear zone (SSZ), a deep suture formed in a continent–continent-type orogen around 1.6–1.5 Ga. The polymetallic metal associations of Cu–Fe sulfides with trace metals like Ni, Mo, Se, Te, Au, etc., U–(Ni–Mo) and Fe–P along the shear zone (cf. Fig. 3.32), with the regional metamorphic character of the ores (Sarkar and Deb 1974) pose a challenge to any ore genetic model. Significant REE mineralization has been reported recently from Moinajharia-Bagjata-Kanyaluka (M-B-K) sector in the eastern part of Singhbhum Shear Zone (SSZ) in Jharkhand. The available geochronologic data indicate that all mineralization along SSZ is pre- to early syn-shearing. According to Pal and Rhede (2013) the principal phase of uranium mineralization at Jaduguda, the most important U-bearing zone in the belt, took place sometime in ca. 1.80–1.90 Ga. There was a HREE metasomatism at ~1.66 Ga. A late phase of hydrothermal activity affected the Jaduguda ores at

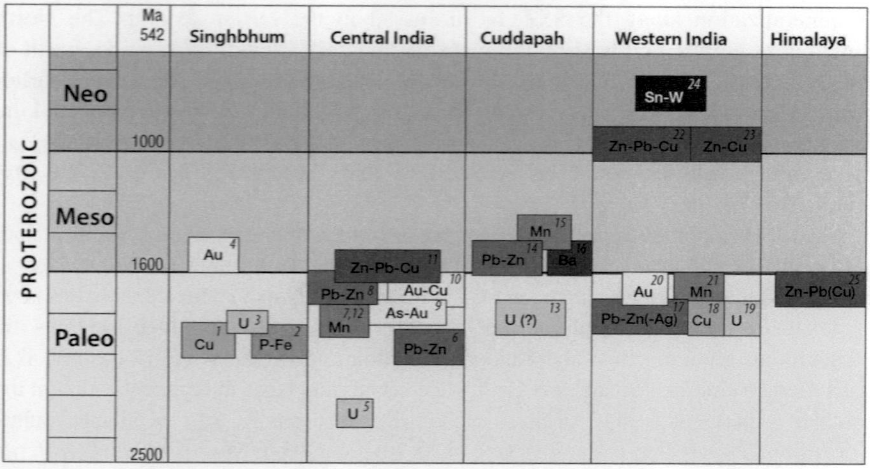

Fig. 5.17 Space–time distribution of Proterozoic ore deposits of India. Numbers in boxes refer to: *1–3* Singhbhum Shear Zone deposits; *4* Lawa, Mysera; *5* Bodal-Bhandaritola; *6* Imalia; *7* Gangpur; *8* Sargipalli; *8a* Kolari-Bhaonri; *9* Imalia; *10* Pular-Parsoli; *11* Betul; *12* Sausar; *13* Tummalapalle; *14* Agnigundala-Zangamrajupalle; *15* Pakhal; *16* Mangampet; *17* Rajpura-Dariba, Rampura-Agucha; *17a* Zawar; *18* Khetri; *19* Umra; *20* Bhukia; *21* Banswara; *22* Ambaji-Deri; *23* Basantgarh; *24* Tosham-Balda-Degana; *25* Askote-Rangpo (after Deb 2014)

~1.0 Ga. Age dating of allanite and co-existing monazite from Bagjata uranium mine by LA-ICPMS technique (Pal et al. 2011a) has also established three stages of hydrothermal activity over a protracted period of time. The LREE mineralization took place around 1.88 Ga, followed by the main uranium mineralization phase at 1.66 Ga and a minor phase at 1.02 Ga. While the LREE and main phase of uranium mineralization as well as the Cu sulfide mineralization at ~1.77 Ga (Johnson et al. 1993) are clearly pre-development of SSZ, the last phase of hydrothermal activity is consistent with the prograde amphibolites facies metamorphism – deformation that produced the NSZ in North Singhbhum mobile belt. The predominant, variously altered basic volcanic host rocks along the SSZ explain satisfactorily the source of Cu and P, as also of the by-product Au. However, for the U-(Mo) association in the southern and central part of SSZ, as well as signatures of strong albitization, late K-metasomatism in some sectors and high salinity of fluid inclusions (Mishra et al. 2003), an acid magmatic-hydrothermal source during the development of SSZ seems to be warranted. Thus a multi-phase ore-forming process can be surmised on a regional scale, which involved an earlier VMS-type Cu sulfide ore formation followed by magmatic-hydrothermal U-(Mo) emplacement predating the regional metamorphic episode. It may nonetheless be noted that based on the metal association of U–Cu–Fe and presence of REE-containing minerals in the Turamdih-Jaduguda sector, an atypical IOCG-type model for the mineralization has been proposed more recently (Pal et al. 2009, 2010, 2011b). Obviously, further work on the genesis is necessary since there is a clear temporal gap between Cu and

U mineralization along the SSZ, as discussed in the earlier section. The North Singhbhum mobile belt hosting the meta-sedimentary Chandil Formation, north of the Dalma volcanics in SC holds significant promise for gold. There were earlier mines at Lawa and Maysera (presently submerged by the Chandil reservoir) and the recently discovered prospects at Lungtu-Parasi and Babaikundi. Temporally, all these mineralizations could be ascribed to late Paleoproterozoic (~1600 Ma) (Gupta 2010).

The other important repository of copper in India is the 80 km-long Khetri copper belt in the North Delhi fold belt of ADOB. Conspicuous mineralization occurs at Banwas, Madan Kudan, Kolihan, and Chandmari (from N to S) where the ore has been exploited for the past several decades. The host rocks in the first three locations are garnetiferous chlorite schist and banded amphibolites-quartzite while in the last, it is garnetiferous chlorite schist alone. Carbonaceous phyllite hosts the mineralization in the southern part of the belt, such as in parts of Kolihan mine. Cordierite-anthophyllite-cummingtonite rocks are recorded in Madan Kudan but do not form the immediate wall rock to the ore. The chalcopyrite (+pyrite-pyrrhotite)-rich ores are metamorphosed though primary banding is preserved at many places. The ore bodies are in the form of single or compound lenses, peneconcordant with the S-surface in the host rocks. The opinions on the origin of these deposits have varied over the years, from epigenetic hydrothermal related to granitic intrusions in the vicinity (Roy Chowdhury and Dasgupta 1965) to sedimentary diagenetic with later remobilization during metamorphism (Sarkar and Dasgupta 1980) to the IOCG model, most recently (Knight et al. 2002). Some of the features mentioned above do not support the first and the last options. However, the presence of magnetite-apatite bodies, uranium mineralization and extensive sodic metasomatism in zones just south and east of the KCB warrant a closer look, which may establish IOCG-type mineralization in some sectors. It is thus likely that KCB and its adjacent zones might turn out to be a complex ore zone with multi-phase ore-forming processes, just like the SCB.

The stratiform to stratabound major Pb-Zn sulfide deposits of Rajasthan are all of Paleoproterozoic age (~1.8–1.7 Ga: Deb et al. 1989), are regionally metamorphosed from granulite-upper amphibolites facies (Rampura-Agucha: Deb and Sehgal 1997; Mishra and Bernhardt 2009) to middle amphibolites facies (Rajpura-Dariba: Deb and Bhattacharya 1980; Mishra et al. 2006) to upper greenschist facies (Zawar: Sarkar and Banerji 2004) and are hosted by metamorphosed clastic and/or carbonate rocks belonging to Paleoproterozoic supracrustal belts of ADOB. Their tectonic setting varies from intracontinental rift basin (Rampura-Agucha and Rajpura-Dariba deposits) to passive continental margin (Zawar deposits). When compared to MVT, SEDEX, Irish, Penine, and Bleiberg types of Pb–Zn sulfide deposits, the first two deposits match broadly with the SEDEX-type (Deb and Pal 2004) while the Zawar deposits share certain features with the MVT, or more specifically its Irish sub-type (Sarkar and Banerji, 2004). However, Bhattacharya and Bull (2010), based on the tectono-sedimentary setting of the Zawar deposits, considered them to be of SEDEX origin. The sediment-hosted Bhukia Au prospect in southern Rajasthan, in the basal Aravalli sequence of ADOB, has been classified as a Proterozoic orogenic gold deposit (Deol et al. 2012) and is likely to be of the same Paleoproterozoic antiquity. Uranium in carbon phyllites of Umra-Udaisagar area and the bedded Mn ores of Banswara, based on their stratigraphic position in the Aravalli sequence, are also considered to be of late Paleoproterozoic age.

It may not be out of place to mention here that there was isochronous development of Aravalli, Bhilwara (northern Aravalli basin of Heron 1953), and NDFB (north of Ajmer) sedimentary sequences around 1.8 Ga in a variety of tectonic settings, as suggested by the Pb–Pb model ages of galena from stratiform sulfide deposits in each of these belts (Deb and Thorpe 2004; Deb et al. 1989), as well as the recent detrital zircon analyses from them (Kaur et al. 2011; Mckenzie et al. 2013). Two major tectono-thermal events are also recorded in the ADOB that can be broadly classified as late Paleoproterozoic (1.9–1.6 Ga) and early Neoproterozoic (1.0–0.8 Ga) events (Bhowmik et al. 2010; Buick et al. 2006; Deb et al. 1989, 2001).

The polyphase, polymetallic As–Au–Ag–Pb–Zn–Sn association in siliceous and carbonaceous, carbonate sequence of western Mahakoshal belt around Imalia represents the oldest Pb–Zn mineralization in the Indian shield. The metamorphosed Pb–Zn mineralization here is $\sim$2.0 Ga old (Tripathi 2008) while the As–Au mineralization is of a later phase, around 1.7 Ga (Deb 2014). This sparsely studied complex mineralization, explored decades ago by the Geological Survey of India, points to the possibility of both Carlin-type and carbonate replacement-type deposits in this region. The stratiform Sargipalli Pb–Zn deposit in Sundergarh district of Odisha, exploited by Hindusthan Zinc Ltd till the nineties, is isofacially metamorphosed with its host metasediments belonging to the uppermost part of the Gangpur Group (Sarkar 1974). Its Pb isotope composition has a high μ-value suggesting a granitic source for the lead and its Pb–Pb model age is estimated to be 1665 Ma (Viswakarma and Ulabhaje 1991). Several small Zn–Pb–Cu ore prospects of VMS affiliation in Betul belt (Mahakud et al. 2001), the Au-Cu sulfide vein ores at Pular-Parsori (Saha et al. 2001), stratiform and stratabound zinc sulfide mineralization at Kolari-Bhaonri in volcano-sedimentary sequence of Bhiwapur Formation (Bandyopadhyaya et al. 1992) in Sakoli belt do not yet have good age constraints, but from geological considerations appear to be of late Paleoproterozoic age. The Himalayan polymetallic (Zn–Pb–Cu sulfide) deposits are associated with Daling Group of metasediments in Sikkim or with the Pithoragarh Group of metasediments in Garhwal, Uttarakhand. They also fall in the Paleoproterozoic age bracket of 1.89–1.60 Ga (Sarkar et al. 2000) and interestingly, have very similar isotopic constitution as the Aravalli deposits of Rajasthan. The stratabound cherty dolostone-hosted Pb–Zn $\pm$ Cu mineralization in the Bandalamottu deposit ($\sim$10 Mt reserve with 6.59% Pb, 1.79% Cu mined by HZL in the past; Raghunandan et al. 1981) in the Agnigundala belt within the Cumbum Formation of the Nullamalai Group of Cuddapah basin, possibly belongs to the MVT clan, rather than a remobilized ore. The Pb isotope data on the Cuddapah sulfides available in the Canadian Geological Survey open file (Thorpe 2008), indicates an average age of 1600 Ma for the Agnigundala (Bandalamottu) and Zangamrajupalle

Pb–Zn ores, though with different μ-values. The Mangampet area in the southernmost part of the Nullamalai rocks is highly potential for similar Pb–Zn mineralization and hosts a very large bedded barite deposit of comparable Mesoproterozoic age (Deb and Pal 2015).

The largest repository of Mn ores in India (~142 Mt) is the 200-km long belt of Sausar Group of rocks in central India constituting a part of CITZ. These regionally metamorphosed Mn ores are hosted by gonditic rocks (Roy 1966) whose protoliths were laid down in shallow shelves, above the Mn^{4+}/Mn^{2+} redoxcline during a transgressive phase of the sea. Manganese deposition took place on the oxygenated basin margins by upwelling of anoxic deep sea water with a high content of dissolved Mn metal (Dasgupta et al. 1992; Roy 2006) and was subsequently affected by regional metamorphism in the terrain. Though depositional age estimate for Sausar Mn ores is still not available, the geological continuity of Sausar rocks into the manganiferous Gangpur sequence in the east through the manganiferous Chilpi rocks suggests their stratigraphic equivalence. The gonditic Mn ores of Sausar (cf. Sect. 3.2.2.4) may, therefore, be considered to be Paleoproterozoic in age. Mn ores of sedimentary affiliation are mined in the Penganga beds at Adilabad in Andhra Pradesh and appear to be, on geological grounds, of Mesoproterozoic age.

The oldest uranium concentration in the Indian shield (other than the Mahagiri detrital uraninite of uncertain economic worth) is represented by the epigenetic mineralization along deep crustal faults within metabasalts and meta-rhyolites of the Dongargarh Supergroup in Bodal-Bhandaritola and Jangalpur-Baghnadi-Parsori belt in BC of central India. Hydrothermal solutions generated during emplacement of Dongargarh Granite, considered to be 2.4–2.2 Ga old, were enriched in uranium during circulation through the felsic volcanic and deposited the U-ores along zones of dislocation (Mahadevan 1992).

In the Tummalapalle-Rachakuntapalle-Giddankipalle sector of the Papagni sub-basin of Cuddapah Supergroup in Andhra Pradesh, stratabound U mineralization with a large tonnage (65,000 t eU_3O_8) and a mineable grade (~0.045% U_3O_8) has been established by the Atomic Minerals Directorate in the stromatolitic and phosphatic dolostones of the Vempalle Formation (Vasudeva Rao et al. 1989). The mineralization is traceable for a total length of ~160 km with a width of ~100–150 m, out of which 5 blocks have been developed, spreading over 15.2 km from west to east (Atomic Minerals Directorate News, Feb, 2012). The uranium mineralization and stromatolitic dolomite from the Vempalle and Tadpatri formations have been dated at 1756 ± 29 Ma (Pb–Pb method) (Zachariah et al. 1999). The stratabound U ores at Tummalapalle formed by sedimentary-diagenetic processes with the metals leached from the fertile basement rocks of the Cuddapah basin. Metallogeny in the Cuddapah and Bhima basins represented by unconformity-proximal U-veins both in the basement and overlying cover sequence, formed by hydrothermal ore fluids at Lambapur and Gogi respectively.

The Neoproterozoic saw very few ore-forming pulses in the Indian shield. A conspicuous exception were the polymetallic VMS-type (Zn–Pb–Cu in Ambaji and Deri and Zn–Cu in Basantgarh) sulfide deposits (Deb 2000a) in the western fringe of the ADOB, named Ambaji-Sendra belt by Deb et al. (2001). Subsequent metamorphism of these ores (Deb 1979) is related to the 'Pindwara orogeny' around 900 Ma (Deb 2000b). Bhowmick et al. (2010) determined monazite chemical ages and found evidence of a Grenvillian age (M2) metamorphism between 946 ± 6 and 865 ± 10 Ma and interpreted them to indicate a collisional orogeny in ADOB. The extensive Erinpura-Malani magmatism of Sirohi and Nagaur districts in western Rajasthan and Bhiwani district of Haryana has concentrated sub-economic deposits of W–(Sn) in several locations such as, Balda, Degana, Sewariya and Tosham around 0.8–0.7 Ga (cf. Deb 2000b; Murao et al. 2000).

5.9.3 Concluding Remarks

The spatiotemporal distribution of ore deposits in the Indian shield, as outlined above, highlights the fact that metallogeny in the Indian shield continued over a large time span of almost 2900 My (~3.6–0.7 Ga) with marked intensity during four specific epochs: (I) Mesoarchean (~3.3–3.1 Ga), (II) Neoarchean (~2.8–2.5 Ga), (III) Paleo-Mesoproterozoic (~2.2–1.5 Ga), and (IV) Neoproterozoic (1.0–0.7 Ga) (Fig. 5.18). The two peaks of metallogenic epochs in terms of economic value are mainly in Neoarchean and Paleo-Mesoproterozoic times respectively.

A substantial part of the Mesoarchean metallogeny in the Indian shield is intrinsically linked to the evolution of the older Sargur greenstone belt in WDC,

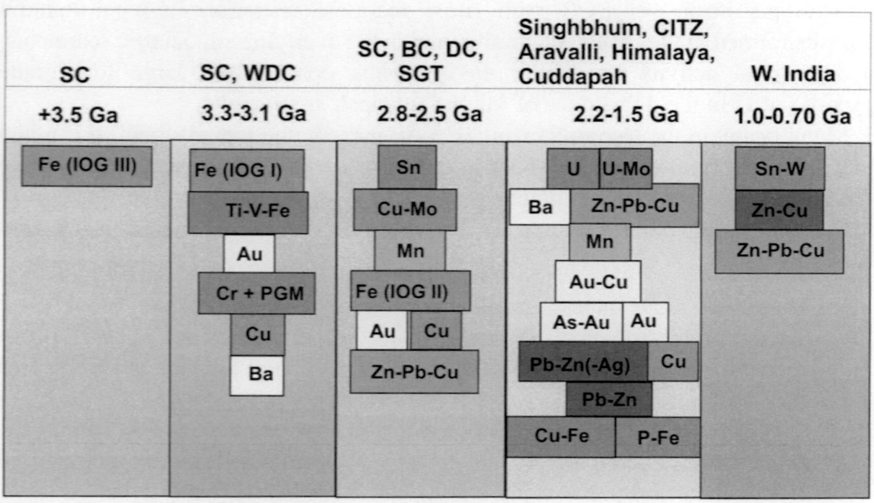

Fig. 5.18 A graphical summary of metallogenic provinces and epochs in the Indian shield (after Deb 2014)

through magmatism in a subduction setting, related to partial melting of an oceanic slab. Mn-oxide ore association in the Tumkur area of WDC probably reflect localized availability of sufficient oxygen in the southern proto-continent. A parallel metallogeny in this early stage of continental growth of the Indian shield is represented by the large IOG I and some gold in the suspected greenstone belt environment or Cr (±PGM) mineralization related to mafic-ultramafic intrusive complexes emplaced in a supra-subduction zone setting in the Singhbhum-North Odisha craton.

The Neoarchean metallogeny of Fe and Mn in SC and DC, that of Au in DC and Cu–Mo and Sn in BC was related to crustal growth and stabilization of the early shields through cratonization. The 2.7 Ga superplume that preceded the assembly of Kenorland was responsible for the extensive development of greenstone belts in the DC. Algoman-type BIF developed in the greenstone belts of WDC during periods of volcanic quiescence while orogenic gold in both EDC and WDC was generated during accretionary collision and related suturing (Naqvi 2005). Superior-type BIF in large masses, closely associated with Mn ores in both SC and EDC, formed along the first continental shelves of the Archean cratonic margins, implying availability of sufficient oxygen in the late Archean atmosphere.

Though the origin remains contentious, Fe–P, Cu–U–REE associations formed in late Paleoproterozoic in SC and NDFB (ADOB) in rifted intra-cratonic (volcano-) sedimentary basins. Orogens with subduction-accretion characteristics, such as in southern ADOB and parts of CITZ, saw the development of Proterozoic orogenic gold deposits. SEDEX-type Pb–Zn concentration and atypical MVT deposits formed in intra-continental abortive rifts and passive continental margin, respectively. Large Mn accumulations took place along shallow cratonic margins of central India through upwelling of bottom waters along redoxclines (Roy 2006). In the intracratonic basins of SCB with rifted margins, unconformity-type uranium deposits formed in the basement granites and in the overlying siliciclastic sediments. Hydrothermal activity in similar environments concentrated large low grade resource of U in the dolostones of lower Cuddapah stratigraphy.

Metallogeny in the Neoproterozoic is very sparse in the Indian shield. It is noted only in the western edge of ADOB in the form of Zn–Pb–Cu or Zn–Cu-bearing VMS deposits or anorogenic granite associated W–Sn deposits.

Some other interesting inferences that could be drawn in conclusion (cf. Deb 2014) include:

(i) There was more intense metallization of the southern cratonic block compared to the north, with the bulk of the Fe, Mn, Cr, Au, Cu, and U ores concentrated here. The NCB saw the concentration of the bulk of Pb–Zn ores only.
(ii) Archean metallogeny was restricted to the cratons of SCB only, while the major part of Proterozoic metallogeny was confined to the Proterozoic mobile belts and partly to the intracratonic basins.
(iii) This points to the different paths of evolution of the SCLM underneath the two cratonic blocks of the Indian shield in the Archean times, which is also

supported by geological features unique to the SCB. These include: (i) development of komatiites in the Sargur belt; (ii) profusion of late Archean auriferous greenstone belts in the EDC; (iii) widespread development of granulite facies rocks in the SGT and EGMB. The Mesoarchean metallogeny in the WDC and SC can be ascribed to the coupled evolution of the relatively pristine Archean crust and SCLM under these cratons.

The periodicity of Precambrian metallogeny in India can be identified with major tectonic cycles, which in turn are manifestations of cycles of amalgamation and dispersal of supercontinents (Barley and Groves 1992; Nance et al. 2014). The metallogenic epochs of Neoarchean (2.8–2.5 Ga), Paleoproterozoic ($\sim$1.8 Ga), and Neoproterozoic ($\sim$1.0–0.7 Ga) mentioned above coincide closely with the supercontinent assembly of Kenorland ($\sim$2.7 Ga; Rogers and Santosh 2004), and Columbia (1.9–1.8 Ga; Evans 2013) and the assembly (1.1–1.0 Ga) and dispersal (0.8–0.7 Ga) of Rodinia; Li et al. 2008) respectively. They also match well with the approximate 750 Ma interval between supercontinent assemblies (Meert 2012). The metallogenic epochs of most intense metallization in Neoarchean and Paleoproterozoic also coincide with the major peaks of juvenile crustal growth worldwide (cf. Condie 2000). Similar correlation of metallogeny with supercontinent cycles have been recorded from other cratonic areas too (e.g., Teixeira et al. 2007; Zhai and Santosh 2013 amongst others).

The enormous concentration of Superior-type Fe-ores in DC, BC, and SC in the late Archean can be related to the first extensive development of passive margins on the edges of the early continental shields during the assembly of Kenorland supercontinent at $\sim$2.7 Ga (Rogers and Santosh 2004). The VMS-type pyritic Cu deposit at Ingaldhal in WDC can also be attributed to the mafic volcanism during Kenorland assembly. The relative importance of plume tectonics for the evolution of the EDC with its conspicuous greenstone belts in the Neorchean has been emphasized by several workers (e.g. Choukroune et al. 1997) and fits well with the earlier-discussed gold endowment model as well as with the evidence of a superplume during the assembly of Kenorland (Kerrich et al. 2010). The Sn and RM-bearing pegmatites and granites in the Bastar craton as also the Cu–Mo mineralization of Malanjkhand can be related to the late stages of assembly of Kenorland.

The oldest volcano-sedimentary basin in SC, the Dhanjori basin, developed on the Singhbhum Granite Basement (SGB) during the assembly of Kenorland supercontinent at $\sim$2.8 Ga. However, mineralization in the Dhanjori basin took place in its uppermost part only during the assembly of Columbia around 1.9 Ga. The Paleoproterozoic metallogeny in ADOB of sediment-hosted Zn–Pb deposits of SEDEX type and stratiform Cu deposits was related to intra-continental rifting that coincided with the assembly of Columbia and continued as MVT(?) Pb–Zn ores in the passive continental margin during the initiation of its dispersal. The Mn ores of Sausar Group in CITZ may have also formed during the assembly of the Paleoproterozoic supercontinent, Columbia.

The assembly of Rodinia saw the development of continent–continent orogens in the western fringe of ADOB as well as amalgamation of the SCB and the NCB of

Table 5.2 Relationship of Indian ore deposits to supercontinent cycles (as defined by Evans 2013)

Assembly	Between assembly and break-up	Dispersal
Kenorland (~2.7–2.5 Ga)		
Fe deposits (IOG II in SC, EDC, BC) often assoc. with Mn ores		
VMS Cu (Ingaldhal, WDC)		
Porphyry(?) Cu–Mo (Malanjkhand)		
Orogenic Au (EDC)		
Sn–RM deposits (BC)		
Columbia (1.9–1.8 Ga)		(1.4–1.2 Ga)
Cu and REE-rich apatite-magnetite (SSZ, SC)		
Mn ores (Sausar and Gangpur)		
SEDEX Pb–Zn ores (Aravalli)	MVT(?) Pb–Zn ores (Aravalli)	
Orogenic gold (Aravalli)	Pb–Cu ores (Sargipalli, Gangpur)	
Black shale hosted U (Aravalli)	U ores (Tummalapalle, Cuddapah)	
Stratiform Cu (NDFB)		
As–Au ores (Imalia, W. Mahakoshal)		
Rodinia (1.1–1.0 Ga)		(0.8–0.7 Ga)
VMS Zn–Pb–Cu ores (Ambaji-Sendra belt)		W ores (Western Rajasthan)
VMS Zn–Cu ores (Ambaji-Sendra belt)		Sn polymetallic ores (Haryana)

the Indian sub-continent along the CITZ. This Grenvillian event was marked by a high grade upper amphibolite facies metamorphism between 1.0 and ~0.9 Ga in the northern and eastern part of Chotanagpur Granite Gneiss (CGC) as well as the eastern and central part of CITZ (Chatterjee and Ghosh 2011). Similar metamorphic event has been recorded earlier from the western fringe of ADOB (Deb et al. 1989, 2001; Bhowmik et al. 2010). The development of VMS-type Zn–Pb–Cu and Zn–Cu deposits in southern ADOB with the age of these ores and their bimodal volcanic host rocks at ~990 Ma conform to the last stages of Rodinia assembly. Dispersal of Rodinia started around 800 Ma when Erinpura-Malani anorogenic magmatism formed Sn–W mineralization to the western fringe of the ADOB. This anorogenic magmatism took place in the time window between the dispersal of Rodinia and the assembly of Gondwanaland, around 700 Ma.

A summary of correlation between Precambrian ore deposits of Indian shield and the supercontinent cycles is presented in Table 5.2.

References

Abbot DH, Isley AE (2002) The duration, magnitude and intensity of mantle plume activity over the last 3.8 Ga. J Geodyn 34:265–307

Acharyya SK, Roy A (2000) Tectonothermal history of the Central Indian Tectonic Zone and reactivation of major faults/shear zones. J Geol Soc Ind 55:239–256

Anand R, Balakrishnan S, Mezger K (2005) Geochemical and geochronological studies on titanites from the granitoid rocks of the eastern Dharwar craton, South India. Geophys Res Abstr Eur Geosci Union 7:1565

Aubouin J (1965) Geosynclines. Elsevier, Amsterdam, p 335

Banerjee PK (1972) Geology and geochemistry of the Sukinda ultramafic field, Cuttack district, Orissa, vol 103. Geol Surv Ind Mem, 171 p

Bandyopadhyaya BK, Roy A, Shukla RS, Guha Sarkar TK (1992) Sedex-type copper and zinc deposits in the Proterozoic Sakoli Group, Nagpur and Bhandara district, central India. In: Sarkar SC (ed) Metallogeny related to the tectonics of the Proterozoic mobile belts. Oxford and IBH Co Pvt Ltd, New Delhi, pp 53–102

Barrie CT, Hannington MD (1999) Classification of volcanic-associated massive sulfide deposits based on host rock composition. Rev Econ Geol 8:1–11

Barley ME, Bekker A, Krapez B (2005) Late archean to early paleoproterozoic global tectonics, environmental change and the rise of atmospheric oxygen. Earth Planet Sci Lett 238:156–171. doi:10.1016/j.epsl.2005.06.062

Barley ME, Krapez B, Groves DI, Kerrich R (1998) Late Archean Bonanza: metallogenic and environmental consequences of the interaction between mantle plumes, lithospheric tectonics and global cyclicity. Precambr Res 91:65–90

Barley ME, Pickard AL, Sylvester PJ (1997) Emplacement of a large igneous province as a possible cause of banded iron formation 2.45 billion years ago. Nature 385:55–58

Barley ME, Groves DI (1992) Supercontinent cycles and distribution of metal deposits through time. Geology 20:291–294

Bell K (2001) Carbonatites: relationship to mantle plume activity. Geol Soc Am Spec Pap 352:267–290

Bhattacharya HN, Bull S (2010) Tectono-sedimentary setting of the Paleoproterozoic Zawar Pb-Zn deposits, Rajasthan, India. Precambrian Res 177:323–338

Bhowmik SK, Bernhardt H-J, Dasgupta S (2010) Grenvillian age high-pressure upper amphibolite-granulite metamorphism in the Aravalli-Delhi Mobile Belt, Northwestern India: new evidence from monazite chemical age and its implication. Precambr Res 178:168–184

Bilibin YuA (1955) Metallogenic provinces and metallogenic epochs. Geosgeoltekhizdat, Moscow, p 88 (in Russian)

Binda PL (1995) Stratigraphy of Zambian copper belt orebodies. J African Earth Sci 19:251–264

Bonatti E, Honndrez-Guerstein MB, Honnorez J, Stern C (1976a) Hydrothermal pyrite concretions from Romanche trench (Equatorial Atlantic): metallogenesis in oceanic fracture zones. Earth Planet Sci Lett 32:4–10

Bonatti E, Zerbi M, Kay R, Rydell HS (1976b) Metalliferous deposits from Apenine ophiolites: Mesozoic equivalents of modern deposits from oceanic spreading centres. Bull Geol Soc Am 87:83–94

Buick IS, Allen C, Pandit M, Rubatto D, Hermann J (2006) The Proterozoic magmatic and metamorphic history of the Banded Gneiss Complex, central Rajasthan, India: LA-ICP-MS U-Pb zircon constraints. Precambr Res 151:119–142

Burke K, Dewey JF, Kidd WSF (1976) Dominance of horizontal movement, arc and micro-continent collisions during the later permobile regime. In: Windley BF (ed) The early history of the earth. Wiley, New York, pp 113–129

Burke KC, Wilson JT (1976) Hot spots on the earth's surface. J Geophys Res (B) 93:7690–7708

Cawood PA, Hawkesworth CJ (2014) Earth's middle age. Geology 42(6):503–506. doi:10.1130/G35402.1

Chadwick B, Vasudev VN, Hegde GV (2000) The Dharwar craton, southern India, interpreted as the result of Late Archean oblique convergence. Precambr Res 9:91–111

Chakraborty KL, Chakraborty TL (1984) Geological features and origin of the chromite deposits of Sukinda valley, Orissa, India. Mineral Deposita 19:256–265

Chatterjee N, Ghosh NC (2011) Extensive early neoproterozoic high-grade metamorphism in north chotanagpur gneissic complex of the central Indian tectonic zone. Gondwana Res 20:362–379

Choukroune P, Ludden JN, Chardon D, Calvert AJ, Bouhallier H (1997) Archean crustal growth and tectonic processes: a comparison of the Superior province, Canada and the Dharwar craton, India. In: Burgand JP, Ford M (eds) Orogeny through time. Geol Soc London Spec Publ No. 121, pp 63–98

Condie KC (2004) Supercontinents and superplume events: distinguishing signals in the geological record. Phys Earth Planet Inter 146:319–332

Condie KC (2000) Episodic continental growth models: after thoughts and extensions. Tectonophys 322:153–162

Condie KC (1998) Episodic continental growth and supercontinents: a mantle avalanche connection? Earth Planet Sci Lett 163:97–108

Dasgupta S, Banerjee H, Bandopadhyay G (1992) Manganese deposition in the proterozoic global perspective and Indian scenario. In: Sarkar SC (ed) Metallogeny related to tectonics of the proterozoic mobile belts. Oxford & IBH Publ Co Pvt Ltd, New Delhi, pp 163–176

Deb M, Pal T (2015) Mineral potential of proterozoic intra-cratonic basins in India. In: Mazumder R, Eriksson PG (eds) Precambrian basins of India: stratigraphic and tectonic context, vol 43. Geol Soc London Mem, pp 309–325. doi:10.1144/M43.21

Deb M (2014) Precambrian geodynamics and metallogeny of the Indian shield. Ore Geol Rev 57:1–28

Deb M, Pal T (2004) Geology and genesis of base metal sulphide deposits in the Dariba-Rajpura-Bethumni belt, Rajasthan, India in the light of basin evolution. In: Deb M, Goodfellow WD (eds) Sediment-hosted Lead-Zinc Sulphide Deposits. Narosa Publ House, New Delhi, pp 304–327

Deb M, Thorpe RI (2004) Geochronological constraints in the Precambrian geology of Rajasthan and their metallogenic implications. In: Deb M, Goodfellow WD (eds) Sediment-hosted lead-zinc sulphide deposits. Narosa Publ House, Delhi, pp 246–263

Deb M, Thorpe RI, Krstic D, Corfu F, Davis DW (2001) Zircon U-Pb and galena Pb isotope evidence for an approximate 1.0 Ga terrane constituting the western margin of the Aravalli-Delhi orogenic belt, northwestern India. Precambr Res 108:195–213

Deb M (2000a) VMS deposits: geological characteristics, genetic models and a review of their metallogenesis in the Aravalli Range, NW India. In: Deb M (ed) Crustal evolution and metallogeny in the northwestern indian shield. Narosa Publ House, New Delhi, pp 329–363

Deb M (2000b) Metallogeny in the Aravalli-Delhi orogenic belt, northwestern India. Geol Surv Ind Spec Publ 55:333–350

Deb M, Sehgal U (1997) Petrology, geothermobarometry and C-O-H-S fluid compositions in the environs of Rampura-Agucha ore deposit, Bhilwara district, Rajasthan. In: Proceedings of the Indian Academy of Science, vol 106(4) (K. Naha comm vol), pp 343–356

Deb M, Hoefs J, Baumann A (1991) Isotopic composition of two Precambrain Stratiform barite deposits from the Indian shield. Geochim Cosmochim Acta 55:303–308

Deb M, Thorpe RI, Cumming GL, Wagner PA (1989) Age, source and stratigraphic implications of lead isotope data for conformable sediment-hosted base metal deposits in the Proterozoic Aravalli-Delhi orogenic belt, northwestern India. Precambr Res 43:1–22

Deb M, Bhattacharya AK (1980) Geological setting and conditions of metamorphism of Rajpura Dariba polymetallic ore deposit, Rajasthan, India. In: Proceedings of V IAGOD symposium 1, Schweiserbart, Stuttgart, pp 679–697

Deol S, Deb M, Large RR, Gilbert S (2012) LA-ICPMS and EPMA studies of pyrite, arsenopyrite and loellingite from the Bhukia-Jagpura gold prospect, southern Rajasthan, India: implications for ore genesis and gold remobilization. Chem Geol 326–327:72–87

Evans DAD (2013) Reconstructing pre-Pangean supercontinents. Geol Soc Am Bull **25** (11/12):1735–1751. doi:10.1130/B30950.1

Goldfarb RJ, Groves DI, Taylor RD, Deb M (2010) Controls on the global distribution of orogenic gold and their significance in relation to India. In: Deb M, Goldfarb RJ (eds) Gold metallogeny: India and beyond. Narosa Publ House Pvt Ltd, New Delhi, pp 48–57

Goldfarb RJ, Groves DI, Gardoll S (2001) Orogenic gold and geologic time: a global synthesis. Ore Geol Rev 18:1–75

Groves DI, Bierlein FP, Meinert LD, Witzman MW (2010) Iron oxide copper gold (IOCG) deposits through Earth history: implications for origin, lithospheric setting and distribution from other epigenetic iron oxide deposits. Econ Geol 105:641–667

Groves DI, Bierlein FP (2007) Geodynamic settings of mineral deposit systems. J Geol Soc London 164:19–30

Groves DI, Vielreicher RM, Goldfarb RJ, Condie KC (2005a) Controls on the heterogenous distribution of mineral deposits through time. In: McDonald I, Noyce AJ, Butter IB, Herrington RJ, Polya DA (eds) Mineral deposits and Earth evolution, vol 248. Geol Soc London Spec Publ, pp 71–101

Groves DI, Condie KC, Goldfarb RJ, Hronsky JMA, Vielreicher RM (2005b) Secular changes in global tectonic processes and their influence on the temporal distribution of gold-bearing mineral deposits. Econ Geol 100:203–224

Groves DI, Goldfarb RJ, Robert F, Hart CJR (2003) Gold deposits in metamorphic belts: overview of current understanding, outstanding problems, future research, and exploration significance. Econ Geol 98:1–29

Gupta A (2010) Gold mineralization in the eastern segment of Indian Precambrian shield: a review. In: Deb M, Goldfarb RJ (eds) Gold metallogeny: India and beyond. Narosa Publ House, New Delhi, pp 256–280

Gurney JJ, Helmstaedt HH, Richardson SH, Shirey SB (2010) Diamonds through time. Econ Geol 105:689–712

Gurney JJ, Helmstaedt HH, Le Roex AP, Nowicki TE, Richardson SH, Westerlund KJ (2005) Diamonds: crustal distribution and formation processes in time and space and an integrated model. In: Hedenquist JW, Thompson JFH, Goldfarb RJ, Richards JP (eds) Economic geology 100th anniversary volume, pp 143–177

Haldar D, Mukhopadhyay AK, Mukherjee P (2005) Petrotectonic environment of chromitites associated with the Precambrian ultramafic rocks of Jharkhand and Orissa. In: Proceedings of seminor mineral and energy resources Eastern and Northeastern India. MGMI, Kolkata, pp 203–225

Hannington MD, de Ronde CEJ, Peterson S (2005) Sea floor tectonics and submarine hydrothermal systems. In: Hedenquist JW, Thompson JFH, Goldfarb RJ, Richards JP (eds) Economic geology 100th Anniversary volume, pp 111–141

Hargraves RB (1986) Faster spreading or greater ridge length in the Archean? Geology 14:750–752

Haywood N (2008) The geodynamic setting and magmatic controls on genesis of world class IOCG ore deposits. Geol Soc Aust Abstr 89:130

Heron AM (1953) The geology of central Rajputana. Mem Geol Surv Ind 78, 389 pp

Hildreth W, Moorbath S (1988) Crustal contributions to arc magmatism in the Andes of Central Chile. Contrib Mineral Petrol 98:455–489

Hoffman PE (1988) United States of America, the birth of a craton. Ann Rev Earth Planet Sci 16:543–604

Isley AE, Abott DH (1999) Plume-related mafic volcanism and the deposition of banded iron formation. J Geophy Res 104B:15461–15477

Jaques AL, Webb AW, Fanning CM, Black LP, Pidgeon RT, Smith OB, Gregory GP (1984) The age of the diamond-bearing pipes and associated leucite lamproites of the West Kimberley region, Western Australia. BMR J Aust Geol Geophy 9:1–7

Jayananda M, Kano T, Peucat JJ, Channabasappa S (2008) 3.35 Ga komatiite volcanism in the western Dharwar craton, southern India: constraints from Nd isotopes and whole rock geochemistry. Precambr Res 162:160–179

Johnson PT, Dasgupta D, Smith AD (1993) Pb-Pb systematic of copper sulphide mineralization, Singhbhum area, Bihar. Indian J Geol 65:211–213

Kaur P, Chaudhuri N (2014) Metallogeny associated with the Paleo-Mesoproterozoic Columbia supercontinent cycle: a synthesis of major metallic deposits. Ore Geol Rev 56:415–422

Kaur P, Zeh A, Chaudhuri N, Gerdes A, Okrusch M (2011) Archean to Paleoproterozoic crustal evolution of the Aravalli mountain range, NW India, and its hinterland: the U-Pb and Hf isotope record of detrital zircon. Precambr Res 187:155–164

Kay M (1951) North American geosynclines. Mem Geol Soc Am 48:1–131

Kerrich R, Goldfarb RJ, Cline J, Leach D (2010) Metallogenic provinces of Laurentia in a superplume-supercontinent framework with a focus on gold. In: Deb M, Goldfarb RJ (eds) Gold metallogeny: India and beyond. Narosa Publ House, New Delhi, pp 1–29

Kerrich R, Goldfarb RJ, Richards JP (2005) Metallogenic provinces in an evolving geodynamic framework. In: Hedenquist JW, Thompson JFH, Goldfarb RJ, Richards JP (eds) Economic geology 100th anniversary volume, pp 1097–1136

Klein C, Beukes NJ (1993) Sedimentology and geochemistry of the glaciogenic late Proterozoic Rapitan Iron Formation in Canada. Econ Geol 88:542–565

Knight J, Lone J, Joy S, Gameron J, Merrilee SJ, Nag S, Shah N, Dua G, Jhala K (2002) The Khetri copper belt, Rajasthan: iron oxide-copper-gold terrane in the Proterozoic in India. In: Porter TM (ed) Hydrothermal Iron oxide-copper-gold and related deposits: a global perspective, vol 2. Australian Mineral Foundation, Adelaide, pp 321–341

Kolb J, Hellmann A, Rogers AA, Sindern S, Veneemann T, Bottcher ME, Meyer FM (2004) The role of a transcrustal shear zone in orogenic gold mineralization at the Ajjanahalli Mine, Dharwar Craton, South India. Econ Geol 99:743–759

Kosticin N, Krapez B (2004) SHRIMP U-Pb detrital zircon geochronology of Late Archean Witwatersrand Basin of South Africa: relationship between zircon provenance age spectra and basin evolution. Precambr Res 129:141–168

Krienitz M-S, Trumbull RB, Hellmann A, Kolb J, Meyer FM, Wiedenbeck M (2008) Hydrothermal gold mineralization at the Hira Buddini gold mine, India: constraints on fluid evolution and fluid sources from boron isotopic compositions of tourmaline. Min Deposita 43:421–434

Kusky TM, Kidd WSF (1992) Remnants of an Archean oceanic plateau, Belingwe greenstone belt, Zimbabwe. Geology 20:43–46

Large RR, Bull SW, McGoldrick PJ, Walters S (2005) Stratiform and stratabound Zn-Pb-Ag deposits in Proterozoic sedimentary basins, Northern Australia. In: Hedenquist JW, Thompson JFH, Goldfarb RJ, Richards JP (eds) Economic geology 100th anniverary volume, pp 931–963

Leach DL, Sangster DF, Kelly KD, Large RR, Garven G, Allen CR, Gutzmar J, Walters S (2005) Sediment-hosted Pb-Zn deposits: a global perspective. In: Hedenquist JW, Thompson JFH, Goldfarb RJ, Richards JP (eds) Economic geology 100th anniverary volume, pp 561–607

Li ZX, Bogdanova SV, Davidson A, Collins AS, De Waele B, Ernst RE, Fitzsimons ICW, Fuck RA, Gladkochub DP, Jacobs J, Karlstrom KE, Lu S, Natapov LM, Pease V, Pisarevsky SA, Thrane K, Vernikovsk V (2008) Assembly, configuration, and break-up history of Rodinia: a synthesis. Precambr Res 160:179–210

Mahadevan TM (1992) Uranium metallogeny in the Proterozoic mobile belts in India in relation to tectonic developments. In: Sarkar SC (ed) Metallogeny related to tectonics of the proterozoic mobile belts. Oxford & IBH Publ Co Pvt Ltd, New Delhi, pp 177–208

Mahakud SP, Raut PK, Hansda C, Ramteke FP, Chakraborty U, Praveen MN, Sisodia DS (2001) Sulphide mineralization in the central part of Betul belt around Ghisi-Mauriya-Koparpani area, Betul district, Madhya Pradesh. Geol Surv Ind Spec Publ 64:377–385

Martin H (1986) The effects of steeper Archaean geothermal gradients on geochemistry of subduction zone magmas. Geology 14:753–756

McGowan RR, Roberts S, Boyce AJ (2006) Origin of the Nechanga copper-cobalt deposits of the Zambian copper belt. Min Deposita 40:617–638

McKenzie NR, Hughes NC, Myrow PM, Banerjee DM, Deb M, Planavsky NJ (2013) New age constraints for the Aravalli-Delhi successions of India and their implications. Precambr Res 238:120–128

Meert JG (2012) What's in a name? The Columbia (Paleopangea/ Nuna) supercontinent. Gondwana Res 21:987–993

Meyer C (1988) Ore deposits as guides to the history of the Earth. Ann Rev Earth Planet Sci 16:147–172

Meyer C (1981) Ore forming processes in geologic history. In: Skinner BJ (ed) Economic geology 75th anniversary volume, pp 6–41

Mishra B, Bernhardt HJ (2009) Metamorphism, graphite crystallinity and sulfide anatexis of the Rampura-Agucha massive sulfide deposit, northwestern India. Min Deposita 44:183–204

Mishra B, Pal N (2008) Metamorphism, fluid flux, and fluid evolution relative to gold mineralization in Hutti-Maski greenstone belt, Eastern Dharwar Craton, India. Econ Geol 103:801–827

Mishra B, Upadhyay D, Bernhardt H-J (2006) Metamorphismof the host and associated rocks at the Rajpura-Dariba massive sulfide deposit, Northwestern India. J Asian Earth Sci 26:21–37

Mishra B, Pal N, Ghosh S (2003) Fluid evolution of the Mosaboni and Rakha copper deposits, Singhbhum district, Jharkhand: evidence from fluid inclusion study of mineralized quartz veins. J Geol Soc Ind 61:51–60

Mishra B, Panigrahi MK (1999) Fluid evolution in the Kolar Gold Field: evidence from fluid inclusion studies. Min Deposita 34:173–181

Mitchell AHG, Garson MS (1981) Mineral deposits and global tectonic settings. Acad Press, London, p 405

Mondal SK, Frei R, Ripley EM (2007) Os isotope systematics of mesoArchean chromitite-PGE deposits in the Singhbhum Craton (India): Implications for the evolution of lithospheric mantle. Chem Geol 244:391–408

Mondal SK, Ripley EM, Li Chusi, Frei R (2006) The genesis of Archean chromitites from Nuasahi and Sukinda massifs in the Singhbhum craton, India. Precambr Res 148:45–66

Morgan JW, Stein HJ, Hanna JL, Markey RJ, Wiszniewska J (2000) Re–Os study of Fe–Ti–V oxide and Fe–Cu–Ni sulfide deposits, Suwalski anorthosite massif, north east Poland. Min Deposita 35:391–401

Morgan WJ (1972) Plate motions and deep mantle convection. Geol Soc Am Mem 132:7–22

Mukherjee R, Mondal SK, Rosing MT, Frei R (2010) Compositional variations in the Mesoarchean chromites of the Nuggihalli schist belt, Western Dharwar Craton (India): potential parental melts and implications for tectonic setting. Contrib Mineral Petrol 160:865–885

Mukhopadhyay J, Misra B, Chakrabarti K, De S, Ghosh G (2016) Uraniferous paleoplacers of the Mesoarchean Mahagiri quartzite, Singhbhum craton, India: depositional controls, nature and source of >3.0 Ga detrital uraninites. Ore Geol Rev 72(2):1290–1306. doi:10.1016/j.oregeorev.2015.05.020

Mukhopadhyay J, Crowley QG, Ghosh S, Ghosh G, Chakrabarti K, Misra B, Heron K, Bose S (2014) Oxygenation of the Archean atmosphere: new paleosol constraints from eastern India. Geology 42:923–926. doi:10.1130/G36091.1

Mukhopadhyay J, Beukes NJ, Armstrong RA, Zimmermann U, Ghosh G, Medda RA (2008) Dating the oldest greenstone in India: a 3.51 Ga precise U-Pb SHRIMP zircon age for dacitic lava of the Southern Iron Ore Group, Singhbhum craton. J Geol 116:449–461

Murao S, Deb M, Takagi T, Seki Y, Pringle M, Naito K (2000) Geochemical and geochronological constraints for tin polymetallic mineralization in Tosham area, Haryana, India. In: Deb M (ed) Crustal evolution and metallogeny in Northwestern Indian Shield. Narosa Publ House, New Delhi, pp 430–442

Murphy JB, Nance RD (1992) Supercontinents and the origin of mountain belts. Sci Am 266 (4):84–91

Nair RVG, Nair MM, Maji AK (2005) Gold mineralization in Kottathara prospect, Attapady valley, Kerala, India. Gondwana Res 8(2):203–212

Naldrett AJ (2010) Secular variation of magmatic sulfide deposits and their source magmas. Econ Geol 105:669–688

Nance RD, Murphy JB, Santosh M (2014) The supercontinent cycle: a retrospective essay. Gondwana Res 25(1):4–29. doi:10.1016/j.gr.2012.12.026

Naqvi SM (2005) Mantle plumes, accretion-collision processes and the orogenic gold deposits with special reference to the Indian plate: a review and discussion. J Him Geol 26(1):139–142

Nelson DR, Bhattacharya HN, Thern ER, Altermann W (2014) Geochemical and ion microprobe U-Pb zircon constraints on the Archean evolution of Singhbhum craton, eastern India. Precambr Res 225:412–432

Nelson DR, Chivas AR, Chappel BW, McCulloch MT (1988) Geochemical and isotopic systematic in carbonatites and implications for the evolution of ocean island sources. Geochim Cosmochim Acta 52:1–17

Nutman AP, Chadwick B, Ramakrishnan M, Viswanatha MN (1992) SHRIMP U-Pb ages of detrital zircons in Sargur supracrustal rocks in western Karnataka, southern India. J Geol Soc Ind 39:367–374

Page NJ, Banerjee PK, Haffty J (1985) Characteristics of the Sukinda and Nuasahi ultramafic complexes, Orissa, India by platinum group element geochemistry. Precambr Res 30:27–41

Pal DC, Rhede D (2013) Geochemistry and chemical dating of uraninite in the Jaduguda uranium deposit, Singhbhum Shear Zone, India—implications for uranium mineralization and geochemical evolution of uraninite. Econ Geol 108(6):1499–1515

Pal DC, Chaudhuri T, Mcfarlane C, Mukherjee A, Sarangi AK (2011a) Mineral chemistry and in situ dating of allanite, and geochemistry of its host rocks in the Bagjata Uranium Mine, Singhbhum Shear Zone, India—implications for the chemical evolution of REE mineralization and mobilization. Econ Geol 106:1155–1171

Pal DC, Sarkar S, Mishra B, Sarangi AK (2011b) Chemical and sulfur isotope compositions of pyrite in the Jaduguda U (-Cu-Fe) deposit, Singhbhum shear zone, eastern India: implications for sulfide mineralization. J Earth Sys Sci 120:475–488

Pal DC, Trumbull RB, Wiedenbeck M (2010) Chemical and boron isotope compositions of tourmaline from the Jaduguda U(-Cu-Fe) deposit, Singhbhum shear zone, India: implications for the sources and evolution of mineralizing fluids. Chem Geol 277:245–260

Pal DC, Barton MD, Sarangi AK (2009) Deciphering a multi-stage history affecting U-Cu-(Fe) mineralization in the Singhbhum shear zone, eastern India, using pyrite textures and compositions in the Turamdih U-Cu-(Fe) deposit. Min Deposita 44:61–80

Panigrahi MK, Mookherjee A (1997) The Malanjkhand copper (+molybdenum) deposit, India: mineralization from a low temperature ore fluid of granitoid affiliation. Min Deposita 32:133–148

Pereira J, Dixon CJ (1965) Evolutionary trends in ore deposition. Trans Inst Mining Metallurgy 74:B506–B527

Perkins WG (1984) Mount Isa "silica-dolomite" and copper orebodies: the result of a syntectonic hydrothermal alteration system. Econ Geol 79:601–637

Pirajno F (2004) Hot spots and mantle plumes: global intra-plate tectonics, magmatism and ore deposits. Min Petrol 82:183–216

Pirajno F (2000) Ore deposits and mantle plumes. Kluwer, Dordrect, p 556

Polat A, Kerrich R (2004) Precambrian associations: boninites, adakites, magnesian andesites and N-enriched basalts. In: Kusky TM (ed) Precambrian ophiolites and related rocks. Developments in precambrian geology, vol 13. Elsevier, Amsterdam, pp 567–569

Polat A, Kerrich R, Wyman D (1999) Geochemical diversity in oceanic komatiiites and basalts from the Late Archean Wawa greenstone belts, Superior province, Canada. Trace element and Nd isotope evidence for a heterogenous mantle. Precambr Res 94:139–173

Radhakrishna BP, Naqvi SM (1986) Precambrian continental crust of India and its evolution. J Geol 94:145–166

Raghunandan KR, Dhruba Rao BK, Singhal ML (1981) Exploration for copper, lead and zinc ores in India. Bull Geol Surv Ind A Econ Geol 47, 195 pp
Rai KL, Venkatesh AS (1990) Malanjkhand copper deposit—a petrological and geochemical appraisal. Geol Surv Ind Spec Publ 28:563–584
Ramesh Babu PV (1999) Rare metals and rare earth pegmatites of Central India. Special issue on rare metal and rare earth pegmatites of India. Expl Res Atomic Min 12:7–52
Ram Mohan M, Sarma DS (2010) Geochemistry and evolution of the Hutti-Maski greenstone belt: link between Archean crustal evolution and gold mineralization. In: Deb M, Goldfarb RJ (eds) Gold metallogeny: India and beyond. Narosa Publ House, New Delhi, pp 191–205
Richards JP, Mumin AH (2013) Magmatic-hydrothermal processes within an evolving earth: iron oxide-copper-gold and porphyry Cu ± Mo ± Au deposits. Geology 41:767–770. doi:10.1130/G34275.1
Richards JP (2009) Post-subduction porphyry Cu-Au and epithermal Au deposits: Products of remelting of subduction modified lithosphere. Geology 37:247–250
Richards JP, Boyce AJ, Pringle MS (2001) Geologic evolution of the Escondida area, northern Chile: a model for spatial and temporal localization of porphyry Cu mineralization. Econ Geol 96:271–305
Rogers AJ, Kolb J, Meyer FM, Vennemann T (2013) Two stages of gold mineralization at Hutti mine, India. Min Deposita 48:99–114
Rogers JJW, Santosh M (2004) Continents and supercontinents. Oxford Univ Press, Oxford, p 289
Rogers JJW (1996) A history of continents in the past 3 billion years. J Geol 104:91–107
Rona P (1980) Global plate motion and mineral resources. Geol Assoc Can Spec Pap 20:607–622
Rosing MT, Rose NM, Bridgewater D, Thompson HS (1996) Earliest part of Earth's stratigraphic record: a reappraisal of the >3.7 Ga Isua (Greenland) supracrustal sequence. Geology 24:43–46
Roy S (2006) Sedimentary manganese metallogenesis in response to the evolution of the Earth System. Earth Sci Rev 77:273–305
Roy S (1966) Syngenetic manganese formations of India. Jadavpur Univ Press, Calcutta, p 219
Roy Chowdhury MK, Dasgupta SP (1965) Ore localization in the Khetri copper belt, Rajasthan, India. Econ Geol 60:69–88
Saha AK, Bhoskar KG, Chattopadhyay S (2001) Ore mineralogy, fluid inclusions, and sulphur isotope studies of Parsori gold-copper deposit, Sakoli fold belt, Maharashtra and its genetic modeling. Geol Surv Ind Spec Publ 58:163–180
Santosh M, Maruyama S, Yamamoto S (2009) The making and breaking of supercontinents: some speculations based on superplumes, super downswelling and the role of tectonosphere. Gondwana Res 15:324–341
Santosh M, Nadear S, Javoy M (1995) Stable isotopic evidence for the involvement of mantle-derived fluids in Wynad gold mineralization, south India. J Geol 103:718–728
Sarkar SC, Gupta A (2012) Crustal Evolution and Metallogeny in India. Cambridge Univ Press, Delhi, p 840
Sarkar SC (2010) Gold mineralization in India: an introduction. In: Deb M, Goldfarb RJ (eds) Gold metallogeny: India and beyond. Narosa Publ House, New Delhi, pp 95–122
Sarkar SC, Gupta A (2005) Nature and origin of iron ores of Eastern India, a subject of scientific interest and industrial concern. In: Proceedings of seminor mineral and energy resources of Eastern and North-eastern India. MGMI, Kolkata, India, pp 79–101
Sarkar SC, Banerji S (2004) Carbonate-hosted lead-zinc of Zawar, Rajasthan in the context of world scenario. In: Deb M, Goodfellow WD (eds) Sediment-hosted lead-zinc sulphide deposits. Narosa Publ House, Delhi, pp 328–349
Sarkar SC, Chernyshev IV, Banerjee H (2000) Mid-Proterozoic Pb-Pb ages for some Himalayan base metal deposits and comparison to deposits in Rajasthan, NW India. Precambr Res 99:171–178
Sarkar SC, Kabiraj S, Bhattacharyya S, Pal AB (1996) Nature, origin and evolution of the granitoid-hosted early Proterozoic copper-molybdenum mineralization at Malanjkhand, Central India. Min Deposita 31:419–431

Sarkar SC (1988) Genesis and evolution of the ore deposits in the early Precambrian greenstone belts and adjacent high grade metamorphic terrains of peninsular India—a study in similarity and contrast. Precambr Res 39:107–130

Sarkar SC (1985) Concepts of tectonic control of metallisation—yesterday and today. In: Proceedings symposium. Megastructures and plate tectonics and their role as a guide to mineralisation. Geol Min Met Soc Ind Bull 53:181–204

Sarkar SC, Dasgupta S (1980) Geological setting and transformation of sulfide deposits in the northern part of the Khetri copper belt, Rajasthan, India: an outline. Min Deposita 15:117–137

Sarkar SC (1974) Sulphide mineralization at Sargipalli, Orissa, India. Econ Geol 69:206–217

Sarkar SC, Deb M (1974) Metamorphism of the sulphides of the Singhbhum copper belt, India—the evidence from the ore fabric. Econ Geol 68:1282–1293

Sarma DS, Fletcher IR, Rasmussen B, McNaughton NJ, Ram Mohan M, Groves DI (2011) Archaean gold mineralization synchronous with late cratonization of the Western Dharwar Craton, India: 2.52 Ga U-Pb ages of hydrothermal monazite and xenotime in gold deposits. Min Deposita 46:273–288

Sarma DS, McNaughton N, Fletcher IR, Groves DI, Mohan MR, Balaram V (2008) Timing of gold mineralization in the Hutti gold deposit, Dharwar craton, south India. Econ Geol 103:1715–1727

Sawkins FJ (1984) Metal deposits in relation to plate tectonics. Springer, Berlin, p 325

Sawkins FJ (1990) Metal deposits in relation to plate tectonics, 2nd edn. Springer, Berlin, p 461

Seedorf E, Dilles JH, Profett JM Jr, Einaudi MT, Zurcher L, Stavast WJA, Johnson A, Barton MD (2005) Porphyry deposits: characteristics and origin of hypogene features. In: Hedenquist JW, Thompson JFH, Goldfarb RJ, Richards JP (eds) Economic geology 100th anniversary volume, pp 251–298

Shelly D, Broughton D, Scott R, Hitzman M, Bull S, Large RR, McGoldrick P, Croaker M, Pollington M, Barra F (2005) A new look at the geology of the Zambian copper belt. In: Hedenquist JW, Thompson JFH, Goldfarb RJ, Richards JP (eds) Economic geology 100th anniversary volume, pp 965–1000

Sikka DB, Nehru CE (2002) Malanjkhand copper deposit, India: is it not a porphyry type? J Geol Soc Ind 59:336–339

Sillitoe RH (2010) Porphyry copper systems. Econ Geol 105:3–41

Sillitoe RH (1999) VMS and porphyry copper deposits: products of discrete tectono-magmatic setting. In: Stanley CJ et al (eds) Mineral deposits: processes to processing. In: Proceedings of 10th IAGOD symposium, 2001, pp 7–10

Stanton RL (1972) Ore petrology. McGraw Hill Book Co, New York, p 931

Stein HJ, Hanna JL, Zimmerman A, Markey RJ, Sarkar SC, Pal AB (2004) A 2.5 Ga porphyry Cu-Mo-Au deposit at Malanjkhand, Central India: implications for late Archean continental assembly. Precambr Res 134:189–226

Stille H (1936) Wege und Ergebnisse der Geologisch-Tektonischen Forschung. Gesell Wiss Forh 25 Jahre Kaiser Wilhelm, Bd 2: 84–85

Storey M, Mahoney JJ, Kroenke LW, Saunders AD (1991) Are oceanic plateaus sites of komatiite formation? Geology 19(4):376–379

Teixeira JBG, Misi A, da Silva MG (2007) Supercontinent evolution and the Proterozoic metallogeny of South America. Gondwana Res 11:346–361

Tembo F, Kampunzu AB, Porada H (1999) Tholeiitic magmatism associated with continental rifting in the Lufilian fold belt of Zambia. J Afr Earth Sci 28:403–425

Thorpe RI (2008) Release of lead isotope data in 4 databases: Canadian, Western Superior, foreign, and whole rock and feldspar. Geol Surv Can, Open File 5664. doi:10.4095/224837

Tomlinson KY, Stevenson RK, Hughes DJ, Hall RP, Thurston PC, Henry P (1998) The Red Lake greenstone belt, Superior Province: evidence of plume-related magmatism at 3 Ga and evidence of an older enriched source. Precambr Res 89(1–2):59–76

Tosdal RM, Richards JP (2001) Magmatic and structural controls on the development of porphyry Cu $\pm$ Mo $\pm$ Au deposits. Rev Econ Geol 14:157–181

Tripathi S (2008) Geological characterization of the carbonate-hosted polymetallic prospect at Imalia, Mahakoshal Belt, Central India. Unpublished Ph.D thesis, University of Delhi, 315 pp

Turneaure FS (1955) Metallogenic provinces and epochs. Economic geology 50th anniversary volume, Pt1, pp 38–98

Uvarov EB, Chapman DR (Revised by A. Issacs) (1977) A dictionary of science, ELBS & Penguin Books

Van der Voo R, Spakman W, Bijwaard H (1999) Tethyan subducted slabs under India. Earth Planet Sci Letts 171:7–20

Vasudeva Rao M, Nagabhusanna JC, Jatyagopal AV (1989) Uranium mineralisation in the Middle Proterozoic carbonate rocks of the Cuddapah Supergroup, Southern Peninsular India. Expl Res Atomic Min 2:29–38

Veizer J, Laznicka P, Jansen SL (1989) Mineralization through geologic time; recycling perspective. Am J Sci 289: 484–524. doi:10.2475/ajs.289.4.484

Viswakarma RK, Ulabhaje AV (1991) Sargipalli galena—unusual lead isotope date from eastern zone. Min Deposita 26:26–29

Wang R, Richards JP, Hou Z-Q, Yang Z-M, Gou Z-B, Dutrane A (2014) Increasing oxidation state from Paleocene to Miocene in the eastern Gangdes belt, Tibet: implication for collision-related porphyry Cu–Mo $\pm$ Au mineralization. Econ Geol 109:1943–1965

White RS (1992) Magmatism during and after continental break-up. In: Storey BC, Alabaster T, Pankhurst RJ (eds) Magmatism and the causes of continental break-up. Geol Soc London, pp 1–15

Windley BF (1995) The evolving continents, 3rd edn. Wiley, New York, p 526

Windley BF (1981) In: Kroner A (ed) Precambrian plate tectonics. Elsevier, Amsterdam, pp 1–20

Zachariah JK, Bhaskar Rao YJ, Srinivasan R, Gopalan K (1999) Pb, Sr and Nd isotope systematic of uranium mineralized stromatolitic dolomites from the Proterozoic Cuddapah Supergroup, south India: constraints on age and provenance. Chem Geol 162:49–64

Zhai M, Santosh M (2013) Metallogeny of the North China Craton: link with secular changes in the evolving Earth. Gondwana Res. doi:10.1016/j.gr.2013.02.007

Zhao G, Sun M, Wilde SA, Li S (2004) A Paleo-Mesoproterozoic supercontinent: assembly, growth and break-up. Earth Sci Rev 67:91–123

Ziauddin M, Narayanaswami S (1974) Gold resources of India. Bull Geol Surv Ind, Sr A, Econ Geol no 38 (Pt. I), 168 pp

Part II
Allied Natural Resources

Chapter 6
Energy Resources

Abstract Energy resources are essential for the sustenance of the human society as much as it is needed for the life and existence of an individual. Energy resources are of two kinds: renewable (solar, hydro-power, wind, tidal, biofuels) and the non-renewable energy sources comprise coal, natural gas and petroleum, and nuclear raw materials. Geothermal energy is a continuous form of energy source. The world derives 2/3 of its produced energy from the fossil fuels and 1/3 from the nonfossil type, i.e., hydro-energy + nuclear energy. While the exhaustion of the nonrenewable energy sources is slated, that of the renewable variety appears inexhaustible. Coal (±lignite) provides 2/5 of the fossil fuels going for the production of electrical energy and there lies the main root of the global warming problem. Coal is a combustible sedimentary rock having both organic ($C_{org} \geq 50\%$) and inorganic components. Humic coal forms from different mixtures of macroscopic plant debris, whereas sapropelic coal forms from selected types of microscopic plants. Coalification: plant debris (biochemical changes) peat → (geochemical changes) → coal. Petrographic components: vitrain, clarain, durain, and fusain. Of the world's total reserves of 848 Gt, India's share is 284 Gt. However, it is generally high ash and noncoking. Bulk of India's coals was deposited in Gondwana basins of Permo-carboniferous age. Add to these a proven reserve of 6.1 Gt of lignite in the Tertiary basins. Crude oil or petroleum is a mixture of hydrocarbons that remains liquid at atmospheric pressures. It is lighter than water. Chemically crude oil is made up of carbon ($\sim 85\%$), and hydrogen (13%) and the rest made up of S, O, N along with some trace metals. The constituent hydrocarbon molecules are alkanes, napthanes, and aromatic compounds including benzene. Oil and natural gas are produced from phytoplanktons deposited on water-sediment interface and modified by (bio-) geochemical and geochemical processes. Oil and gas formed at infinite number of points in the host rock and then migrated and accumulated in the low pressure zones (reservoirs). Although India identified 26 oilfields on its rightful areas, it produces about 1/3 of its requirements from its wells. India has a very good reserve of gas hydrates within its EEZ. After the WW II around the middle of the last century, the world saw the advent of another kind of fuel for energy (electrical) production. It was nuclear energy, accounting for about 16% of world energy today. In fact, it is the heat produced by fission of nuclei of suitable radioactive elements,

M. Deb and S.C. Sarkar, *Minerals and Allied Natural Resources and their Sustainable Development*, Springer Geology,
DOI 10.1007/978-981-10-4564-6_6

such as U and Th that is utilized in heating water, creating steam that would rotate turbines and produce electricity. The positive thing about it is that the whole operation is "clean" and a plant runs for about 30–40 years with an initially high but later moderate cost. But two recent disasters in such plants, one at Chernobyl, Ukraine (1986) and the other at Fukushima, Japan (2011) have shaken the faith in this kind of energy source for many prospective users. Toward the end of 2014 India was having 20 nuclear power plants in operation in the country. India is at an enviable position with respect to the provision of Th resources in the form of monazite in the beach sands of Kerala, Tamil Nadu, and Odisha. But its own reserve of U ores is not encouraging. India's main source of indigenous U ores lies in the Singhbhum Cu–U belt in eastern India. The Cuddapah basin in Andhra Pradesh is now showing promising potential in this respect.

Keywords Renewable and nonrenewable mineral resources · Varieties of coal · Distribution of coal in time · Coking coal · Coal petrography · Coal-bearing basins · Coal bed methane · Petroleum and natural gas · Gas hydrates · Oil and natural gas fields in India · Fissionable metals · Radioactive ores · Nuclear energy · Thorium resources of India · Geothermal energy

6.1 Introduction

Energy is the capacity to do work. Energy exists in different forms, such as mechanical energy, light, sound, heat, electrical, chemical, and nuclear energy. In this chapter, we will not go into any details about the nature of these energies, how they are produced or utilized to get work done, except what is barely needed. We shall rather discuss the natural material that is used raw or after processing, to produce energy for various uses in the society. Much of the energy we use today is in the form of electrical energy which is produced by rotating a turbine with the help of a steam jet or water jet. Steam is produced by burning coal, natural gas, and oil, or with the help of heat produced in a nuclear reactor with the disintegration of highly fissionable metals, such as uranium and thorium. Water jet does the job in a hydro-power generator, same as what a steam jet does in thermal power. Add to this solar energy, bioenergy, wind energy, tidal energy, etc., the full potential of which remains to be utilized in the days to come. Of these energy sources, solar energy, hydro-power energy, wind energy, tidal energy, and bioenergy sources are renewable, and the rest are all nonrenewable. Geothermal energy is a continuous form of energy source. Two thirds of the natural raw materials, utilized in producing electrical power, are fossil fuels (coal, natural oil and gas, oil shale) (Table 6.1).

People have been worried for sometime at the fast dwindling reserves of fossil fuels in the world as a whole. At the same time, they have been concerned that continued use of fossil fuel is the cause of fast deterioration of atmospheric quality bringing about climate change. The recently concluded International Climate

Table 6.1 Average world scenario of energy sources for electrical power generation

Coal	39%
Natural oil	10%
Natural gas	15%
Nuclear energy	16%
Hydel power	19%
Miscellaneous	1%

Conference (November–December, 2015) at Paris, France, adopted a few corrective resolutions. It is difficult to forecast right now how long the known reserves of energy sources will last because (i) reserves may increase on further exploration, (ii) we are not sure of the rate at which we will consume energy in the future or replace conventional energy sources with unconventional ones. Moreover, we may reasonably expect that human ingenuity will discover before long cheap and easy methods of producing such renewable energies, such as solar energy, bio-energy, wind energy, tidal energy, etc.

In this chapter we will discuss the major energy resources.

6.2 Coal

In spite of a continuing international debate on coal-based emissions from thermal power plants, considered to be responsible for global warming, coal still remains the major energy source in most countries of the world. The annual world consumption of coal is about 7–8 billion tons. There is a rough estimate of proven global coal reserves of over 848 billion tons, with the largest repositories in North America, followed by former Soviet Republics, China, Australia, Africa, and India. Thus, at the current rate of consumption, that is 6.6 giga tons of hard coal and 1 giga ton of brown coal, there is enough coal to last for the next century or more.

6.2.1 *What is Coal?*

Coal is a black to brownish black combustible sedimentary "rock," having both organic and inorganic constituents (in some anthracitic coal the inorganic component may be negligible). This carbonaceous solid fuel, coal, that burns readily, or rather with little difficulty, is usually found stratified with other sedimentary rocks. To be a "coal" its carbon content should be $\geq 50\%$. Coal differs from charcoal in that the former is a naturally occurring commodity while the latter is prepared artificially from wood by controlled burning, although both have vegetative matter as the initial raw material.

6.2.2 *Use of Coal*

Coal, as the available information suggests, has a long history of use, at least in certain regions. 4000–3000 years ago (in the Bronze Age) some tribes in Wales reportedly used coal in funeral pyres. The Chinese used coal during 1100–1000 BC and the Greeks during 300–200 BC. The British interest reportedly remained live during the intermediate period, before it became intense with the advent of the Industrial Revolution. This was about the time when the rest of Europe and North America also took to the exploitation of coal. India, though a very old civilization, also having large reserve of coal, is not known to be using coal before the entry of the British colonizers in the eighteenth century.

The different types and varieties of coal have different industrial uses. This is shown schematically in Fig. 6.1. Besides the three main uses of power generation, steel and cement production, the other industrial uses of coal include alumina refineries, paper manufacturing, chemical and pharmaceutical industries, and in the production of liquid fuel in some countries. Other important use is in the metallurgy of oxidic ores (as a reductant). Refined coal tar is used for the manufacture of certain industrially useful chemicals, such as benzene, phenols, etc.

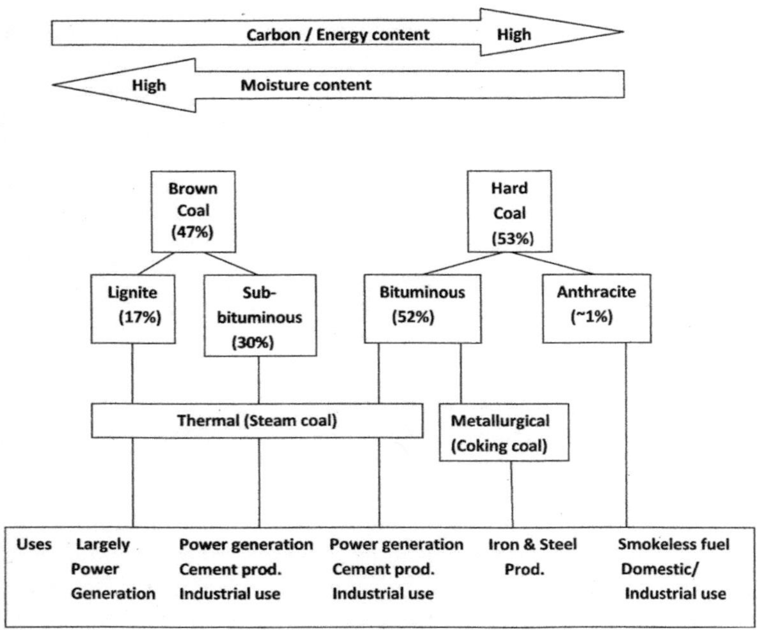

Fig. 6.1 A schematic representation of different varieties of coal and their different uses based on their moisture and carbon content. Percentages in parenthesis refer to their relative proportion in the world

6.2.3 Distribution of Coal

Land plants indeed appeared first in Lower Paleozoic Era (Silurian and Devonian periods). Thus, the greatest period of coal formation globally was in the Carboniferous and Permian when abundant plant material, suitable climate, and basins of accumulation were available. However, there have been three major coal accumulations in geological history (Fig. 6.2). These are the following:

Paleogene–Neogene Period: Worldwide (Lignite—anthracite)
Jurassic-Cretaceous Period: Coals of USA, Canada, China
Carboniferous—Early Permian Period:

Northern hemisphere: USA, Canada, Europe, CIS, Far East
Southern hemisphere: South America, Africa, Indian sub-continent (Gondwanaland), SE Asia, Australia, Antarctica.

The principal areas of prolific coal deposition in Carboniferous-Permian times are shown in a paleogeographic reconstruction of this period in Fig. 6.3.

Estimate for the world reserve of proven black coals (bituminous coal and anthracite) and brown coals (sub-bituminous and lignite) were, respectively,

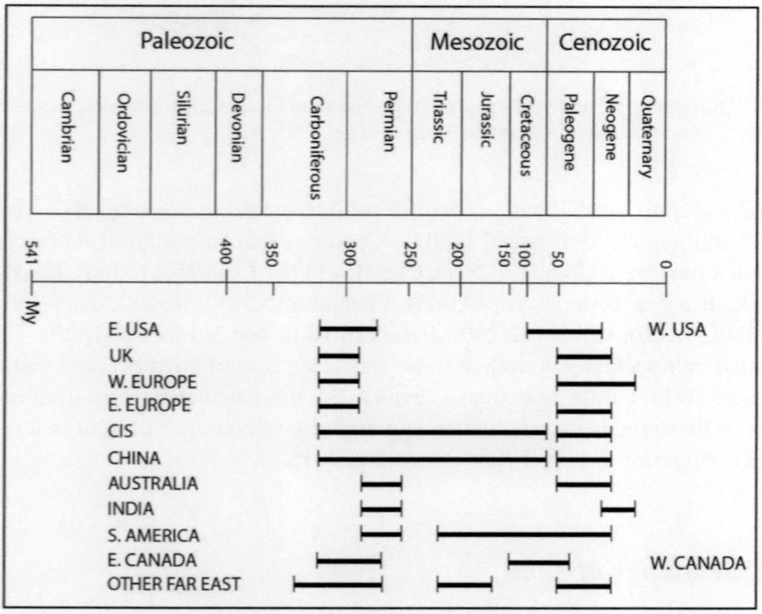

Fig. 6.2 Time-space distribution of coal deposits (coal + lignite) in the world (after Walker 2000 in Thomas 2013)

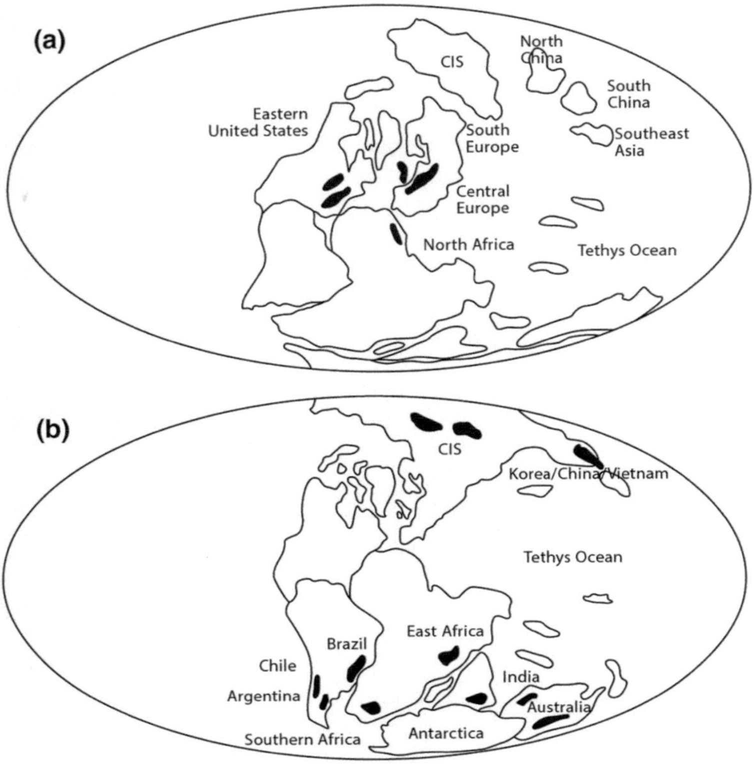

Fig. 6.3 Distribution of principal areas of coal deposition in a paleogeographical reconstruction of **a** Late Carboniferous and **b** Permian-Triassic times

404,762 million tons and 456,176 million tons, respectively, totaling 860,938 million tons during 2012–2013. Countries that have significant reserves of coal at that time were USA (108,501), Canada (3474), Columbia (6366), Kazakhstan (21,500), Russian Federation (49,088), Ukraine (15,351), South Africa (30,156), Australia (37,100), China (62,200), India (56,100), and Indonesia (1520) – all figures in parenthesis being in million tons. This reserve, theoretically could sustain the world's needs for a little more than a century. But this calculation is no simpler in the context of the agreements reached among majority of countries present in the world climate conference in November–December, 2015.

6.2.4 *Varieties of Coal*

Coal is divided into two broad groups: *humic coal* and *sapropelic coal*, on the basis of their source material, physical and chemical properties. Humic coals are formed from different mixtures of macroscopic plant debris and they generally give a banded look. In humic coals the remains of wood tissues and bark are a common feature. Sapropelic coal, on the other hand, formed from a selected type of microscopic plant debris and is generally richer in resins, waxes or fats and therefore, richer in hydrogen compared to the humic coals. They look generally homogeneous. They are generally of lenticular form and localized in extent and thus, economically less important. *Boghead* and *Cannel coals* are varieties of sapropelic coal. The former, also known as *Torbanite*, is brownish, comprising algal and fungal material, and yields tar and oil upon distillation. The latter is nonbanded, very fine grained, black coal of bituminous rank with a conchoidal fracture. It is high in volatile constituents and burns with a smoky flame. It does not soil the finger.

Humic coals are sub-divided into several varieties depending on the degree of maturation. These are peat, brown coal (including lignite and sub-bituminous coal), bituminous coal, anthracite. They are further sub-divided into several groups, as shown in Fig. 6.4.

Peat is the initial decay product of partially carbonized vegetal matter in a swamp or in a shallow basin. Most physical feature of the initial wood virtually remains intact there. It may be light or relatively dark in color, depending on the degree of transformation. To cite an example of typical peat in India, one generally refers to the Nilgiri peat, on the Western Ghats. It has a market too. In and around Kolkata, in the flood plains of the Hoogly River, drillings for tube-wells often encounter a layer of peat. Peat is of common occurrence around the world.

Brown coal is generally considered as one step forward in the passage to typical coal, i.e., bituminous coal. However, it is sub-divided into two: the earlier sub-stage product is called *lignite* while the latter sub-stage is referred to as *sub-bituminous* coal. The color changes from brown (lignite) to black (sub-bituminous), fossilized "wood"-content decreasing accordingly. The woody fibers are embedded in an amorphous macerated groundmass of vegetal origin. Volatile-content is in the range of a little above 60% to about 43% (dry and mineral matter-free basis).

Bituminous coal is the common banded variety of coal with jet black color. It is the most abundant variety of coal. Name notwithstanding, it does not contain bitumen. It derives its name from the fact that it burns with a yellow smoky flame, like bitumen. Carbon content varies from 40 to 70% while moisture drops drastically from 10 to 5%. It is divided into several sub-types depending on parameters. These are: high volatile bituminous, medium-volatile bituminous and Low volatile bituminous. Semi-anthracites are included in this and some further divide High volatile bituminous into A, B, and C.

Anthracite is the product of the final stage of transformation of coaly matter. Nowadays it is sub-divided into semi-anthracite, anthracite and meta-anthracite, the last one representing the highest state of transformation where C-content may be even 98–99%. This on further transformation would be converted to graphite.

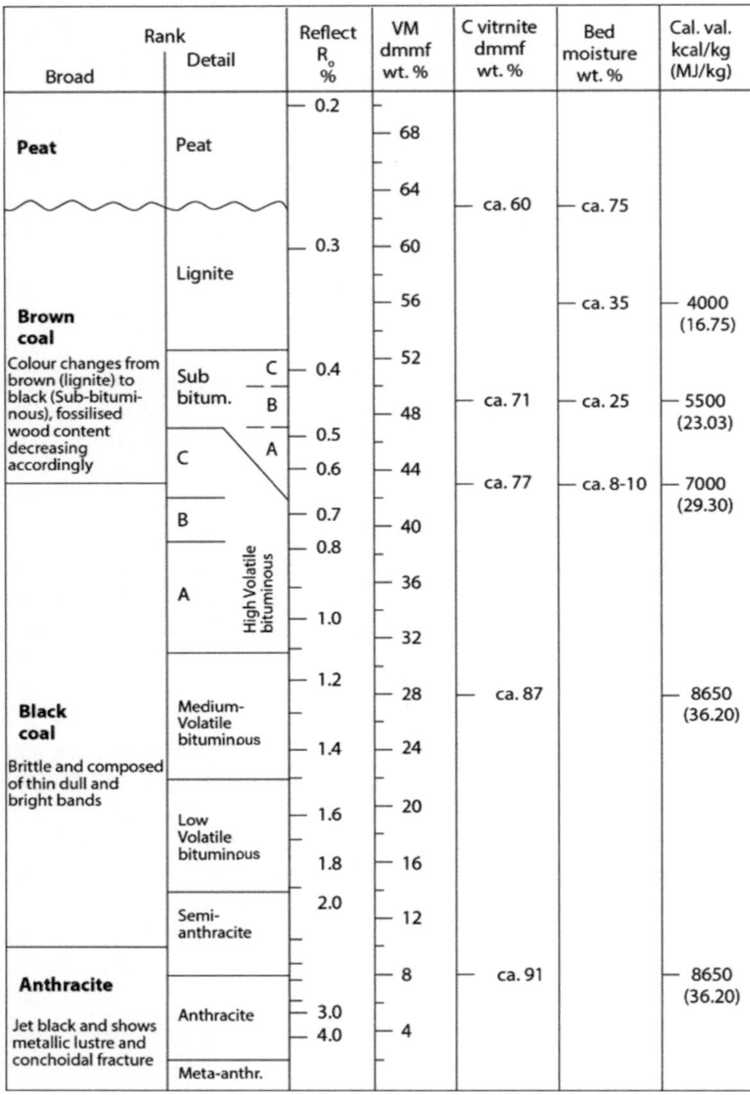

Fig. 6.4 Classification of coal into commonly used divisions and sub-divisions, based on principal physical and chemical properties. *Refl* reflectance; *VM* volatile mater; *Cal.Val* calorific value (heat-producing capacity); *dmmf* calculated to dry and mineral matter-free, i.e., pure coal content basis (after Taylor et al. 1998)

Anthracite is jet black color, is shining, does not soil the hand and burns with a smokeless blue flame. It breaks with a conchoidal fracture. Some important properties of different coal varieties mentioned above are given in Fig. 6.4.

6.2.5 Properties and Composition of Coal

In the above four types of humic coal, the properties are mainly controlled by moisture, volatile matter, ash, and carbon content. These four parameters, measured on weight percent basis, provide the *proximate analyses* of coal. *Moisture* held within the coal is its inherent moisture. It may comprise surface moisture, hygroscopic moisture held within the micro-fractures, moisture due to decomposition of organic matter, and mineral moisture held within the structure of silicate minerals like clay minerals. *Volatile matter* refers to the components of coal, except moisture, that are liberated at high temperature. This is commonly a mixture of short and long chain hydrocarbons, aromatic hydrocarbons, and some sulfur. *Ash* in coal is the solid residue that remains after burning. This is essentially contributed by the oxide and silicate minerals that exist in raw coal as the worst deleterants. High ash coal inevitably will have low calorific value and add to slag in metallurgy. There are different methods of coal washing to reduce the ash. *Fixed carbon* is the carbon found in the material which is left after the volatile materials are driven off. It is used to estimate the amount of coke that will be produced from a sample of coal. *Coke* is a porous and brittle solid containing about 80% carbon, obtained after heating the coal to a temperature of 950 °C in the absence of air in coke ovens. Some coals have intrinsic coking-caking properties. A lot of chemicals are obtained as by-products when coke is produced. *Sulfur* in coal is also unwanted. It generally occurs in three forms: sulfate sulfur, sulfide sulfur, and organic sulfur. Sulfur-rich coal, even if low in ash content, is unwanted in metallurgy, particularly in making iron and steel. *Phosphorus* is also unwelcome. During the making of coke, phosphorus gets into it and ultimately forms "weak spots" in the blast furnace.

The *ultimate analysis* provides an element-by-element composition of the coal's organic fraction, namely carbon, hydrogen, oxygen, and sulfur, all on a weight percent basis. Hydrogen by oxygen ratio in coal plays an important role in determining the caking properties of coal. When H/O ratio is 59% or more, coal shows strong caking qualities. When it is between 59 and 55% it is fairly caking, but when the ratio is below 55% the coal is noncaking. With the change of wood to anthracite through several stages, we note changes in chemical composition of each phase and ultimately in the *calorific value,* which is a measure of the heat produced after complete combustion of a given weight of coal. Peat formation is considered as the *biochemical stage* of coal forming process. During the geochemical stages of *coalification* successively higher *ranks of coal* are formed from lignite through sub-bituminous and bituminous to anthracite, there is a gradual elimination of oxygen and when the fixation of carbon is sufficiently high there is a significant change in the hydrogen content (Table 6.2) (Fig. 6.5).

Spontaneous combustion of coal is moderately common, though a fully satisfactory explanation is not yet available. However, it is established that high volatile coal is more susceptible to spontaneous combustion, aided and abated by some higher temperatures and O_2-supply. It may mean a disaster in a coalfield, as has happened in the Jharia belt (again discussed later).

Table 6.2 Average ultimate analyses of wood, peat, and different varieties of coal

	Carbon (%)	Hydrogen (%)	Oxygen (%)	Nitrogen (%)	Calorific value (kJ/kg)
Wood	50.0	6.0	43.0	1.0	14,400–17,400
Peat	57.0	6.0	35.3	1.7	13,800–20,000
Lignite	65.0	5.2	28.3	1.5	23,000–29,000
Bituminous coal	84.0	5.2	28.3	1.5	25,000–29,000
Anthracite	93.5	2.8	2.8	0.9	35,000–35,420

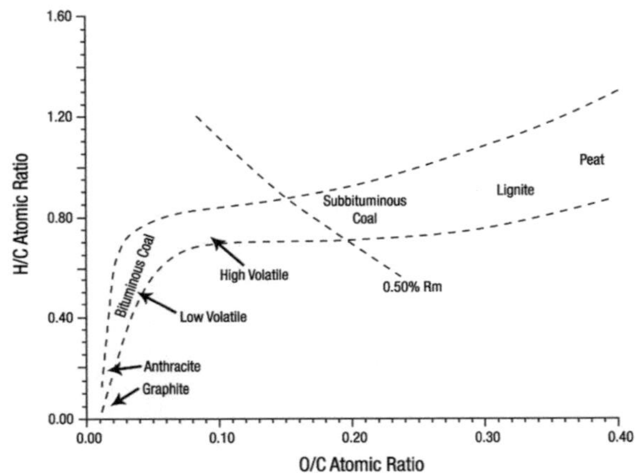

Fig. 6.5 Van Krevelan's diagram showing the variation of H/C and O/C ratios in different varieties of coal

6.2.6 *Mode of Occurrence and Structure of Coal Seams*

Coal generally occurs as sheet-like bodies, commonly low dipping and extensive over an area of variable dimensions, length varying from a few meters to, rarely, hundreds of kilometers. One such unit is called a *coal seam*. The thickness of a coal seam generally varies between a few to a few tens of meters. Indian coal seams, however, hardly exceed a few tens of kilometers in length. Their thickness may vary from less than a meter to more than 100 m. Examples of some thick Indian coal seams are:

A Singrauli coalfield seam in Madhya Pradesh	131.36 m
An Ib river coalfield seam in Orissa	180.0 m
A Birbhum coal seam in West Bengal	180.0 m

These coal seams are constituents in the geological sequence of a sedimentary basin in which they formed. They are also conformable with the associated sedimentary rocks, dominated by sandstone, shale, and clays. In the Gondwana basins the deposition of sediments is often cyclic (Fig. 6.6); the sequence starting with

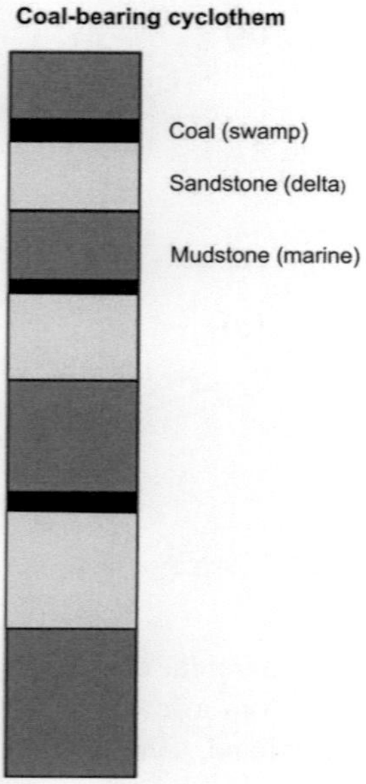

Fig. 6.6 A sketch of the cyclic sedimentation pattern of coal

sandstone may be overlain by sand-shale intercalations, sandy shale, siltstone, carbonaceous shale, and then coal. In other words, it is an upward fining sequence, sans coal. However, the sequence may and do vary in details depending on the variation in the environment of deposition. In the Jharia coalfield, eastern India, 60 such cycles were identified in a thickness of 1130 m of Barakar sediments. A coal seam itself may be split into two or more units by included shaly layers. Structurally, coal seams are more often affected by faults than by folds particularly in the intra-continental basins, and amongst faults, they are affected more often by normal rather than other types of faults. Thin bands/strata of terrestrial or volcano-clastic sediments may cause splits within a coal seam (Fig. 6.7). They are unwanted in the industrial context. Discordant seams may occur above a stratigraphical unconformity.

Humic coals are commonly stratified and comprise layers or bands of variable physical properties, such as luster and texture and thickness. *Cleats* or joints are also very important features in coal (cf. Fig. 6.18), particularly from the exploitation point of view. Generally, three sets of cleats are found in coal seams. One set is parallel to the bedding plane. The other two occur at right angles to each other and the first one. Of course there may be less regular cleats due to tectonic reasons. Cleats when not naturally sealed by mineralization help in coal mining, as well as in withdrawing coal bed methane (CBM) provided it is there.

Research has indicated that cleating decreases from lignite to bituminous coal. In anthracite, it may increase or decrease which may be explained by a competition between fracture formation and annealing. More studies are needed on this. Coal type and ash content apparently affect cleating.

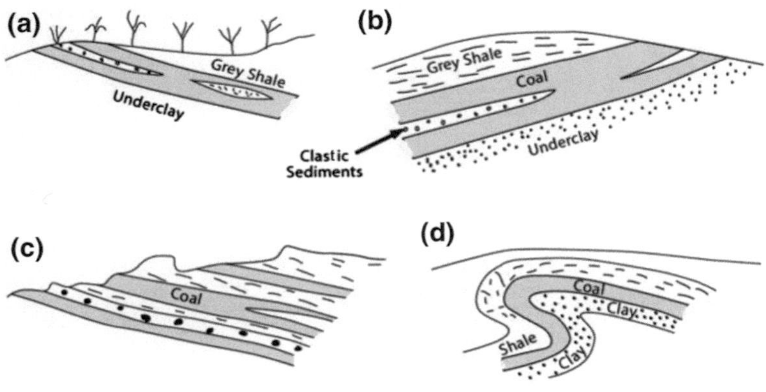

Fig. 6.7 Sketches (**a**–**d**) showing some structural dispositions of coal seams including splitting

Spheroidal or oval-shaped bodies of coal are known as *ball coal* which vary in size from a few to 50–60 cm and are found embedded in the coal seams. On the other hand, spheroidal or ovoid, sometimes even angular concretions and fossilized plant remains, occur within coal seams and are known as *coal balls*. They may contain even significant amounts of organic matter.

6.2.7 Coal Petrography

Humic coals, particularly the bituminous variety, are banded in nature. The bands, varying in thickness from a few mm to a few cms, are defined by variable physical properties, such as luster and texture. They are commonly represented by bright and dull bands (Fig. 6.8). The components of these bands are vitrain, clarain, durain, and fusain, according to the original nomenclature of Marie Stopes (1919). *Vitrain* or anthraxylon constitutes thin bands of bright, glassy-looking jet black coal with conchoidal fracture in which no woody matter can be identified megascopically. The brightness of vitrain varies considerably with the rank of coal. *Clarain* is characterized by bright color and silky luster. It is composed of translucent, finely divided plant residue which is more resistant. *Durain* is the dull material in bituminous coal. It has no luster and has an earthy appearance. It is hard, black to lead gray in color and can contain cuticles, spores, etc. *Fusain* is the mineral charcoal. It is high in ash. These coal *lithotypes* (Table 6.3) are important as determinants of coal properties, such as calorific values, liquefaction potential, and coal bed methane (CBM) potential.

Fig. 6.8 Banded constituents of Bituminous coal: *bright shining bands* are vitrain and the *dull bands* are durain

In coal petrography, coal is a rock and consists of several macerals, i.e., mineral equivalents, such as vitrinite (or huminite), exinite (or liptinite), and inertinite which are studied under incident as well as transmitted light (Fig. 6.9). The macerals are distinguished on the basis of morphology, optical properties as well as polishing hardness. The analogy with minerals in rocks is not really correct because minerals have definite chemical compositions while macerals range widely in their composition and physical properties even with the same rank of coal. Macerals are formed during the decomposition of plant material in either aerobic or anaerobic state and depend on the condition of the swamps. The three petrographic constituents have the following properties: (1) Vitrinite, the main constituent of coal is a translucent, light brown colloidal jelly. (2) Exinite consists of clear, translucent spores and

Table 6.3 Lithotypes of humic and sapropelic coals (modified after Thomas 2013)

Lithotype	Macerals	Physical properties	Types of coal
Vitrain	Vitrinite with 20% exinite	Black, very bright lustre, thin layers break cubically, thick layers have conchoidal fractures	Humic coal
Clarain	Variable	Finely stratified layers of vitrain, durain and in cases, fusain. Medium lustre	Humic coal
Durain	Mainly inertinite and exinite	Black or gray, dull, rough fracture surface	Humic coal
Fusain	Mainly fusinite	Black silky lustre, friable and soft	Humic coal
Cannel coal	Fine maceral particles, usually dominated by sporinite	Black, dull, lustre greasy, breaks with a conchoidal fracture	Sapropelic coal
Boghead coal	Dominated by alginate	Black or brown, dull, homogeneous, breaks with conchoidal fracture. Lustre may be greasy	Sapropelic coal

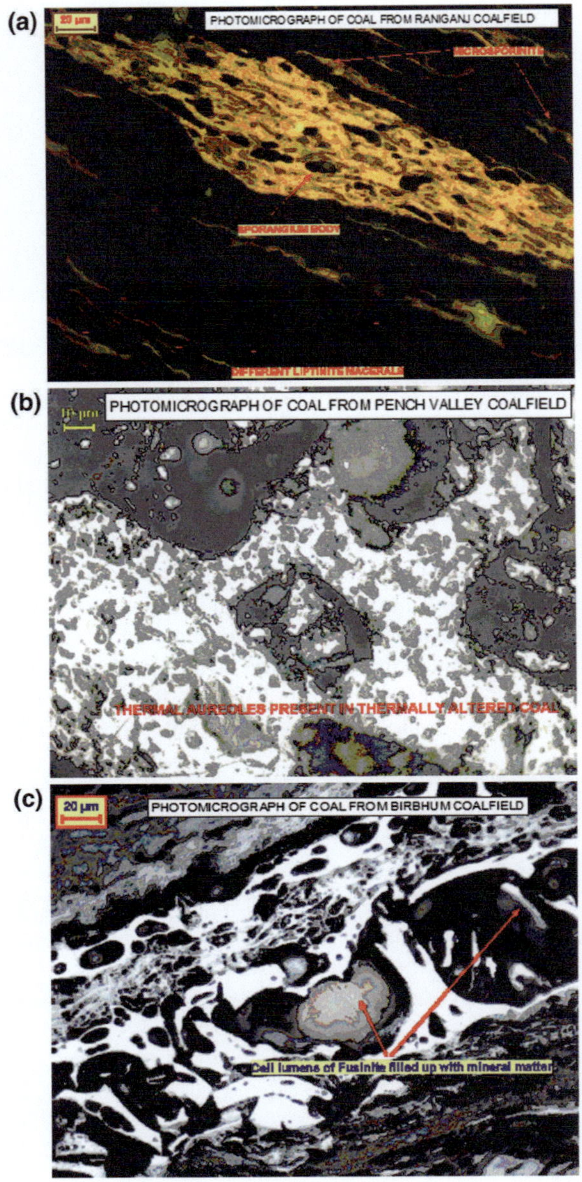

Fig. 6.9 **a** Different liptinite macerals under transmitted light using a thin section of coal from Ranigunj coalfield. **b** Vitrinite mass in thermally altered coal from Pench valley coalfield. Reflected light. **c** Coal sample from Birbhum coalfield under reflected light. Bright mass is vitrinite. Cell lumens of fusinite filled with mineral matter can also be seen (Courtesy: GSI)

smaller plant bodies. (3) Inertinite includes micrinite, semi-fusinite, fusinite, and some opaque minerals and is the major maceral in sapropelic coal. According to Deb (1980) vitrinite is ortho-hydrous, exinite is per-hydrous and micrinite or fusinite is sub-hydrous.

Table 6.4 Microlithotypes and their maceral composition (after Taylor et al. 1998)

Microlithotype	Maceral composition
Vitrite	Vitrinite (V) > 95%
Liptite	Liptinite (L) > 95%
Inertite	Inertinite (I) > 95%
Clarite	V + L > 95%
Durite	I + L > 95%
Vitrinertite	V + I > 95%
Duroclarite	V > L, I (each > 5%)
Clarodurite	I > V, L (each > 5%)

Maceral associations at microscopic scale are known as *micro-lithotypes*. The common micro-lithotypes in hard coal are listed below (Table 6.4).

6.2.8 Evaluation of the Quality of Coal

Quality of coal is determined mainly by two parameters:

1. Type of coal
2. Rank of coal

The type is principally determined by the relative proportion of the three macerals: vitrinite, liptinite (formerly exinite), and inertinite, as they have different physical and chemical properties. The vitrinite group is rich in oxygen, leptinite is rich in hydrogen, and inertinite is rich in carbon. Liptinite contains almost twice as much volatile matter as that of vitrinite whereas inertinite contains half as much volatile matter as that of vitrinite.

Rank is the level of maturity of coal. The increase in rank from lignite to anthracite is manifested by a rise in carbon content relative to hydrogen and oxygen. It is also recorded by the reflectance of coal which is estimated under a reflected light microscope with oil immersion objectives, from vitrinite maceral of coal after careful polishing. The light used should desirably be monochromatic. The relationship between vitrinite reflectance and chemical composition of coal is shown in Fig. 6.10 (cf. Sharma and Ram 1966).

6.2.9 Coal-Bearing Basins

We already know that coal formed in sedimentary basins. But we do not know yet their exact tectonic settings. However, these settings are generally divided into two principal types: paralic, located on sea coasts, and limnic that formed in intra-continental and even intra-montane swamps. Wetland, rather than a lake,

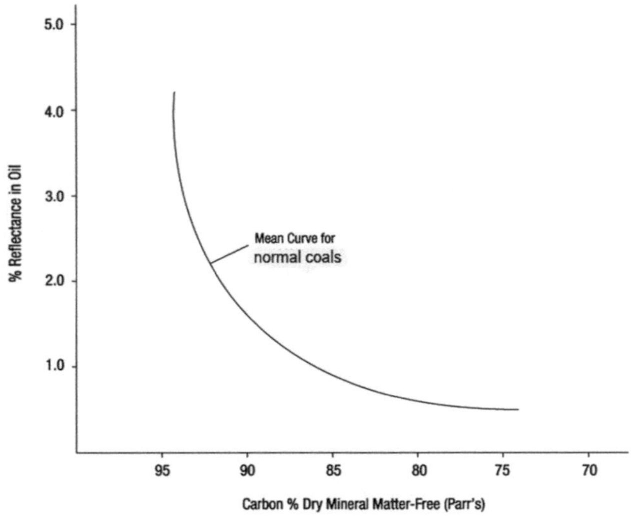

Fig. 6.10 Relationship of reflectance to carbon content of coal

because of its wide expanse and shallow depth, should be a more appropriate term for such a basin.

Tectonically, paralic coal basins are located in foreland downflexures produced by subduction tectonics. The sediments, mostly of continental derivation, may be considerably thick. The sequence, together with the included coal measures, is generally deformed, the intensity of which decrease from the orogenic front. Coal seams are numerous but generally thin (1–2 m). Best examples of this type of deposits occur in the Appalachian foredip in North America and in the North European foredip.

Intra-continental and platformal coal-bearing sequences are generally not thick, but exceptions exist. Coal seams are few (compared to the paralic environment), but may be thick. Coal seams are sub-horizontal to low dipping and sheetlike. Examples of this type of coal-bearing sequences are the Gondwana deposits of Indian sub-continent, South Africa, and South America. The brown coal deposits of Germany also belong to this type (Thomas 2013).

Another type of coal basins is called intra-montane basins that form as transtensional, transpressional, and pull-apart basins as effects of late stage tectonism in an orogen. Sediments are limnic to alluvial. Coal seams are a few but generally thick and their deformation is variable. Early Tertiary Raton and Powder River deposits of North America are examples of this type.

6.2.10 *Genesis of Coal/Coalification*

Coal forms from organic (plant) debris in two principal stages: peatification and coalification. The former is essentially a biochemical process, whereas, the latter is essentially a geochemical process of conversion. In peatification biologic activities involving the plant materials are most important. In coalification, geogenic pressure and temperature, as well as time, play the most important roles. To attain the stage of typical coal, the plant debris has to go through the process of peatification.

Plant debris cumulates Biochemical changes ⟹ ***Peat*** Geochemical changes ⟹ ***Coal***

Peatification

About 3% of the Earth's surface, totaling ~310 million hectares, is covered by peat, which means it is widespread. Peat developed in wetlands.

Dissel (1992) divided the peat producing wetlands into *ombrogenous* peatlands or *mires* that formed due to heavy rainfalls and *topogenous* peatlands caused by surface/groundwater regimes. Ombrogenous peatlands are greater in extent and less varied in composition, whereas topogenous peatlands are more varied. The term *Mire* is now used for peat-forming eco-systems. Development of a peat deposit requires the satisfaction of the following conditions (Pohl 2011):

(1) A profuse growth of higher plants.
(2) Dead plant debris is covered by water to restrict oxidation and microbial decomposition.
(3) A minimum quantity of clastic sediments enters the peatland.

Climate has an essential control on plant growth which provides the necessary raw material (plant debris) for the ultimate product which is peat or coal. Sufficient precipitation must have characterized the area. To be precise, it should not fall below 40 mm/month even in dry seasons and the accompanying temperature remaining above 10 °C to favor a prolific plant growth. This qualifies tropical situations most as regions of higher plant growth and subsequent coal formation. We need not, however, write off other climatic zones. Significant coal resource developed in temperate, even cool climates. The Late Carboniferous, i.e., Pennsylvanian coals of Europe, North America and China were deposited in swamp forests dominated by Sigillaria and Lepidodendron at tropical latitudes. The Permian flora of southern continents, characterized by Gangamopteris-Glossopteris species formed during a cool and humid climate, possibly with tropical summer temperatures (Ziegler et al. 1997; Rayner 1996).

One may ask how a mire or coal (rather peat)-forming ecosystem develops. According to Thomas (2013) peat formation may be initiated by:

(a) *Terrestrialization* in which a waterbody (pond, lake, lagoon) is replaced by a mire.
(b) *Paludification.* It is the opposite of terrestrialization, i.e., a dryland is replaced by a mire. This happens due to the rise of the ground water table.

Peat is relatively impermeable. Therefore, its growth may obstruct drainage over extensive areas. In such areas, if annual rainfall exceeds evaporation, a raised mire may develop. Such a mire can develop upwards, because they maintain their own water table. The progression of a peat-forming environment from the infilling of a water body or lake to a low-lying mire is shown in the accompanying diagram (Fig. 6.11). Note that the peat thus formed will be zoned.

Majority of coals are formed from plant aggregates where they grew. Such coals are underlain by "seat earths" or rootlet beds and are known as autochthonous coals. On the other hand, coal that has formed from plant remains that came floating from

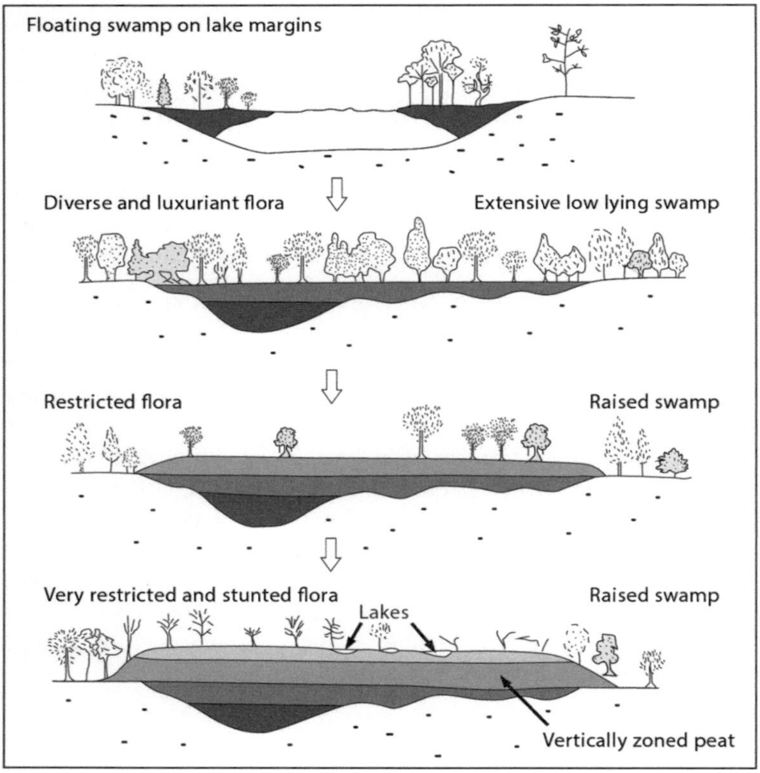

Fig. 6.11 Sequence of mire development, showing vertical zonation of peat (after Thomas 2013)

a distance is called allochthonous. They lie directly on the bed below. Locally, as in the Cooper basin, Australia both the evidences are present. Allochthonous coal is usually high in ash and varies in thickness. This variety is relatively more common in Gondwana coal.

Biochemical Peatification

A peat-forming wetland has an oxidized near-surface zone, below which occurs a reducing zone. A fully developed vertical redox zonation consists of oxic, sub-oxic, sulfidic, and methane zone, from top downwards.

Dead plant cumulate in the mire is decomposed by bacteria, archae, actinomycetes, and fungi. Synthesis of humic macromolecules occurs by microorganisms proliferating on plant residues (Kögel-Knabner 1993). Sugars, starch, pectin, and protein are rapidly metabolized, as is cellulose. Lignin is preferentially attacked by fungi, transforming into solid humic substances. Lipids, such as pollens, spores, and cuticles preserve both the structure and material composition. The decay of plant material within the peat-profile is called *humification*. Humification may be followed by biogenic gelification. Microbial activity causes more intense decomposition. High microbial activity is principally controlled by (i) a basic pH of the swamp; (ii) availability of sufficient nutrients, such as Ca, P, K, and N; (iii) elevated temperatures and (iv) free oxygen in surface water (Pohl 2011). Burial of the peat by marine sediments or alluvial or limnic sediments, as the case may be, indicates the end of the biochemical stage of the coalification process, which is then taken over by the geochemical stage.

Geochemical Coalification

Though lignite is supposed to be the next stage to peat (and sub-bituminous to lignite), their difference is not very clear-cut. Covered by sediments, coalification proceeds forward. Humins evolve during coalification by increasing aromatization, condensation, and clustering (Hatcher and Clifford 1997). During coalification ratios like H/C and O/C change and decomposition reactions release H_2O, CH_4, and CO_2 or CO (Kopp et al. 2000). Increase in the rank of coal is reflected in physical, chemical, and petrographical properties.

Effect of Pressure, Temperature and Time on Coalification

Direct evidence of the rise of coal rank due to temperature-rise is obtained close to the igneous intrusion. Moreover, a higher rank/grade is obtained when the thermal gradient is higher at shallower depths. But in the case of lower thermal gradient, it will need greater depths to attain the same rank (Stach 1982; Teichmuller 1987). Even anthracite formation is facilitated by proximity to a magmatic heat source.

Pressure result in compaction, decrease in porosity and loss of moisture in the plant mass undergoing coalification. According to Stach (1982), pressure causes “physico-structural coalification,” whereas temperature causes chemical coalification. Pressure effect will be less in case of confining pressure, where the loss of gasses will be difficult. In contrast, coal seams affected by disjunctive structures, such as pervasive shears, well developed cleats, will favor rank increase by letting out volatile matters in a favorable temperature regime.

Table 6.5 Major stages of coal formation (after Thomas 2013)

Coalification stage	Approximate ASTM rank range	Predominant processes	Predominant physic-chemical changes
Peatification	Peat	Maceration, humification, gelification, fermentation, concentration of resistant substances	Formation of humic substances, increase of aromaticity
Dehydration	Lignite to sub-bituminous	Dehydration. Compaction loss of O_2-bearin groups, expulsion of COOH	Decreased moisture contents and O/C, CO_2 and H_2O ratios, increased heating values, cleat growth
Bituminization	Upper sub-bituminous A to high volatile A to low volatile bituminous	Generation and entrapment of hydrocarbons, depolymerization of matrix, increased hydrogen bonding	Increased vitrinite Ro, increased fluorescence, increase in extract yields, decrease in density and sorbate accessibility, increase in strength
Debituminization	Uppermost high volatile A to low volatile bituminous	Cracking expulsion of low molecular weight hydrocarbons, specially methane	Decrease in fluorescence, molecular weight of extract, H/C ratios, strength; cleat growth
Graphitization	Semi-anthracite to anthracite	Coalescence and ordering of pre-graphitic aromatic lamellae, loss of hydrogen, loss of nitrogen	Decrease in H/C ratio, stronger XRD peaks, increased sorbate accessibility, anisotropy, strength ring condenzation and cleat healing

Time is important in case of high temperatures when a shorter time will cause coalification (if there is a release of volatile matter). But when the temperature is low then it is generally of little help, as is shown by the Lower Carboniferous lignite of Moscow basin, Russia.

The major stages of coal formation are shown in a tabular form in Table 6.5.

6.2.11 Coal in India

Indian coal resources are confined to two distinct geological periods and basinal set-ups: (i) bulk of Indian coal (>99%) is Permian in age and occurs in intra-cratonic Gondwana basins of peninsular India and a few minor occurrences as thrust sheets (overriding Siwalik sediments) in the foothills of Darjeeling and Arunachal Pradesh Himalaya. These are spread over the states of West Bengal, Jharkhand, Madhya Pradesh, Chattisgarh, Maharastra, Andhra Pradesh, and Orissa (Fig. 6.12). (ii) A very minor part ($\sim$1 Bt) of Indian coal occurs in the Early Tertiary sediments deposited in the near-shore pericratonic basins and shelves of

Fig. 6.12 The distribution of coal fields in India. Gondwana coalfields: 1. Damodar-Koel valley coalfields (*1–15*); 2. Satpura coalfields (*31*, *35*, *36*); 3. Son valley coalfields (*17–26*, *32–34*); 4. Mahanadi valley coalfields (*27–30*, *37–38*); 5. Wardha valley coalfields (*39–42*); 6. Godavari valley coalfields (*43*); 7. Rajmahal—Birbhum belt coalfields (*14*). Tertiary coalfields: Assam-Nagaland-Arunachal Pradesh (*44–49*); Meghalaya (*50–53*); Lignite fields: Jammu & Kashmir (*54*, *58*); Gujarat (*56*); Rajasthan (*57*); Tamil Nadu (*55*)

Jammu & Kashmir, Gujarat, Rajasthan and Tamil Nadu and the north-eastern region of Assam, Arunachal, Meghalaya, Nagaland, respectively.

The Gondwana sediments are represented by thick sequence of glacial, marine, fluvial, and lacustrine facies with characteristic floral–faunal content within a geological time span of about 200 My. The Gondwana sequence has a cumulative stratigraphic thickness of 4–6 km (see Table 6.6). The basins are generally of half graben type and their prominent structural element is the boundary faults. Interestingly, the faults show parallelism with the dominant structural trend of the basement. All these features attest to simultaneous sedimentation and subsidence (Fig. 6.13), leading to the formation of the coal-bearing Gondwana basins. The coal seams are flexed and faulted in different measures, causing at places some correlation problems.

Gondwana coalfields are found in the following linear belts, mostly along major river valleys in the eastern and central parts of the country (Fig. 6.12):

1. Damodar-Koel valley coalfields
2. Satpura coalfields
3. Son valley coalfields
4. Mahanadi valley coalfields
5. Wardha valley coalfields
6. Pranhita-Godavari valley coalfields
7. Rajmahal-Birbhum belt coalfields.

Table 6.6 A version of Gondwana stratigraphy, with the coal-bearing formations shown in bold

Supergroup	Group	Formation	Maximum thickness (m)	Age
Upper Gondwana	Jabalpur	Umia	915	Lower Cretaceous
		Jabalpur	305	Upper Jurassic
	Rajmahal	**Kota**	610	Middle Jurassic
		Rajmahal	30 (Inter-trap)	Lower Jurassic
	Mahadeva	Maleri		Uppermost Triassic
		Panchmarhi	760	Upper Triassic
Unconformity				
	Panchet	Panchet	610	Lower Triassic
Lower Gondwana	Damuda	**Ranigunj**	915	Upper Permian
		Barren Measures	610	Middle Permian
		Barakar	760	Lower Permian
		Karharbari	120	Lower Permian
	Talcher	Talcher Boulder bed	275	Upper Carboniferous
Unconformity				
	Basement rocks			Archean

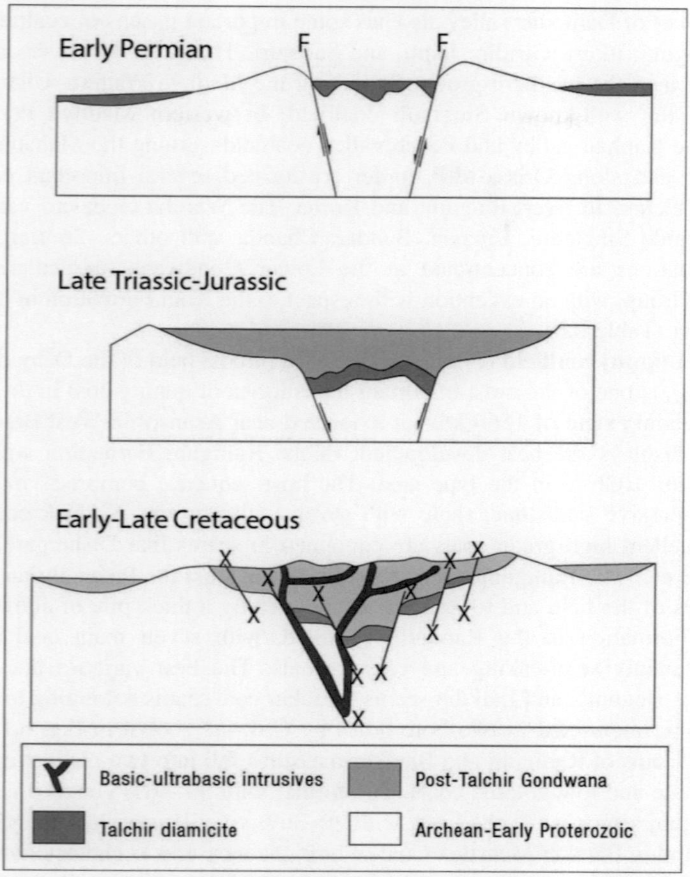

Fig. 6.13 Sedimentation and subsidence through faulting in coal-bearing Gondwana basins (after Casshyap et al. 1993)

The belts are disposed either along E–W or NW–SE in general, except the Rajmahal-Birbhum belt that trend N–S. The main repository of coal-bearing sediments of Early Permian overlying the Talchir, is the Barakar Formation. The Gondwana coal-bearing formation continues sub-surface across the Rajmahal – Garo gap up to the Sungrimari coal basin in Meghalaya. Gondwana coal in sub-economic quantity has developed in several places in the Lesser Himalaya. The thickness of Barakar Formation in parts of Damodar valley coalfield is more than 1000 m while it is less than 500 m in all other basins, reflecting the variation in subsidence histories. The southern part of Damodar valley has some of the most important coal-bearing areas in India which include Hutar, Auranga, North and South Karanpura, Ramgarh, Bokaro, Jharia, and Ranigunj coalfields. These fields have almost all the caking coals and superior quality noncaking coal reserves. The northern part of Damodar valley also has some important groups of coalfields, such as Daltongunj, Itkuri, Giridih, Jainti, and Sahajuri. These are, however, much less important than the southern group. Straddling the Madhya Pradesh–Uttar Pradesh border is the well-known Singrauli coalfield. In western Madhya Pradesh are located the Kanhan valley and Pench valley coalfields. Along the Mahanadi valley in Orissa, and along Orissa-M.P border are located several important coalfields, such as Talcher, Ib river, Raigarh, and Korba. The Wardha-Godavari valley coalfields include Singareni, Lingala, Bandar, Chanda, and others. Stratigraphically, coal formations are concentrated in the Lower Gondwana, particularly in the Damuda Group, with an exception with respect to the Kota Formation in the Upper Gondwana (Table 6.6).

The **Ranigunj coalfield** (Fig. 6.14), the easternmost field of the Damodar valley (Fig. 6.12), is one of the most important repositories of quality coal in the country. With an aerial extent of 1560 km^2, it is located near Asansol in West Bengal. This coalfield displays the best development of the Ranigunj Formation with a total thickness of 1036 m in the type area. The host sequence comprises mainly fine grained, massive sandstone, shale with seven major groups of thick coal seams. Some excellent high grade coals are contained in seams like Dishergarh, Poniati, Sanctoria, etc. The Ranigunj sequence is faulted against the Precambrian rocks on three sides of the field and toward east is covered by a thick pile of alluvium. The Barakar Formation in the Ranigunj coalfield, with seven main coal horizons, contains mainly semi-caking and caking coals. The best varieties are found in Ramnagar, Begunia, and Laikdih seams. Barakar coal seams belonging to Salanpur "A" Group, intersected in two bore holes by GSI, are shown in Fig. 6.14 (inset). The coal seams of Ranigunj and Barakar measures fall into two distinct categories, high volatile and low volatile coals. The former with 30–36% volatiles is found in the Ranigunj seams while the latter with 20–30% volatiles and low moisture of 1–3% is found in Barakar Measure. Coal of both the measures is characterized by low sulfur (<0.1 to 1%) and low phosphorus (0.02–0.25%).

The **Jharia coalfield** is sickle shaped, 19 km long and 38 km wide, covering an area of 450 km^2 around the city of Dhanbad in Jharkhand. The Gondwana sequence here contains a major proportion of Barakar ($\sim$50%) and a smaller proportion of Ranigunj rocks, the former hosting at least 25 coal seams, 1.22 m or above in thickness, and out of which 18 seams (numbered from I to XVIII) are regionally extensive and of low volatile caking quality. Seams from X and above are of exceptionally high quality. Many of these seams are of caking type coal. The best caking coal reserves in India are stored in this field. Ranigunj Measure in this field has 3–4 major seams of low volatile, semi-caking quality. Along its southern margin the coalfield is faulted against the Archean basement. Normal mining operations are often hindered by dislocations in coal seams by extensive faulting. A number of dolerite and mica-peridotite dykes traverse the coalfield destroying some parts of the important coal seams.

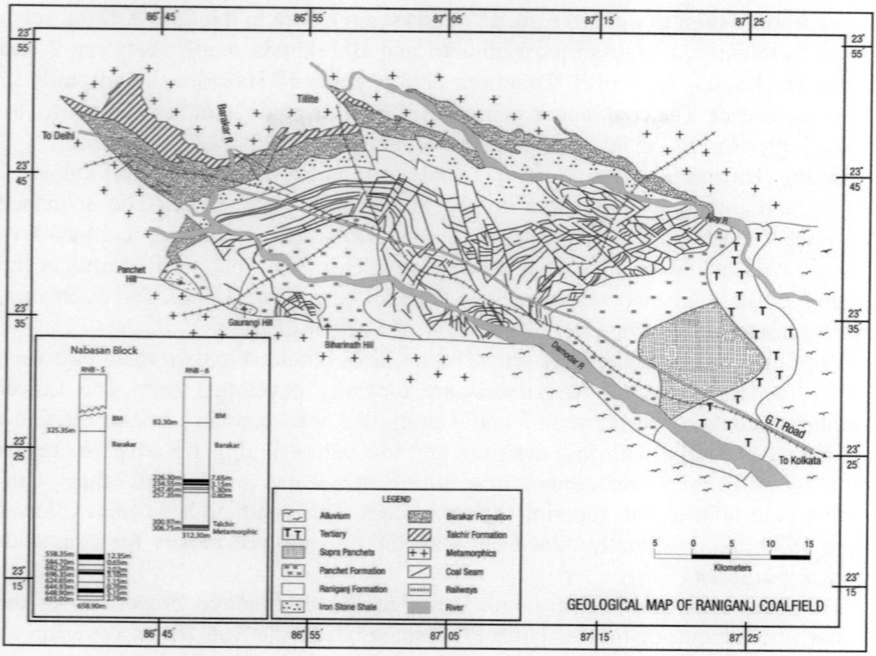

Fig. 6.14 Geological map of Ranigunj coal field and *inset* showing intersections of coal seams (8 and 5) in two bore holes in Nabasan block, varying in thickness from 0.40 to 12.35 m and occurring between 226.30 and 657.70 m depths (after GSI)

After Ranigunj and Jharia, **Bokaro coalfield** comes next in importance amongst the Damodar valley coalfields. It is located just to the west of Jharia coalfield and measures 64 km from east to west. Its width is variable between 6.5 and 18 km. This field, which is in two parts, east (207 km^2) and west (154 km^2) Bokaro, hosts coal mainly in the lower Damudas (Barakar). There are 29 coal seams in the Barakar Formation of East Bokaro. The famous 34–41 m thick Kargali seam occurs in the middle sandstone and shale stage of the lower Damudas. The other important seams are Karo (21–30 m thick) and Bermo (12–14 m in thickness). The coal found in the western part is not as good as in the eastern part of Bokaro coalfield. There are nine major seams within the Barakars in west Bokaro. They mostly belong to Class III with ash content between 15 and 26%. The Ranigunj Formation occurring in Bokaro coalfield contains only a few thin coal seams.

The **Karanpura coalfield** shows the full development of Lower Gondwana sequence. It is divided into north and south fields with a total area of about 1400 km^2. The latter in the south-east is much smaller, where the Barakar Formation assumes an elongated shape. In south Karanpura, which lies in Hazaribagh district, there are 24 coal seams of which eight seams produce a cumulative width of 18.8 m while the other 16 produce 54 m. Some important seams from bottom to top are named Argada seam (33 m in thickness), Sirka seam (13 m in thickness), Bansgarha, Hathidari and Bhurkunda seams (between 2 and 5 m in thickness). The north Karanpura field is partly in Hazaribagh and partly in Palamau district. The coal here is generally of inferior grade with high moisture (8–13%), high ash (14–35%), and is generally noncoking.

In the **Rajmahal area**, Lower Gondwana rocks extend for 160 km in a north-south direction from Siuri to the south of Ganges River. The sequence comprises Talchirs, Barakar measures, and Dubrajpur sandstone (=Mahadeva Group), overlain by basaltic lava flows toward east. The coal seams occur in the Barakars. There are two seams each at Hura, Jilbari, Chuparbhita, and Pachwara, and Brahmani areas. The coal is generally of inferior quality.

The **Giridih coalfield** covers only 18 km^2 area, bounded by two main east-west faults. The Barakars and Karharbaris are typically developed here. The Lower Karharbari coal seams, between 3 and 7 m in thickness, contain some of the finest coking coal in India with low moisture and low ash. The upper Karharbari seams are 1.2–3 m in thickness. Above these with a barren gap are the Hill seams, containing both inferior and superior variety of coal. This coalfield has been exploited since 1870 and is virtually mined out with the coal utilized mostly for nonmetallurgical purposes.

The **Singrauli coalfield** in south Rewa region of Madhya Pradesh is in the border of Mirzapur district of Uttar Pradesh, south of the Son River, covering an area of 2300 km^2. Talchir, Barakar, and Ranigunj measures are developed here, of which Barakar is the thickest (300–360 m). An important seam in this field is the Naunagar seam, 5.5–10.7 m in thickness and extending for a distance of 6 km. It has variable ash (12–36%) and high moisture (about 8%) and is of the noncoking type. Six other seams have been reported from this field out of which the Turra seam contains coal of superior quality. In the Mohar area, a very thick (134 m) coal seam (Jhingurda) occurs. The **Pench valley coalfield** has four main seams with high volatile noncaking type coal. The **Kanhan valley coalfield** also has high volatile caking type coal with high ash content (up to 22%).

The **Singareni coalfield** in Andhra Pradesh includes the Tandur, Yellandu, and Kothagudem regions along the Godavari valley. There are three main seams with coal of noncaking type having low ash content and high moisture content. The **Talchir coalfield** in Orissa has three main coal horizons. The bottom most seam is of better quality. Generally the coals are noncaking type with high moisture content (8–10%).

In all the above-mentioned Gondwana belts, there are obvious similarities of sedimentation, but there are some differences too. The Gondwana succession started with early Permian and ended up with lower Cretaceous. They were in general deposited as fluviatile and fluvio-lacustrine sediments within rift-related intra-cratonic basins. The sequence is best developed in Son-Damodar basins (Fig. 6.12) where at the bottom there are glacial/glaciogenic sediments called Talcher Formation. Coal first appeared in the overlying Karharbari Formation. The environment appears to be near-shore as suggested by marine beds at Umaria and Daltongunj coalfields. Karharbari Formation is overlain by the Barakar Formation which is not only geologically well developed but contains several thick coal horizons. The Ranigunj Formation that also overlies Barakar, is interrupted by another Formation which is characteristically free of coal, called Barren Measures. The Ranigunj Formation is coal-bearing and contains several coal seams. It may be pointed out that commercial grade coal appeared in the Son-Damodar basin earliest in time and disappeared much later. However, the Barakar Formation contains coal in all the belts but not the Ranigunj Formation or its time-equivalents. There are thin and impersistent coal seams at most. Gondwana coal seams have been flexed and faulted in the post-depositional stage. Lamprophyric and doleritic dykes penetrated the coal seams, spoiling the coal at the contact.

A very small part of Indian coal is **Tertiary** in age, occurring in the eastern states of **Assam**, **Arunachal Pradesh**, **Meghalaya**, and Jammu & Kashmir in the west (Fig. 6.12). These coal reserves are located in Upper Assam (east of Dibrugarh), Garo Hill region and also in Khasi-Jaintia hills (north of the Surma valley) and Mikir hills (just north-east of Shillong plateau). Geologicaly, the area is characterized by marine transgression and regression, close to the Meghalaya-Assam crystalline massif. The two stratigraphic units that host workable coal seams in the Tertiary sequence of the region are the Jaintia Formation (915 m thick) of Lower Eocene age and the Barail Formation of Upper Eocene to Oligocene age. In the Jaintia Formation, the lower Tura Sandstone unit contains workable coal seams including two workable seams of good quality lignite. The Kopili unit on the top of the formation has thin coal seams in Khasi and Jaintia hills (Fig. 6.15). The upper unit of the Barail Formation, the Tikak Parbat Unit contains thick workable coal seams enclosed in a sandstone-shale sequence. The Makum coalfield, 28 km long and 5 km wide, is the best developed coalfield in Assam in the southeastern corner of Lakhimpur district near Margherita. This field has five workable coal seams, of which two are of 18 and 6 m thickness. The coals of the Makum coalfield, Assam, show tangible coking properties. This qualifies the coal to be used with prime-coking coals for a mix. These Tertiary coals are low in ash but high in sulfur,

Fig. 6.15 Small-scale mining of Tertiary coal in Khasi hills near Cherapunji (Photo: M. Deb)

varying between 3 and 6%. Sulfur in coal disqualifies it for many industrial uses. The Assam coals, although of low rank, have high caking properties with low ash content (3–4%).

The **coals of Jammu & Kashmir state** (mainly in the Riasi district) occur within Eocene Subathu sediments, unconformably overlying the Proterozoic Jammu Limestone. The coal is highly crushed and has developed a foliated structure. It has developed semi-bituminous to anthracitic character. The best occurrence is at Kalakot where there is a semi-bituminous coal seam, 5.2 m in thickness. Lignite also occurs in the Kashmir valley within loamy, sandy clays and soft sandstones of Karewa Formation of post-Tertiary (Pleistocene) age. The main occurrences are in the Shaliganga and Handwara areas.

The lignite deposit of Neyveli in **South Arcot district of Tamil Nadu** (Fig. 6.12) extends over an area of 330 km^2 hosted by Cuddalore Sandstone of Upper Tertiary (mid-Miocene) age. The lenticular lignite deposit with a maximum thickness of 20 m is underlain by plastic and sandy clays and Cretaceous limestone and marl, overlying Archean basement rocks (Fig. 6.16). The entire sedimentary sequence is under extreme artesian conditions. The lignite appears dark brown to black in color and on the average has moisture about 53%, ash 2.5%, volatile matter 24%, fixed carbon 20%.

In the **Palana area of Bikaner district of Rajasthan**, about 2.6 m thick seams of lignite occur in middle to lower Eocene sequence underneath an overburden of 60 m thickness. The lignite field has NW–SE trend and is of 3.2 km in length and 1 km width.

Lignite also occurs in the Tertiary formations on the west coast, in the **Kutch and Bhavnagar districts of Gujarat** (Fig. 6.17). The seams vary in thickness from 3 to 8 m. The most significant deposit is at Umarsar, 145 km NW of Bhuj.

Lignite deposits of economic significance have been located by GSI in Barmer, Jaisalmer, Bikaner, and Nagaur basins of Rajasthan and in Kachch, Bavnagar, Bharuch, and Surat areas of Gujarat where lignite is present in the form of highly irregular, discontinuous and isolated bodies. Lignite deposits at deeper level (beyond 600 m depth) have been identified in the Sanchor basin of Rajasthan and Cambay basin of Gujarat.

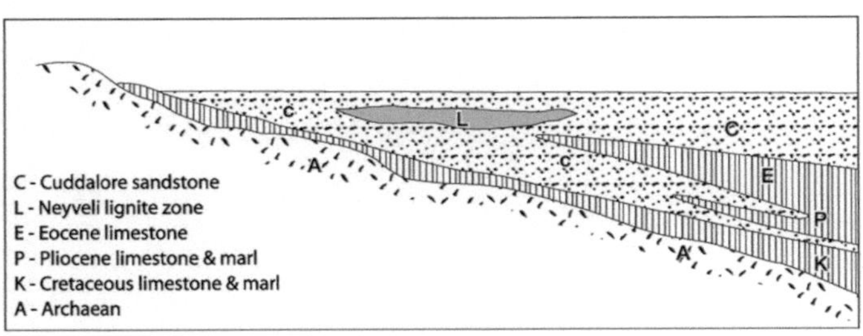

Fig. 6.16 A section through Neyveli lignite deposit geology (after Deb 1980)

Fig. 6.17 Lignite occurrence in Gujarat: **a** Panandhro mine of GMDC; **b** Valia mine of GIPCL (Photo: Courtesy T. Pal)

6.2.12 Reserves of Coal and Lignite in India

There is no denying the fact that coal is a necessary evil today. We need it for many essential uses, but at the same time are concerned that we continue to increase greenhouse gasses in the atmosphere by burning coal. India is a large country and its unsatisfied energy needs are still very large. In this context, we see our position with respect to bituminous coal reserves in India, as follows.

Coal resources in India, although there is nothing to be complacent about, are not too small. We have a recent (1.4.2012) geological estimate of 293.5 billion tons of coal for coal seams of thickness ≥ 0.9 m and down to 1200 m depth, including the mined out reserves. The total lignite resource of the country presently stands at 41.9 billion tons (GSI, PINSA, 2012) (cf. Tables 6.7 and 6.8 and the text thereafter). The principal use of coal in India is in raising steam in thermal power stations and in metallurgy, particularly in iron and steel industry. Indian Gondwana coals are rich in ash and poor in sulfur, except locally. Ash content is reduced by "washing," particularly for the coal used for metallurgy. Coal washing is a specialized

Table 6.7 Bituminous coal reserves (Mt) of India (as on 01.04.2009/GSI)

State	Proved	Indicated	Inferred (exploration)	Total
West Bengal	11652.84	11603.25	5070.70	28326.79
Jharkhand	39479.33	30894.31	6338.32	76711.96
Bihar	0.0	0.0	160.00	160
Madhya Pradesh	8041.18	10294.58	2645.25	20981.01
Chattisgarh	10910.64	29191.79	4380.67	44483.10
Uttar Pradesh	866.05	195.75	0.00	1061.80
Maharashtra	5255.36	2907.21	1992.17	10154.74
Odisha	19943.63	31484.05	13799.18	65226.86
Andhra Pradesh	9193.61	6748.04	2985.27	18926.92
Sikkim	0.00	58.25	42.98	101.23
Assam	348.65	35.85	3.04	387.52
Arunachal	31.23	40.11	12.89	84.31
Meghalaya	89.04	16.51	27.58	116.62
Nagaland	8.76	0.00	8.60	17.36
Total	105820.32	123469.70	37464.11	266754.13

Table 6.8 Coking coal resources in different states (GSI 2009)

Jharkhand	Prime coking	7.31 Bt
	Medium coking	25 Bt
	Semi-coking	0.43 Bt
Madhya Pradesh	Medium coking	2.20 Bt
Chattisgarh	Semi-coking	0.15 Bt
West Bengal	Medium coking Semi-coking	0.23 Bt 0.79 Bt
Total		34.11 Bt

technology though. Presently there exist in India 19 coal washeries (15 in public sector and 4 in private sector) with 33.28 Mt annual capacity. Washing is a very important step in the sustainable use of coal.

Indian coal resources are of two types: (i) Known resource, established through different stages of exploration. Based on character such resource is classified as "coking," "noncoking," and "high sulfur" types. (ii) Prognosticated resource: coal likely to be present mostly at deeper levels beneath the cover of younger rocks in areas identified through geological and geophysical studies, subject to confirmation by exploratory drilling. With continued exploration the resource base is consolidated and upgraded continuously. We present here the most recent available data.

Within a depth of 0–1200 m from the surface, India's coal reserve is as follows (as on 01.04.2011) (2/3rd of these figures should be available within 300 m). This resource figure does not take into account the mined out reserve. However, it needs to be mentioned that all the coal resources are not mineable.

Proved category	114 Bt
Indicated category	137.5 Bt
Inferred category	34.4 Bt
Total	285.9 Bt

Typewise break-up of Gondwana coal reveals that coking and noncoking coal of the country are 33.4 Bt (14%) and 250.9 Bt (86%), respectively. Jharkhand state is practically the lone contributor of coking coal in India. Prime-coking coal of metallurgical specifications occurs in seams (IX–XVIII) of Barakar Formation in Jharia coalfield of Damodar valley basin.

India's reserve of lignite is encouraging. Till 1.4.2010 its reserve stood at:

Proved category	6146 Mt or 6.1 Bt
Indicated category	25,343 Mt or 25.3 Bt
Inferred category	8407 Mt or 8.4 Bt

Till the above date, 636 Mt of lignite has been mined in India. More recently (01.04.2011), the total lignite resource of the country stands augmented to 40.9 Bt, which did not take into account the mined out reserve. Lignite deposits in India contain high moisture (40–55%) and variable ash (<5 to 45%). Based on Gross Calorific value Indian lignite is classified into A, B, and C grades.

Bulk of Indian lignite (31,976 Mt) is located in Tamil Nadu, followed by Rajasthan (4804 Mt), Gujarat (2663 Mt), Pondicherry (416.6 Mt), Jammu & Kashmir (27.6 Mt), Kerala (9.7 Mt), and West Bengal (1.15 Mt). Out of the total of 6.14 Bt of country's "proved" resources of lignite, Neyveli field in Tamil Nadu contains 3.74 Bt. A total of 8.5 Bt of country's lignite resources are confined within 150 m depth from the surface. In Mannargudi lignite field most of the resources occur below 300 m depth. From the reserve figures of coal and lignite that we record here, it is difficult to forecast how long these may satisfy India's needs. No doubt we need to increase our energy production fast if we would like to maintain the pace of industrial development. But as yet we do not know how much of coal and lignite we will need for that. Moreover, a part of the "indicated" and "inferred" category may be upgraded to "proved" category. Apparently there is no cause for panic. We will come back to this at a later section.

Total (coking + noncoking) production of coal in India during 2014 was 462.4 Mt. In India 77% coal output is consumed in the power sector. Coal is mainly imported from Indonesia (49%), Australia (33%) and South Africa (12%). Coke was mainly imported from China (85%). India is the third largest producer of coal after China and USA.

6.3 Oil and Natural Gas

6.3.1 Introduction

Petroleum and its use as a source of energy were reportedly known to some people (Chinese, Burmese) for a long time. But it became an essential and hence commercial commodity when Americans successfully drilled a well close to an oil show in USA in 1859. Inevitably, it was a booster to industrial development. Mineral oil caught up with people's interest throughout the world before long. As a result, an annual production of 20 million tons of petroleum (mtpa) became as large as 3590 Mtpa in 1997. However, this spectacular rise in petroleum production was due not only to the fact that it was an important source of thermal energy but also because it was an important source of many chemicals (petrochemicals) obtained during fractionation (fractional distillation) of petroleum. These are the raw material for many of the downstream industries today.

A modern man cannot now imagine of a life without the transport system consisting of automobile, airplane and ship, gas for cooking and heating, and electricity for almost everything. We are very largely dependent on fossil fuels (coal, natural oil, and gas) for all our energy needs. This makes people also worry about how long the world's known reserves will last.

We have already discussed the position of coal in an earlier section. World's proven reserves of crude oil and natural gas (in 2015) were estimated at 226.9 billion tons and 187.3 trillion cubic meters respectively. The world's crude oil production in 2014 remained stable at 4.2 billion tons and natural gas at 4.3 trillion cubic meters.

India's reserves of crude oil in 2015 was 0.8 billion tons and that of natural gas (oil related) was 4.2 trillion cubic meters. The world picture of the reserve of crude oil and natural gas is not quite encouraging, but that of India is much worse. Production of crude oil and natural gas in India in 2014–2015 was 37.461 million metric tons (Mt) and 33.656 billion cubic meters (BCM), respectively, while the consumption of oil and natural gas in India for the year 2011 was 182 Mt and 61.1 BCM, respectively.

However worrisome is the global resource position of oil and natural gas today, human ingenuity cannot surrender to it without accepting the challenge. What is advisable under the given situation, inter alia, are:

1. Use of improvised exploration methods
2. Increasing efficiency in the increase of oil recovery from the well
3. Conservation through efficient use
4. Adaptation/renovation of the refining process
5. Replacement by biofuels/mixing with biofuels
6. Going for alternative cost-effective energy sources.

6.3.2 Chemical Composition of Natural Hydrocarbons

Oil and gas are referred to as natural hydrocarbons that grade from gasses to liquid to semi-solids and solids. The gas includes the dry gas (methane) and the wet gasses (ethane, propane, butane, etc.), classified thus according to the amount of liquid vapor they contain. Liquid hydrocarbons are referred to as crude oil, thus distinguishing it from the refined products. Semisolid to plastic hydrocarbon includes asphalt while solid includes coal and kerogen. A relatively new energy resource of the future is the gas hydrate which comprises frozen water (ice crystals) with the atomic lattice containing molecules of methane and other gasses. The ice and gas molecules are also known chemically as clathrates. These are stable under high pressure and low temperatures.

The major constituents of the natural gas are the hydrocarbons of the paraffin series in which the heavier members decline in abundance with increasing molecular weight (Selley 1998). Methane (CH_4) also known as marsh gas is the most abundant followed by ethane (C_2H_6), butane (C_4H_{10}), and propane (C_3H_8). Paraffins with molecular weight more than pentane are least common. Methane can come from three known sources: (i) from mantle degassing, (ii) from thermal maturation of buried organic matter (known as thermogenic methane), and (iii) bacterial degradation of organic matter at shallow depths (known as bacteriogenic methane). All the other gasses mentioned above form by thermogenic processes only.

Crude oil is a mixture of hydrocarbons that remain liquid at atmospheric pressures. They are lighter than water and have widely varying viscosities. Chemically crude oil is mainly made up of carbon ($\sim 85\%$) and hydrogen ($\sim 13\%$) with much less content of heterocompounds, sulfur, oxygen, and nitrogen along with traces of vanadium, nickel, and other elements. Although the chemical composition is relatively simple, crude oil can have a very large number of molecular compounds whose number and proportions give a chemical signature to particular oil. These molecular compounds may be grouped under alkanes (C_nH_{2n+2}), naphthanes or cyclo-alkanes (C_nH_{2n}), and aromatic compounds which include benzene (C_6H_6). Geologists classify the crude oils according to the ratio of paraffins, naphthenes, and aromatics which give indications of their source and geologic history.

6.3.3 Surface Indications of Oil and Gas

Surface indications are of great significance in the discovery of prospective oil fields. Most of the producing fields are marked by pronounced seepages of oil or methane gas on the surface. Where oil oozes out on the ground and collects in a pool, it is recognized easily by its smell, appearance, and inflammability. Usually oil from seepages floats on stagnant water in a brownish or greenish film. Gas seeps on the other hand are easier to detect. If they occur under water, bubbles will be produced on the surface of the pool. Some large gas seepages may also burn spontaneously. It is necessary to ascertain if the gas seeps are superficial, originating from the decay of organic matter near the surface, or whether it is derived from deep-seated reservoir rocks, leaking through faults, joints, and fractures. Gas seeps are promising surface indications if they are issued out from bed rocks. When the seeps contain heavy hydrocarbons and burn in flames, they clearly indicate presence of oil in the area.

6.3.4 Origin, Migration and Accumulation of Oil and Natural Gas

Oil and natural gas form from phytoplanktons deposited on the water–sediment interface in marine environments and subjected to (bio-) geochemical and geological processes. This whole process of conversion of phytoplanktons to oil and natural gas is divided into three stages: diagenesis, catagenesis, and metagenesis (metamorphism) (Tissot 1977).

Diagenesis: Takes place below the surface of deposition of sediments down to a few hundred meters characterized by temperatures ≤ 50 °C. Much of the organic matter is oxidized or is consumed by burrowing organisms or bacteria. The upper part of this zone is characterized by intense activity of anaerobic methanogenic bacteria and production of biogenic gas. Kerogen is a common organo-chemical product, disseminated solid hydrocarbon formed from buried organic matter. It is the fossilized insoluble (in normal petroleum solvents) organic material found in sedimentary rocks that convert to oil, natural gas, and coal under suitable conditions. Chemically, kerogen comprises carbon, hydrogen, and oxygen with minor amounts of nitrogen and sulfur. Kerogen is classified into three types depending on their source materials and proneness to oil and/or gas production: Type I (algal), Type II (liptinitic from phyto- or zooplankton), and Type III (humic from land plants). While types I and II produce oil, type III tends to generate mostly gas. Oil generation takes place between 60 and 120 °C while gas requires 120–225 °C to be produced (Selley 1998).

Catagenesis: Characteristic of the depth-zone of 2–5 km in the sediment where the ambient temperature ranges between 50 and 150 °C and the pressure does not exceed 1500 bar. Kerogen that starts developing in the diagenetic stage largely converts into oil and then wet gas (mainly methane with a little of ethane, propane, and butane).

Metagenesis: Oil is no more stable. Instead it is all thermogenic gas.

In Fig. 6.18, the role of pressure-temperature in the formation of oil and natural gas has been depicted. But time and the composition of the original organic matter do also play their role. For example, at higher temperature it needs less time to complete the process. Marine sediments with lipid-rich organic matter deposited at a given site will yield oil and wet gas. Cellulose-rich terrestrial organic matter under

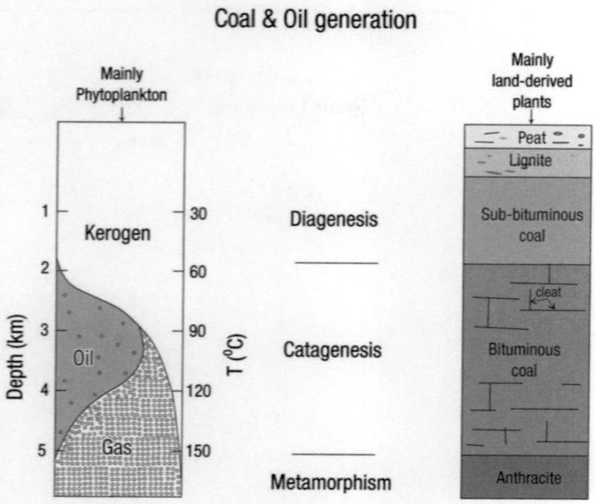

Fig. 6.18 Diagrammatic presentation of coal and natural oil (-gas) generation in terms of pressure and temperature in the crust (after Hunt 1979)

the same conditions will produce coal and dry gas (i.e., nearly pure thermogenic methane (see coal bed methane).

Why do we have fossil fuels confined to the Phanerozoic rocks? The principal answer should be the prolific growth of phytogenous material in the Phanerozoic that worked as a raw material bank for the fossil fuels. The advanced forms of plants with cellulose-rich tissues did not become prolific until the Devonian period. It is not impossible, however, that some oil and natural gas developed in the Neoproterozoic, such as the oil in Nonesuch Shale horizon in the mid-continental rift region of north-central USA, or in the early Phanerozoic, but were lost due to geologic reasons, such as tectonism, weathering, erosion, etc.

It is interesting to note that actually oil and natural gas did not form where they are found today. They formed at infinite number of points as minute entities within the primary host rock. Then they migrated through porous and permeable rocks due to tectonic and lithostatic pressure and accumulated in low pressure areas. The rocks in these three stages are, respectively, called source rocks, reservoir rocks (sandstones, porous carbonate rocks) and traps. The traps may be structural or stratigraphic (Fig. 6.19). Oil and gas are trapped in the highest point (structural culmination, stratigraphic pinch out, etc.) of a permeable rock unit suggesting that they are free to migrate vertically and laterally. When all three of the fluids are present in the system, they occur stratified due to differential density, with water at the bottom and gas at the top. Migration of oil and gas is further distinguished into primary and secondary migration. In primary migration the hydrocarbons are seen to migrate from primary source rocks, such as clays and shales into the permeable reservoir beds, such as sands and limestone. In secondary migration the hydrocarbons move within the reservoir. It happens due to buoyancy that is differential densities of the fluids and in response to differential pressures. It is generally conceived that periodic flushing of water from overpressured shale by hydraulic

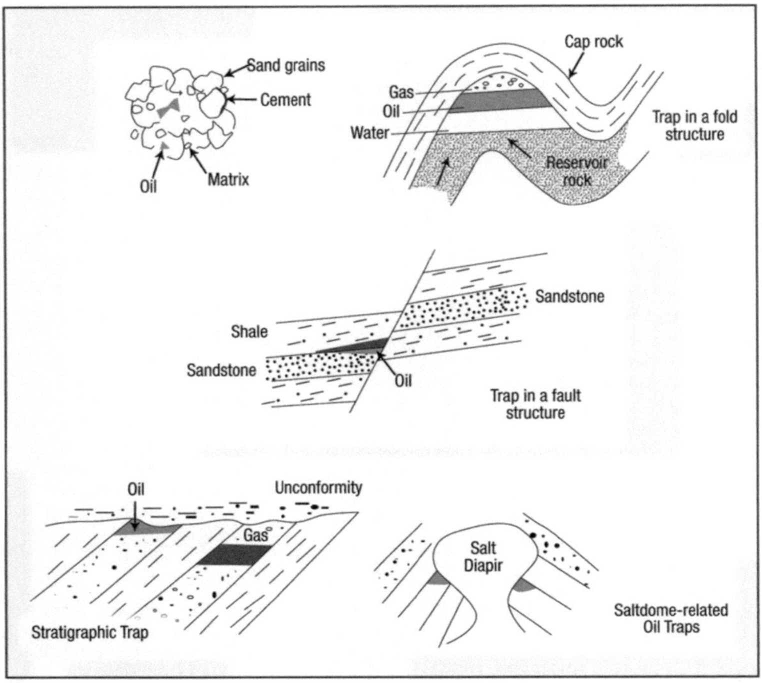

Fig. 6.19 A sketch of different types of structural and stratigraphic traps for oil and natural gas

fracturing is an important process that allows petroleum to migrate from the source rocks. It is generally believed that oil migrates from the source bed not in any solution but as a discrete oil phase associated with pore waters. Oil can migrate to great distances. A 1000 km migration has been recorded in Alberta, Canada. Today oil and gas are recovered through drilled wells, some several kilometers deep.

6.3.5 Oil and Natural Gas Fields of India

In India, hydrocarbon accumulations are found in sedimentary basins of varying ages throughout the Phanerozoic and having an aerial extent of about 3.14 million square km on land and sea including deep waters. However, most of the commercial accumulations are within Cenozoic successions both on land and offshore. The Mesozoic comprising a shallow marine Jurassic succession in Kachchh and Rajasthan and poorly developed shallow shelf facies of the Cretaceous in most parts of the eastern coast are characterized by relatively lesser hydrocarbon accumulations. The Gondwana succession in different parts of the country contains rich coal deposits where exploration for coal bed methane is underway in a big way.

A total of 26 sedimentary basins are identified which are categorized according to their hydrocarbon potential, as under (Fig. 6.20) (Bharadwaj and Saini 2008):

Category I: Proven commercial productivity (Mumbai offshore (Fig. 6.21), Cambay and Rajasthan basins; Assam shelf and Assam-Arakan fold belt; Cauvery, Krishna-Godavari basins).

Category II: Identified prospectively (Kachchh, Andaman-Nicobar, and Mahanadi basins).

Category III: Prospective basins (Himalayan foreland, Ganga Vindhyan, Konkan-Kerala-Lakswadeep (Fig. 6.20) and Bengal basins).

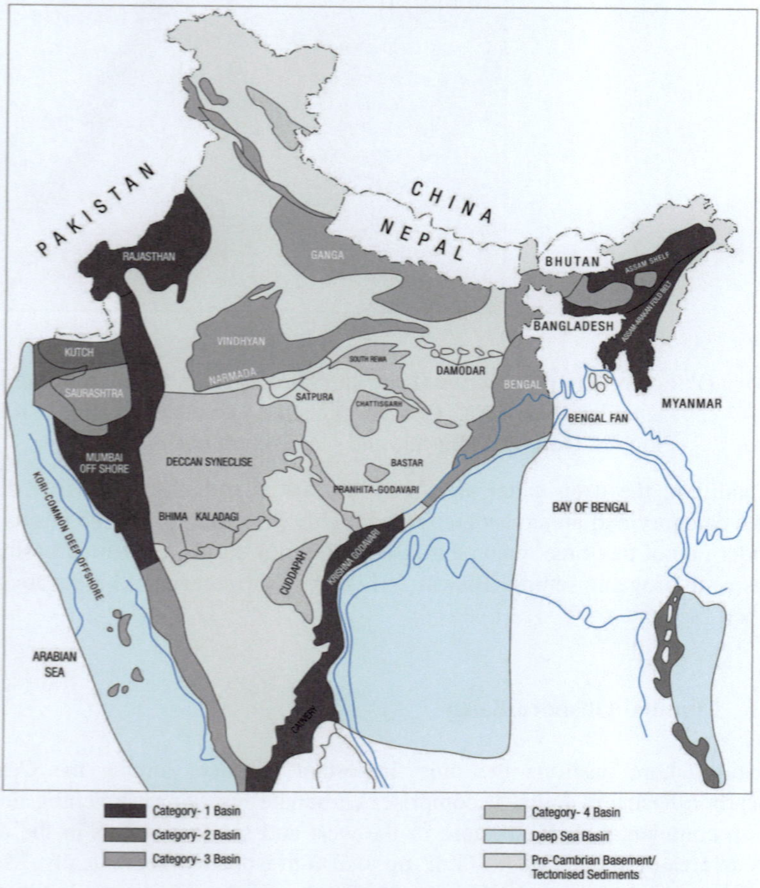

Fig. 6.20 Categorization of sedimentary basins in India based on their hydrocarbon potential (after Biswas 2012)

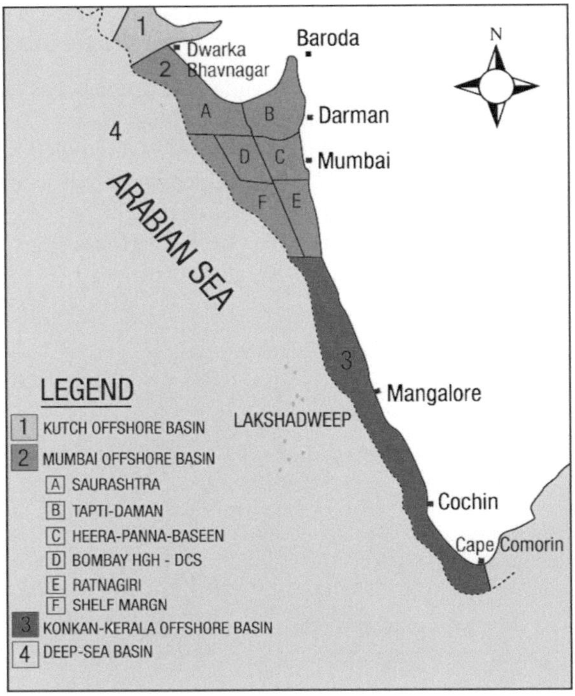

Fig. 6.21 Different offshore basins with hydrocarbon potential in the west coast of India (after Bhowmick 2008)

Category IV: Potentially prospective (Karewa, Spiti-Zanskar, Satpura, South Rewa, Damodar, Narmada, Deccan syneclise, Cuddapah, Bhima-Kaladgi, Pranhita-Godavari, Bastar, and Chattisgarh basins).

In addition, the deep water areas beyond 200 m isobaths up to EEZ form a separate category and are considered to be highly prospective. All the basins listed fall under one of the three major settings: (1) Interior basins, (2) Rifted basins, and (3) Basins in orogenic setting (Biswas 2012). Only the category I basins are briefly described below.

6.3.5.1 Mumbai Offshore Basin

Mumbai offshore basin is the most important producer among the Cenozoic hydrocarbon basins in India. It comprises carbonate reservoirs stretching between the deep continental shelf structure in the west and shallow waters in the east. It covers an area of about 1,48,000 km^2 up to 200 m isobaths. Tectonically, Mumbai basin is a continuation of the Cambay rift to the south with a lateral shift and is dissected by ENE–WSW grabens. The Mumbai High arch divides this basin into two sub-basins: Surat depression in the north and the Ratnagiri basin in the south (Fig. 6.22). Based on the tectonic evolution of the basin, two sequences have been

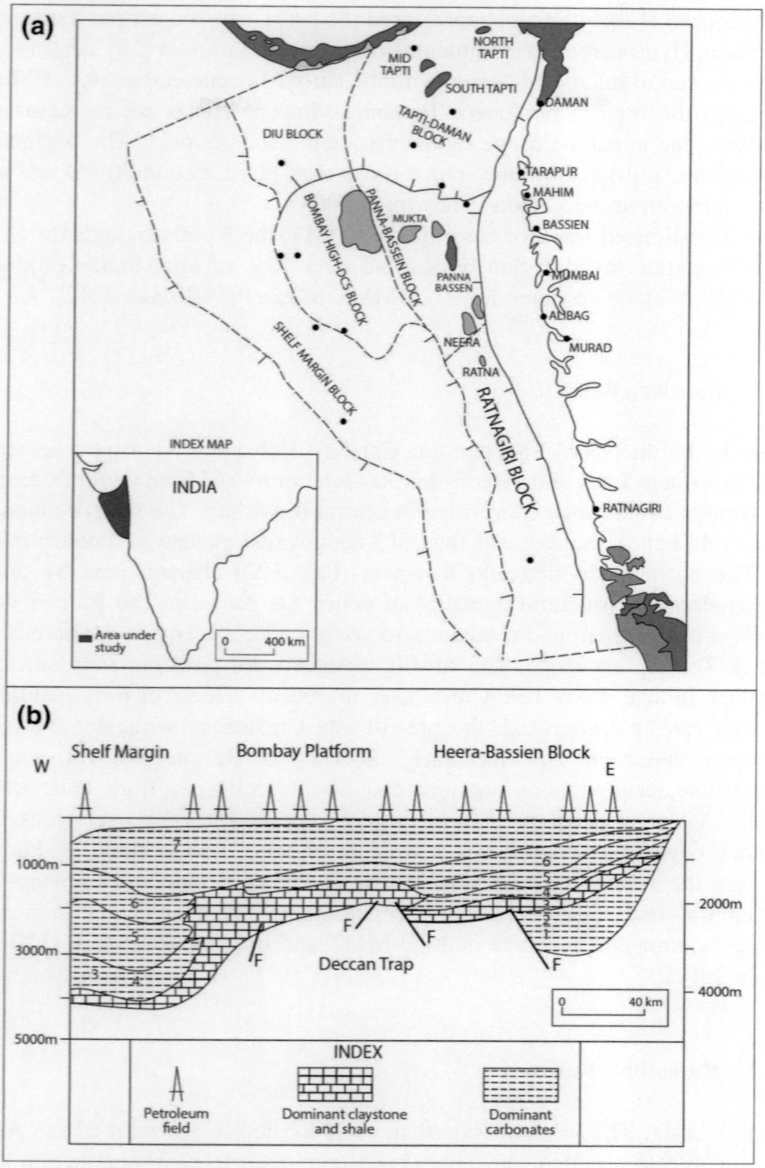

Fig. 6.22 **a** Disposition of the various sub-basins in Mumbai offshore; **b** E–W section across Bombay platform (after Bhowmick and Misra 2008)

identified in the Cenozoic succession: The lower Paleocene sediment is syn-rift while the upper post-rift sedimentary succession is late Paleocene to recent. The lower succession is dominantly fluvial inter-tidal to shallow inner shelf clastics.

Late Paleocene to early Eocene represented the first Cenozoic marine transgression in the basin. Hydrocarbon accumulations occurred here in reservoirs ranging in age from Paleocene to middle Miocene, with the early Miocene carbonates of Mumbai High being the biggest producers. Post-mid-Miocene shales act as regional cap rock. Paleocene to early Eocene sediments are the source rocks. The beginning of hydrocarbon expulsion from the source rocks took place around 30 Ma ago with a critical migration around 5 Ma (Bhowmick 2008).

With an estimated resource base of 9190 MMT, this basin accounts for 1/3rd of the total resource of all Indian basins and over 50% of total Indian production. Mumbai High alone accounts for 2122 MMT of reserve (Biswas 2012).

6.3.5.2 Cambay Basin

Cambay basin: the NNW–SSE trending Cambay rift basin is a conspicuous tectonic feature in western India and extends for 500 km northward from Gulf of Cambay in south Gujarat to Jaisalmer-Mari ridge in central Rajasthan. The rift is contained by Aravalli fold belt in the east and Deccan Trap covered plateau of Saurashtra in the west. The basin is divided into 8 sectors (Fig. 6.23) characterized by different sedimentation and structural domains, of which the Narmada and Barmer depressions form the southern and northernmost sectors. Deccan Trap is the basement for the thick Tertiary sequence. The rift-fill sediments are predominantly siliciclastic and range in age from late Cretaceous to recent. Three distinct sedimentary sequences can be recognized: the pre-rift late Cretaceous sequence, the syn-rift Palaeocene sequence and the early Eocene to Recent post-rift sequence. Hydrocarbon accumulations are recorded in all sequences from Palaeocene to Miocene. Major accumulations are located in middle Miocene over block uplifts and block edge folds. Two major petroleum systems are identified in this basin. These are the Cambay-Kadi/Kalol system in the north and the Cambay-Hazad system in the south. Early to middle Eocene Cambay Shale is the main source rock. The latest estimate of resource is 2050 MMT and in place reserve is 1140 MMT (Biswas 2012).

6.3.5.3 Rajasthan Basin

Rajasthan basin: The western Rajasthan shelf located to the west of the Aravalli range underneath a part of the Thar desert comprises three important sub-basins: Jaisalmer, Bikaner-Nagaur, and Barmer-Sanchor, covering an area of about 1,20,000 km^2. This is the only basin in India where hydrocarbon potential has been reported from the Mesozoic and Infra-cambrian sediments (Bhowmick and Misra 2008). Of these three sub-basins, it is the Barmer-Sanchor basin, a northward extension of the Cambay rift, which has commercial hydrocarbon reserves presently being exploited by Cairn.

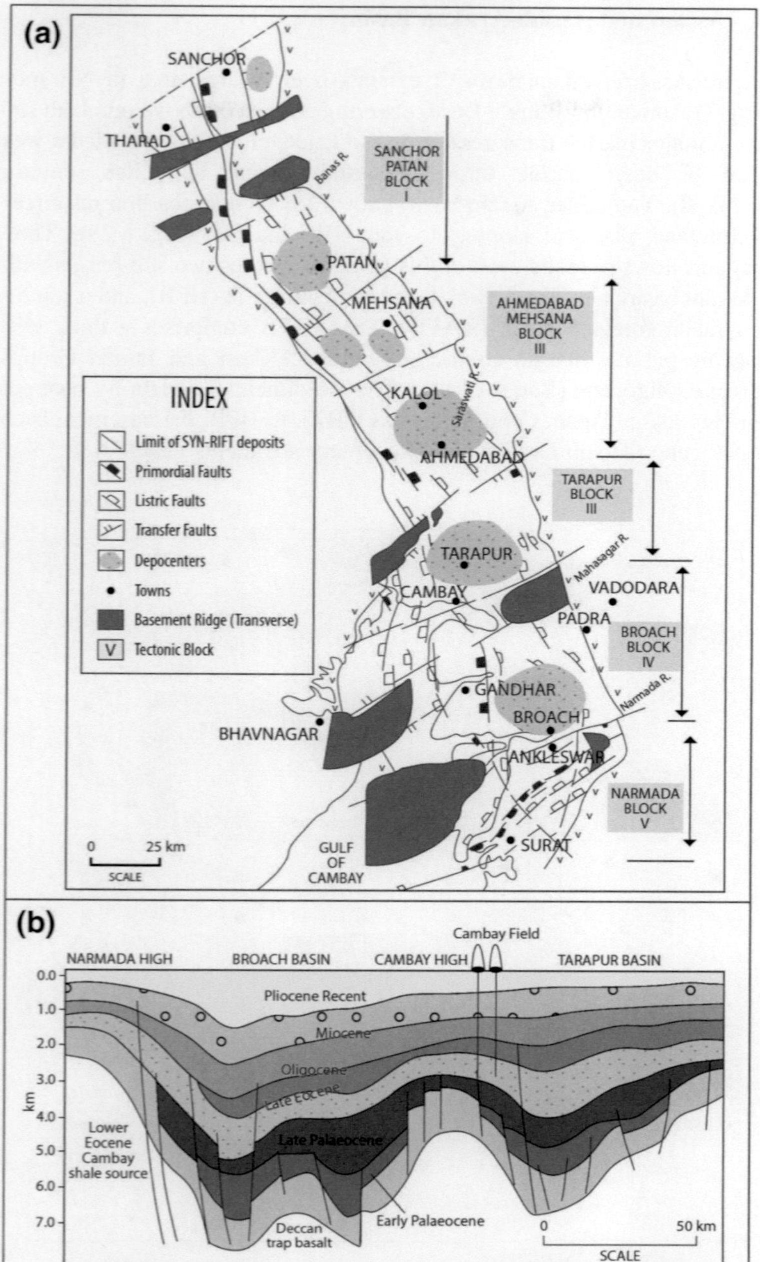

Fig. 6.23 Map and section of Cambay rift basin (after Bhowmick and Misra 2008)

6.3.5.4 Assam and Assam-Arakan Basin

Assam and Assam-Arakan basin: it extends over a large area in NE India and adjoining Myanmar and Bangla Desh, covering an area of above one lakh sq km. It had seen a major marine transgression in the Paleogene which entailed a westward migration of deep marine through coastal marine to deltaic environment (Fig. 6.24a–d). The upper Assam basin known for its hydrocarbon resources is an arched foreland platform sloping to the NW and SE (Fig. 6.25). The River Brahmaputra flows over the crest of this arch and creates two sub-basins within the upper Assam basin, a northern Himalayan foreland basin (HFB), and a southeastern Assam-Arakan foreland basin (AAFB). The AAFB comprises a thick wedge of pre-orogenic passive margin Cretaceous-Eocene (Khasi and Jaintia Groups) and deep marine Oligocene (Barail Group) flysch sediments overlain by post-orogenic Neogene mollasse (Tipam Group) (Biswas 2012). In HFB, the basement is covered by the Neogene (Tipam Group) and Quaternary sediments (Table 6.9).

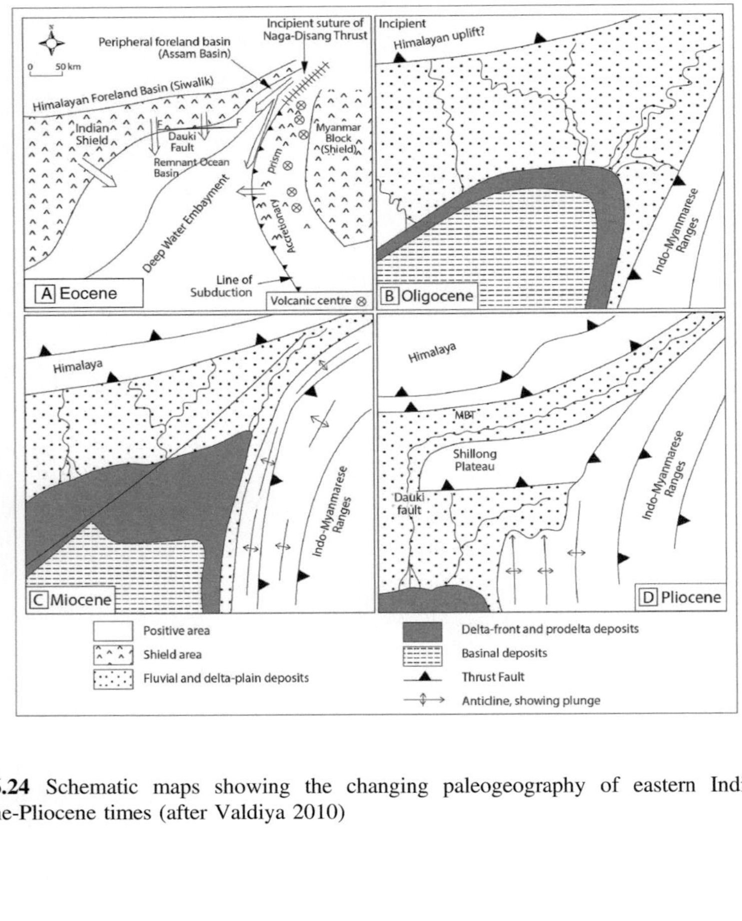

Fig. 6.24 Schematic maps showing the changing paleogeography of eastern India during Eocene-Pliocene times (after Valdiya 2010)

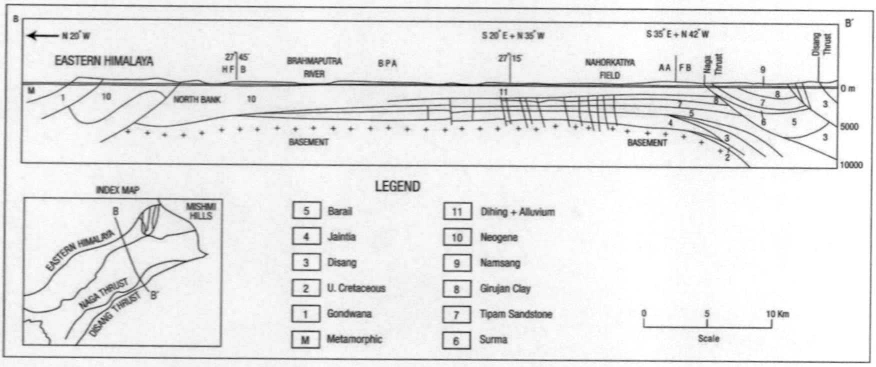

Fig. 6.25 A section across upper Assam basin showing the location of Naharkatia oil field (after Dasgupta and Biswas 2000)

Table 6.9 Petroleum and natural gas-bearing broad tertiary stratigraphic sequence of the Assam and Assam-Arakan basins (after Dasgupta and Biswas 2000)

Mid-Pleistocene unconformity	Multiple uplifted terraces
Upper Pliocene unconformity	Upper Dupitilas, Kimins, Dihings
Mio-Pliocene	Subansiris, Lower Dupitilas, Namsangs
Oligo-Miocene	Dalfas, Surmas, Tipams, Jirujans
Early Cretaceous to Late Eocene	Barails Local unconformities at multiple levels

All the oil/gas fields are confined to the southeastern foreland close to the Naga thrust (Fig. 6.26). Sandstones of the Oligocene Barail and Miocene Tipam Formation are the major reservoirs while geochemical studies indicate that Barail coal-shale sequence is the main source rocks. The Miocene Girujan Clay is the regional cap rock for the Tipam reservoir. Oil accumulations are also known in the Eocene sequence of Sylhet and Kopili Formations. In most cases, the accumulations are in structural traps and also in strato-structural combination traps (Biswas 2012) (Fig. 6.26 inset). Digboi oil field is the oldest with the first historic discovery here in 1889. Major output comes from Naharkatya, Hugrijan and Moran fields. Sibsagar, Rudrasagar, and Lakwa are other important fields.

The total in place geological reserve of oil and gas in this basin are 1000 MMT and 221 BCM, respectively.

6.3.5.5 Tripura-Mizoram Fore-Arc Basin

Tripura-Mizoram Fore-arc basin: it is also a category I basin. It is an accretionary prism, being an western extension of Assam-Arakan-Andaman fore-arc trench. Mio-Pliocene Surma and Tipam Groups are extensively exposed in this basin with N–S trending narrow ridges which represent steep anticlines. The thick Bhuban

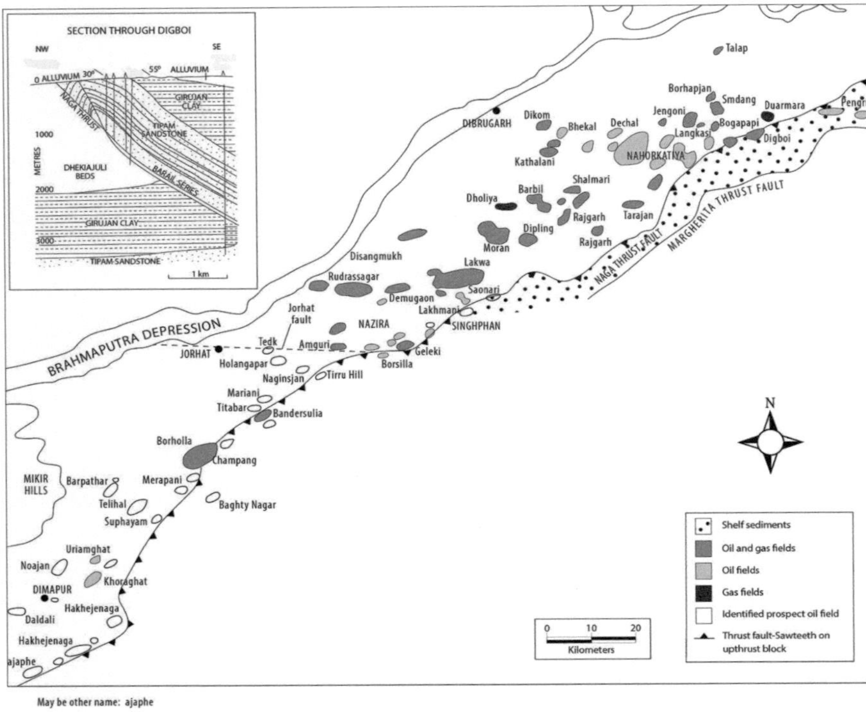

Fig. 6.26 Various oil and gas fields in between the Brahmaputra depression and the Naga thrust, Assam (after Bhowmick and Misra 2008). *Inset* shows Digboi structural trap (after Valdiya 2010)

Shales of Lower Miocene age in the Surma Group are considered to be the source beds for gas and the upper Bhuban Sandstones in Bokabil Formation are considered to be the reservoirs capped by interbedded shales. Presently gas is being produced from Agartala dome, Baramura, and Rokhia anticlines at the rate of 1.5 Mm^3/day.

6.3.5.6 Cauvery Basin

Cauvery Basin is defined by NE–SW horst and grabens in the southern part of the Eastern Pericratonic rift basin at an angle to the N–S trending coastline and shelf. Extensive sandstone deposition took place in all the grabens/half-grabens during the syn-rift phase of Cauvery basin evolution. The common hydrocarbon traps are structural, stratigraphic and combination which drape over the basement highs. Hydrocarbon potential of this basin is not as high as the K-G basin. Gulf of Mannar, a pull-apart basin between India and Srilanka appears prospective. The estimated in place reserve is 92.29 MMT. The latest deep water discovery and in Mannar sub-basin may raise the reserve base (Biswas 2012).

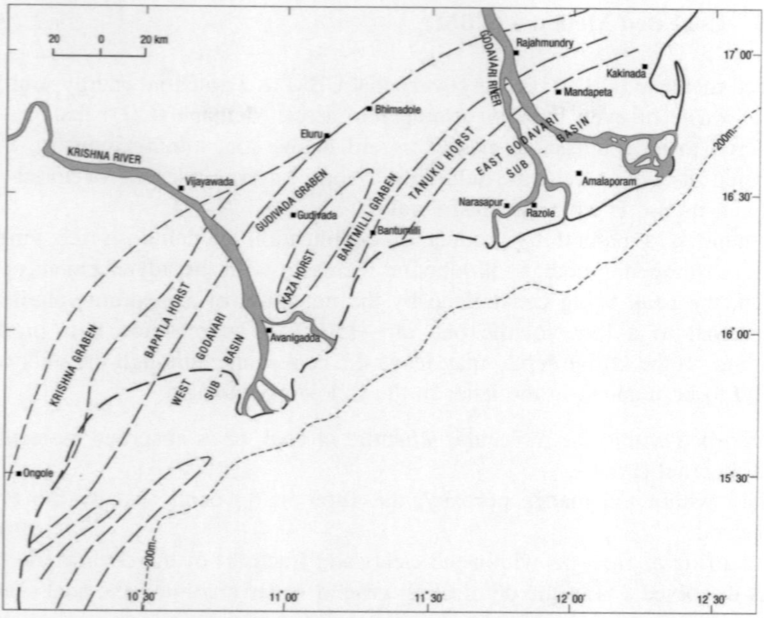

Fig. 6.27 Depiction of the horst-graben structure in the hydrocarbon-bearing Krishna-Godavari basin on the east coast (after Biswas 2012)

6.3.5.7 Krishna-Godavari (K-G) Basin

Krishna-Godavari (K–G) basin is located in the central part of the eastern passive margin of India and is defined by NE–SW trending horst-graben structures sub-parallel to the coast (Fig. 6.27). Lower Gondwana rocks are present below the Cretaceous-Tertiary sediment-fill of this coastal rift basin. This basin has potential reservoirs ranging in age from Permian to Pliocene. 42 oil and gas-bearing traps are identified in this field. Of these the Palaeogene and the Neogene systems are important producers. The prognosticated resource of the on land and shallow offshore (up to 200 m isobaths) is 1130 MMT. The gas reserve in this basin is expected to be 15 TCM (Biswas 2012).

6.3.6 Nonconventional Hydrocarbon Resources

There are few other natural fuels or fuel sources that are potential supplements to conventional fossil fuels.

6.3.6.1 Coal Bed Methane (CBM)

Coal bed methane (CBM): the discovery that CBM as a potential energy source is a story of waste or even liability turning into asset. Methane (CH_4) that was long considered to be a nuisance, even a hazard in the coal mining industry, is now considered to be a welcome addition to our known reserves of fossil fuels. Moreover, its use is environment-friendly.

Methane is a natural by-product of coalification of cellulose-rich terrestrial plants. Thermogenic methane production increases with the advancement of coalification, the peak being constrained by the transition of a medium-volatile bituminous coal to a low volatile one at ~ 150 °C. The methane thus produced, depending on the burial depth, may leave the coal seam, although the bulk of it is believed to be retained in the latter in the following manner:

(a) Absorbed within the molecular structure of coal, or as absorbed molecules on the internal surface.
(b) Held within the matrix porosity, the latter being controlled by the type of maceral
(c) Occurring as free gas within the cleats and fractures of the coal seam.
(d) As dissolved gas in the permeating ground water involving the coal seam.

Coal is thus both the source and reservoir rock for CBM. CBM production is done by simple depressurization and dewatering process. A few countries like USA, Australia, and China are already commercially exploiting CBM. The production of CBM started in India in 2007 through the efforts of Great Eastern Energy Corporation Ltd. with its base near Asansol, West Bengal. The total coal-bearing area in India is 35,400 km^2 consisting of 44 major coal and lignite fields distributed in 12 states which gives a prognosticated CBM resource of 162 TCF (4.6 TCM). According to some estimates CBM could be able to contribute up to 15% of India's natural gas production. The Government has so far awarded 26 coal bed methane blocks for exploration. These are spread over 13,600 km^2 and the total methane resource in these blocks are estimated at 1374 bcm which translates into a production potential of 38 mscmd (Singh and Saxena 2008).

6.3.6.2 Oil Shales (OS)

Oil shales are hydrocarbon-bearing "mud rocks" or shales that yield oil on heating. Oil shales form by accumulation of microorganisms from the overlying water mass on to the anaerobic bottom mud and protected from oxidation by younger sediments. Some oil shales are rich in hydrocarbons (yielding up to 240 L of oil per ton of rock. Although some countries (Russia, China, Estonia) produced oil from oil shales commercially, there are some inherent problems of production: (a) non-availability of cheap source of heating; (b) disposal of a huge volume of rock waste;

(c) nonavailability of a large volume of water; (d) other environmental issues. Oil shale reserves are generally believed to be high in the world.

In India, oil shale deposits in Assam and Arunachal Pradesh are under study.

6.3.6.3 Tar Sands and Heavy Oils

Tar sands are sandstones in which intergranular space is occupied by bitumen or asphalt. Since both the bitumen and asphalt are heavier than the normal variety of oil, these are sometimes referred to as heavy oils.

Large resources of this variety of hydrocarbon occur in nature. Reportedly, Venezuela alone has 1.2 trillion barrels. The global total may not be less than the conventional crude oil reserve. In Venezuela, oil from the tar sands is extracted by emulsification with water. Hydrogenation could be an alternative technique, provided sufficient hydrogen is available at an affordable cost.

In India, heavy oil is reported to occur in the Bikaner basin and low wax heavy crude occurs in Assam.

6.3.6.4 Gas Hydrates

Gas hydrates are crystalline ice-like material composed of water and gas (predominantly methane) molecules, stable at low temperatures (10–20 °C) and high pressures (8–30 MPa). Gas hydrates are essentially water clathrates of natural gas molecules, held captive below water at depths of ≥ 800 m, mostly in the continental rise and partly in the continental slope in the permafrost and outer continental margins of the world (Fig. 6.28). Gas hydrates are potential source of hydrocarbons in future, with one unit of gas hydrate broken down to 164 units of gas and 0.8 units of water at STP. The potential methane availability from gas hydrates is estimated to be 2×10^4 trillion cubic meters (TCM). An estimated natural gas reserve of 300 TCM, appears insignificant compared to the above. Further the total carbon locked in the gas hydrates is twice the total carbon locked in all types of conventional fossil fuels (coal, oil and natural gas) taken together. Gas hydrates occur in three principal regions: polar, shallow arctic, and sub-arctic offshore and deep oceanic regions of the tropics (Fig. 6.29).

There are a number of potential offshore areas of gas hydrates accumulation within India's exclusive economic zone. A national gas hydrate programme (NGHP) exists and several R & D projects are also there to develop vast resources of gas hydrates in western, eastern and Andaman offshore areas. NGHP has shortlisted Mahanadi, Krishna-Godavari and Kerala-Konkan basins, and Andaman Sea for drilling/coring of gas hydrates in deep waters. Approximately, 1900 trillion cubic meter of methane has been prognosticated within the vast exclusive economic zone (EEZ) of India. This volume of gas is more than 1500 times of India's present natural gas reserve (Sain et al. 2012).

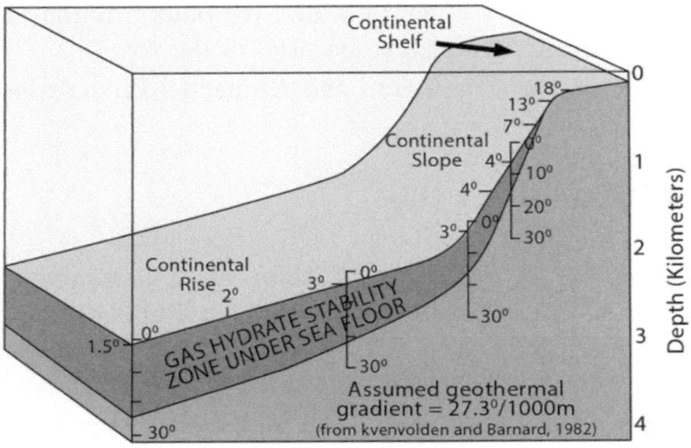

Fig. 6.28 Stability thickness zone of gas hydrates in the outer continental margin (after Sain and Gupta 2008)

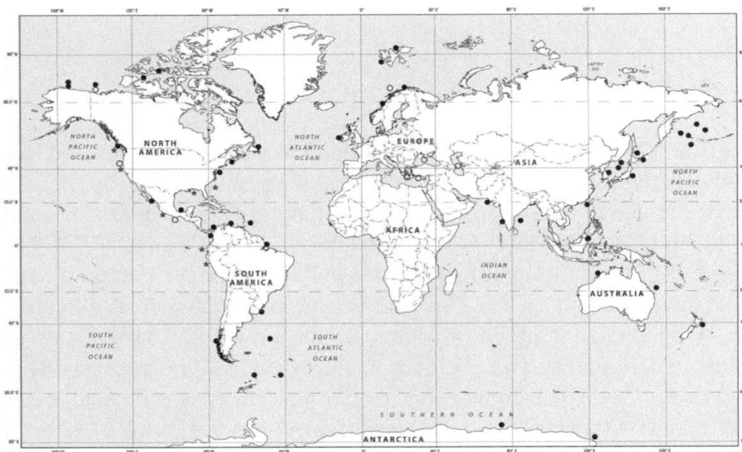

Fig. 6.29 Global occurrence of gas hydrates in the permafrost and outer continental margins of the world. *Open circles* location of recovered naturally occurring gas hydrate samples; *closed circles* inferred location of gas hydrate from seismic reflection; *stars* location of gas hydrates in DSDP and other programs (after Sain and Gupta 2008)

India is the third country after USA and Japan to have taken up seriously R & D studies on the development of gas hydrate resources.

6.3.6.5 Hydrogen

Hydrogen is fast gaining world attention as a "clean" fuel and at the same time energetically efficient as an automobile fuel. It is believed to be capable of replacing or supplementing oil used in transport vehicles. The source material can be a fossil fuel or a renewable resource. As of today, hydrogen is costlier than the conventional fuel. The world is eagerly waiting for the discovery of a cheap and safe method of manufacturing hydrogen as a transportation fuel.

However, scientists of the Rutherford-Appleton Laboratories, Oxford Cellar Energy, England have developed micro beads of densely packed hydrogen that can be used in a gasoline or petrol vehicle without engine modification. A tankfull of this artificial fuel is expected to last 300–400 miles.

6.3.6.6 Biodiesel

It is chemically processed vegetable oil (or animal fat) which can be mixed with conventional diesel and used in transports. It is commonly derived from the seeds of jatropa, mohua, karanja, kusum, dhupa, undi, sal, pilu, simarouba, jojoba, tumba, tung, neem, mango, carnel, nahor, and kokum. Biodiesel–diesel blend is environmentally superior as it is nontoxic and biodegradable. Besides causing less pollution, it increases the life of the engine.

6.3.6.7 Ethanol

Ministry of Petroleum and Natural Gas allowed 5% blending of ethanol with petroleum from 1st November, 2006. But the problem was that India does not produce sufficient ethanol for the purpose. Some countries like USA and Brazil mix it with petroleum in greater proportions.

6.4 Nuclear Energy

6.4.1 Introduction

What is the difference in science between burning of fossil fuels such as coal and petroleum that emit heat and the generation of heat due to the breakdown of a large nucleus, such as uranium into smaller nuclei, e.g., barium and krypton? In the first the process involved is a chemical reaction with the transfer or sharing of electrons so that the nuclei remain unaffected and the elements retain their identity. In the second case nuclear energy is produced because of changes in the bonding that holds the nucleus together. It creates an enormous amount of energy, perceivable by Einstein's (1905) landmark postulation:

$$E = MC^2$$

where E is the energy produced; M is the mass of the matter involved; C is the velocity of light.

In nuclear reactions, energy may be produced in two ways: *fission* and *fusion*. In nuclear fission a large nucleus such as that of uranium breaks down into two smaller nuclei, barium and krypton when bombarded by a neutron. In fusion reaction, energy is released by fusion or joining together of light elements such as hydrogen and lithium. In nuclear reactors the fission principle is commonly followed.

6.4.2 Radioactivity and the Principal Radioactive Elements

Way back in 1896, a French chemist, Henry Becquerel discovered that some elements such as uranium and thorium underwent spontaneous disintegration producing energy in the form of rays or particles. The phenomenon was subsequently called *radioactivity*, a misnomer, since no radio waves are involved in the disintegration. It was later found that all elements with atomic number exceeding 83 are radioactive.

The most eminent radiations from the radioactive elements are α, β, and γ *rays*:

α *rays*: He ion with a double positive charge. Ejected from radioactive atoms with velocities up to 10% of that of light. They have smaller penetrating power compared to other radiations.
β *rays*: Electrons carrying negative charge and weighing about 1/1837 amu. They move with velocities in cases exceeding 98% of that of light. They are faster than α rays and their direction is changed markedly by magnetic field.
γ *rays*: Electromagnetic radiation of the same nature but with a short wave length than the X-rays. They have no mass in the ordinary sense and they are not deflectable in a magnetic field. They carry no electrical charge and travel with the velocity of light. They thus have the greatest penetrating power.

Besides α, β, and γ rays there are other rays such as neutrons (mass 1 amu and 0 charge), protons (mass 1 amu and single positive charge), deuterons (ions of H^2 having a mass of 2 amu and single positive charge), and positrons (positive electrons of mass 1/1837 amu and single positive charge). As the electron generally comes from the K level, the process is called K-capture. L-capture may also take place. As mentioned already, finally there is fission which results in the fragmentation of the whole nucleus.

Radioactivity is the specific property of specific isotopes. While certain isotope (s) of a chemical element may be radioactive, others may not be. A radioactive element may also have more than one decay mode and each radioactive element has a characteristic *half-life*, which is the amount of time required for half the mass of a radioactive element to decay, e.g., the half-life of ^{238}U is 4.468 billion years.

There are only three naturally occurring elements which show radioactivity. These are U (Z = 92), Th (Z = 90) and K (Z = 40). They generate 0.73, 0.20, and 27×10^{-6} calories/gm/year of radioactive heat in the earth's interior, thought to be responsible for some of the partial melting in the mantle.

6.4.3 *Geochemistry of Uranium and Thorium*

Uranium and thorium being members of the Actinide series are highly electropositive metals. The former is the last member of the Group VIB, while the latter is the last element in Group IVB. Both uranium and thorium are highly oxyphile in nature, the former occurring as oxides, hydroxides, silicates, phosphates, vandates, molybdates, carbonates, sulfates, arsenates. The latter occurs mainly as oxides, silicates, and phosphates. Their elemental states are not reported from nature. Affinity of these elements toward organic matter such as humus, coaly matter, petroleum and bitumen are known. In nature U (IV and VI) and to some extent U (V) are common as also Th (IV). Needless to say U (IV) is stable in reducing and U (VI) is stable in oxidizing conditions. Results of the study of the $U–O_2–CO_2–H_2O$ system at $T = 25$ °C and $P_{CO2} = 10^{-2}$ atm show that for most meteoric waters in near neutral pH range, the dominant aqueous species of uranium are expected to be oxide or carbonate complexes of U^{6+}. Uraninite or pitchblende (UO_2) will precipitate with the lowering of Eh (Fig. 6.30).

In nature, uranium is composed of three principal isotopes in the following proportions:

^{234}U	0.0054%
^{235}U	0.720%
^{238}U	99.275%

The above-mentioned isotopes, ^{235}U and ^{238}U, decay following finite rates into ^{207}Pb and ^{206}Pb, respectively. ^{235}U undergoes fission when bombarded with slow neutrons while ^{238}U absorbs slow neutrons to form ^{239}U which decays to form the fissionable ^{239}Pu. When the content of ^{235}U is increased from 0.72% to about 2.5–3.0%, it gives the enriched fuel used in light water nuclear reactors. There are some energetic similarities (ionic size, charge, electronegativities, etc.) between uranium and several other elements in the Periodic Table, i.e., Y, Zr, Th, Ca, and Ba. Hence these elements are partially replaced by uranium in the mineral structure (diadochy). For example, U^{4+} replaces Ce in monazite, Y in xenotime, Zr^{4+} in zircon and other zirconium minerals, Th^{4+} in thorianite and Ca^{2+} in apatite and fluorite.

Uranyl $(UO_2)^{2+}$ solutions reacting with soluble, colloidal, or gel silicates, arsenates, phosphates, molybdates, vandates, carbonates, sulfates, selenites, tellurites, and metal hydroxides precipitate a large number of insoluble uranyl

Fig. 6.30 Eh–pH diagram showing the stability field of uraninite and other U-species (after Langmuir 1978)

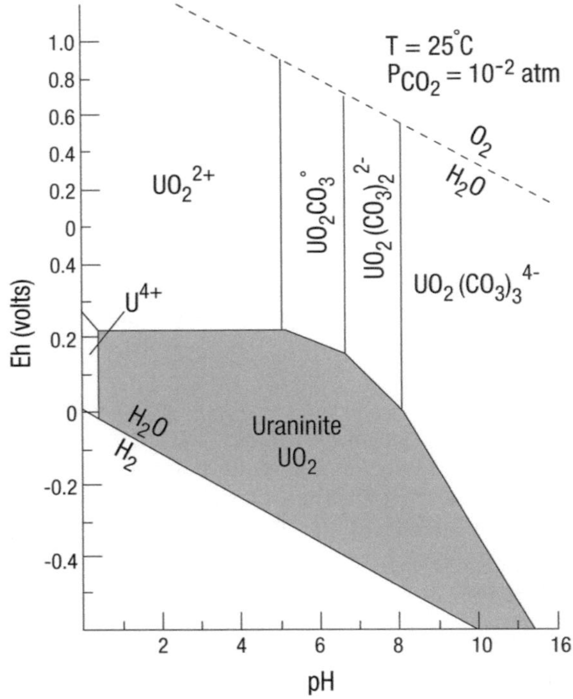

compounds, many of which are common in nature. Similarly, uranyl solutions reacting with soluble or colloidal organic (humic) complexes in nature precipitate a host of uranium-organic compounds.

Because of their large size and high charge density both the U^{4+} and $(UO_2)^{2+}$ ions are strongly adsorbed by a large number of natural substances, both organic and inorganic, including gels of hydrous iron and manganese oxides, silica-alumina gels, clay minerals, clay-humic complexes, zeolites, and humic substances. The sorption capacity increases with the increase of pH up to about 8.5.

Although in the crustal rocks thorium is more abundant than uranium (Table 6.10), the former hardly develops a mineral deposit through a primary geological process. It forms exploitable secondary (placer) deposits though, as we will see later in the chapter.

Thorium (Th) with its atomic number of 90 has 6 isotopes in nature, of which the one with longest half-life is ^{232}Th. This isotope again ultimately decays to ^{208}Pb, via a set of steps. Some of the intermediate isotopes of this series, particularly ^{208}Th are useful in geochemical prospecting. The isotope ^{232}Th absorbs slow neutrons and is usable in Breeder type reactors.

ThO_2 is more basic than the dioxides of the preceding elements of Group IV (Si, Ge, Sn, Pb, Ti, Zr, Hf), its salts exhibit less tendency to hydrolysis and the elements form no oxyions, unlike uranium. The hydroxide $[ThOH_4]_n$ dissolves readily in

Table 6.10 Distribution of uranium and thorium in common crustal rocks (after Boyle 1982)

Rock type	Th (ppm)	U (ppm)	Th/U
Igneous rocks			
Ultrabasics (peridotite, dunite)	0.1	0.02	5
Kimberlites	12.0	4.5	2.6
Lamprophyres	15.0	5.0	3
Basic intrusive and extrusives	3.0	0.6	5
Acidic intrusive and extrusives	15.0	4.5	3.3
Alkali syenites and granites	Up to 100	Up to 100	1
Sedimentary rocks			
Arenites and rudites (Sandstone, arkose, greywacke, conglomerate)	5	1.5	3.3
Lutites (shale, argillite)	12	3.5	3.4
Sapropelites (carbonaceous pyritic shales, oil shales)	Up to 20	Up to 1200	-
Chemical precipitates (limestone, dolomite)	1	1.5	0.6
Chemical precipitates (chert, cherty iron formations)	3	3	1.5
Tuffs (intermediate-acidic compositions)	6	2	2
Phosphorites (oceanic)	Up to 12	Up to 300	-
Metamorphic rocks			
Quartzites, metagreywacke	5	1.5	3.3
Quartz-pebble conglomerate (pyritiferous)	Up to 435+	Up to 2000+	–
Phyllite, meta-argillite, shale	10	2.5	4
Marble, crystalline dolomite	1	0.5	2
Schists (sedimentary parentage)	10	2	5
Gneisses and granulites	10	3	3.3
Hornfels (sedimentary parentage)	10	2	5
Skarn (sedimentary parentage)	15	3	5

strong acids. In water, thorium hydroxide is very insoluble. The common soluble salts of thorium are nitrate, sulfate and chloride. Th-tetrachloride is apparently volatile at elevated (>500 °C) temperatures. Chloride and other soluble salts undergo extensive hydrolysis in aqueous media.

The average abundance of uranium in Earth's crust is ~2 ppm while that of thorium is about 8 ppm, making the Th/U ratio vary between 3 and 4. Th/U ratio in some cogenetic igneous rocks exhibit decrease with the increase of SiO_2 at total alkali content. But in other situations the relationship may be reverse or constant. In normal igneous rocks most of the thorium is present in the accessory minerals, particularly thorite, monazite, allanite, zircon, sphene, anatase, rutile, and perovskite. In some sedimentary rocks thorium-bearing minerals, such as monazite, zircon, and sphene may be present in considerable proportion.

The Th-content in soils is highly variable, from 0.1 to 50 ppm, the average being about 5 ppm. Unusually high content of thorium (200–1000 ppm or more) may be present in certain soils and clays, laterites, bauxites developed on alkaline rocks and carbonatites and in the vicinity of Th-bearing pegmatites and veins. During soil formation Th/U tends to increase due mainly to the differential solubility of U and Th.

The aqueous chemistry of thorium in soils is largely dependent on hydrolytic, colloidal and absorption reactions (Boyle 1982). Redox reactions are unimportant as the element can exist in nature in only one oxidation state.

In neutral waters, Th is stable as sulfate, phosphate, and fluoride complexes at pH = 5.5–7.5 and $[Th(OH)_4]^0$ above 7.5. The carbonate complex is decomposed in acid solutions (pH < 5) and breaks down in highly alkaline (pH > 10) with the precipitation of hydroxide, or hydroxide gels. Langmuir and Herman (1980) stressed that Th-organic complexes probably predominate over inorganic complexes. The difference in geochemistry of U and Th, particularly in low temperatures, explains the differential distribution of the two elements in nature.

Radioactive anomaly, i.e., higher than threshold-concentration of U or Th or both, may occur in some parts of the earth's crust due to specific crustal processes. Such anomalies are located by detecting the γ-radiation of long range penetration by using instruments like Geiger Müeller (GM) Counter and Scintillometer in the field and Gamma-ray spectrometers in the laboratory. Since it is not known if the radioactivity measured is from U or Th or both, or from some other radio-elements, it is expressed as "equivalent (e)U_3O_8," which means that the sample measured should contain that much actual U_3O_8 had U been present in radioactive equilibrium with its daughter elements and no Th was present.

6.4.4 An Outline of Ore Mineralogy

Uranium occurs in a large number of minerals (>200) and thorium in a few. Only a few of them form ores. The principal ore minerals of uranium are uraninite and pitchblende. Coffinite $[U(SiO_4)_{1-x}(OH)_{4x}]$ and brannerite (a combined Ti- and U-oxide) are also included in the ores if they occur not in trivial proportions.

Although the two mineral names, uraninite and pitchblende are sometimes used interchangeably, a distinction between them is generally discernible. Uraninite is macrocrystalline, higher temperature in origin and generally contains up to several percent thorium and REE. Ideally UO_2, uraninite occurring in nature is partially oxidized and should more accurately be referred to as UO_{2+x}, rather $(U^4_{1-x}U^6_x)\ O_{2+x}$. The name pitchblende may better be confined to the cryptocrystalline variety of uranium oxide. Its composition is also referable to UO_{2+x}, although the value of x is a little larger than that of uraninite, i.e., pitchblende is more oxidized. That makes the cell-edge of pitchblende somewhat smaller than that of uraninite. Pitchblende contains little or no thorium or rare earths. Cuney (1978) however, reported from Noirs-Limouzat, France, a CaO-content of up to 8.39%. This and the Pb-content of >6% reported in literature, appear to have been structurally accommodated.

A uranium deposit exposed to surface gets oxidized and in a moist/wet climate the metal is leached out as soluble complexes from the top zone. Part of the uranium is stabilized in the form of secondary uranyl minerals, autunite, meta-autunite

(Ca-bearing phosphates), torbernite (copper-bearing phosphate), schoepite and meta-schoepite (U-hydroxides), carnotite (K-bearing vanadate), uranophane (Ca-bearing silicate), gummite (mixture of uraninite and secondary u-minerals of variable composition), etc., which develop on the sub-outcrop (Sarkar 1970, 1984). Most of these secondary uranium minerals have bright yellow, green or greenish yellow color and fluoresce variably under ultraviolet light. Somewhat analogous to gummite, "thucolite" refers to oxidized hydrocarbons with varying amounts of U and Th.

Th-mineralogy, in contrast, is rather simple due to only one principal oxidation state of thorium (Th^{4+}) The most common endogenic (primary) minerals of thorium are thorianite [$(Th, U)O_2$], thoro-uraninite, and thorite ($ThSiO_4$).

U and Th in significant amounts ($\geq 1\%$) may occur in a large number of rock-forming accessory minerals. These include zircon, monazite, xenotime, allanite, apatite, sphene, rutile, and anatase. The lattice of some of these minerals may get partially or totally destroyed by α-emission of uranium to produce a metamict. Uranium may also be present as a substitution ion in several niobates, tantalates, and titanates. These include minerals, such as betafite, davidite, samarskite, columbite-tantalite, and pyrochlore.

6.4.5 *Radioactive Ore Deposit Types*

Uranium and thorium are the primary fuel-feeds in nuclear reactors. Plutonium is a secondary feed. Both uranium and thorium as we have already discussed occur in nature primarily in the form of minerals. Uranium occurs as finite ore bodies, such as in:

Quartz-pebble conglomerate
Vein complexes in Proterozoic unconformity surfaces
Beds and rolls in sandstone
Franceville and "Copper belt" type mineralization of Africa
Hematite breccias in granitoid rocks
Albititic metasomatites/IOCG type mineralization
Beaverlodge district type vein deposits in metamorphic tectonites

Besides, there are economic/sub-economic deposits in ferricrete, calcrete, carbonatite, pegmatite, phosphorite, and black shales.

The first three types of deposits have been discussed in reasonable details in Chap. 2 and are not repeated here.

Franceville (Gabon) and "Copper belt-type" (Shaba – Zambia) Deposits
The early Proterozoic Franceville deposit of Gabon, West Africa, occurs in deltaic sediments overlying fluviatile deposits of coarse sandstone and conglomerates.

Uranium as pitchblende occurs as disseminations or as veins when occurring in tectonic structures. In the latter, U-content may be as high as 10%.

The "Copper belt" or Shinkolobwe deposit contains 25.5−31.5 × 10^3 t U (0.36–0.8% U). Besides copper and uranium, cobalt and nickel occur as important associate metals in the belt, though not always maintaining proportionality. It seems logical to conclude that uranium originally distributed in sediments was later collected in diastropic structures during deformation and greenschist metamorphism (Menegel 1981).

Uranium Mineralization Associated with Hematite Breccias

The only deposit of this type known to date, and that too extraordinarily large, is the Olympic Dam deposit, Roxby Downs, South Australia. Host rock of the ores here is hematitic breccias, occurring as steeply dipping bodies within the fractured syeno-granite. The estimated metal reserve in the deposit is: U 1088 × 10^3 t (0.054%), Cu 32 × 10^3 t (2.6%), Au 1200 t (0.6 g/t), Ag 7 × 10^3 t (3.5 g/t), La 3.4 × 10^6 t, and Ce 5.4 × 10^6 t. It is a polyphase hydrothermal deposit in which meteoric water also participated.

Albititic Uranium Deposits

Albitites are metasomatic rocks, associated with tectonic zones, characterized by repeated mylonitization, cataclasis, and brecciation. They are lenticular to anastomosing in shape. These rocks often, though not invariably, contain U, Th, Be, Zr, P, Ta, Nb. A new nomenclature of ore type IOCG (Iron oxide-copper-gold) at places comes close to it but hardly coincides.

Beverlodge District Type Veins in Shear Zones in Metamorphic Tectonites

The geological framework is made up of Archean gneiss, migmatite, and amphibolites with subordinate metasediments. Mineralization is controlled by a Lower Proterozoic ductile-brittle shear and mylonite zone. Pitchblende is accompanied by the gangue minerals such as calcite, quartz, hematite, pyrite and albite. Reserve: 27 × 10^3 t U (0.17% U). Genetic history of the Beverlodge mineralization in Saskatchewan, Canada, is complex (Sarkar 1995). Uranium was introduced in granitoid plutons during Kenoran orogeny and was further concentrated in sedimentary rocks which underwent mylonitization and faulting during Hudsonian orogeny. Metals from the sediments were remobilized and redeposited by hydrothermal process which was accompanied by sodic metasomatism, hematitization, chloritization, and carbonatization. Trembley (1978) suggested a main stage of mineralization at 1780 Ma, with later rejuvenation at 1140 Ma. Similar deposits are known in the eastern and western parts of the Bohemian Massif in the Czech Republic and Slovakia.

Other types of uranium deposits, usually sub-economic, are ferrocrete, calcrete, carbonatite, pegmatite, phosphorite, black shale, etc.

6.4.6 *Uranium Mineralization in India*

Uranium occurs in different parts of India (Fig. 6.31), but its economic, rather the strategic concentration is far below its projected or perceived need. The present position of uranium resource in India stands at 1,75,010 t U_3O_8 under 'indicated' and 'inferred' categories (as on Feb., 2012, AMD). The principal repository of the metal in the country is the Singhbhum copper-uranium belt and presently, all the working mines of uranium are located along this belt, e.g., Bagjata (Moinajharia), Jaduguda,

Fig. 6.31 Map of India showing the location of different radioactive deposits and prospects in India (after AMD 2012)

Bhatin, Naroapahar, Turamdih – Bandhuhurang – Mohuldih, Singridungri-Banadungri, Bangurdih going from east to west (vide Fig. 3.34). Besides there is minable uranium mineralization at a couple of places in and around the Cuddapah basin and Bhima basin in South India and Domiasiat – Gomaghat – Pdengshapak area and Lostoin-Wahkut-Umthongkut area within the Cretaceous Mahadek Sandstone of Meghalaya. Mining is withheld here presently for one or the other reason. Some areas in Rajasthan (Vein type deposit at Rohil central and north), in central India (vein type mineralization at Bodal-Jajawal, Chattisgarh) and Himalaya look prospective, but there should be more rigorous investigation there. While exploratory mining at Gogi in Yadgir district of Karnataka has already started, permission for exploratory mining has been sought recently for Rohil deposit, Rajasthan and Wahkyn in Meghalaya.

In the Singhbhum copper-uranium belt above-background isorads are obtained virtually at every point, but it is only at certain locales that the concentration has been high enough to be economic. Coming from the SE along the shear zone, the first economic mineralization is encountered at Bagjata, about 10 km away from the former copper mine at Mosabani. The ore zone, 600 m along the strike and 600 m along the dip are being mined by the Uranium Corporation of India (UCIL) since 2008. In the Mosabani-Rakha Mines sector, the Cu-enriched and U-enriched zones occur close to each other. When the copper mines were in operation, uranium was separated from the copper ore tailings.

Jaduguda is the best deposit in the belt, being mined since 1967. The host rocks of mineralization at Jaduguda comprise apatite-magnetite-biotite-chlorite-bearing quartz schists ("granular rocks"), brecciated quartzites, and conglomerate and biotite-chlorite schists. The ore zone is split into two zones, 60 m apart. The better developed footwall lode continues below, to a vertical depth of 800 m (Fig. 6.32).

3 km to the NW of Jaduguda, there is another mine, Bhatin, geologically very similar to the former. Naroapahar deposit occurs further 9 km to the west, along the belt. The ore grade is somewhat lower than the Jaduguda-Bhatin deposits, being in the 0.04–0.05% U_3O_8 range, but the total reserve is moderately good. Here too there are four parallel orebodies, the lowermost one being the best. Banadungri may be considered as the western extension of the Naroapahar mineralization. Turamdih – Bandhuhurang – Keruadungri zone, located 5–6 km SW of Tatanagar Railway station, is mineralized and presently being mined. Mohuldih, adjacently to the west, is in an advanced state of development. The host rocks in the Naroapahar-Mohuldih sector consist of sericite, chlorite, quartz, and albite in varying proportions and the ore grade is usually ≤ 0.05 U_3O_8.

The uranium-bearing minerals in the Singhbhum belt comprise uraninite, shooty pitchblende, complex U–Ti oxide ("Pronto ore"), allanite, xenotime, monazite, davidite, pyrochlore-microlite, clarkeite, autunite, torbernite, schoepite, uranophane. Jaduguda, and Bhatin produce nickel and molybdenum as by-products. Xenotime in Kanyaluka may be mineable (Sarkar 1984; Sarkar and Gupta 2012).

Age of Singhbhum mineralization is 1500–1600 Ma according to Rao et al. (1979). However, in a recent interesting chemical dating work on uraninite from Jaduguda deposit, Pal and Khede (2013) obtained ages that indicate initial uranium

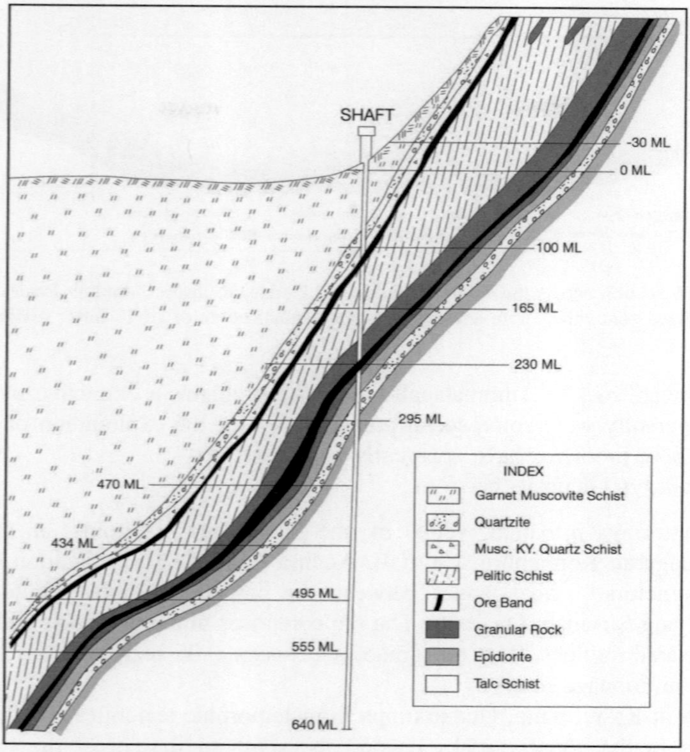

Fig. 6.32 Transverse geological section of Jaduguda mine (after UCIL Jaduguda)

mineralization to have taken place at ~1.80 to 1.90 Ga, modified at places by later events. The age obtained by the pioneer workers Rao et al. (1979) apparently relate to one such modified value.

The ore genesis is controversial, but people in general agree about the "hydrothermal" origin. But hydrothermal fluids may form in many ways (vide Chap. 2). What better fits here is not certain at the moment. Keenly contesting models refer to those of "Proterozoic unconformity" and IOCG origin. We wait for more work in the future, particularly on the genesis of IOCG deposits.

The next important uranium province in India is the Cuddapah basin and its surroundings (vide Fig. 4.8). The deposits are both stratabound and vein type. The best stratabound deposit occurs in the phosphatic siliceous dolomitic rocks of Vempalle Formation (Tummalapalle area, Rachakuntapalle, Madyalabodu) of the Cuddapah Supergroup in Andhra Pradesh. The mineralization extends for a total length of ~160 km and a width of ~100–150 m. The overall grade is low (~0.045% U_3O_8) but with large tonnage. It is the largest deposit in India with more than 65,000 t U_3O_8 (AMD, as on Feb., 2012). The ores comprise pitchblende and coffinite, associated with collophane. An underground mine is being developed at Tummalapalle in the richest section of a ~15 km zone (Fig. 6.33) by UCIL. The

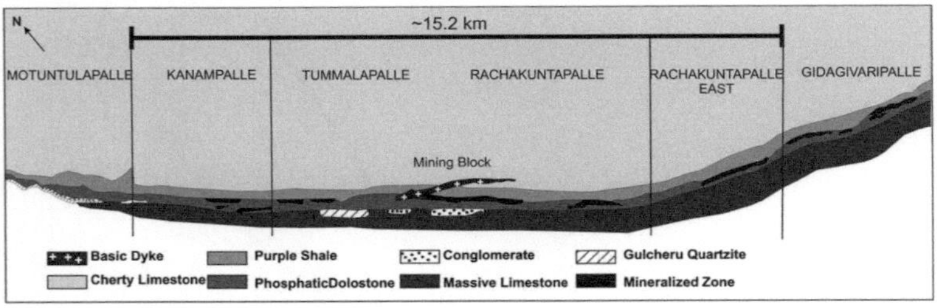

Fig. 6.33 A section across the Tummalapalle area, Cuddapah district, Andhra Pradesh showing the mineralized zone, host strata and the proposed mining block of UCIL (after AMD 2012)

host carbonate rock at Tummalapalle offered a problem in extractive metallurgy. There apparently were some social problems too for the exploitation of the ores. Most of these problems have apparently been resolved.

The vein type deposits occur as:

(a) Unconformity-proximal veins in the basement granitoids at Lambapur (Peddagattu, Koppunuru, Chitrial), Andhra Pradesh. Mineralization occurs in the structurally weak zones between the basement biotite granite and the overlying Srisailam Quartzite. The ore comprises uraninite, lesser coffinite and associated sulfides. The ore grade is between 0.05 and 0.1% U_3O_8 with a medium tonnage reserve.
(b) Veins at Kasturigattu, Gudarikuppa in metamorphic tectonites.
(c) Veins in the country rocks beyond the southern margin of the basin near Rayachoti.

Uranium mineralization is proven at Ukinal and Gogi (Yadgir district) in the Bhima basin, occurring within sedimentary rocks (phosphates, carbonates). Grade is better than most uranium deposits in India (up to 1.68% U_3O_8), but the resource potential is not high ($\sim$2500 t). The mineralization occurs in both the biotite granite forming the basement and the overlying limestone, above the unconformity. The ore mineralogy includes pitchblende and coffinite, associated with organic matter and sulfides, such as pyrite, chalcopyrite, galena, etc.

Another economically potential uranium mineralization belt occurs at Domiasiat-Wakhyn – Gomaghat – Pdengshakap (Lostoin, Tyrnai, Umthongkut) area within the Mahadek sandstone of Upper Cretaceous age ($\sim$100 Ma) in Meghalaya. Although this area has been handed over to UCIL, mining is reportedly withheld due to physical (a very high rainfall area) and some social reasons.

In Rajasthan, some moderately good deposits occur at Umra, Udaisagar, Kalamagra, and Haldighati (Udaipur district) and Putholi, Salera, Dadiya (Chittorgarh district) and Maonda, Bandhaki Dhani, Kerpura, Dhaneswar, Ghateswar-Rohil (Sikar district). However, these are not yet considered to be adequately economic or strategic.

6.4.7 *Resources of Thorium and Associated Metals in India*

Thorium and zirconium share one common feature in nature – they get concentrated in placer deposits in the exogenic environment. Thorium is located mostly in monazite. Larger part of REE[1] (70–75%) occurs in monazite (Ce, La phosphate in which Th is present as substitution). A fair quantity of titanium in the form of ilmenite and rutile occurs in such placers. Placer sands containing ilmenite (520.38 Mt), rutile (29.11 Mt), monazite (10.70 Mt), zircon (32.28 Mt), sillimanite (195.85 Mt), garnet (154.26 Mt), kyanite, etc., is a common feature along the beaches of peninsular India in the states of Maharashtra, Goa, Karnataka, Kerala, Tamil Nadu, Andhra Pradesh, and Odisha, inland placers in parts of Jharkhand and West Bengal and Teri red sand in Tamil Nadu (AMD 2012). However, their best concentration is at Chavara and Manavalakurchi in south Malabar coast and Chatrapur in the Odisha coast (Fig. 6.31). The monazite content in these beach sands may be up to 10–11% (by wt.). ThO_2 in Indian monazite may be as high as 10% or a little more. Average ΣREEO of Indian monazite is about 60% (vide Table 3.8).

The estimated reserve of monazite in the country is 10.70 Mt, which is expected to contain $\geq 7 \times 10^5$ tons of ThO_2. Bulk of this reserve is located at Chavara and Manavalakurchi. The best of such beach placers developed particularly in the sectors where the continental wash pouring out into the sea was derived from the Precambrian country rocks consisting of granites, granitic pegmatites, granite gneisses, migmatites, and granulites. Warm and humid climate with abundant rainfall has been conducive to the decomposition and disintegration of the rocks that originally contained the minerals that now have formed the placer deposits. Action of waves, tides, current and wind did the necessary winnowing and deposition along the coast (Sarkar et al. 1995).

6.4.8 *Nuclear Energy: The World and Indian Scenario*

World's fossil fuel reserves are not expected, at the present state of knowledge, to last more than a few decades, even at the present rate of consumption. Naturally we have to look for alternative resources, deservedly renewable, such as solar energy, wind energy, tidal energy, hydel energy, biofuels, etc. All these are already in use, but generally not in sufficient quantities because of uneven distribution of facilitating situations in different parts of the world and generally high production cost. The world was first horror-struck seeing the devastation caused by atom bombs in

[1]REEs in fact are not extremely rare and are even more abundant than metals such as, Cr, Ni, Cu, Zn, Mo, Sn, W, or Pb in the Earth's crust. Only they do not show a tendency to form large mineral accumulations of economic interest and resembled the oxides of alkaline earths, such as, CaO, BaO etc. (Vide Chap. 3).

Hiroshima and Nagasaki but soon became attracted to nuclear energy seeing its great potential for use in civilian enterprise. With time it earned the sobriquet "clean energy." Before the decade of fifties of the last century, there were a few nuclear reactors for experiments or energy production in a small number of countries including India. But a few years ago, some of the major users of nuclear energy for power generation came to be (Newton et al. 2006):

France	77%	Switzerland	40%
Lithuania	72%	Japan	35%
Slovakia and Belgium	55%	Germany	32%
Sweden	52%	USA	20%
Ukraine	51%		

The history of development of nuclear energy for civilian use is divisible into four stages:

First stage: Some scattered research and production units by the end of fifties.
Second stage: It started in the early seventies as a fall out of the Arab–Israeli war, when there was a sudden rise in the price of petroleum and natural gas and many people decided to go for nuclear energy. There was a frantic search for uranium. Soon a new type of uranium deposit, i.e., "Proterozoic unconformity type" was discovered. Then came suddenly the Chernobyl reactor disaster in Ukraine (former USSR) on April 26, 1986. This was a knee-jerk event for the entire world. USA stopped setting up of 22 reactors out of 26 in the process. Many uranium mines closed down due to reduced ore-demand.
Third stage: The human memory being what it is, the disaster of the Chernobyl incident was largely forgotten. International Atomic Energy Agency (IAEA) again sanctioned 60 nuclear energy plants by 2020. This included eight in India. Presently 27 nuclear energy plants, including eight in India, are in various stages of development.
Fourth stage: Nature is hardly in peace for too long. On March 11, 2011 occurred the triple disaster of tsunami, earthquake (RS = 9) and the reactor leakage at Fukushima Daichi on the NE coast of Honshu, Japan. The Japanese were stunned. They exhausted their high technology, and also the whole world came forward to their help. The death toll, however, remained indefinite, and life became hell for hundreds of thousands of people who were alive. Houses with all belongings were washed away for some; others had to run away 30 km or so from the site of the accident to save themselves from the radiation hazard. Some believe, in total effect, the Fukushima Daichi disaster was comparable to the Chernobyl disaster of 1986, if not worse.

As it seems, the world suffers from a love-hate relationship with respect to nuclear energy. Love, because a small mass of nuclear fuel is needed to produce a large amount of energy, the involved technology is generally environmentally clean, the energy cost is initially somewhat high but evens out before long and a modern plant is expected to last 30–40 years. Hate, because the people still suffer from the trauma of Hiroshima and Nagasaki from the last days of World War II. The memory of the last two events has been refreshed by reactor disasters, occasionally though not frequently. Some nuclear reactor accidents that preceded Chernobyl and Fukushima are Khistim disaster, USSR (1957) leading to cancer death of 200 people; Windscale accident in England (1957) which also caused cancer deaths of 200 people; workers in the Tokaimuk nuclear plant, Japan were exposed to severe radiation in 1999. The disaster of generally believed “full proof” Fukushima Daichi reactor led many people of the world to think of their energy security problem anew. Germany will be happy to see all their nuclear reactors shut down within the next 20 years. France and Japan, presently two of the major users of nuclear energy are on the revise-mode. France has already reduced its nuclear power dependence by 1/3rd, i.e., to 50%.

The challenge of Fukushima, as it seems, has already been taken up by the industry. The holding company has reportedly devised some safety features that operate in the absence of electricity or human intervention that was lacking in Fukushima. The March 2011 earthquakes at Fukushima knocked out the plants connection to the grid and the Tsunami that followed damaged the backup generators and electrical equipment, crippling the cooling system, leading reactor-cores to heat up. In the proposed system a giant water tank will be placed above the reactor core. In the event of a potential melt down, the heat building will trigger a valve, allowing the water to flow into the reactor. But till about the end of 2014, a large proportion of civil population in Japan has been averse to nuclear power plants in general and against resumption of activities for production in the devastated Daichi plant.

India’s Position

The first Prime Minister of India, Pandit Jawaharlal Nehru, who himself was a student of Science once, was a great appreciator of Science and Technology. He set up the Atomic Energy Commission of India immediately after Independence under direct supervision of Prime Minister’s Office (PMO). The renowned Indian scientist, Homi Bhaba, was given the full charge of this development. But unfortunately he died immaturely in an air crash in 1966. With him India’s atomic energy program did not die, though got a halted growth. India’s refusal to sign the Nuclear Proliferation Treaty (NPT) was a particular impediment to her quicker development of nuclear industry and technology. A few years back (2008), India entered into a treaty with USA, according to which USA would help India in developing her nuclear industry for civil (peaceful) uses. In the changed situation some Nuclear Supply Group (NSG) countries have also agreed to help India in this respect. This is necessary as the USA itself is a poor producer of uranium ores (2% of world production). While Canada and Australia share half of world production of uranium ores between them in sub-equal proportion, followed by Kazakstan Russia and Namibia—each producing 8–10%.

Toward the end of 2014 India was having 20 nuclear power plants in operation within the country. Of these, 10 (KGs 1–4 at Kaiga, Karnataka; NAPs 1–2 at Narora, U.P; MAPs 1–2 at Kalpakkam, Tamil Nadu, and TAPs 3–4 at Tarapore, Maharashtra) with a gross installed capacity of 2840 MWe, use uranium obtained from inside the country. Supply from domestic mines increased by 8% during 2013–14, reaching a total of 812 metric tons. They are beyond the supervision of International Atomic Energy Agency (IAEA). 10 others (RAPs 2–6 at Rawatbhata, Rajasthan; KAPs 1–2 at Kakrapar, Gujarat, and TAPSI 1–2 at Tarapore, Maharashtra) run with imported fuel, supplied by Russia, Kazakhstan, and France, are under IAEA supervision. Two more units at Narora were to come under the IAEA supervision before the end of 2014. Two more could be added to this category.

India is hopeful since Bhaba's time that she will utilize a large part of her thorium reserve in making Fast Breeder reactors. In this type of reactor, fast neutrons from a uranium + plutonium mix impinge on a thorium shell and ^{232}Th is converted into fissile ^{233}U. Presently we have an experimental type of Fast Breeder Reactor at Kalpakkam which is producing 40 MW electricity for some time now. The 500 MWe Prototype Fast Breeder Reactor at Kalpakkam is under commissioning, and it will become operational in a few months. But one difficulty with this type of reactor is that they are cooled by liquid sodium instead of light and heavy waters in some others. This liquid sodium is highly explosive when coming in contact with air and water. Reportedly, Russia has recently been able to offset this problem by using liquid lead.

Any reactor site should be chosen with the background knowledge and necessary wisdom. The site should be seismically safe, not densely populated and the technology adopted must be proven at the first place. The controversy regarding the proposed Jaitapur nuclear power project in Maharashtra must be looked into in this context.

Recently Government of India informed the Indian Parliament that after the Fukushima disaster several task forces were set up to review the safety features in Indian nuclear power plants and it was found that Indian reactors have sufficient margins and features in design to withstand high intensity earthquakes and tsunami. The task force recommended that the reactors should shut down automatically on sensing abnormal seismic activity and additional sources of power will come into operation through hook up with external sources, such as generators, batteries. Nuclear reactors at Tarapore and Madras power stations will be given additional shore protections. The safety measures will continuously be reviewed. However, the Government refuses to compromise on its present nuclear energy program. Government of India may be right in drawing up its present nuclear energy program, but it must be holistic in its perspective and be ready to change if the situation so demands.

6.5 Geothermal Energy

This is the vast reservoir of heat energy in the Earth's interior which is manifested by volcanoes, hot springs, geysers, etc. Historically, Romans and Greeks are known to have developed cities around hot springs which they used extensively for health and recreation purposes. The city of Bath, near Oxford in England is a classic example in this regard. The tradition still continues in Europe and the famous hot spring at Karlovy Vary in Czech Republic attracts thousands of people every year for spa treatment. Many mineral waters from thermal springs have enough soluble salts and sometimes a minute amount of radioactivity, imparting the waters therapeutically important properties. The principal ingredients are sulfate of magnesium and sodium, bi-carbonates, and chloride of alkalis and sometimes, alkaline earths. This renewable energy is also technically and economically viable for generation of electricity and has minimum negative environmental impact. It is continuously regenerated by mainly the decay of radioactive elements that occur in all rocks. From the surface down through the crust, the normal temperature gradient—the increase of temperature with increase of depth—in the Earth's crust varies from 17 to 30 °C per kilometer depending on the location and heat flow. Geothermal energy can be used in various fields, such as space heating, cooking, cold storage, in greenhouse cultivation, crop drying, in tourism and in health resorts.

Major geothermal fields are situated in the circum-Pacific margin, along East African rift, North Africa, around Mediterranean Sea, Iceland, Yellowstone Park, USA amongst other places. Oldham and Oldham (1882) were the first to record the location of 300 known thermal springs in India. In recent years, Geological Survey of India has identified 350 geothermal energy locations in India. The most promising of these are the geysers in Puga valley of Ladakh and at Manikaran in the Kulu Himalaya. Well hot springs include those at Gauri kund, Badrinath pilgrimage in Himalaya, Tattapani in Chhattisgarh, Rajgir in Jharkhand, and Bakreswar in West Bengal. The estimated potential for geothermal energy in India is about 10,000 MW.

There are 10 geothermal provinces identified in India. They are shown in Fig. 6.34 and the first three in the list below are marked:

(i) Himalayan province
(ii) Son-Narmada-Tapi (SONATA) province
(iii) West Coast province
(iv) Cambay graben province
(v) Aravalli (Sohna) province
(vi) Naga-Lushai province
(vii) Andaman-Nicobar island province
(viii) Godavari province
(ix) Mahanadi province
(x) South Indian cratonic province.

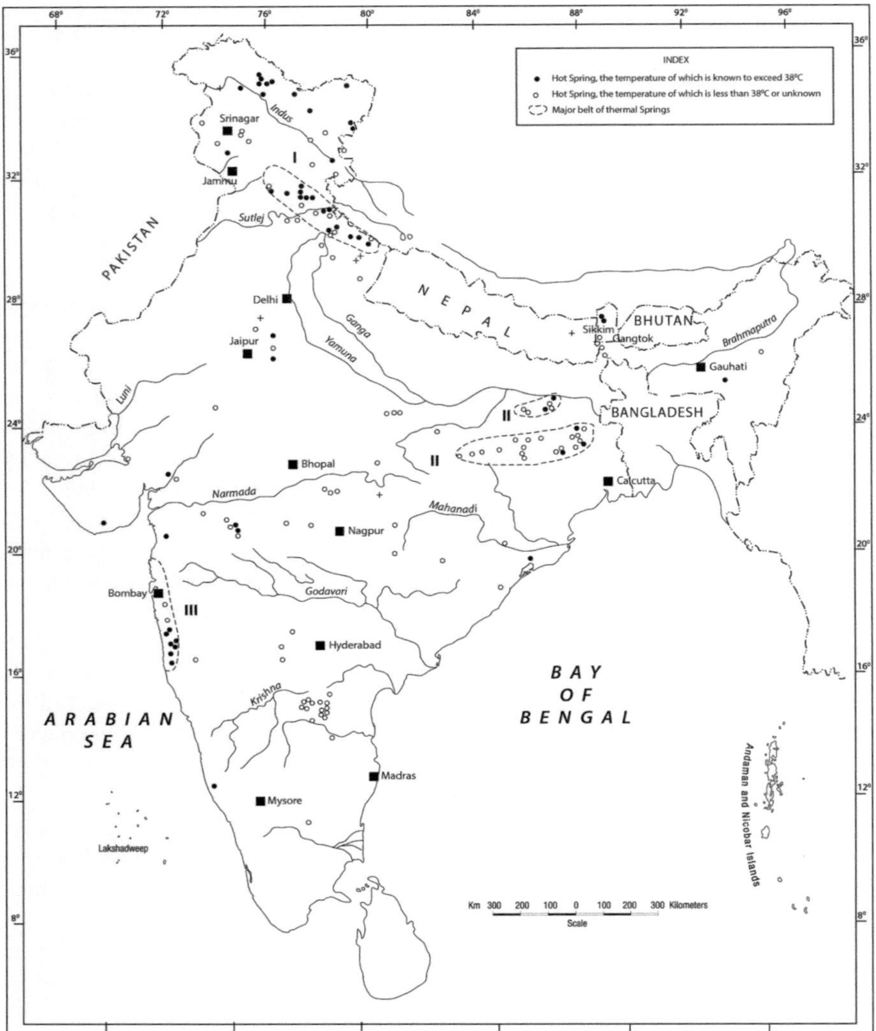

Fig. 6.34 Map showing the distribution of hot springs in India (after Deb 1969)

These provinces are associated with major rifts or subduction zones and registered high heat flow and high geothermal gradient. The reservoir temperatures estimated are 120 °C (West coast), 150 °C (Tattapani) and 200 °C (Cambay). The depth of the reservoirs in these provinces is between 1 and 3 km. These geothermal systems are liquid-dominated and steam dominated systems prevail only at Puga and Tattapani. The issuing temperature of water at Puga is 98 °C discharging >190 tons/h of thermal water. Post-Tertiary granite intrusive is responsible for the high temperature gradient (>100 °C/km) and heat flow (>468 mW/m^2). The Cambay province is 200 km long, bracketed by deep-seated faults. Older granite intrusive, such as those at Tuwa (Gujarat) produce 90 °C issuing temperatures. Also Mio-Pliocene basic intrusives contribute partly to the high geothermal gradient (>60 °C). More than 15 thermal discharge sites are located in this province. The SONATA province extends from Cambay in the west to Bakreswar in West Bengal in the east in an area characterized by very high heat flow and geothermal gradient. It encloses the well-known Tattapani geothermal field spreading over an area of 80,000 km^2. This field has 23 thermal discharge sites with temperatures varying from 60 to 95 °C and flow rate >4000 l/m. The estimated reservoir temperatures are as high as 217 °C at 3 km depth. The Bakreswar-Tantloi field in West Bengal marks the junction between SONATA and the Singhbhum shear zone. High helium gas discharge (4 l/h) is encountered in the thermal springs here. The group of hot springs passing through Rajgir in Jharkhand lie on a lineament sub-parallel to the Son-Narmada lineament to its north (GSI unpublished report).

None of the thermal springs are being used for electricity generation in India as yet. Only a pilot project has been carried out at Puga valley, Ladakh with UNDP aid. However, several of the Himalayan springs are being used for bottling table mineral water. In India, a major problem of contamination of the spring water emanates from the fact that wherever there is a hot spring, the place turns into a pilgrimage attracting large crowds who invariably bathe in these waters.

References

Bharadwaj VK, Saini SS (2008) Hydrocarbon potential of India. In: Singhvi AK, Bhattacharya A, Guha S (eds) Glimpses of Geosci research in India, 2004–2008. Ind Nat Sci Acad, pp 251–252

Bhowmick PK (2008) Phanerozoic petroliferous basins of India. In: Singhvi AK, Bhattacharya A, Guha S (eds) Glimpses of Geosci research in India, 2004–2008. Ind Nat Sci Acad, pp 253–268

Bhowmick PK, Misra R (2008) Indian oil and gas potential. In: Singhvi AK, Bhattacharya A, Guha S (eds) Glimpses of Geosci research in India, 2004–2008. Ind Nat Sci Acad, pp 218–228

Boyle RW (1982) Geochemical prospecting for thorium and uranium deposits. Elsevier, Amsterdam 498 p

Biswas SK (2012) Status of petroleum exploration in India. Proc Ind Nat Sci Acad 78(3):475–494

Casshyap SM, Tewari RC, Khan A (1993) Alluvial fan origin of the Bagra Formation (Mesozoic Gondwana) and tectonostratigraphic implications. J Geol Soc Ind 42:269–279

Cuney M (1978) Geologic environment, mineralogy and fluid inclusions of the Bois Noirs—Limouzat uranium vein, Forez. France. Econ Geol 73(8):1567–1610

Dasgupta AB, Biswas SK (2000) Geology of Assam. Geol Soc Ind, Bangalore 169 pp

Deb S (1980) Industrial minerals and rocks of India. Allied Publishers Pvt Ltd, New Delhi 603 pp

Deb S (1969) Thermal springs in India. J Asiatic Soc XI:1–4

Dissel CFK (1992) Coal-bearing depositional systems. Springer, Berlin 721 pp

Hatcher PG, Clifford DJ (1997) The organic geochemistry of coal: from plant materials to coal. Org Geochem 27:251–274

Hunt JM (1979) Petroleum geochemistry and geology. WH Freeman & Co, 617 pp

Kögel-Knabner I (1993) Biodegradation and humification processes in forest soils. In: Bollag JM, Stotsky G (eds) Soil Biochemistry. Marcel Dekker, New York, pp 101–135
Kopp OC, Bennet ME, Clark CE (2000) Volatiles lost during coalification. Int J Coal Geol 44:69–84
Langmuir D (1978) Uranium solution-mineral equilibria at low temperatures with applications to sedimentary ore deposits. Geochim Cosmochim Acta 42: 547–570
Langmuir D, Herman JS (1980) The mobility of thorium and natural waters at low temperatures. Geochim Cosmochim Acta 44:1753–1766
Menegel L (1981) The occurrence of uranium in the Katanga system of Northwestern Zambia. Econ Geol 76:56–68
Newton FT, Collings SP, Little BC (2006) Nuclear power update. SEG Newslett 67(1):8–15
Oldham T, Oldham RD (1882) Thermal springs of India. Mem Geol Surv Ind, XIX
Pal DC, Khede D (2013) Geochemistry and chemical dating of uraninite in the Jaduguda uranium deposit, Singhbhum shear zone, India—implications for uranium mineralization and geochemical evolution of uraninite. Econ Geol 108:1499–1515
Pohl WL (2011) Economic geology, principles and practice. Metals. Minerals, coal and hydrocarbons—introduction to formation and sustainable exploitation of mineral deposits. Wiley-Blackwell, Sussex, p 663
Rao NK, Aggarwal SK, Rao GVU (1979) Lead isotope ratios of uraninites and the age of uranium mineralization, Singhbhum shear zone, Bihar. J Geol Soc Ind 20:124–127
Rayner RJ (1996) The paleoclimate of the Karroo: evidence from plant fossils. Palaeogeogr Palaeoclimatol Palaeoecol 119:385–394
Sain KC, Gupta H (2008) Gas-hydrates—future potential source of energy in India. In: Singhvi AK, Bhattacharya A, Guha S (eds) Glimpses of Geosci research in India, 2004–2008. Ind Nat Sci Acad, pp 244–250
Sain KC, Gupta H, Mazumdar A, Bhaumik AK, Bhoumik PK (2012) Geoscientific investigations of gas-hydrates in India. Proc Ind Nat Sci Acad 78(3):503–511
Sarkar SC, Gupta A (2012) Crustal evolution and metallogeny in India. Cambridge University Press, Delhi 840 pp
Sarkar SC (1995) Modelling Singhbhum uranium mineralization in the light of Proterozoic uranium metallogeny. Expl Res Atom Minerals 8:81–93
Sarkar SC (1984) Geology and ore mineralization of the Singhbhum copper-uranium belt, Eastern India. Jadavpur University, Calcutta 263 pp
Sarkar SC (1970) A study of mineralization of radioactive elements in the Singhbhum shear zone, Bihar. Proc Ind Nat Sci Acad 36A:246–261
Sarkar SC, Dwivedy KK, Das AK (1995) Rare earth deposits in India - an outline of their types, distribution, mineralogy, geochemistry and genesis. Global tectonics and metallogeny 5:53–61
Selley RC (1998) Elements of petroleum geology, 2nd edn. Academic Press, New York 470 pp
Sharma NL, Ram KSV (1966) Introduction to the geology of coal and Indian coalfields, 2nd edn. Oriental, Jaipur
Singh MP, Saxena R (2008) Status of coal bed methane investigations in India. In: Singhvi AK, Bhattacharya A, Guha S (eds) Glimpses of Geosci Res in India, 2004–2008. Ind Nat Sci Acad, pp 229–240
Stach E (1982) Text book of coal petrology, 3rd edn. Gebruder Borntraeger, Berlin 535 pp
Stopes MC (1919) On the four visible ingredients in banded bituminous coal. Proc R Soc London 90:470–487
Taylor GH, Teichmuller M, Davis A, Diessel CFK, Littke R, Robert P (1998) Organic petrology. Gebruder Borntraeger, Berlin
Teichmuller M (1987) Coalification studies and their application to Geology. In: Scott AC (ed) Coal and coal-bearing strata—recent advances. Geol Soc London, Spec Publ 32:127–169
Thomas L (2013) Coal geology, 2nd edn. Wiley Blackwell, 444 pp
Tissot BP (1977) The application of the results of organic chemical studies in oil and gas exploration. In: Hobson GD (ed) Dev in petroleum geol, vol 1, Appl Sci Publ, London, pp 53–82

Trembley LP (1978) Uranium provinces and types of uranium deposits in the Precambrian rocks of Saskatchewan. Geol Surv Can Paper 78-1A:427–435

Valdiya KS (2010) Making of India, geodynamic evolution. MacMillan Publishing India Ltd, Delhi 816 pp

Ziegler AM, Hulver ML, Rowley DB (1997) Permian world topography and climate. In: Martiny IP (ed) Late glacial and post-glacial environmental changes: quaternary, carboniferous-permian and proterozoic. Oxford University Press, Oxford, pp 111–146

Chapter 7
Soil: An Essential but Somewhat Neglected Natural Resource

Abstract Soil is an essential natural resource for the existence of life on Earth. The Science dealing with soil is Pedology. Soil is a complex aggregate of inorganic $\pm$ organic matter occurring at or close to the Earth's surface. Soil is, however, defined differently for different purposes. Soil formed even in the Archean time, but was modified subsequently. The soil section starting from the surface layer to the protolith is called *soil* profile. Texturally, soil is referred to three end members: sand, soil, and clay. Major composition of the soil and their relative proportions are given as follows: O > Si > Al > Fe = C = Ca > K > Na > Mg > Ti > N > S. Soil essentially forms by the decomposition (with/without transport of the products) of the exposed rocks. Commonly organic matter is added to the upper horizons. The following factors play important roles in the formation of soil: climate, activity of organisms, nature of the protolith, topography, and duration of the exposure time. There are a number of systems of classification of soils, including one proposed by FAO. However, many countries prefer to use their own system(s), based on the ground reality. Soil erosion is the worst hazard in the case of soils. Erosion means removal of the upper fertile part of the soil, by natural or human interference. A grass cover is a natural protection for the soil. Despande and Sarkar (2009) estimated the different soil types of India as follows: alfisols 13.55%, inceptisols 39.74%, aridisols 4.3%, entisols, 28.1%, vertisols 8.5%, and ultisols 2.5%. Conditions of soil formation being different in different parts of the country, nature of soil in India varies widely from place to place.

Keywords Pedology · Soil profile · Textural and chemical compositions of soils · Soil nutrient and their supply · Acidification of soil · Salination of soils · Black soil · Desert soil · Soil classifications

7.1 Introduction

Soil is an absolutely necessary natural resource for the existence of all life on land. Food grains for us could not be grown unless the land surface was covered by soil. In fact the food chain would not have existed but for soil. Quality of life is also

M. Deb and S.C. Sarkar, *Minerals and Allied Natural Resources and their Sustainable Development*, Springer Geology,
DOI 10.1007/978-981-10-4564-6_7

dependant on the quality of soil and the kind and quality of plants grown on them for humans and animals. Plants and trees would not have been there without soil to provide oxygen and absorb carbon dioxide exhaled by us. In turn, without plant and trees, the climate would have been much more aggressive, even hostile to life, with little or no rainfall and higher temperatures. All great river valley civilizations have had good soil as one of their chief natural resources. Human society is indebted to soils as one of their greatest natural assets, but it has not been mindful of the latter as much as it should have been. Agriculture itself reduces soil mass through increased erosion and mindless use of inorganic fertilizers and pesticides degrade soil. Desertification is caused by overgrazing, large-scale felling of trees and mismanagement of groundwater. Development of soil is a slow process: good topsoil may accumulate at the rate of 10 tons/ha/year, which translates to 1 mm of soil cover in a hectare of land. Under poor conditions, the latter may take thousands of years. Thus, soil is a renewable but slowly produced resource. It needs to be used (or abused) carefully, with an eye for conservation as much as possible.

The scientific study of soils falls under the discipline of *Pedology,* while the process of soil formation is referred to as *Pedogenesis*. Soil studies form an essential part of environmental problem, mitigation and management efforts in terms of land use planning, waste disposal and natural hazard management, and for tracing signatures of regional and global climate change, amongst others.

Soil is a complex aggregate of inorganic + organic matter covering the uppermost zone of the continental crust, except in rugged mountains where weathering is slower than erosion, or on sediment cover of continental basins. It contains matter in all three states: solid, liquid, and gaseous (Fig. 7.1) and forms from rocks involving the participation of components of atmosphere, hydrosphere, and biosphere, thereby establishing a link between climate and the vegetation that grows on soil. For example, a silty loamy soil, good for plant growth, would normally contain half solids (mineral + organic matter) and half pore space filled by water + air. The latter, at optimum condition for plant growth, would have half-and-half pore space for air and water. This proportion would vary from place to place depending on weather

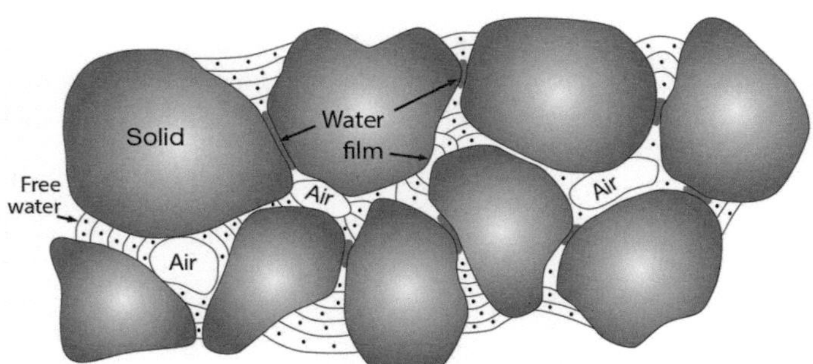

Fig. 7.1 Presence of three states of matter in soil: mineral grains (*solid*), free water in intergranular space, and water films on mineral grains (*liquid*) and micro-pockets of air in the intergranular space (*gas*) of soil

and other factors. In majority of places soil develops in situ, implying hardly any transportation of soil material from the site of generation, although the overall proportion of transported soil is not insignificant either, particularly in temperate regions. Sediments in contrast are materials that are transported and redeposited by the action of wind, water, or ice in different sedimentary environments.

Soil is defined differently for different purposes. To the engineer, it is all unconsolidated material above the bed rock which can be removed without blasting, while to the soil scientist it is the material on the surface of the earth supporting crops and plant growth. However, what is good soil for road building (gravel with not much clay) to the engineer may be a poor quality soil for growing crops. Similarly, organic-rich soil is good for agriculture but unsuitable for road construction. Montmorillonite-rich black soil, developed over the Deccan traps, is good for cotton cultivation, but is unwanted by engineers as it damages the overlying roads by swelling after prolonged rains during monsoon season. Soil is now being increasingly used as the recipient of undesirable municipal, industrial, and mining waste. Bricks made of silty soil have been an important building material since long.

There is a common belief (and reasonably so) that the surficial zone of the continents was first colonized by photosynthetic bacteria and the multicellular land plants first appeared in the Silurian period (~425 Ma ago) and prolific growth of plants took place by the middle of Devonian to early Carboniferous times (400–350 Ma ago). Sedimentary rocks containing soil horizons similar to those found today (Wild 1993) are reported from even older stratigraphies. Paleosols of Precambrian age are reported from several parts of the world, including India. Amongst the very few Archean paleosols known worldwide, the pyrophyllite-bearing paleosol near Keonjhar in Singhbhum craton formed between 3.29 and 3.02 Ga is the oldest reported from India (Mukhopadhyay et al. 2014). Proterozoic paleosols in India are reported from Rajasthan, from Meghalaya, Odisha, and also from the Himalayan region (cf. Bhargava et al. 2011). These metamorphosed paleosols do not retain the textures and structures of original soils, and are mostly identified on the basis of the chemical changes in the perceived weathering profiles. Fire clays, below some coal seams, are sort of paleosols or fossil soils. Their primary characters are not largely modified, however.

7.2 Soil Profile

The soil horizons that occur roughly as a blanket are hardly ever thicker than 2 m. In situ unconsolidated material (regolith) on the surface has a gradational boundary with the bed rock (protolith) and this vertical column, known as *soil profile,* can be divided broadly into two zones, the topsoil and the subsoil (Fig. 7.2a). The uppermost part of a soil profile is commonly gray in color being rich in leaf litter and partially decomposed organic matter, designated as **O** horizon. This is followed downward by the **A** horizon, the plowed horizon or Top soil, which comprises both mineral and organic matter as well as living organisms. It is followed by the zone of

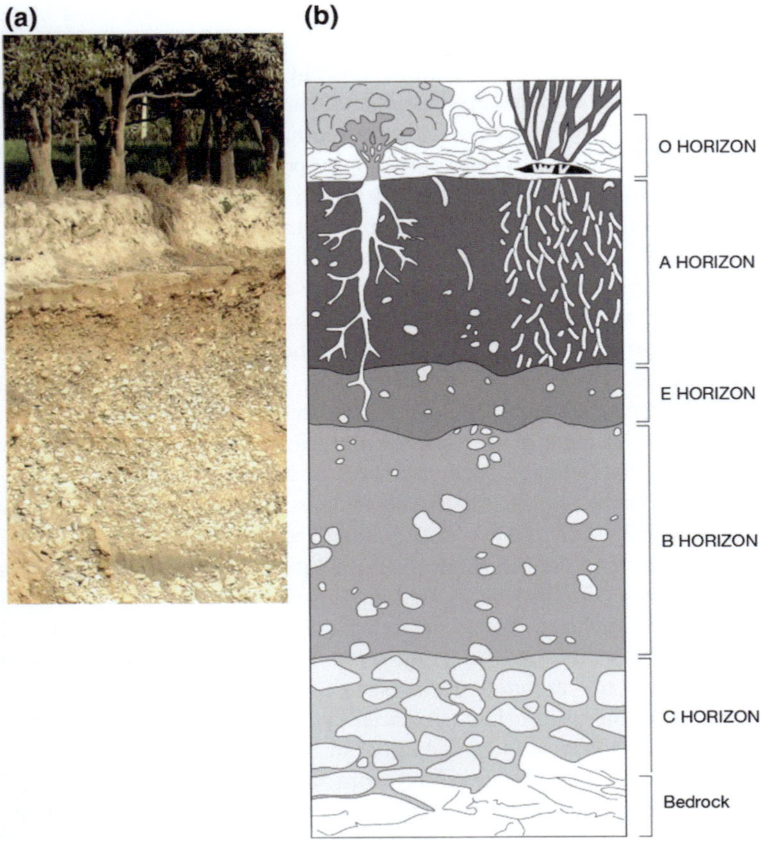

Fig. 7.2 **a** Actual soil profile in a river terrace (Photo: M Deb); **b** a hypothetical soil profile showing the different soil horizons

leaching, **E** or eluviated horizon, which is generally acid or slightly basic, light colored, with the leaching of clays and other minerals resulting in iron and calcium carried below to the **B** horizon, the zone of accumulation. This illuvial B horizon below forms the subsoil and is characterized by the accumulation of clays, iron oxide, carbonates, and other humic materials leached from the A and E horizons above. It is underlain by the **C** or the lowest little altered horizon above the bed rock, sometimes containing pieces of coarsely broken weathered bed rock. A hypothetical soil profile is presented in Fig. 7.2b, in which the actual number of horizons, their thickness, and composition may vary from place to place in different soil types.

The profile described above is represented by soils that are predominantly mineral-rich or inorganic in composition. Although organic matter (OM) is present in such soils, it ranges commonly between 1 and 10%, even in the upper layers. In contrast, soils that are present in swamps, bogs, and marshes are very rich in organic matter, having between 80 and 95% OM. All soils containing >20% OM are classified as *organic* soils. However, since the mineral soils cover much larger land area they are of much greater importance.

7.3 Characteristics of Soil

Soils have many properties but those that are commonly studied are color, textures, mineral and chemical compositions, ion exchange properties and pH of the soil solution, as well as organic content of the soil and interactions amongst the soil components.

Color of soils can vary widely from white to black, which provides a lead to understand how a soil was formed and what it is made up of. It is normally defined with the help of Munsell color chart, an internationally accepted system of color designation. The dark colors such as black or dark brown commonly reflect the high humus content in the soil, such as those found in cool humid areas. On the other hand, humus is present scantily in desert or semi-desert areas rendering the soil light brown or gray. Reddish color of soil is related to the presence of ferric oxides and hydroxides.

Texturally soil is referred to three end members: sand, silt, and clay fractions defined in terms of grain size. According to the International scale,

Gravel: > 2 mm
Coarse sand: 0.2–2 mm
Fine sand: 0.02–0.2 mm
Silt: 0.002–0.02 mm
Clay: < 0.002 mm.

Based on texture, soil is divided into several types and subtypes depending on the relative proportion of sand, silt, and clay. Sand is dominated by quartz (SiO_2) and clay by kaolinite, halloysite, illite, and montmorillonite. Silt may consist of fine quartz, smectite, illite, and chlorite. Textural classification of soil and its nomenclature is presented in Fig. 7.3. Sand particle sizes are determinable instrumentally but an experienced person can get an idea of it by rubbing the soil between fingers. Coarse sand feels gritty; fine sand gives the feel of the individual grains, but not gritty. Silt gives a silky feel and clay sticks to fingers. The importance of clay in soil lies in the fact that its mineralogy and content controls the cation absorption property of the soil from solution. Clay-rich soil can retain water longer. Clay particles stick to each other except in "Alkali soils". The most stable aggregates are found in neutral and calcareous soils and also in some tropical soils in which the mineral particles are cemented by iron oxides. Laterite is a common variety of that.

Minerals in soil are dominantly quartz (SiO_2) and clays. Two groups of clays are recognized: (a) silicate clays, characteristic of temperate regions and (b) iron and

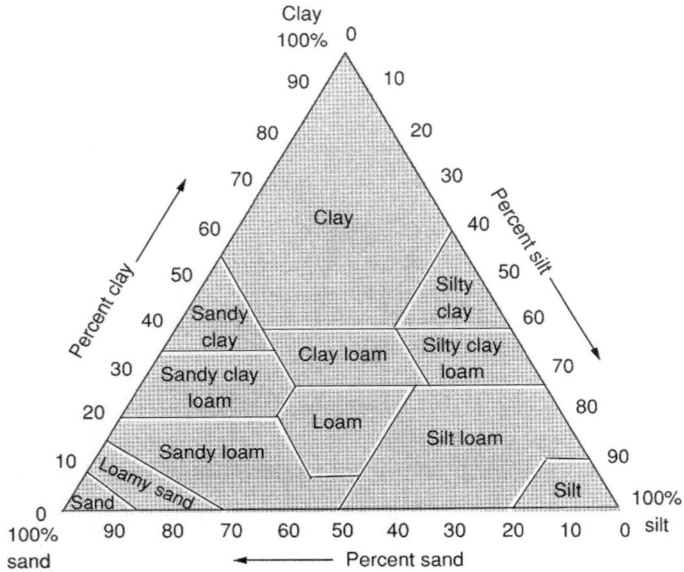

Fig. 7.3 Textural classification of soils based on the proportion of sand, silt, and clay

aluminum oxide clays which are prominent in the well-weathered soils of the tropics and semi-tropic. The properties of red and yellow soils in these regions are largely controlled by these clays. The silicate clays are divided into the following groups (Table 7.1) with the members included in parentheses: Kaolinite/Kandite (kaolinite, dickite, nacrite, halloysite), Illite (illite, hydromica, phengite, glauconite), Smectite (montmorillonite, nontronite, saponite, hectonite), Chlorite (chlorite, chamosite, clinochlore). The first group minerals are hydrated aluminosilicates, and the second group is K-bearing hydrated aluminosilicates. The third group contains Na instead of K and the fourth group contains both Mg and Fe. Otherwise they are also aluminosilicates. Attapulgite which includes the 'wool cleaner' 'Fuller's earth' contains Mg in a fair proportion, but no Al. Vermiculite is an expanding-type mineral with structural characteristics similar to that of smectite. These minerals may form hydrothermally, by supergene alteration of rocks, or by sedimentation (Table 7.2). Feldspars and micas may be present in soils that developed on not fully decomposed rocks. Accessory minerals such as, ilmenite, rutile, and zircon are resistant to alteration. Top soil being in the atmospheric oxidation zone will fix iron in Fe (III) state, i.e., hematite and Mn in several oxidation states. Fe and Al oxide clays are represented by gibbsite [$Al(OH)_3$] and goethite (FeOOH). Most Fe and Al oxides are not as sticky, plastic, and cohesive as the silicates.

Fine-grained micas (illite), chlorite, and vermiculite are produced by rather mild weathering of primary aluminosilicate minerals, whereas kaolinite and oxides of Fe and Al are products of much more intense weathering. Conditions of intermediate weathering encourage the formation of smectite. In each case, the formation of

Table 7.1 Mineralogical classification of silicate clays

Distinctive properties	Groups				
	Kaolinite group (kaolinite, dickite, nacrite, halloysite)	Illite group (illite, hydromicas, phengite, glauconite)	Smectite group (montmorillonite, nontronite, saponite, hectorite)	Chlorite (chlorite, chamosite, clinochlore)	Attapulgite
Composition	$Al_4[Si_4O_{10}](OH)_8$ $Al_4[Si_4O_{10}]$ $(OH)_8.8H_2O$ For halloysite	$K_{1\text{-}1.5}Al_4[Si_{7\text{-}6.5}\ Al_{1\text{-}1.5}O_{20}]$ $(OH)_4$	$Na_{0.7}(Al_{3.3}Mg_{0.7})[Si_8O_{20}]$ $(OH)_4.nH_2O$ In hectorite Mg and some Li in place of Al	$(Mg,Al,Fe)_{12}$ [(Si, $Al)_8O_{20}]$ $(OH)_{16}$	$Mg_5Si_8O_{20}(OH)_2.4H_2O$
Structure	One layer of SiO_4 tetrahedra bonded to a layer with Al in octahedral coordination (1:1)	A layer of octahedrally coordinated Al atoms sandwiched between two tetrahedral $(SiAl)_4$ layers (2:1 type). K between these layers. Non-expanding type	Also a (2:1) sheet or layer structure with interlayer Na or H_2O. Noted for interlayer expansion when the minerals are wetted. High plasticity and cohesion	A (2:1) sheet with another sheet of octahedrally coordinated Al, Fe, Mg, i.e., a (2:1:1) sheet	A chain-like structure, where a double silica is linked together by octahedral groups of oxygen and hydroxyls, containing Al and Mg atoms
Origin	May be any of hydrothermal, supergene-residual or sedimentary origin	Generally sedimentary in origin	Montmorillonite or smectite form by the alteration of mafic volcanic rocks	Chlorite is a supergene alteration product or hydrothermal alteration product of mafic rocks. A common mineral in a variety of soils	Deposits are of sedimentary origin
Use	Porcelain, paper and many other industries	Associated commonly with other clay minerals. It is used in making bricks, tiles etc.	Used in foundry clays and drilling muds. Also used as bleaching muds and absorbents		An important component of Fuller's earth, which has cleansing properties of wool etc.

Table 7.2 Some rocks and rock-forming minerals and their supergene alteration products

Rocks and primary rock-forming (silicate) minerals	Alteration product
Tuff (mafic)	Vermiculite (Ca, Al, Mg, Fe)$_3$ [(Si, Al)$_4$O$_{10}$](OH)$_2$.4H$_2$O Montmorillonite (Ca, Na, Al, Mg, Fe)$_2$[(Si,Al)$_4$O$_{10}$] (OH)$_2$.nH$_2$O
Olivine (Mg, Fe)$_2$SiO$_4$	Limonite, 2Fe$_2$O$_3$.3H$_2$O Serpentine, Mg$_6$Si$_4$O$_{10}$(OH)$_8$
Pyroxenes and amphiboles (complex silicates of Mg, Fe, Ca)	Chlorite, (Mg, Fe)$_5$Al[(Si$_3$, Al)O$_{10}$] (OH)$_2$Talc, Mg$_3$[Si$_4$O$_{10}$](OH)$_2$ Vermiculite Serpentine Limonite
K-Feldspar, KAlSi$_3$O$_8$ plagioclase feldspar (Na, Ca)AlSi$_3$O$_8$	Kaolinite, Al$_4$[Si$_4$O$_{10}$](OH)$_8$ Bauxite minerals, Al$_2$O$_3$.nH$_2$O Illite, KAl$_2$[Si$_{3.5}$Al$_{0.5}$O$_{10}$](OH)$_2$
Biotite muscovite KAl$_3$Si$_3$O$_{10}$(OH)$_2$ magnetite (FeO.Fe$_2$O$_3$), hematite (Fe$_2$O$_3$) Quartz is very resistant to both chemical and physical weathering	Vermiculite, smectite, Fe oxides, illite Illite Hematite, limonite (Fe$_2$O$_3$.nH$_2$O)

silicate clays is accompanied by the removal of soluble elements, such as K, Na, Ca, and Mg.

In cultivated soils, weakly acidic-alkaline (pH 6.5–7.5), the ions that are commonly present are Ca^{2+}, Mg^{2+}, K^+, Na^+, NO_3^-, Cl^-, and SO_4^{2-}. Soil solutions also contain silicic acid [Si(OH)$_4$]. $(NH_4)^+$ is present sometimes.

The major elements in soil and their relative proportions are as follows:

$$O > Si > Al > Fe = C = Ca > K > Na > Mg > Ti > N > S.$$

Most of these elements are derived from the protolith (original parent rock) of the soil. However, in most soils the proportions of C, N, and S are much higher. This is due to biological activity.

The charges associated with soil particles attract simple and complex ions of opposite charge. Thus, in a 1:1-type silicate clay particle, the adsorbed cations and anions may be represented as $=/Ca^{2+}$, $+/HSO_4^-$, $-/K^+$, $=/Mg^{2+}$, $++/HPO_4^{2-}$, $-/H^+$, $+/NO_3^-$, $-/Al(OH)_2^+$, and $+/H_2PO_4^-$. The various cations adsorbed by soil colloids are subject to replacement by other cations through a process called *cation exchange*. Thus, hydrogen ions formed by the decomposition of organic matter can displace Ca and other metallic cations from the colloidal complex. However, these reactions are reversible. When lime is added to soil, the Ca replaces hydrogen and other cations. The clay becomes higher in exchangeable Ca and lower in adsorbed

H^+ and Al, and the soil pH is raised. The capacity of soils to hold exchangeable cations increases with the increase of pH. The cation exchange capacity of a soil is defined as the sum total of the exchangeable cations that a soil can absorb, expressed in terms of moles of positive charge per unit mass (Brady 1984). The exchangeable cations are generally available to both higher plants and microorganisms as nutrients. Any water moving through the soil will lose many of its soluble cations to the soil and pick up those cations that were replaced from the exchange sites by the cations being absorbed. The cation affinity toward an absorbing surface is greater for (a) divalent than for monovalent ions and (b) large cations than for smaller ones of the same charge. The adsorption strength of cations increases approximately in the following order:

$$Na^+ < K^+ = NH_4^+ < Mg^{2+} = Ca^{2+} < Al(OH)_2^+ < H^+.$$

There is commonly an enormous population of bacteria in soils – several million of them per gram. Add to these the larger organisms: earthworms, insects, etc. They play important roles in recycling of nutrient elements and thereby maintaining the soil fertility. In the top 10 cm or so of the soil, generally 1–3% carbon is present as organic compounds, products of decomposition of grass, leaves and other parts of plants, organisms, etc. Soil is enriched in carbon compounds as products of photosynthesis in roots, stems, leaves, etc. As a result of decomposition of organic matter by soil organism, the structure of the former is obliterated and the remaining carbonaceous matter is called *humus*.

7.4 Soil Nutrients and Their Supply

A minimally quantitative and qualitative presence of nutrients in soils is essential for the healthy development of agriculture and horticulture at a place. These may be naturally derived or naturally derived + artificially supplemented. Naturally derived nutrients are sourced from

1. Atmosphere: rain water, snow, aerosols, ammonia, nitrogen oxides; C, H, O
2. Soils: Macronutrients: N, P, K, Na, Ca, Mg, S
   Micronutrients: Fe, Mn, Cu, Zn, Mo, Co, B, Cl

A major part of the nitrogen and sulfur is derived from the atmosphere. Bulk of the nutrients in soil is otherwise derived by the chemical/biochemical degradation of rocks. There is recycling of the nutrients by the assimilation of dead and decomposed plants and plant materials. Microbiological activity helps in fixing nitrogen in the soil. Water in the soil may add or remove nutrients depending on its chemistry and mobility.

Fertility of soils is augmented by the application of fertilizer, organic or inorganic. The organic one is mostly compost. Organic fertilizers can better hold water in the soil. The inorganic list is rather long. Commonly used manure and fertilizers are as follows: Sulfate of ammonia (20.5% N), urea (45% N), ammonium nitrate (35% N), rock phosphate (11–15% P) (at least 25% P_2O_5), superphosphate (7–8% P), basic slag (2–8% P), bone meal (7–13% P), and potassium chloride (39–42% K).

7.5 Soil Formation

Soil essentially forms by the decomposition of exposed rocks, generally with the addition and decomposition of organic matter, with subsequent accumulation in situ, or elsewhere through transportation. Transportation (erosion) of the soil material formed leads to new soil formation. Faster rise of mountains in tropical climates lead to faster weathering and erosion. The eroded soil forms alluvial fans, large alluvial basins, such as, the Indo-Gangetic-Brahmaputra basin, the largest of the type in the World.

Soil formation is essentially a product of several natural processes which are largely controlled by five major factors (Brady 1984):

1. Climate (mainly temperature and precipitation)
2. Living organisms (native vegetation, biota and humans)
3. Nature of parent material (Texture, structure, chemical and mineralogical composition)
4. Topography of the area
5. Time that parent materials are subjected to chemical weathering.

Climate is by far the most dominant factor because of the effects of precipitation and temperature on the physical and chemical processes that operate during soil formation. It controls the nature of weathering itself. For every 10 °C rise in temperature, the rate of chemical reaction appears to double. Temperature and moisture increase also impacts biochemical changes by soil organisms. The contrast between thin soil profile in arid climates and deep weathered profile in humid tropical regions is clearly an evidence of climate control in soil formation. Extreme weather condition like very high temperature during drought can remove all the moisture from the soil allowing it to crack up into polygons rendering it useless for growth of any crop (Fig. 7.4).

Fig. 7.4 Mud cracks in soil during drought conditions. Millions of people in 13 states are affected by severe drought in 2016, Maharashtra being worst affected (Photo: *Indianexpress.com*)

The nature of vegetation is also an important factor that determines soil character. Soils formed under grassland and forest cover differ markedly, for example. Organic matter content is generally much higher in soil under grassland with higher moisture and cation-holding capacities. In a forest, the cycling of nutrients from the litter falling from the trees will also differ markedly depending on the type of trees in the forest. Nitrogen is added to the soil system by microorganisms interacting with the plant litter. Vegetative cover also reduces the erosion rates of soils. Further, human activities like cutting of trees, cultivation, irrigation, and adding fertilizers to the soil may have significant effect on soil-forming processes.

The nature of the parent material, as mentioned above, has strong influence on the effectiveness of the weathering process, mainly by allowing differing rates of water percolation and partially controlling the nature of vegetation. It also controls the type of clay minerals that develop in a profile. Illite, for example, develops by weathering of K-rich micas, whereas smectite develops from minerals high in Ca and Mg. Chemical weathering or decomposition may take place by chemical decomposition of orthoclase feldspar and other rock-forming minerals (Table 7.2), producing kaolinite (Fig. 7.5) as a step in the process of bauxitization:

$$\underset{\text{(Orthoclase)}}{4KAlSi_3O_8} + 4H^+ + 18H_2O \rightarrow \underset{\text{(Kaolinite)}}{Si_4Al_4O_{10}(OH)_8} + 4\,K^+ + 8\,Si(OH)_4.$$

Ferrous iron converts to ferric iron on oxidation throwing out one electron from the outer shell and combining with oxygen forms ferric oxide and (OH), ferric hydroxide:

$$Fe^{2+} \rightarrow Fe^{3+}(+\ \text{electron}).$$

Ferric hydroxide [$Fe(OH)_3$] forms a coating on the weathering products or 'micro'-nuggets.

Also, silica will be hydrated through the reaction: $SiO_2 + 2H_2O \rightarrow Si(OH)_4$(Silicic acid).

Topography of the land, whether flat or rolling, will affect the rate of removal of the water of precipitation and thus modify both the effects of climate and vegetation on soil formation. In a rolling topography, the nature of the soil in the depressions will be significantly different from those in the uplands and so will be the accompanying vegetation.

The length of time that rock materials have been subjected to weathering also controls soil formation. It also has its effect on the other factors. The soil, for example, in the alluvial plains does not get as much time to develop as in the upland areas.

Some other factors also play varying roles in soil formation. These include the following (Wild 1993):

Addition and partial decomposition of organic matter
Formation of structural units

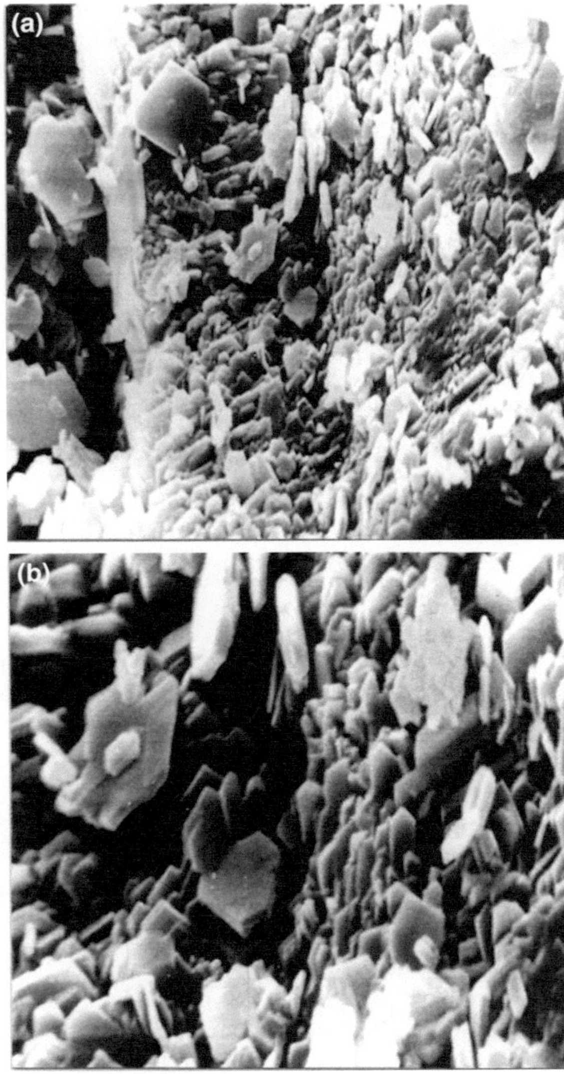

Fig. 7.5 Scanning electron microscope (SEM) photographs of well-crystalline kaolinite mass in altered khondalite at Galikoda, Andhra Pradesh. **a** 2400x; **b** part of the same field at 6000x

Leaching and acidification
Clay eluviations
Podzolization
Desilication
Reduction
Salinization and alkalization
Erosion, transport, and deposition, if the soil is a transported one.

Decomposition is augmented by lichens, i.e., association of fungi and algae, soil microorganisms, and roots of higher plants. Wet and hot tropical climate is favorable for soil formation.

Factors that determine soil property can be put into an equation for soil formation (cf. Wild 1993):

$$S = f\ (L_0, P_x, t),$$

where S, the soil produced, is a function (f) of L_0, the state of the system when soil formation starts and is inclusive of mineral and organic materials and relief; and P_x includes climate and organisms and t is the time involved.

7.6 Soil Classification

A large number of systems of classification of soils have been proposed to date, including one proposed by FAO (UNESCO soil map of the World). The US Department of Agriculture has proposed one (Table 7.3) based on soil taxonomy.

Table 7.3 Soil orders used in soil taxonomy (after US Department of Agriculture)

Soil order	Some properties of the order	Mha	%
1. Aridisols	Soils in dry climates where salts and carbonates (caliche) may have accumulated. Low organic accumulation	2480	18.8
2.. Alfisols	Soils with clay translocation and deposition (high base status)	1730	13.1
3. Inceptisols	Young soils with few diagnostic features. Most common in humid climates but may range from Arctic to the tropics	1170	8.9
4. Mollisols	Temperate grassland soils	1130	8.6
5. Oxisols	Highly weathered soils of the tropics and subtropical regions. Leached of bases, hydrated	1120	8.5
6. Entisols	Recently formed soils with limited development of horizons	1090	8.2
7. Ulfisols	Soils with clay translocation and deposition (low base status). Characterized by an argillic horizon, restricted to humid climates	730	5.6
8. Spodosols ('podzols')	Soils with subsoil accumulation of humus and iron/aluminum oxides	560	4.3
9. Vertisols	Swelling clay soils, that expand or contract with changing moisture content in regions of pronounced wet and dry season	230	1.8
10. Histosols	Organic soils (peat, muck, bog)	120	0.9
Andisols	Soils derived primarily from volcanic material, relatively rich in chemically active minerals		
Gelisols (proposed new order)	Soils with permafrost within a depth of about 1–2 m		
TOTAL		13,170	

The hierarchy of division is somewhat extended, the first level of division being called the order, which are ten in number. The orders are characterized by diagnostic horizons.

Generalized classifications are meant for international reference and overall comparisons. Most countries have their own system of soil classification based on the ground reality. In India this job is entrusted to National Bureau of Soil Survey and Land Use Planning. For a large country like India it is a time taking and expensive but a useful exercise.

7.7 Acidification of Soils

Acidification of soil is a neutral process, involving (a) oxidation of NH_3, NH_4^+, S^{2-}, and Fe^{2+}; (b) increase of humus content; (c) removal of basic cations in crop harvests; and (d) secretion of protons by plant roots. The acids produced are H_2CO_3, HNO_3, H_2SO_4, and HCl. Atmosphere contributes to acidification of soils through pouring of acid rains. Mine waters and, locally, industrial discharge may be acidifying for the local soil.

High soil acidity has a severe adverse effect on plant growth. It hinders activity of micro- and meso-organism. There is an increase in phytotoxic elements, such as, Mn and Al, in the soil solution. There is also decrease in the availability of P, deficiency of Mo. Liming is the time-tested method of mitigating soil acidity. It supplies Ca, improves the solubility of P, and precipitates Al. Lime amendment is generally in the range of 1.5–1.6 t of $CaCO_3$/ha.

7.8 Contamination of Soil

There are quite a few elements, mostly belonging to the 'heavy metals' group (density >5 or 6 gm cm^{-3}), that occur in soils and are hazardous for plants and animals, including human kind. They enter the biological systems through the food chain and water. This group includes Cd, Cr, Cu, Pb, Hg, Ni, Zn, Tl, and As. Sources of these deleterious elements are industries utilizing them, mines and smelters excavating and processing the same ores, municipal waste not properly treated. Nature itself may be the source of some such elements, e.g., arsenic in the Bengal basin and Tl in the groundwater system southwest Guizhou Province, China (Tangfu Xiao et al. 2003). As, F, Hg, and Se are sometimes contributed by a dying volcano. Some hazardous elements and compounds in soils are contributions by inorganic fertilizers and pesticides. It must be mentioned here that all the above elements are not deleterious for human health under all circumstances. Some of them are beneficial if kept within limits. One may recall here an old dictum, "everything is a poison, nothing is a poison, the dose alone makes the poison" (Paracelsus, 1493–1541 in Ross 1999). There is a great amount of truth in it, but possibly not the whole truth.

Soil also gets contaminated from the following agricultural sources:

Fertilizers: As, Cd, Mn, U, V, Zn in some phosphatic fertilizers
Manures: As and Cu in some pig and poultry manure
Pesticides: As, Cu, Mn, Pb, Zn, persistent organic chemicals (DDT)
Corrosion of metals: galvanized metal objects, e.g., fencing, troughs etc.
Fuel spillages: hydrocarbons
Burial of dead livestock: pathogenic microorganisms.

7.9 Salination of Soils

Irrigation is a common practice where there is limited rainfall. But a common problem in such cases is the accumulation of salts, particularly Na salts, which may adversely affect the plant growth. Irrigation water drawn from surface or ground water sources typically contain 200–2000 ppm of salt. Thus, application in a single season of 1000 mm of water with 500 ppm of salt would add 5 tons of salt to a hectare of land. White encrustrations on soil of saline salt are generally composed of chlorides and sulfates of Na, Ca, and Mg (Fig. 7.6). As a remedy more water

Fig. 7.6 White encrustations of saline material spread extensively on soil in Churu district, Rajasthan (Photo: M Deb)

should be added to the system, so that the soils are washed into the subsoil and discharge into a river via drains. Excess water from soil or subsoil is drained through ditches, perforated pipes, by gravity flow or pumping.

7.10 Soil Erosion

Erosion is the worst hazard in case of soils. Soil erosion means removal of the upper fertile part of the soil profile by surface run-off (Fig. 7.7) or blowing air. This part is played by nature, but aided and abetted by mankind in most cases. A grass and/or plant cover is a natural protection for soil. But people cut trees and plants for domestic use or to increase their croplands. Soils, softened in the process of cultivation, are blown away by strong wind. Herbivorous domestic animals (cows, sheep, and goats) remove the grass cover by overgrazing, in some places leading to desertification. Steep topography and high rainfall promote erosion. Eroded lands are poor crop growers not only because their fertile top soil is removed but also for the fact that their water penetration and retention capacity is low. The top 15–20 cm of the soil profile is most productive.

Factors that affect erosion are included in the *Universal Soil-Loss equation*:

$$A = RKLSCP,$$

Fig. 7.7 Soil erosion by surface run-off in a sloping topography at Ghora Katora hill, Rajgir, Bihar (Photo: courtesy Jyothi Kandula)

where A, the predicted soil loss in metric tons per hectre per year, is the product of

R Rainfall and run-off
K Soil erodibility
L Slope length
S Slope gradient or steepness
C Cover and management
P Erosion control practice.

In conclusion, soil is important to all lives on land, next only to water, for their sustenance. Without the soil there would be no trees, no plants, no crops—and no food for people and animals, and so no oxygen—carbon dioxide balance. So we must conserve its volume and preserve its quality. Soil profiles must be maintained minimizing erosion, salination, and acidification. Deforestation in the region must be checked so that required rainfall is there. Hazardous elements must be identified and removed as far and as soon as possible. It must be pointed out that today the world's population is about 7 billion and India's population has exceeded 1.2 billion. A strong demand in India would be to appropriately manage tropical and subtropical soils for their restoration and preservation. If they are not cared for, crops will fail even with good rainfall. By the end of the century (21st) the world population is likely to peak at 11 billion before it stabilizes on a plateau. The world has to arrange for their sustainable living, in which at least food and water must be secured. Besides, the people today need so many industrial products for a reasonable living. As a result, there is now a raging controversy regarding distribution of land between agriculture and industry. Hence, land use planning is an advisable practice for any government today, particularly in the developing world.

7.11 Soil in India

The beginning of soil study in India was made by Voelcker in 1893 and by Leather in 1898. They categorized the soils in India into four major groups: the Indo-Gangetic alluvium, the black cotton soil or *regur*, red soil, and laterite soil. Since then various workers have categorized soils in India into 16 groups, 16 major and 108 minor soil regions, 27 major soil groups (Raychaudhuri and Govindarajan 1971), 25 broad soil classes, 24 major soil groups, and most recently into 11 major soil groups (Table 7.4) (Bhattacharyya et al. 2013). Regarding soil nomenclature, these authors draw attention to the problem of classifying soils as red, laterites and hilly since at many places these names overlap. Moreover, recent findings on shrink–swell soils (generally known as black soils) have shown them to be red in color, which makes the common name of red and black soils redundant.

In a recent soil survey, the areal extent of different soil orders, suborders, and great group in India has been estimated (Despande and Sarkar 2009). The data indicate that 13.55% of Alfisols, 39.74% of Inceptisols, 4.3% of Aridisols, 28.1% of

Table 7.4 The extent and distribution of the different soil classes of India along with their equivalents according to United States Department of Agriculture, USA nomenclature system (modified after Bhattacharyya et al. 2013)

Major soils (traditional name)	Extent (percentage)	Distribution in states	Soil orders US soil taxonomy
Alluvial	30.4	J&K, H.P., Punjab, Haryana, Delhi, U.P., Gujarat, Goa, M.P., MS, A.P., Karnataka, TN, Kerala, Puducherry, Bihar, Odisha, WB, A.P., Assam, Nagaland, Manipur, Mizoram, Tripura, Meghalaya, A&N	Inceptisols, entisols, alfisols, aridisols
Coastal alluvial	3.1	A.P., Karnataka, TN, Kerala, WB, Gujarat, Odisha, Puducherry, Lakshadweep, A&N	Aridisols, inceptisols, entisols
Red	26.8	A.P., Karnataka, Kerala, TN, Puducherry, Rajasthan, M.P., MS, Gujarat, Goa, ArP, Assam, Manipur, Meghalaya, Nagaland, Mizoram, Tripura, Delhi, U.P., H.P., A&N	Alfisols, ultisols, entisols, inceptisols, mollisols, aridisols
Laterites	5.5	A.P., Karnataka, Kerala, TN, Puducherry, MS, Odisha, WB	Alfisols, ultisols, inceptisols
Brown forest	0.2	Karnataka, Maharashtra	Mollisols, inceptisols
Hill	0.7	Manipur, Odisha, WB, Tripura, Nagaland	Inceptisols, entisols
Terai	0.1	U.P., Sikkim	Mollisols, entisols
Mountain meadow	–	J&K	Mollisols
Sub-montane	–	J&K	Alfisols
Black	16.6	M.P., MS, Rajasthan, Puducherry, TN, U.P., Bihar, Odisha, A.P., Gujarat	Vertisols, mollisols, inceptisols, entisols, aridisols
Desert	8.0	Rajasthan, Gujarat, Haryana, Punjab	Aridisols, inceptisols, entisols
Others[a]	8.6	–	–
Total	100	–	–

[a]Includes glaciers (0.4%), sand dunes (0.01%), mangrove swamps (0.04%), salt waste (0.01%), water bodies (0.1%), rock land (0.25%) and rock outcrops (7.8%). *MP* Madhya Pradesh; *MS* Maharashtra; *UP* Uttar Pradesh; *J&K*, Jammu and Kashmir; *TN* Tamil Nadu; *AP* Andhra Pradesh; *ArP* Arunachal Pradesh; *WB* West Bengal; *HP* Himachal Pradesh; *A&N* Andaman and Nicobar Islands

Entisols, 8.5% of Vertisols, 2.5% of Ultisols, 0.40% of Mollisols, and 2.92% Others occur in India.

Soils in India are diverse and differ from area to area. The wide variety of geological formations of India is responsible for the development of a wide variety of soils on them under different climatic, topographic, and vegetative conditions as well as the nature and duration of the weathering processes they have undergone. Consequently, the characteristics of soil mentioned above vary over a wide range and can be related broadly to the main physiographic divisions, such as the plateaux of the peninsular region, the Indo-Gangetic plains, and the Himalayan terrain to the north of it. Wadia and his co-workers (1935) compiled a soil map of India with emphasis on geological formations and classified the soils as red, black (regur), laterite, and lateritic soils of Peninsular India, delta, desert, bhabar, terai, and alkali soils of the Indo-Gangetic Plains (IGP). The soils of peninsular India, south of the

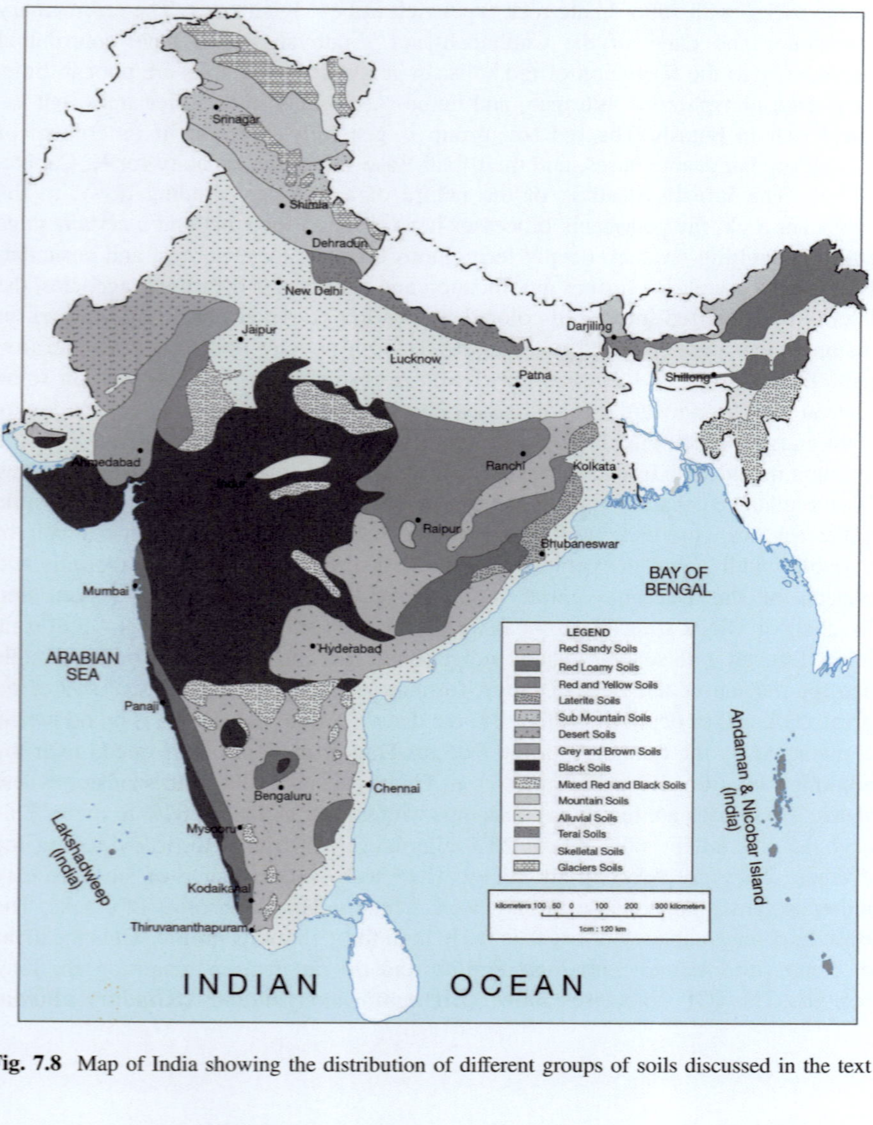

Fig. 7.8 Map of India showing the distribution of different groups of soils discussed in the text

Vindhyan range, are very mature soils and are of great antiquity and stand out in contrast to the drift soils in the great plains of north India. The alluvial soils of IGP are largely immature and have not developed any prominent profile. All these groups of soils and more are shown in a soil map of India (Fig. 7.8). Readers interested in a modern soil map of India in terms of sub-order associations are referred to the NBSS & LUP (1985) map in a scale of 1:7,000,000.

The **red soils** form a major group in India and can be categorized under "Ferruginous soils" of the tropics. They are variously termed by their color as "red", "brown", "yellow", "gray", "lateritic", "ferralitic", or "latosols", in the absence of a precise definition. The parent rocks are granites and gneisses, quartzitic, or feldspathic rocks, with subordinate rock types rich in Fe–Mg silicates. The sedimentary sandstones and clays of the Cuddapah and Vindhyan basins have contributed secondarily to the formation of red soils. In general, the red soils are poor in lime, magnesia, phosphorous, nitrogen, and humus, especially in the drier areas, but are fairly rich in potash. The red soil group is generally deficient in its content of soluble exchangeable bases, and their total Base Exchange capacity for K, Ca, etc. is low. The **laterite** itself is of the nature of a soil cap, grading down to the subjacent rock, the pedogenic processes have stopped short beyond a certain stage and the resulting product, deeply ferruginous compact clay, porous, and vesicular, is not really a soil. By further modification and action of soil-forming agencies, the laterite is converted into a red-colored soil with a common profile of 1–3 feet, of disintegrated loam speckled with segregated iron nodules and sometimes, an iron pan. The laterite cap, however, may be of great thickness, 100–200 feet, on some plateau tops. Downward this soil passes into soft, easily sectile, yellow, or buff clay (lithomarge). Often the clay base of lateritic soil is not aluminum silicate but free alumina mixed with iron sesquioxides. Because of the intense leaching and the low Base Exchange capacity, typical lateritic soils are lacking in fertility and are of little value for crop production. Regionally, the lateritic soil group forms a belt of variable width round the peninsula, in its best development capping the hills and plateau of the Deccan, central India, Bengal, and extending to Assam and Meghalaya. **Black soils** (*Regur or black cotton soil*) are another large group of soils in the Deccan with several distinct subvarieties. The most characteristic black soils cap the plateau of the Deccan Traps, forming a mantle of rich residual soil of no great thickness or depth of profile. But the distribution of black soils is by no means constrained by the boundary of the Deccan Trap formation, being spread over the adjoining granitic and mafic gneisses or Cuddapah and Vindhyan sandstones and slates. These soils are highly argillaceous, with large clay factor, 62% or more. This clay factor contains 60% silica, 25% alumina, and 15% of ferric oxides on the average. They are exceedingly sticky when wet. A dominance of smectite clay minerals causes appreciable shrink–swell, which induces formation of cracks. The Base Exchange capacity of *regur* is fairly high throughout its profile. Thus the areas of *regur* are credited with high fertility and do not require manuring for long periods. The IGP comprises **older (Bhangar)** and **younger (Khadar) alluvia**.

The older alluvium, marked by a higher degree of pedogenesis, covers the Gangetic basin in U.P., Bihar, and Bengal. It is represented by darker brown argillaceous loam with a considerably large clay factor, less sand, and greater proportion of moisture. The lime content is higher with thickly diffused nodular *kankar*. The *Bhangar* is assigned a minimum age of 120,000 years (Pal 2008). The younger *khadar* soil is common in the north-west, in Punjab and adjoining areas. It is represented by pale brown, sandy clays and loams, less calcareous and carbonatic, i.e., with less *kankar*. The water table in the Bhangar is always relatively higher than in the north-west alluvial tract of khaddar where the subsoil water may be as low as or lower than 50 feet. Salt-impregnated or **alkaline soils** form a negatively important group in the drier parts of north Bihar, U.P., Punjab, and Rajasthan. These soils have many undecomposed rock and mineral fragments which on weathering liberate Na, Mg, and Ca salts and sulfurous acids. These soils are notably impervious and have impeded drainage and thus, low agricultural value. The salts are confined to the top layer of the soil being transferred from below by capillary action. The *nitre-impregnated* soils of Bihar and Punjab is a variant of the above, the difference being that the salt (potassium nitrate) is introduced into the soil cap through human activities in densely populated cultivated districts in a warm humid climate. In the great deltas of Indus and Ganges, and the rivers draining to the east coast, are the **Delta soils** as a continuum of the loamy alluvial soils upstream, but generally richer in organic matter. These soils are periodically renewed by floods and are as a rule very fertile.

Mountain and Forest soils of the Himalayan region are found in the depressions within the mountains, valleys and on the less steeply inclined slopes. Generally, it is the north-facing slopes of the Himalayan ranges which support a considerable soil cap. The larger extent of these soils is under heavy growth of perennial forests of conifers. The soil caps on the broad zones of Tertiary sandstones and clays, constituting the Siwalik foothills, are primary soils, shallow and immature, containing a large proportion of undecomposed mineral grains. In the broad lateral valleys – the *doons* – as in Dehra Dun, Kashmir, and Nepal are alluvial soils of high fertility.

Desert soils: the Thar area of Rajasthan, covering an area of 40,000 sq. miles is generally devoid of a soil cap of agricultural utility, soil-forming conditions being absent in these extremely arid region. Its place is taken by a mantle of dry barren sands, derived by the mechanical disintegration of the ground rocks or sand blown in from the Indus basin or the coast. The Runn of Kuchchh in the south and southern Punjab in the north are extensions of this desert. Both Kuchchh and Sind are overspread by a surface mantle heavily impregnated with common salt and alkaline salts and dotted with salt pans. Barren sandy soils without a clay factor are also common in the coastal dunes of Odisha, Nellore, and Tamil Nadu as well as in the Malabar Coast.

References

Bhargava ON, Gurmeet Kaur, Deb M (2011) A Paleoproterozoic paleosol horizon in the lesser Himalaya and its regional implications. J Asian Earth Sci 42:1371–1380

Bhattacharyya T, Pal DK, Mandal C, Chandran P, Ray SK, Sarkar D, Velmourougane K, Srivastava A, Sidhu GS, Singh RS, Sahoo AK, Dutta D, Nair KM, Srivastava R, Tiwary P, Nagar AP, Nimkhedkar SS (2013) Soils of India: historical perspective, classification and recent advances. Curr Sci 104(10):1308–1323

Brady NC (1984) The nature and properties of soils. MacMillan Publishing Company, USA 750 p

Despande SB, Sarkar D (2009) Reviews of soil surveys in India. In: Bhattacharyya T, Sarkar D, Pal DK (eds) Soil survey manual, NBSS & LUP, Nagpur, pp 5–16

Mukhopadhyay J, Crowley QG, Ghosh S, Ghosh G, Chakrabarti K, Mishra B, Heron K, Bose S (2014) Oxygenation of the Archean atmosphere: new paleosol constraints from eastern India. Geology 42(10), August: 923–926. doi:10.1130/G36091.1

Pal DK (2008) Studies on Indian soils. In: Singhvi AK, Bhattacharya A, Guha S (eds) Glimpses of geoscience research in India, 2004–2008, pp 61–63 (Ind Nat Sci Acad)

Raychaudhuri SP, Govindarajan SV (1971) Soils of India. Tech Bull (Agr) 25:1–45

Ross M (1999) The health effect of mineral dusts. In: Plumplee GS, Logsdon MJ (eds), Environmental geochemistry of mineral deposits. reviews in economic geology, 6A: 339–356

Tangfu X, Boyle D, Guha J, Rouleau A, Yetang H, Basshan Z (2003) Ground water related Tl transfer processes and their impact on the ecosystem: southwest Guizhou province, China. Appl Geochem 18:675–691

Wadia DN, Krishnan MS, Mukherjee PN (1935) Introductory note on the geological formation of the soils of India. Rec Geol Surv Ind 68:363

Wild A (1993) Soils and the environment—an introduction. Cambridge University Press, Cambridge, UK, p 287

Chapter 8
Water

Abstract Water is an inalienable resource for all kinds of life on Earth. Fast-developing urban population and ever-increasing industrial activities added with negative effects of climate change have made availability of adequate and suitable quantity of water a subject of intranational and international strife in many parts of the world, solution of which, at times, appear elusive. The oceans contain 96.5% of the water in the hydrosphere in the saline state. Only 2.4% is fresh water, but 87% of it resides in ice caps and glaciers. This means much less than 1% is available for the biosphere. Availability of this water is highly variable in space and time. Penetration of ground water and accumulation of surface water in drainage basins and wetting of the land are effects of a complicated *hydrologic cycle*. It is notable that 95% of the liquid fresh water available is ground water in reality. Natural and anthropogenic pollution of ground water (and locally surface water also) is a menace in different parts of the world. Saving water resource is a loud cry today. Best way of doing it is by minimizing domestic and industrial consumption, recycling used water after proper treatment, making direct use of rain water, going for more scientific irrigation and reducing loss during transport through pipes. Desalination of sea water is a far cry for general use at the moment. Water crisis is an acute problem in India. We should resort to as many of the above means as possible to restrain it before long. The usefulness of the river-linking proposal in India is still considered dubious.

Keywords Hydrologic cycle · Global water distribution · Watershed · Effluent/influent streams · Aquifer · Aquiclude · Water pollution · Water conservation · Desalination of sea water · River-linking proposal in India

8.1 Introduction

Water is a basic resource for all kinds of life on earth. Civilizations grew on the banks of rivers where water was easily available. Cities, such as, Fatehpur Sikri, had to be abandoned when there was protracted water scarcity. There are many

M. Deb and S.C. Sarkar, *Minerals and Allied Natural Resources and their Sustainable Development*, Springer Geology,
DOI 10.1007/978-981-10-4564-6_8

other similar cases throughout the world. Therefore ensuring adequate supply of water, both in terms of quantity and quality, to the ever-increasing population in the developing world is a major contemporary challenge, satisfaction of which has far-reaching socioeconomic and even political implications. Due to spatial and temporal variation of rainfall, there is considerable skewness in the availability of water from region to region. Surging urban population in megacities and ever-increasing industrial and agricultural production, along with apparent effects of climate change have made water one of the major resource issues of the twenty-first century. Large countries like China and India, where there are floods in some parts and droughts in others, almost every other year, have planned to implement a more equitable distribution of water by linking the rivers involved in flood with those having water stress. This could ensure adequate water availability in water-deficient areas for irrigation, for producing hydropower, navigational facilities, etc., bringing about considerable economic growth. But this approach also is fraught with serious long-range ecological and environmental consequences (discussed in a later section), besides being very costly. Thus we need to understand varied aspects of source and distribution of water, movement of water through the hydrological cycle, its quality and quantity, withdrawal and conservation, etc., to develop a sustainable water management capacity for every population.

It hardly needs to be emphasized that water is in many ways a unique matter. It occurs in three different states: solid, liquid and gaseous, an overwhelmingly larger portion of it occurring in liquid state though. It is a rare compound in which the solid form, ice, is lighter than the liquid form because its formation is accompanied by a volume expansion of 8%, making it less dense than water and hence it floats on it. If ice were denser than water and the lower parts of ocean, lakes, and rivers were frozen, aquatic life in the biosphere would have been severely affected. Sunlight can penetrate water bodies to variable depths allowing photosynthesis below the surface which is also vital to the biosphere. Compared to other common liquids, water has the capacity to absorb and store heat which has a strong influence on climate. Solar energy heats ocean water, which evaporates as water vapor and forms the cloud of moisture droplets in the atmosphere by condensation to start the *hydrologic cycle*. Warm ocean currents move to higher latitudes making those regions suitable for human habitation. Water is also a universal solvent and can dissolve a variety of compounds, including many minerals in rocks, taking them out of the system through running waters. It has thus the power to shape topography. Water also has high surface tension which helps it to fill tiny pores in rocks and also move through them. It can also react with many organic complexes, including some of those in the human body which contains 50–75% water mostly as intracellular fluid. Water availability may restrict ecosystem development because most organisms need large amount of water to survive. All these qualities, amongst others, make water indispensable to all forms of life on Earth.

8.2 Global Water Distribution

About 70% of Earth's surface is covered with water. The oceans hold 96.5% of this water, largely in saline and non-potable state (Fig. 8.1). Another 1.1% of total global water is also saline, found in lakes and ground water, while only 2.4% is fresh water, that is, water with less than 1000 mg/L of salts or total dissolved solids. However, not all of this water is available for use and consumption. Out of the total fresh water budget, ice caps and glaciers account for 87% (that is, 2.09% out of 2.4%). Thus the amount of water available to the biosphere (human beings, plants, and animals) is much less than 1% of the total. The reservoirs of this liquid fresh water are the subsurface water accumulations, i.e., *ground water* (95%), lakes, rivers, and streams (3%) and soil moisture (2%). It is important to note that all this fresh water resource is linked to the process of precipitation from the atmosphere which holds only about 0.001% of the total water on earth at any one time with an average residence time of only about 9 days (cf. Kundzewicz 2008).

These figures give the impression that available fresh water is really scarce and insufficient for the growing human population on the globe which has already exceeded 7 billion and may increase by about 50% at the close of the present century. However, except for Saharan and sub-Saharan nations, the per capita water availability round the world is still satisfactory on a per country basis. The global average for the total actual renewable water resource (TARWR) per capita is 6000 m^3/inhabitant/year while this figure for India is 1582 (UN Report, 2010). Human populations are not distributed evenly across the planet and tend to concentrate in specific urban patches. Similarly, fresh water resources in the world are far from being evenly distributed. Also the quantity and quality of water available at any particular time and place is highly variable. It is therefore difficult to determine if water is truly scarce in the physical sense at the global or regional scale or it is mainly a problem of inequitable distribution and usage. Further, global population growth in urban centers is projected to double by 2025 (UN Report, 2000) leading to a tremendous increase in urban water demand. Thus water shortages are bound to occur with increasing frequency causing human suffering and socioeconomic disruption. This can also lead to international tensions across borders, particularly

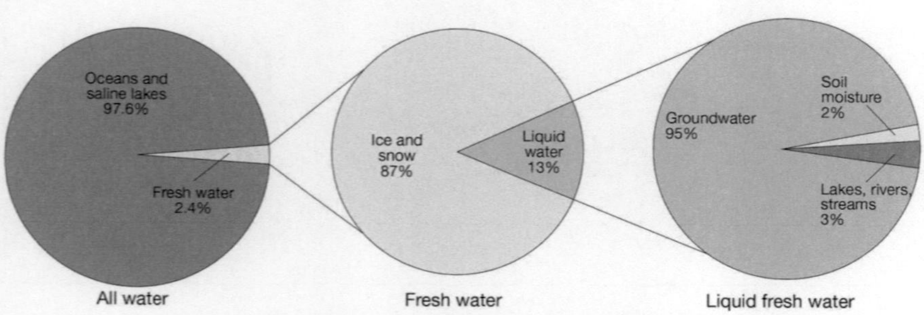

Fig. 8.1 Schematic depiction of global water distribution with emphasis on fresh water

where an important river is passing through more than one country. This added pressure on world's fresh water resources may redefine our understanding of global water resource and its distribution in the near future.

8.3 The Hydrologic Cycle

The hydrosphere, atmosphere, and the uppermost part of the lithosphere constitute the three spheres of our environment through which water of the earth circulates. This global water cycle (Fig. 8.2) is driven by the energy from the sun and involves three major processes: evaporation, precipitation, and transpiration. The solar energy changes part of the water in oceans, lakes and rivers, and also the soil, from the liquid to the gaseous state by evaporation, and it transforms from the gaseous to the liquid phase by condensation in the atmosphere. This water is drawn by gravity back to the earth's surface and enters the hydrologic system as precipitation, in the form of rainfall or snowfall and dew. It leaves the system as stream flow or surface runoff, and evaporation from open bodies of water, soil surfaces, and evapotranspiration from the surfaces of plants (Fig. 8.2). Over land, about 90% of water that reaches the atmosphere comes from the last process and also from the soil.

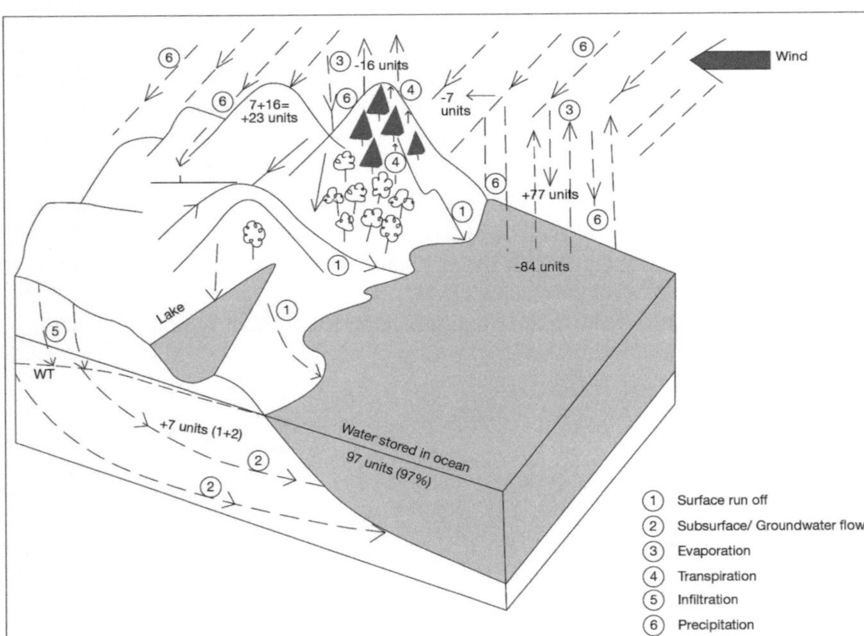

Fig. 8.2 The hydrologic cycle. A rough mass balance is shown between evaporation, precipitation, infiltration, and transpiration

In addition, living organisms emit the water and moisture they have consumed through respiration and perspiration. A large part of this water of precipitation infiltrates into soil and bed rocks of the subsurface.

The water returning to the earth from the atmosphere thus takes various paths in the hydrosphere and lithosphere and gets into different reservoirs (Fig. 8.3). The major part, about 110,000 km^3 falls on land as precipitation each year and after plants have used a part of it, and another part entering subsurface, about 40,000 km^3 is released into streams, rivers, and lakes as *surface runoff* and eventually gets released to the ocean. A part of the precipitation is converted to ice and remain stored in ice caps and glaciers usually for long periods of time. The hydrology of an area is essentially determined by the *watershed* or catchment (also known as the *drainage basin*), defined by the total area that hosts all the surface runoffs to a particular river. Obviously, the watershed can vary in size greatly from a small area for a stream to thousands of square km for large river systems like the Indus or Ganges. Streams are classified as *effluent* or *influent*. The former is maintained in the dry season by ground water seepage into the stream channel, whereas the latter is above the water table and flows only as a response to precipitation. Because of these reasons, an effluent stream could be *perennial* while an influent stream could be *ephemeral*, not flowing all through the year.

The part of the precipitation that percolates through the soil fills up the porous spaces between soil particles while a major part infiltrates into underground through faults, fractures, and shear zones filling intergranular spaces in porous and

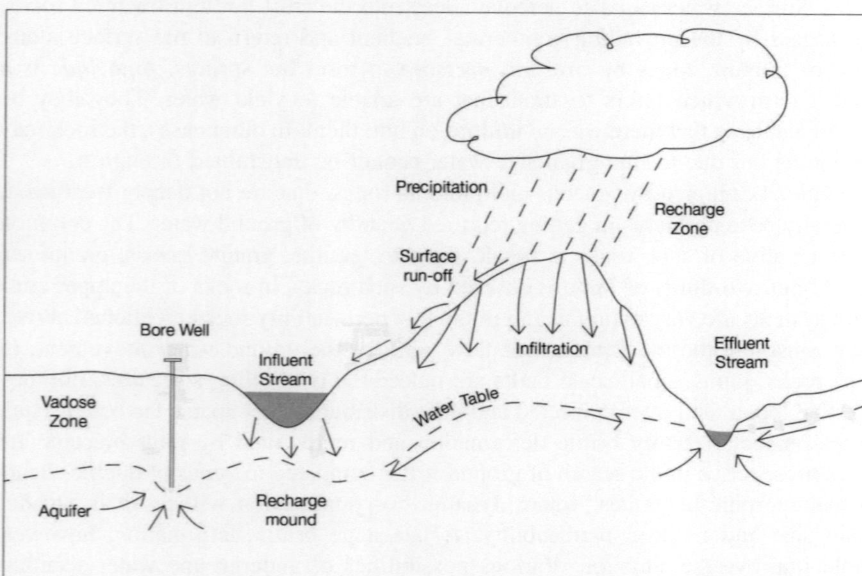

Fig. 8.3 A sketch of two main water reservoirs, surface water and ground water, and the link between the two

permeable rock formations where all spaces are filled with water to define *the zone of saturation*, whose upper limit delineates the *ground water table*. In places where the ground water table intersects the topography a *discharge zone* is formed, comprising water seeps, or springs. Otherwise, the ground water needs to be withdrawn through suitable means, such as dug and borewells.

Ground water accumulation, as can be easily understood, is not uniform over large tracts. They occur selectively in some rocks—in a sedimentary rock sequence or along some structures, fractures, and/or breccias in hard massive rocks, such as basalts, granites, granite gneiss, etc. The rocks and structures which contain more water than the surrounding rock masses are called *aquifers*. The boundaries of the water-filled rock formation delimit the aquifer which may be confined or unconfined, depending on whether or not the aquifer is sandwiched between impervious strata like clay or shale. In sedimentary sequences, coarse grains sandstones, gravel or pebble beds are generally better aquifers, particularly when not characterized by a large content of matrix or cement. The water storage capacity of these materials is relatively high, approximately equivalent to their porosity. Karstic aquifers are commonly met in carbonate rocks, where they form by chemical dissolution of rock matter by water flowing along fractures. Cavities thus formed may at places be large, particularly at the intersection of fractures. Structure-controlled aquifers are found in hard rock formations made up of crystalline, volcanic, or sedimentary strata that allow the flow of ground water through geological structures in them (faults, fractures, and shears). The porosity of these aquifers is relatively low but interconnected openings in these structures often permit a significant flux of ground water. Surface water can also percolate deep into the crust through fractured rocks, get heated by the prevailing geothermal gradient and return to the surface along fault or fracture zones by *artesian pressure* to form hot springs. *Aquiclude* is a related term which refers to strata that are unable to yield water. They may be impermeable so that there was no infiltration into them. In other cases, the rock may be porous but due to fine grain size water cannot be transmitted through it.

Regions occupied by igneous and plutonic rocks, that are not deeply weathered, generally pose problems in getting required quantity of ground water. The common representative of such rocks is basalt, rhyolite, granite, granite gneiss, granulites, etc. About two-thirds of India is covered by such rocks. In rocks of the upper crust joint systems are very common and in the low permeability rocks mentioned above, they constitute the most prominent flow path for the ground water movement. In such rocks, joints, small scale faults are indeed the controlling structures, but they must be 'open' and interconnected and their distribution less sparse the better. Fault zones characterized by brittle deformation and represented by fault breccias are more prospective in the search of ground water compared to zones of ductile shears in metamorphic tectonites, where dynamic recrystallization will result in smaller grain size and reduced permeability. A late-stage brittle deformation, however, could improve the situation. Various possibilities of underground water accumulation and flow are shown in a few sketches in Fig. 8.4.

Ground water flow varies laterally in an aquifer depending on the type and structure of the geological material that constitutes the subsurface. The flux of

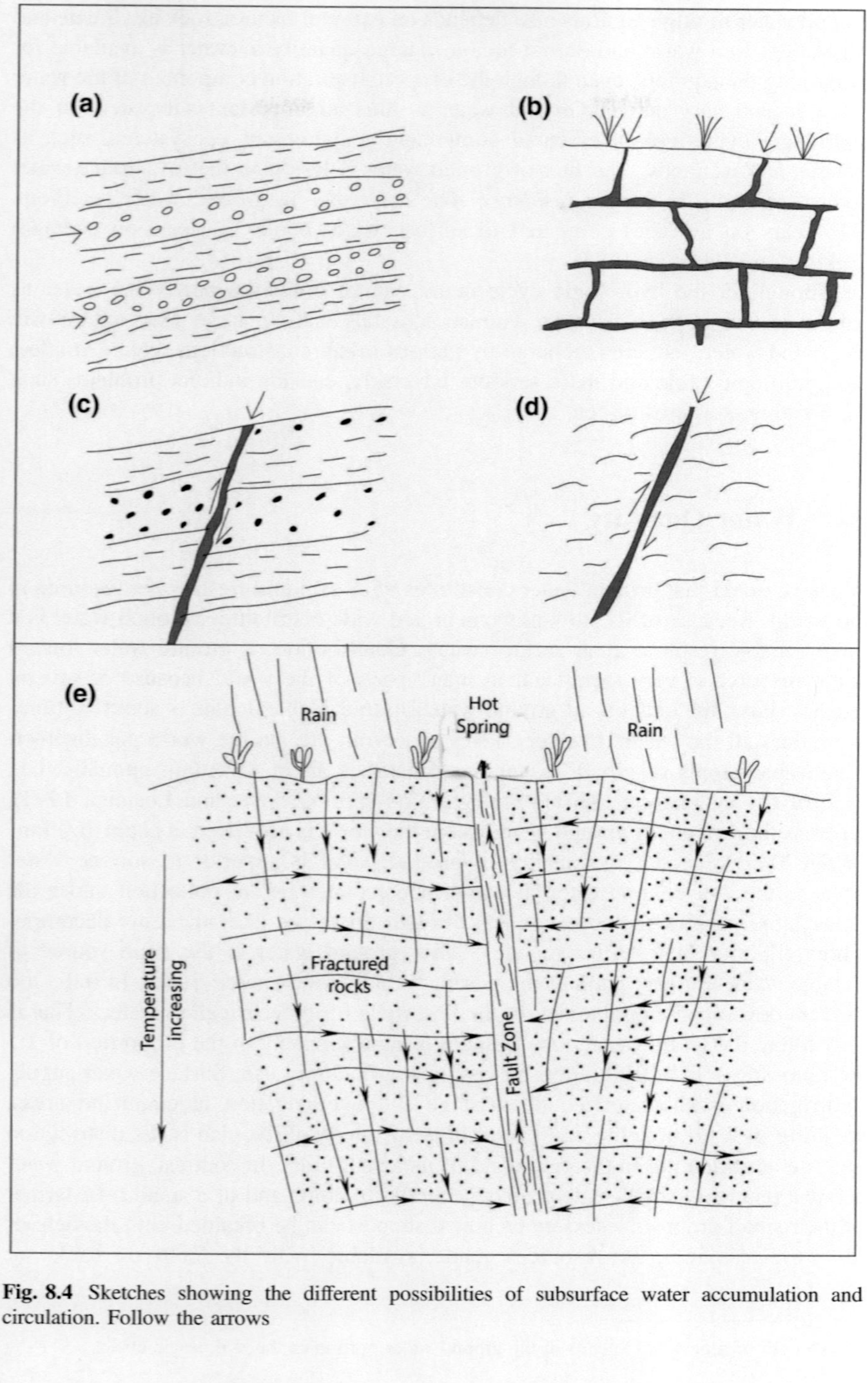

Fig. 8.4 Sketches showing the different possibilities of subsurface water accumulation and circulation. Follow the arrows

ground water through aquifers also depends on external factors, such as climate and vegetation. In a warm and humid region, a large quantity of water is available for recharging the aquifers, even though the evapotranspiration component of the water cycle is also very active. Ground water is thus an important subsystem of the hydrologic cycle and an essential component of numerous ecosystems, such as streams and wetlands. The flux of ground water is less than that of surface water bodies, resulting in longer *residence time*[1] of water in ground water reservoirs (315 years on average) compared to surface water bodies (6 years on average) (Zektser and Loaiciga 1993).

Throughout the hydrologic cycle many natural processes purify the water as natural service in the ecosystem. Human activities such as, large scale withdrawal of ground water, reducing recharge by various urban constructions, etc., can affect the hydrologic cycle and these services adversely, causing indirect problems such as, flooding, soil erosion, etc.

8.4 Water Quantity

We have noted that ground water constitutes 95% of liquid fresh water resource in the world. Because of its slow movement and wide distribution, ground water is a more reliable resource than surface water. Contribution of ground water toward water resource is very significant in many parts of the world because of several reasons. First, the quantity of ground water up to drillable depths is about 70 times more than all the waters in rivers, lakes, reservoir, etc., in the world put together. Renewable supply of ground water is estimated at about 4180 km^3 annually, i.e., 11% of the total runoff (38,000 km^3/year) on earth (Zektser and Loaiciga 1993), whereas the amount of ground water use in the world is estimated at about 760 km^3 (WRI 2000). Second, the quality of ground water is superior to surface water because the soil column through which it percolates before collection under the water table purifies the contaminants through processes like anaerobic decomposition, filtration, ion exchange, etc. Third, ground water is the main source of potable water in many parts of the world, including the United States. In India too farmers developed a preference for the borewells for their irrigation waters. That is why today the ground water use in India is approximately in the proportion of 1:5 for domestic + industrial purposes vis-à-vis agricultural use. Surface water supply to irrigation is characterized by wastage and overirrigation in command areas, resulting in water logging and salinization of low-lying croplands. Its distribution and use are often not in the command of its actual users. In contrast, ground water use is a relatively small-scale project generally in command of a small time farmer or their small groups. Necessary technical support can be obtained with the help of his own resources, or resources made available with the help of banks or

[1]The length of time water spends in the ground water portion of the hydrologic cycle.

government agencies. However, as more and more ground water is extracted for domestic and irrigation purposes, in many countries, including India, water table is going down by 1–3 m per year. The ground water resources therefore need to be monitored carefully for sustainable development both in terms of overextraction and possible pollution (cf. Aswathanarayana 2003).

The quantity of ground water should be evaluated by a regional water budget, over a watershed or an aquifer system. This requires data from a properly designed network of monitoring wells as well as data from all important inputs and outputs to this system. Over a large region (e.g., large river basin), the ground water flow across surface water divides are relatively less important. In these conditions, the following water budget relationship may be applied (Rouleau 2003): $P = Q + E$, where

P total precipitation over the watershed
Q total flow of the draining stream at the outlet of the watershed
E Evapotranspiration over the watershed

The contribution of ground water is included in the total flow at the outlet as: $Q = Q_s + Q_g$,

where Q_s = surface flow and Q_g = ground water discharge. These two components of flow have some basic difference. The surface flow is a short-term component resulting from precipitation moving overland to a stream without percolating to the aquifer. Ground water discharge on the other hand is a long-term persistent component that results from a part of the precipitation infiltrating into the soil and percolating into the aquifer and then flowing slowly to the stream. The relative importance of the two components varies widely depending mainly on the geological material present along the course. In areas covered by clay, surface flow dominates and the streams are often ephemeral. Sand and gravel deposits favor water infiltration and ground water recharge providing a fairly constant base flow to the stream. In the equation above, the difference between the total precipitation over the watershed and the flow at the outlet is the amount of evapotranspiration over the watershed. It varies from a fraction up to the totality of precipitation depending on the factors like temperature, vegetation, air moisture, etc. In most regional water budget estimations, however, the direct ground water discharge to estuary or sea is not considered, although some studies have shown that it can account for about 2400 km^3 annually, i.e., about 6% of total runoff on earth (cf. Zektser and Loaiciga 1993).

In many regions worldwide, a strong demand is made on rivers for supply of water for agriculture or domestic use in large urban centers. This raises the question of water transport from distant rivers to where it is needed, as was done in historical times by the Romans by constructing aqueducts. In recent times, large quantities of water have been transported in California from the water-rich northern part to the water-deficient southern part through canals, mostly for agricultural use but increasingly for urban use as well. Similarly, rapid urban growth of New York City required water to be transported from larger and larger dams in the upstate regions. The Indira Gandhi Nahar Project (IGNP), started in 1958, constructed the longest canal in India and the largest irrigation project in the world. Indira Gandhi Canal is 649 km long and comprises Rajasthan feeder canal and Rajasthan main canal (totaling 492 km) and runs through 167 km in Punjab and Haryana, utilizing the surplus water of Ravi and Beas Rivers. The canal is one of the projects of Green revolution in India and runs through the Great Thar Desert. As with other resources, as the supply of water shrinks or remains constant while the demand increases, stronger conservation measures and obtaining water from distant sources will push water prices up and costlier sources may even be tapped. It is important to note that in countries like the United States, water withdrawals decreased after 1980 while the population kept increasing. Obviously, it points to improved water management and water conservation which other countries also need to follow.

8.5 Water Quality

Water quality changes as it moves through the hydrological cycle and through biological and anthropogenic processes. This puts constraints on the types of use listed earlier, which is elaborated upon in Table 8.1. The highest quality of water should be meant for drinking purposes only, separated from other domestic uses, such as washing, bathing, sanitation, etc., which can take somewhat lower quality of water. In modern houses, a dual plumbing system operates where partially treated waste water is recycled for flushing or for horticulture. Though

Table 8.1 Constraints of water quality on water use (after Aswathanarayana 2003)

Pollutant	Use A	Use B	Use C	Use D	Use E	Use F	Use G
Pathogens	**	0	**	0 to **	*	na	na
Suspended solids	**	**	**	*	*	*	**
Organic matter	**	*	**	* to **	+	*	na
Nitrate	**	*	na	0 to **	+	na	na
Salinity	**	**	na	* to **	**	na	na
Inorganic toxins	**	**	*	*	*	na	na
Organic toxins	*	**	*	* to**	*	na	na
Acidification	*	**	*	*	na	*	na
Aquatic biota	*	*	**	* to **	+	*	*

Note
A = Drinking water; B = fisheries; C = recreation; D = industrial use; E = irrigation; F = power and cooling; G = transport
*Minor impairment for use needing minor treatment
**Presence of the pollutant markedly impairs the use of water for the concerned purpose. Can only be used for this purpose after major treatment
0 = No impairment of use
na = Not applicable
+ = Degradation of the water quality may be beneficial for this specific use

technologically feasible, it is prohibitively expensive to purify municipal waste water to levels required for domestic purposes. Such waters can be used by most industries except those which are food related. Sewage water which has high organic content and nitrates is not only acceptable but actually preferred for use in irrigation.

Pollution of surface and ground water is a major concern, particularly in urban areas, which constrains the usage of available waters. In rural areas, both types of water may get polluted through geogenic, agricultural, and mining processes. Water pollution in the two different environments can result from *point* and *nonpoint sources*. The former may be related to farm runoff or to leachates from waste dumps, mine over burden dumps or from *tailing dumps*. Such sources are more easily identified and are relatively easier to control. Pollution from point sources is mostly under control in the developed world. The latter, however, may come from various urban runoffs or deposition from the atmosphere where the pollutants are introduced by some distant industrial plant. It is even more difficult to identify certain geogenic pollutants, such as fluoride and arsenic in ground water, discussed below.

Ground water in natural conditions is generally of good quality. Dissolved substances in ground water depend on the transit time of water underground and the geological material through which it percolates. Generally, more than 90% of major ions (>5 mg/l) in total dissolved solids in ground water include Ca^{2+}, Mg^{2+}, Na^{+}, HCO_3^{-}, Cl^{-}, SO_4^{2-}. Other minor chemical constituents of ground water include B^{3+}, K^{+}, Sr^{2+}, Fe^{2+}, Fe^{3+}, F^{-}, $Si(OH)_4$, SiO_2^0, CO_3^{2-}, NO_3^{-}. Many of the chemical and isotopic constituents are valuable tracers reflecting processes that have affected ground water.

Contamination of ground water can take place by human activities in various ways. From the surface, polluted waste water can infiltrate down to the water table. Leachates from various waste dumps, stockpiles, sewage sludge, animal feeds, highway de-iceing salts, fertilizers, and pesticides can similarly infiltrate. In the ground but above the water table, contaminants may be introduced to the ground water from cemeteries, septic tanks, leaking tanks of petroleum and other chemical products, sanitary landfills, leakage from waste water sewers, underground pipeline, etc. Pollutants below the water table may come from underground storage of radioactive waste, mines and exploratory wells, etc. Contamination of ground water may come from natural sources as well, as already mentioned. It is now well known that prevention of pollution is much less costly and more efficient than the remediation of an aquifer once contamination has occurred.

A list of various common water pollutants include:

Natural pollutants

Biochemical oxygen demand (BOD), pH, salinity, $CaCO_3$, CH_4, As, Cd, Tl, Pb, Hg, Fe, Ni, Cu, Zn, F, Rn. They are contributed by country rocks through which the water flows. These could also be from abandoned mines or exposed deposits. To these may be added pathogenic microorganisms, and suspended solids.

Anthropogenic pollutants

Industrial and municipal effluents containing such toxic substances as Cu, Zn, Ni, Pb, As, Hg, benzene, toluene, chlorides, phenols, acids;
Agricultural waste (pesticides, fertilizers, and animal sewage);
Radioactive waste (containing As, S, CH_4, Cd, Hg, Pb).

The parameters covered in any common study of drinking water quality assessment along with their acceptable limits are given in Table 8.2.

Table 8.2 Drinking water quality parameters and acceptable limits (BIS 2009)

Physical parameters	Acceptable limit	Chemical parameters	Acceptable limit (mg/l)	Biological parameters
Color (hazen unit)	5	DO		Total coliform count/100 mL
Odor	Agreeable	Total alkalinity As Ca carbonate	200	As many pathogens as possible (e.g., giardia, salmonella, etc.)
Turbidity (NTU)	1	Total hardness (as $CO_{3,}$)	200	BOD
Temperature		Fluoride as F	1.0	COD
EC ($\mu s\ cm^{-1}$)		Chloride as Cl	250	
		Nitrate as NO_3	45	
TDS(mg/l)	500	Sulfate as SO_4	200	
pH	6.5–8.5	Total ammonia	0.5	
		Ca, Mg, Na, K, Ba, Li, B	Ca 75 Mg 30 Ba 0.7 B 0.5	
		SiO_2, Al	Al 0.03	
		Heavy metals (Zn, Cd, Fe, Mn, Cr (as Cr^{6+}), Ni, Cu, Pb, Ag, Se, Mo	Fe 0.3 Mn 0.1 Cu 0.05 Zn 5 Cd 0.003 Pb 0.01 Ni 0.02 Cr 0.05 Ag 0.1 Se 0.01 Mo 0.07	
		As, Hg	As 0.01 Hg 0.001	

Fluorine is an essential element in human diet. Lack of it has long been linked with tooth decay. In small doses (0.3–1.0 mg/l) it inhibits dental caries and hence fluoride toothpastes are commonly recommended. Many countries add fluorine to water supplies. Fluorine is, however, very toxic for both plant and animal life at higher doses (>1.5 mg/l)—it causes dental and skeletal fluorosis (cf. Teotia and Teotia 1994). Health problems caused by fluorine released into the air by burning of domestic coal have been recorded extensively. Excess fluoride in ground water is reported from many (17) states in the country of which Andhra Pradesh, Rajasthan and Gujarat, are severely affected (cf. Brindha and Elango 2011 and the references therein) (Fig. 8.5). Concentration level of fluoride reported from ground water in India varies from 0.5 to 50 mg/l. As per water technology mission, 25 million people residing in 8700 villages in India are affected by fluorosis.

Fluoride contamination of ground water is generally ascribed to the geological environment. Fluorite (48.7% F), apatite (3.5% F), mica (0.14–0.22% F), illite (0.11–0.26% F), and some amphiboles are the main natural sources of fluorine. Thus hard rock granitic aquifers are generally severely affected by fluoride contamination. Water–rock interaction studies in the laboratory (Saxena and Ahmed 2001) under normal temperature and pressure and varying chemical conditions have shown that moderate specific conductivity (750–1750 μS/cm), pH (7.6–8.6), presence of Ca and high HCO_3 (300–450 mg/l) are important chemical parameters for the dissolution of fluoride to ground water from F-rich minerals. Industrial fluorosis is also a serious problem in many developed countries. Industries that use fluoride either as raw material in the manufacturing process or fluoride is formed as a by-product or may even be the end product, include manufacturing or production of aluminum, steel, enamel, fertilizers, chemical, glass, pottery, toothpaste, automobile, polymers, semiconductors, and integrated circuits. Some other concerned industries are oil refinery, refrigeration, and pharmaceutical.

Arsenic pollution of ground water has created a serious problem of public health in the parts of eastern India including Bangladesh, where 50 districts out of a total of 64 are affected by ground water arsenic contamination (Fazal et al. 2001) (Fig. 8.6). In Bangladesh, about 25 million people are drinking As-contaminated (50 μg/L BIS standard and 10 μg/L WHO mpl) shallow tube well water. Nine districts out of 18 are also affected in West Bengal. Here, an estimated +6 million people are living in a region where the top level aquifer (accessed by majority of people through shallow tube well) is contaminated by arsenic (cf. Chakraborty et al. 1994). Several districts of Bihar along the River Ganges have also got affected by arsenic contamination problem. Ghosh et al. (2009) report hot spots of As-contamination of ground water resources of Patna, the most populated district of Bihar. An estimated population of 5.5 lakhs is potentially exposed to serious health risks, with possible rise of morbidity in the next 5–10 years.

While As exposure can occur from food, air and water, all major chronic As poisonings are related to water. Exposure to As leads to an accumulation of As in tissues such as skin, hair and nails, resulting in various clinical symptoms such as hyperpigmentation and keratosis. There is also an increased risk of skin, internal

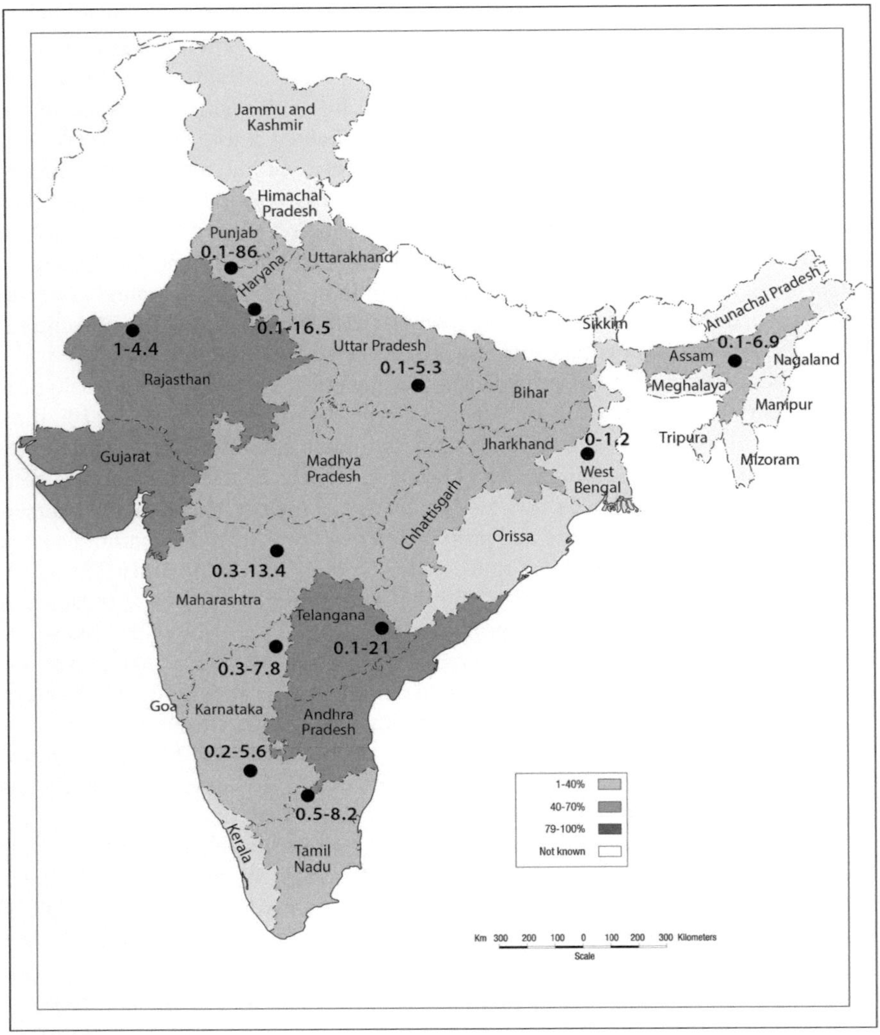

Fig. 8.5 Fluorosis prevalent states in India showing the percentage of population affected. Recorded F-content in ground water (mg/l) also shown (after www.indiawaterportal.org)

organ, and lung cancers. Cardiovascular diseases have also been linked to As consumption. Large intake of arsenic from drinking water over a prolonged period results in keratosis or skin cancer. However, the toxicity of arsenic varies with its state of oxidation in the order, $As^{o} < As^{5+} < As^{3+}$. The very large spread of As-contaminated area suggests it to be related to a geogenic source. Solutions to the problem may lie in drilling deep tube wells, using surface water resources, rain water harvesting, using domestic scale devices that remove arsenic ions, etc. The

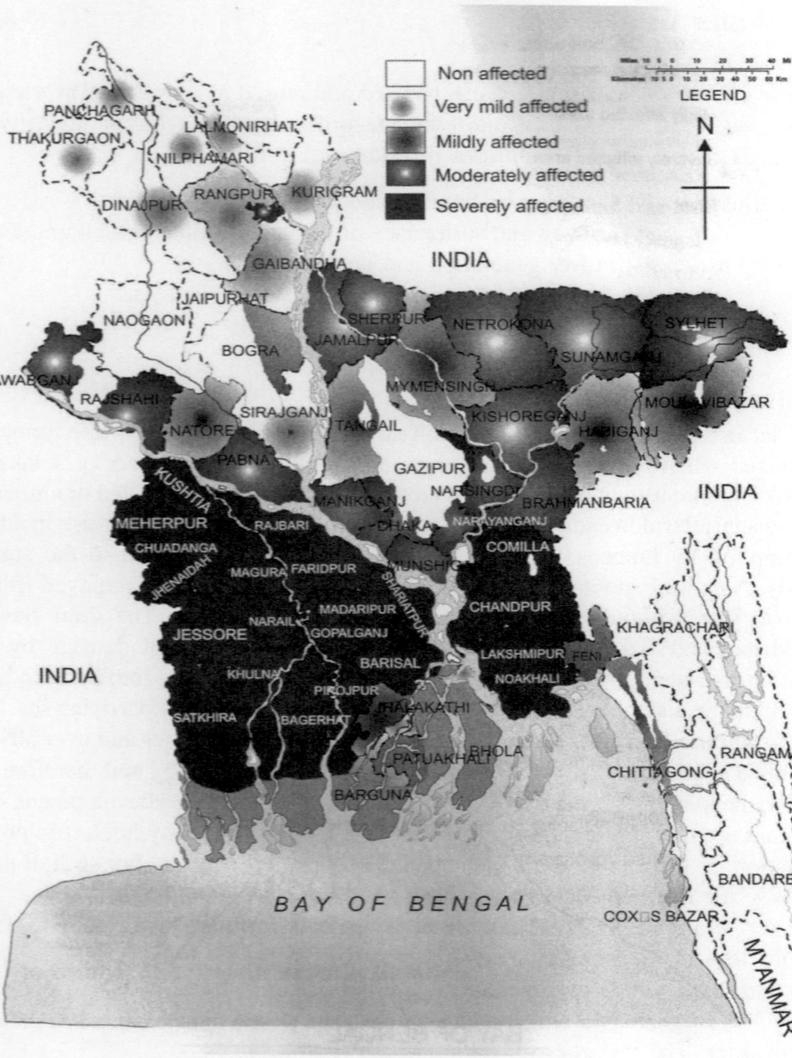

Fig. 8.6 Arsenic affected ground water in different parts of Bangladesh (after Rahman et al. 2003)

problem has been partly remedied following efforts by governments and NGOs in India and Bangladesh.

Punjab is facing severe ground water pollution from radioactive materials, which is linked to rising cases of cancer. A train from the region has been dubbed the "cancer express."

8.6 Water Use

Water is a mobile resource which can be used and reused many times. The various uses of water can be classified into two categories, depending on the intensity of their needs. The most important uses of water are as follows:

Essential uses: (1) Drinking by humans and animals
(2) Growth and sustenance of plants, including agricultural crops
Necessary uses: (3) Power generation and industries
(4) Transport
(5) Pisi-culture (fishery)
(6) Domestic uses
(7) Climate moderation

Water use is also distinguished by off-stream and in-stream uses. The former is recognized when water is removed from its source, such as a river or a lake or reservoir, before use. In most cases it is returned to the source as treated or untreated waste water after domestic or industrial use. A part of this off-stream use includes consumption by humans, plants, and animals and is not returned to the source directly. A major issue in off-stream use is how much water can be removed from a river or lake without damaging the prevailing ecosystems. The Aral Sea in Kazakhstan and Uzbekistan is a glaring case of environmental damage by the diversion of waters for agriculture from the two rivers that flow into it. This large body of inland water has transformed from a vibrant ecosystem into a dying sea. The major water use is in agriculture, over 80% in developing countries and over 50% in developed countries. Much of this water is used very inefficiently and therefore net transfers happen from agriculture to urban sectors with increased development. The in-stream use includes the use of river water for navigation, for hydroelectric power generation, as fish and wild life habitat and for general recreation. For each of these purposes, the requirements of water level and flow rates are different.

Water withdrawals worldwide amount to 4000 km^3 per year equivalent to roughly 10% of the annual global surface runoff. Water use has doubled between 1960 and 2000, which means that average per person use has remained constant. About 70% of global water withdrawals are used for agriculture while 20% is used by industries. The rest goes for domestic use (Fig. 8.7).

We may introduce here the two related emerging concepts: *water footprint* and *virtual water*. The water footprint measures the amount of water used to produce each of the goods and services we use. It can also tell us how much water is being consumed for making a particular product such as, a pair of jeans or to run an enterprise such as the Railways, or by a particular country – or globally – in a specific river basin or from an aquifer. The water footprint has three components: green, blue and gray. Together, these components provide a comprehensive picture of water use by delineating the source of water consumed, either as rainfall/soil moisture (green) or surface/ground water (blue), and the volume of water recovered from effluents. Virtual water is also the water required to produce a particular

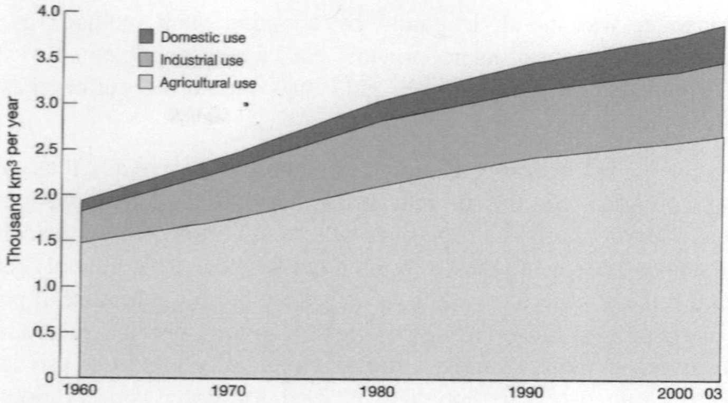

Fig. 8.7 Water use by different sectors

product, be it industrial or agricultural and which is being imported or exported. For example, it takes about 140 l of water to produce a cup of coffee. Sounds incredible, but that amount of water is needed to produce the coffee beans in a far away country like Columbia or Brazil and then used to produce a cup of coffee. Since a large quantity of water is used for agricultural produce, export of grains and other food products from one country to the other involves export of a substantial quantity of virtual water. To mention a few examples, about 1000 l of water is needed to produce a kg of grain while 13,000 l of water was needed to produce a kg of beef. This concept is useful for water resource planning from a local to a global scale. Trading of virtual water will control global water conservation. When products with high water requirement are exported from a country with no water scarcity to one with scarcity, real water conservation happens at a global scale. Politically virtual water can provide greater water security.

8.7 Water Conservation and Its Sustainable Development

Any method that saves water is a water conservation method. Desired methods of water conservation, including the traditional methods, are:

- Minimizing domestic and industrial consumption of water.
- Making better use of rainfall; minimizing its runoff losses; collection and storage of rain water in roof tops, ponds, as well as by putting contour dams, check dams, etc., in suitable locales.
- Recycling of used water after suitable processing.
- Collect ground water without water lifting, through quanats, artesian wells, and horizontal wells.
- Make cultivation climate-specific, e.g., paddy should grow in high-to-moderate rainfall areas and wheat, bajra, and gram in low-to-moderate rainfall areas.

- Control water wastage in irrigation by adopting such methods as, sprinkle irrigation, drip irrigation, microspraying, etc., where possible.
- Loss during transport through pipes and canals should be controlled as much as possible.

These conservation measures are low cost, labour intensive and thus suitable for developing countries. Similar to rain water harvesting, collection, storage and distribution of storm runoff is an age-old practice and is effected by the construction of check-dams or ditches and canals to intercept the flow. This not only reduces the flood hazard but also provides water for irrigation during lean rainfall periods.

Water use can be optimized using pricing as a control, because subsidizing water encourages overuse, though surely, a minimum quantity should be provided to the population free of cost. This approach is used for water conservation in many countries. Research and Development in this field of water conservation may lead to more efficient use of water through: the use of appropriate water dispensing technologies; use of inferior quality of water for specific purposes because water purification is an expensive process; reuse and recycling of water so that wherever there is large water use, such as, in hotels, etc., a zero discharge to the municipal sewerage is ensured; allocation of water among competing users, a very useful though politically sensitive issue.

We have noted earlier that agriculture is the biggest consumer of water resource. It is also well known that the water requirement of different crops vary widely. The water uptake by plants is dependent on (a) characteristics of the plant (rooting density, rooting depth, root extension, etc.), (b) properties of the soil (like porosity and water retention) and (c) on weather conditions (which determine the amount of transpiration). Therefore, the soil–crop–climate system controls the soil water availability. Hence, by encouraging the development of crops that require less water such as, wheat, gram and bajra in a particular climatic zone and providing a tailor-made irrigation system (e.g., surface irrigation, gravity irrigation, sprinkle irrigation, drip irrigation, microsprayers, etc.), water use can be reduced substantially (20–30%), particularly in regions with shortage or in times of scarcity. Agricultural water conservation should also include: lined or covered canals to reduce seepage and evaporation; irrigation when evaporation is minimal, such as, at night or early morning; integrated use of surface and ground water; and if possible, computer monitoring and schedule release of water for maximum efficiency. It will be wise to use surface water for cultivation during the wet season and ground water during the dry.

Providing water to a large population is a complex issue which is likely to become more and more challenging as both the population and the demand for water increases, particularly in the developing world. The problem will be more acute where water is already in short supply. Therefore a whole new approach to water management will be required taking into account geologic, geographic, and climatic factors as well as other economic, social and political factors. This approach will ensure that water resource of a region can be used for a long time in the future without affecting the hydrologic cycle or ecological systems that are dependent on it.

For the supply of water to the urban centers, to croplands for cultivation and for hydropower generation, putting dams across the river valleys is a common practice. The dams in the Damodar valley (DVC), Hirakud, and Bhakra Nangal projects are some Indian examples of these practices. India now has the third largest number (4857 completed; Maharashtra having the highest at 1845) of dams in the world. Experience has shown that large dams inflict severe damage to the river ecosystem, though it is argued that such dams are required in dry and uncertain rainfall zones. Making dams smaller but numerous across a river valley is more advisable. No deforestation should be allowed in the watershed and soil erosion should be prevented within it. The former will assure rainfall in the watershed and the latter will reduce water loss through surface runoff and contribution to ground water will increase.

Sustainable water use and management obviously must have a long-term perspective. It can be achieved through creative alternatives and using variable water sources and not by looking at water issues from only economic and political standpoints. A Master Plan should be drawn up which may include average rainfall over a reasonable period of time, topography, existing and projected land use, including probable industrial development. This should also include importing of water from nearby surplus areas, developing new sources, using reclaimed water and instituting a permanent conservation plan. In wet years the emphasis should be on surface water use, adopt natural and artificial recharge, while in dry years we may resort mainly to ground water withdrawal. In an emergency situation, even the deeper aquifers can be tapped in spite of substandard water quality. Such a plan recognizes that excess and deficient water availability over a period of time is natural and mitigation of its effects can be planned for in advance.

Sustainability of water use would require ensuring enough water resource for maintaining human development and well being and the health of various ecosystems, particularly the aquatic ecosystems. It will also involve maintenance of the standards of water quality for different uses and ensuring long-term renewability of water resources by restricting human intervention. Maximum sustainable yield of water that can be extracted from an aquifer over a year corresponds to the annual quantity of ground water recharge to this aquifer. Summing up, efficient water use technologies have to be promoted and practiced on one hand and policies implemented for withdrawal of subsidies for inefficient water use on the other. A sustainable water supply system should ensure a balance between extraction and replenishment of the water sources, either by natural processes and/or artificial recharge.

8.8 Desalination of Sea Water

A question is often asked: 97% of the hydrosphere is occupied by sea water and even then we are worried about the shortage of useful water today and apprehend a still worse situation in foreseeable future. Why? Can we not find out cheap and simple means of deriving it from such a vast reserve of sea water? The brief answer to this is: it is derivable and even derived locally, but the processes are costly.

There are three processes of desalination of sea water at the moment (Murthy 1992). These are:

1. Distillation
2. Electrodialysis and reverse osmosis
3. Freezing.

The first, a thermal and the second, a membrane technique are the two main competing ones today. Distillation is the oldest desalting process. Story goes that Julius Caeser used a crude distillation apparatus to supply water to armies invading Alexandria. Long sea voyagers used to distil sea water on board their ships in course of time. But evaporating sea water consumes a lot of energy: to evaporate 1 kg of water 540 kcal are required. Today's distillation plants are more advanced. This could be any of the following: multistage flash, multiple effect falling film, or vapor compression types.

Reverse osmosis is a widely used process for the desalination of sea water. In this process the saline sea water is forced under pressure through a specially prepared membrane which allows water molecules to pass faster than the salt ions. Various types of organic polymer films are used as membranes for reverse osmosis; cellulose triacetate is one of the most commonly used materials.

The freezing technique is based on the fact that when sea water is chilled, the ice crystals formed will be salt free. The lower latent heat of fusion, compared to that of vaporization, low corrosion of the equipment used are some of the positive aspects of the process. On the negative side, it is not amenable to multistage operation.

Distillation and reverse osmosis are the most widely used sea water desalination processes in the world today. Currently, there are 12,500 desalination plants in the world, supplying 706 million cubic feet per day, or 1% of the world's production of drinking water (Martin-Lagardette 2001). As we have already mentioned, deriving water suitable for human use (domestic use, industry and agriculture) from sea water on a large scale remains a distant hope for most countries because of the cost factor involved in the known technologies. However, the reduced cost of new desalination equipment is expected to double the world market in this field.

8.9 Water Availability and Water Demand in India

India has a geographical area of 329 million hectares, comprising 2.45% of the earth's landmass, which supports a population of 1.27 billion (2014 estimate), amounting to about 18% of world's population. In comparison, the renewable fresh water resource potential of India is 1869 billion cubic meter (BCM)/year (both surface and ground water) (Table 8.3) which amounts to about 4% of the earth's water resources. However, all this water is not available for use. Due to various constraints of topography and large inequities of the resource in space and time, it is estimated that average annual utilizable quantity of water is only 1123 BCM out of which surface water accounts for 690 BCM and ground water 433 BCM (CWC 2015). River basin-wise water availability in India (Table 8.3) bears out this variability in space. Besides, a storage capacity of 304 BCM will soon be available once the major and medium irrigation projects are completed. The total annual per capita water availability in 2001 corresponded to about 1816 kl which reduced to 1545 kl in 2011 and is projected further to reduce to 1340 kl in 2025 and only

Table 8.3 Water availability of different river basins in India (CWC 2015)

Sl. No.	Name of river basin	Catchment area (km^2)	Average annual flow in the river (BCM)	Utilizable surface water resources (BCM)
1	Indus (up to border)	321,289	73.3	46.0
2	(a) Ganga	861,452	525.0	250.0
	(b) Brahmaputra and Barak	236,136	585.6	24.0
3	Godavari	312,812	110.5	76.3
4	Krishna	258,948	78.1	58.0
5	Cauvery	81,155	21.4	19.0
6	Pennar	55,213	6.3	6.9
7	East-flowing rivers between Mahanadi and Pennar	86,643	22.5	13.1
8	East-flowing rivers between Pennar and Kanyakumari	100,139	16.5	16.5
9	Mahanadi	141,589	66.9	50.0
10	Brahamani and Baitarani	51,822	28.5	18.3
11	Subarnarekha	29,196	12.4	6.8
12	Sabarmati	21,674	3.8	1.9
13	Mahi	34,842	11.0	3.1
14	West-flowing rivers of Kutch, Saurashtra, including Luni	321,851	15.1	15.0
15	Narmada	98,796	45.6	34.5
16	Tapi	65,145	14.9	14.5
17	West-flowing rivers from Tapi to Tadri	55,940	87.4	11.9
18	West-flowing rivers from Tadri to Kanyakumari	56,177	113.5	24.3
19	Area of inland drainage in Rajasthan desert	–	0.0	–
20	Minor river basins, draining into Bangladesh and Myanmar	36,302	31.0	–
	Total		1869.4	690.1

Table 8.4 Projected water demand in India for different uses (CWC 2015)

Sector	Water demand (BCM) in the year		
	2010	2025	2050
Irrigation	557	611	807
Domestic	43	62	111
Industrial	37	67	81
Energy	19	33	70
Others	12	20	35
Evaporation losses	42	50	76
Total	710	843	1180

1140 kl in 2050. If the population of India stabilizes around 1.64 billion by 2050 AD, as predicted, then the per capita water availability will reduce to nearly 60% of the 2001 availability (GOI 2003). The projected water demand in India is presented in Table 8.4. In view of the severity of this problem, optimal water use is essential to ensure comfortable living standard for the people of India.

The concept of seasonality is important in the context of water availability in the Indian subcontinent. Monsoons, a relatively recent phenomenon in geological terms, account for the bulk of precipitation in India, the duration and intensity of which is linked to infiltration and runoff. It is not easily appreciated that 85% of the water in the Ganga Basin is from monsoon rains, although the glacier melts do play a role in maintaining summer flows.

Improved water management in agriculture, industry, and domestic sectors could yield 10–15% of additional availability. Most intensive and indiscriminate water use is in irrigation where improved technology like surface or gravity irrigation, sprinkle irrigation, drip irrigation, and micro-sprayer can reduce water use significantly and should be given incentives. Besides crops like sugarcane which have high water requirement can be replaced by less water intensive crops like wheat, ground nut, grams, etc. The practice of providing free electricity to farmers should be discontinued to stop the misuse of ground water in states like Punjab. In urban areas, treated water suitable for drinking is commonly used for flushing toilets. The quantum of such use often far exceeds the total water supply in a rural area. Industries and cities release untreated or partially treated waste water into surface runoffs causing severe pollution. All such practices need to be reviewed and corrective measures taken. It is important to remember that the cost of reclaiming and reusing polluted water are significantly higher than the care and investment required preventing wastage. It is therefore better to economize in use and protect fresh water. Other water conservation measures such as roof top rain water harvesting, ground water recharge, watershed management through check dams and small dams, also have important roles to play in this regard.

Ground water is a vital resource and is the main source of potable water in many parts of the world. But in India, uncontrolled ground water pumping in urban areas or for irrigation purposes has resulted in marked lowering of ground water level (by tens of meters in many areas) which has brought about concentration of salinity,

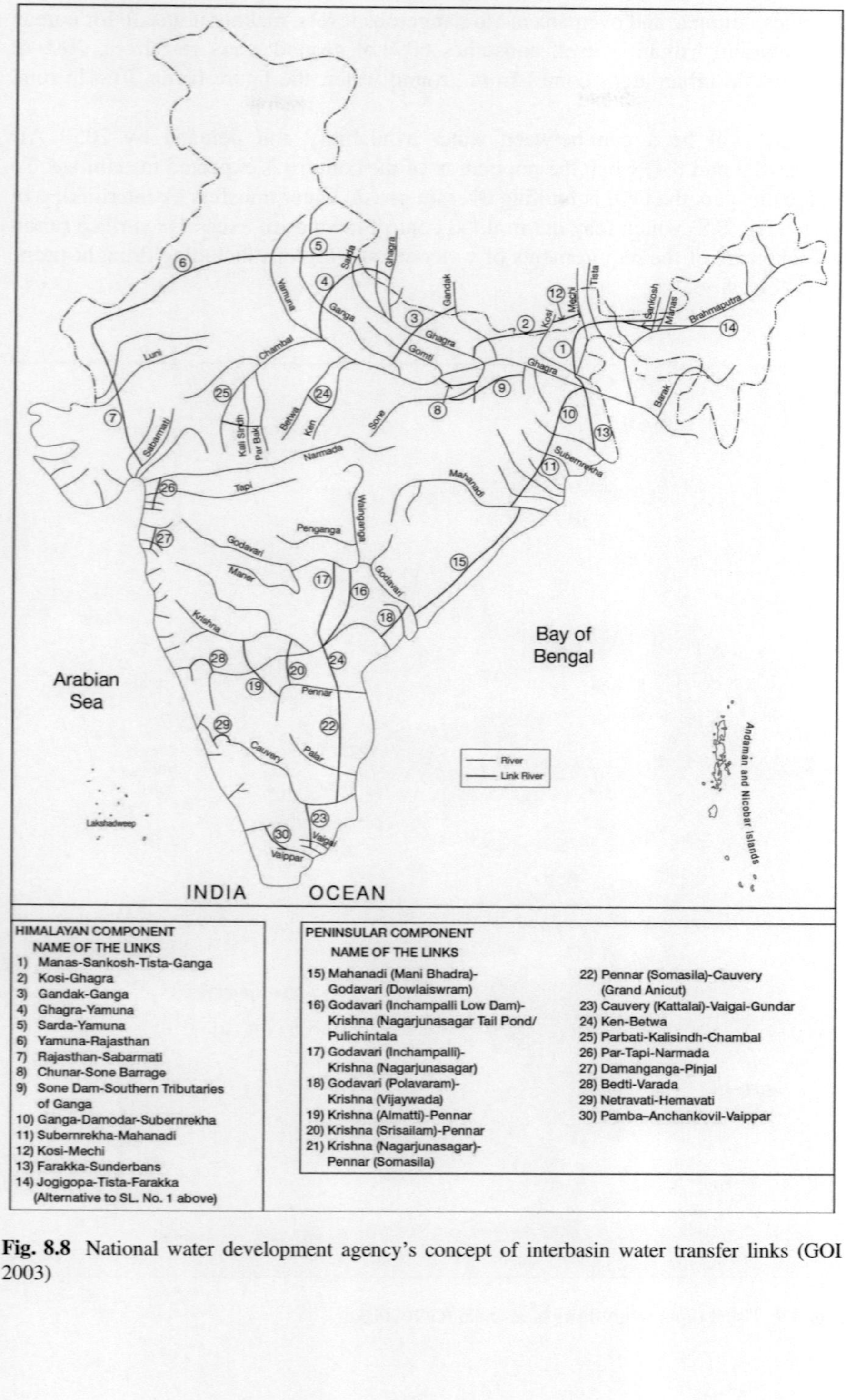

Fig. 8.8 National water development agency's concept of interbasin water transfer links (GOI 2003)

fluorides, nitrates, and even arsenic to dangerous levels, making it unsafe for human consumption. Irrigation itself consumes 60% of ground water resources. 30% of water use in urban areas comes from ground water, the figure being 70% in rural areas.

There will be a gap between water availability and demand by 2050 AD (Tables 8.3 and 8.4) when the population of the country is expected to stabilize. To bridge this gap, the GOI is mulling over interbasin water transfers by interlinking of rivers (Fig. 8.8) which may aid in flood control in areas of excessive surface runoff and take care of the requirements of water-starved regions including drought-prone areas (Fig. 8.9).

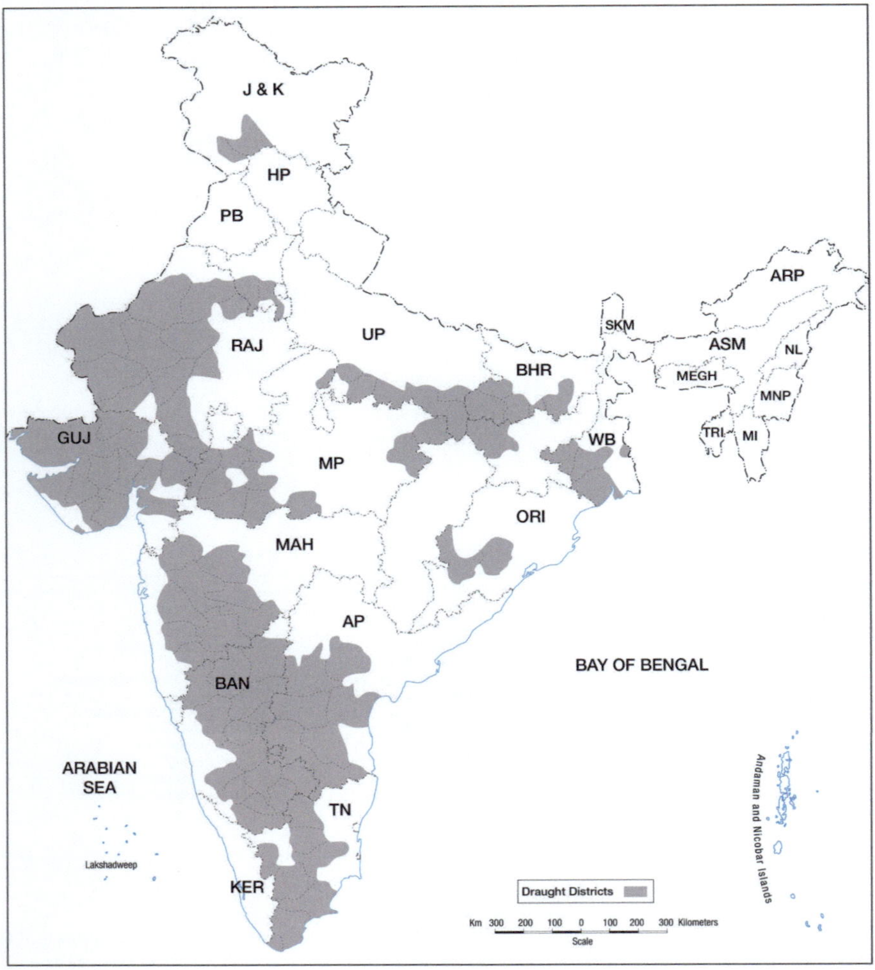

Fig. 8.9 Usual drought-hit districts in India (GOI 2003)

8.10 The River-Linking Proposal

The National Perspective Plan for interlinking of rivers (GOI 2003) comprises two main components.

Himalayan Rivers Development: envisages construction of storage reservoirs on the principal tributaries of the Ganga and Brahmaputra in India, Nepal and Bhutan along with interlinking a canal system to transfer the surplus flow of the eastern tributaries of the Ganga to the west. Besides it plans to link the main Brahmaputra and its tributaries with the Ganga and Ganga with Mahanadi. The Himalayan component has multipurpose benefits of additional irrigation ($\sim$22 million hectares), hydropower ($\sim$30 million kW) and flood control in the Ganges and Brahmaputra basins. It will also augment the flow at Farakka which will eventually help the Kolkata port. Obviously, this development proposal cannot be carried out without international cooperation.

Peninsular Rivers Development: divided into four components.

Interlinking of the Mahanadi-Godavari-Krishna-Cauvery Rivers and building storages at potential sites in the basins.

Interlinking of west-flowing rivers, north of Mumbai and south of Tapti. The scheme includes a water supply canal to the metropolitan areas of Mumbai.

(a) Interlinking Ken–Chambal rivers and provide a water grid for M.P. and U.P. through interlinking canals backed by adequate storage.
(b) Diversion of other west-flowing rivers: heavy rains in the western side of the Western Ghats produce numerous streams that empty into the Arabian sea. Linking them along with adequate storage will meet the water demand of Kerala and also transfer some water to the drought-affected areas to the east. The peninsular component is expected to provide additional irrigation to about 13 million hectares and generate about 4 million kW electricity.

The proposal for interlinking of rivers (ILR) is highly controversial, with strong opinions both for and against the project. Several issues have been raised (Iyer 2014) against the project. The following criticisms are noteworthy:

(i) Instead of finding local or intrabasin solutions to water needs of Rajasthan, Madhya Pradesh, Maharashtra, Karnataka, Andhra Pradesh and Tamil Nadu, the project proponents seem to have looked at a map of India, decided a priori that the rivers of India should be linked and then went ahead to find the modalities of implementation.
(ii) The project mentions a "National water grid" akin to power or road grid. This concept itself is flawed since there cannot be any two-way movement in a water grid. It is argued that rivers are not human artifacts; they are natural phenomena, integral components of ecological systems, and inextricable parts of the cultural, social, economic, and spiritual lives of the communities concerned. They are not pipelines to be cut, turned around, welded, and rejoined.

(iii) If only a small fraction of flood water is diverted from a flood-prone region, it will hardly help in flood mitigation. Most engineers discard the idea of ILR as a flood control measure.
(iv) Linking a river to another will only bring additional water to an area already served by a river. Most of the uplands and dry lands in the country are distant from rivers and therefore will not be helped much by ILR. The problem of drought has to be addressed locally through rain water harvesting and water shed development.
(v) The surplus and deficit concept of ILR is also flawed. Rivers provide many services: support aquatic life and vegetation; provide drinking water to human beings, their livestock and wildlife; influence the microclimate; recharge ground water; dilute pollutants and brings self-purification; sustain a wide range of livelihoods; transport silt and enrich the soil; maintain the estuary in a good state and provide the necessary fresh water to the sea to keep its salinity at the right level; prevent the incursion of salinity from the sea; provide nutrients to marine life. The river is also an integral part of human settlements, their lives, landscape, society, culture, history, and religion. It follows that any diversion of waters from a river, however small, is bound to have some serious consequences, major or minor. Surplus or deficit is thus meaningless in hydrological/ecological context. In many cases, deficit will surely turn out to be bad water management or unsustainable demand.
(vi) Against the main driving force for ILR and large dams, that is irrigation, it is argued that supply-side solutions should be the last resort. Improvement of water use technologies, change of crops and cropping pattern in water-scarce areas, reduction of water demand, etc., should be undertaken as a priority.
(vii) Long distance water transfer with attendant losses due to leakage and evaporation is prohibitively expensive and may not be economically viable.
(viii) Finally, the support to ILR due to generation of hydroelectric power to the tune of 30,000 MW is highly questionable. Also this will require 80+ dams with their reservoirs to be constructed, which will have huge environmental and ecological consequences.
(ix) It will dislodge a large number of people from their accustomed habitats.
(x) It will involve spending an unearthly amount of money for a developing country like India with disproportionate gain potentials.

When international water treaty issues are added to this long list of arguments against ILR, it is but clear that very careful considerations are required before any decision of such national importance can be arrived at.

One other opinion in this debate (Subramanian 2011) supports limited short, localized links after careful economic and environmental cost determination while emphatically declining any grandiose large scale, east to west or north to south links as untenable. The suggestion is worth a serious consideration.

A different but related approach can be followed to take care of better water availability and other water issues in India in place of ILR. It is proper River basin management. The National Ganga basin authority is already in place and a similar Cauvery basin authority with decision-makers from the concerned states is urgently required. Similar authorities for the major river basins in the country would help development in a big way. Besides, with GST, India is set to become a single market and virtual water transfers within the country will become a reality.

References

Aswathanarayana U (2003) Natural resource and environment. Geol Soc Ind 69

Brindha K, Elango L (2011) Fluoride in groundwater: causes, implications and mitigation measures. In: Monroy SD (ed) Fluoride properties, applications and environmental management. www.novapublishers.com, pp 111–136

Bureau Indian Standards (2009) Draft indian standard drinking water specification (2nd rev of IS 10500)

Central Water Commission (2015) Water and related statistics, p 161

Chakraborty D, Samant G et al (1994) Arsenic in ground water in six districts of West Bengal, India: biggest arsenic calamity in the world. Env Geochem Health 18:5–15

Fazal A, Kawachi T, Ichion E (2001) Extent and severity of ground water arsenic contamination in Bangladesh. Int Water Res Assoc Water Int 26(3):370–379

Ghosh AK, Singh SK, Bose N, Roy NP, Singh SK, Upadhyaya AK, Kumar A, Singh S (2009) Arsenic hot spots detected in Patna district, Bihar (India)—a serious health hazard for the estimated human population of 5.5 lakhs. In: Ramanathan AL et al (eds) Assessment of ground water resources and management. ISGSD Spec Publ, IK Int Publ House, New Delhi, pp 62–70

Govt of India, Ministry of water resources (2003) Report of the Task Force on interlinking of rivers. Interbasin water transfer proposals, p 27

Iyer RR (2014) Interlinking of rivers: a plea to the government. Econ Pol Wkly (EPW) XLIX (50):16–18

Kundzewicz ZW (2008) Water cycle. In: Jorgensen SE, Brian F (eds) Encycl ecology. Acad Press, Oxford, pp 3724–3729

Martin-Lagardette JL (2001) Water sources: desalination of sea water. Water Eng Mgmt 18–20

Murthy TKS (1992) Mining the ocean. CSIR Gold Jub Sr, New Delhi, p 98

Rahman MM, Paul K, Chowdhury UK, Sengupta MK, Lodh D, Basu GK, Chanda CR, Roy S, Das R, Quamruzzaman Q, Chakraborti D (2003) Groundwater arsenic contamination and human suffering in Bangladesh and West Bengal, India. Strategic management of environmental and socio-economic issues—a handbook. Guizhou Sci & Tech Publ House, Guiyang, China, pp 102–111

Rouleau A (2003) Groundwater issues in geo-environmental management. A handbook of strategic management of environmental and socio-economic issues (Chap. 12). Guizhou Sci & Tech Publ House, Guiyang, pp 90–101

Saxena VK, Ahmed S (2001) Dissolution of fluoride in groundwater: a water-rock interaction study. Env Geol 40:1084–1087

Subramanian V (2011) Rivers of South Asia: to link or not to link. Capital Publ Co, New Delhi, p 340

Teotia SP, Teotia M (1994) Endemic fluorosis in India: a challenging national health problem. J Assoc Phys India 32:347–352
World Res Inst (2000) World resources 2000–2001: People and ecosystem—the fraying web of life. UN & World Bank, Washington
Zektser IS, Loaiciga HA (1993) Groundwater fluxes in the global hydrologic cycle: past, present and future. J Hydrol 144:405–427

Part III
Sustainable Development of India's Mineral and Allied Resources

Chapter 9
Sustainable Development of Mineral Resources

Abstract The concept of *Sustainable Development* evolved through the efforts of the World Commission on Environment and Development (the Brundtland Commission) during the period 1982–1987 and rested on two basic premises: the issue of "need" for the deprived section of society and "limits" on the ability of the environment to satisfy the need of the present and future generations. Formally, Sustainable Development was defined as: "*the development that meets the needs of the present without compromising the ability of the future generations to meet their own need.*" The scope of this definition was widened in the World Summit of Johannesburg in 2002 to include "economic development, social development and environmental protection at local, regional and global levels." Many interpretations of Sustainable Development were put forward in later years, but they were all based differently on the foundation of these "*three pillars of sustainability.*" More recently, Sustainable Development is understood in terms of complex systems which require nonlinear, organic approach and is not to be considered as a target to be achieved. The concept of sustainable development, however, has had an ambiguous relationship with the extractive industry because of its intrinsic nature, leading to negative consequences. But can our modern societies do without the earth materials for industrial use? Can we really do away with mining? The answer is obvious even to the layman. Hence, the challenge of "Sustainable Development in the Mines and Minerals sector" is to ensure "environmental responsibility" during mining and processing, in which the damage to the environment (including social environment) is to be balanced with the Earth's capacity for accommodating it. Some well-accepted prescribed steps need to be followed in this regard to achieve sustainable development of mineral resources, keeping in mind the valuable guiding principles of *reduce*, *replace*, and *recycle*. "*Free prior and informed consent*" of all the stakeholders must be obtained by the mining industry if development is to be truly sustainable. Further, an alternative interpretation is required for a better understanding of the relationship between sustainable development and mining: a natural resource capital (a mineral or fuel resource) is converted by mining into an economic capital which can be reinvested to create or

M. Deb and S.C. Sarkar, *Minerals and Allied Natural Resources and Their Sustainable Development*, Springer Geology,
DOI 10.1007/978-981-10-4564-6_9

enhance other forms of capital, as exemplified by the “Alaska Fund.” Fast depletion of mineral resources and their long-term availability has been a serious concern for some time now, particularly since ecological footprint of humanity and limits to growth became the focal points of international debate. Though the proponents of “fixed stock paradigm” draw a gloomy picture, discovery of new deposits, advancement of technology and more use of low-grade material have extended the life expectancies of mineral resources. While social costs may limit the use of mineral commodities in future, per capita material throughputs of rich nations must be reduced if sustainable development should become a reality.

Keywords Sustainable development · GDP/GNP · Brundtland report · Three pillars of sustainability · Sustainable development of mineral resources · Environmental responsibility in mining · Free prior and informed consent · Alaska fund · Ecological footprint · Limits to growth · Life expectancy of some mineral commodities

9.1 The Concept of Sustainable Development

With a human-centric view of the environment, natural resource use and development have been at the forefront of western approach to modernization since the days of Industrial Revolution. When a big thrust was made in this direction in the post-World War II years, widespread environmental damage in the wake of unbridled industrial development, urbanization and a consequent desire for enhancement of living standards soon became apparent. The failure of traditional economic approach toward eradication of poverty as well as safeguard of the environment and a growing awareness of the right to social justice, in the world community as a whole, provided the necessary impetus for sustainable development movement worldwide.

The expression “*Sustainable Development*” much in vogue today, thus does not owe its origin to a distant past. It was coined in the last quarter of the last century when the world communities showed some serious concern about growth and economic development that started in the early 1970s, particularly in the Third World countries, often referred to as the “developing countries.” The concept of basic human rights, not very unrelated with the above, made its appearance nearly at the same time. It postulates, inter alia, peoples’ rights to food, shelter, health, and education. It was soon realized that these proposed developments would be impractical unless the environment is properly managed, if not improved. Generally the governments’ concern was mainly with ‘growth’ while the civil societies’ interest was in ‘development’. The former referred to as GDP/GNP (Gross Domestic Product/Gross National Product) as the annual growth in percent, while the latter pertained to wellness of the people. Development is controlled mainly by population, economics, technological development, and a congenial environment and its management. It soon became apparent that the destinies of all people were irrevocably linked and that the increasingly interconnected world cannot be a safe place if a large portion of its population still suffers from poverty, diseases, illiteracy, unemployment, and other barriers to survival. Many countries in the world were articulate in sharing this global concern for alleviation of poverty and finally improving the quality of life on Earth. "Sustainable Development" soon found a respectable position in the lexicons. But what was it actually?

International committees and commissions meeting during this time, deliberated at length on the definition of "*Sustainable Development.*" A series of international conferences took place, starting with the United Nations Conference on Human Settlement in Stockholm in 1972, followed by the World Commission on Environment and Development in 1982–87, the United Nations Conference on Environment and Development (the Earth Summit) in Rio de Janeiro, Brazil in 1992 and the World Summit on Sustainable Development in Johannesburg, South Africa in 2002. The World Commission on Environment and Development (the Brundtland Commission) in their landmark report entitled "*Our common future*" defined "Sustainable Development" as "*the development that meets the needs of the present without compromising the ability of the future generations to meet their own need.*" It reflected a transgenerational outlook of the society for the first time. It covered two key concepts at the same time. The first is the concept of "needs" of the deprived section of the society which had the highest priority. The second is the idea of "limits" on the ability of environment to meet the present and the future needs. Sustainable Development was seen by the Commission as a process of change rather than a fixed state of harmony. The core issues in the report (Brundtland 1987) included the following:

Population and development
Food security
Species and ecosystem
Energy
Industry
Urban challenge

On further consideration, the above list was modified to the following (Dadd-Redalia 1994):

Respecting and caring for the community life
Improving the quality of human life
Conserving Earth's vitality and diversity
Minimizing the depletion of Earth's nonrenewable resources
Working within Earth's carrying capacity
Enabling communities to care for their own environment
Creating a global alliance

Many of these activities are interactive, which will be obvious in the following discussion. The World Summit at Johannesburg in 2002 elaborated upon the definition of Sustainable Development through the inclusion of "economic development, social development and environmental protection – at the local, regional and global levels" as the "interdependent and mutually reinforcing pillars of sustainable development" (Fig. 9.1). In course of time many other definitions of Sustainable

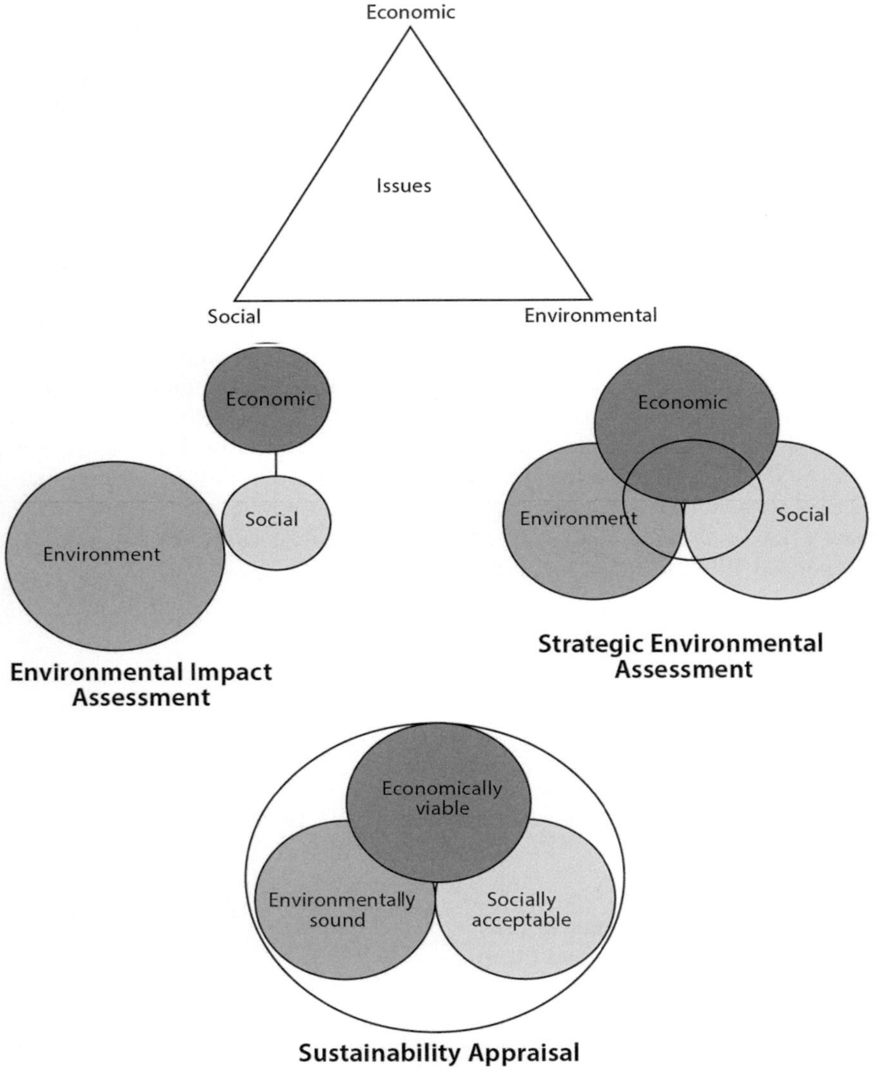

Fig. 9.1 Different grades of relationship between the three pillars of sustainability: environmental, social, and economic issues

Table 9.1 The goals of sustainable development (after Parris and Kates 2003)

What is to be sustained	What is to be developed
1. Nature • Earth • Biodiversity • Ecosystems	1. People • Child survival • Life expectancy • Education • Equity • Equal opportunity
2. Life support • Ecosystem services • Resources • Environment	2. Economy • Wealth • Productive sectors • Consumption
3. Community • Culture • Groups • Places	3. Society • Institutions • Social capital • States • Region

Development emerged from various proponents due to differences in outlook and goals, which have been nicely summarized by Parris and Kates (2003) (Table 9.1). All these definitions have some combinations of the three pillars but with different emphasis. Some suffer from the inherent ambiguity in the definition, which however, can be used positively, by calling for changes in attitude and values to develop the necessary perception for the future. The all-inclusive definition also produces conflicting viewpoints. For example, for the environmentalists, environmental degradation due to development is a basic human rights issue while to the economists, it is an economic problem involving growth and cost–benefit trade-off. But though the term may be looked at as an oxymoron, it is widely considered as a worthwhile goal to pursue because it means economic and social development that endures over the long run, with the issue of intergenerational equity at the core. The latter implies that the current generation is morally obliged to leave sufficient stock of capital for the future generations which desirably should not be less than what it is enjoying now.

A broad range of capital or assets are commonly available to a society. These include: natural capital (both renewable and nonrenewable), man-made capital (hospitals, schools, buildings, machinery, roads, etc.), human capital (skills, talent, etc.), knowledge and social capital (e.g., networks, intellectual wealth, and property). According to some, the different forms of capital are fully substitutable (defined as *weak sustainability*) while others hold that all types of capital are not fully substitutable (defined as *strong sustainability*) because some natural capital or some functions performed by these assets—such as, essential environmental services like global life support—cannot be replaced by others. From a less dogmatic point of view, one may accept that there is potential for substituting different forms of capital over a range, but there are limits to such substitution based on certain thresholds of irreversibility of damage and degradation. These limits, discussed below, may come from two possible constraints—the limited capacity of natural environments to receive the wastes generated by economic systems and the finite nature of the exhaustible resources. The overall scale of economic activity thus must be limited so that it remains within the carrying capacity of the remaining natural capital.

It is by now clear that there are widely different perceptions of the concept of sustainable development. As pointed out by Hjorth and Bagheri (2006), sustainable development is considered by many as a 'project' with an 'end state'. But these authors argue that sustainability is neither the state of the system nor is it a target to be achieved. Sustainability is an ideal to the system which inter-relates different aspects of economy, environment, and society. Sustainable development is an issue of complex systems which cannot be solved by breaking them down into elements, like in classical science, and then focusing on the isolated elements. To deal with it successfully, linear, mechanistic thinking must give way to nonlinear and organic thinking, commonly referred to as systems thinking.

Finally, we also invite the reader's attention to another document called "World scientists' warning to humanity," authored by U Thant, a former UN Secretary General, the report having been endorsed by 1600 scientists, including 102 Nobel laureates from 70 countries. It hardly needs any emphasis to underscore its importance. An excerpt from the document reads as follows:

> Human beings and the natural world are on a collision course. Human activities inflict harsh and often irreversible damage to our current processes, put at serious risk the future that we wish for human society and the plants and the animal kingdoms, and may alter the living world that will be unable to sustain life in the manner we know. Fundamental changes are urgent, if we are to avoid collision that our present course will bring about.

9.2 Sustainable Development of Mineral Resources

Soon after the Brundtland commission gave momentum to the concept of Sustainable Development, James Stevenson, CEO, Rio Tinto Corporation, the world's largest mining enterprise, in 1989 reportedly remarked that *Sustainable Development was an awkward concept for the extractive industry*. The perception that mining itself is unsustainable is still held widely by several social and non-governmental organizations and activists, mainly because of the nature of the extractive industry, which can be summarized as under:

- Projects of nonrenewable resource extraction, with strong negative impact on environment and society, are generally poor contributors to sustainability.
- Energy consumption in the extractive industry is in fact one of its largest environmental impacts. Some estimates attribute 7% of world's energy consumption to mining, mineral processing, and smelting (Richards 2005a).
- Mineral resources are finite and nonrenewable, and therefore, once exhausted, cannot be reproduced as such.

- The life of a mining project is constrained by the geology of the ore body and made uncertain by market fluctuations.
- Often the mining boom in an area ends in a bust, and the lasting local effects, which could be economic, social and ecological, are largely negative.

But does this mean mining should be banned? The answer is obvious to even those who oppose it vehemently with quite a few of them enjoying the benefits of a modern lifestyle. We just need to recall (vide Chap. 1) how in innumerable aspects of our everyday life we are dependent on minerals and metals; how the material fabric of our modern civilization is based on mineral products. The ore body is a dormant asset. As long as it lies in the ground it has no value. Therefore, it suffices to say that mining is a necessary evil and has been one of man's earliest vocations, which is emphasized by the saying in the world of mining: "*If it can't be grown, it has to be mined.*" Some comparisons will make the issue clearer. As we all know, energy use is increasing greenhouse gases in the atmosphere, but will it be a sustainable response if we switch off electricity and gas? Instead, it will be more appropriate to look for cleaner energy and advanced technologies. Similarly, the challenge of sustainable development in the mineral sector is to ensure that mineral resources are developed in an efficient manner with least possible generation of wastes and as Richards (2002) has emphasized, "the damage to the environment (including social environment) by mineral development is to be brought into balance with the planet's capacity for accommodating change."

In the mines and mineral (M & M) sector the concept of intergenerational equity in sustainable development translates into a somewhat complex issue. While mining should contribute to the well-being of the present generation through resource consumption, how does it ensure an equally good quality of life for the future generations? This is possible only through an alternative interpretation of the relationship between Sustainable Development and Mining (Ackerman 1998). The social and economic relationship should be understood in terms of conversion of capital from one form to another. Mining is a process whereby a natural resource capital (the ore body) is converted into products with many uses and sold to generate economic capital which can be reinvested to create or enhance other forms of capital or assets, as mentioned earlier: human (skill, knowledge), social (co-operative networks, community organizations), and physical (roads, schools, hospitals). Such more durable assets will guarantee income for the affected people in a mining project area even beyond the life of the mine. Implementation of this requires a systems approach which involves identification of key stakeholders as well as good governance and management of the mineral sector. A successful example of this approach to sustainability in the M & M sector is the Alaska Fund, established in 1976 for every barrel of oil produced and exported from this US state. Wise use of revenues from the nonrenewable oil has produced a renewable fund now worth more than 40 billion US dollars which can provide benefits for generations to come.

The sustainability of the mining industry also stands on three pillars, discussed above: economic, environmental, and social. Striving for sustainable development in this field involves balancing the inevitable conflicts amongst these three areas (cf. Mohanty and Goyal 2012). Success is clearly dependant on good governance, its efficiency and integrity, which can provide and maintain legal and regulatory framework for achieving the goals of sustainable development. Lack of these attributes in governance can lead to the disastrous situation prevailing in several African countries, such as Democratic Republic of Congo, Angola, Liberia, and Sierra Leone.

The mineral industry commonly offers the only realistic opportunity in developing countries for rural communities to escape the poverty trap. It can provide various economic benefits including employment, particularly to the local community, and generation of wealth only when it makes profit. A mining company, like any other business entity, needs to perform efficiently by minimizing costs and maximizing profits in order to provide competitive return to its shareholders. The profit motive, however, can be counterproductive as it often pushes the management to reduce the social benefits of the workers and cut cost of infrastructure. Profit and growth ultimately depend on how its benefits are used and whether they are acquired at the expense of others (e.g., by externalization of costs, making others pay for the environmental and social impacts or not giving a fair price for the resource extracted). If the benefits of profit are enjoyed by only a few, then ultimately this has to be viewed as unsustainable (cf. Richards 2005b). An argument is often floated that sustainable development incorporates increase in cost. Yes, it does, but on the other hand, when sustainability issues become acute, things may go out of hand and mining itself may not be feasible anymore, as has happened over large areas in east-central India (vide Chap. 11). Sustainable development must involve internalization of various costs of mining and paying a fair price for the commodities extracted. Currently, consumer prices do not include cost of environmental damage and remediation and other social obligations. Ultimately, society must be prepared to pay the full internalized costs for commodities required for a lifestyle it chooses to lead.

Most of the negative effects of mining at the local level relate to environmental impacts as a result of the generation of large volume of wastes (both overburden and tailing) often spilling over beyond the lease area into agricultural land, polluting air and causing noise pollution and discharge of liquid effluents. Drilling and blasting, particularly in open cast mines, commonly add to the atmospheric pollution. There are also problems relating to damage to land and topography and destruction of forests and wildlife habitat. Mitigation of environmental impacts of mining and progressive rehabilitation of the affected area is fundamental to sustainable development in this sector. One of the major technological challenges relates to management and utilization of mine wastes including tailings and mine runoff. Mine closure and reclamation of mined lands after cessation of operations is an established practice in developed mining countries and is now mandatory in India as well. It is possible to return the reclaimed mine sites to their previous natural state, thereby recreating suitable habitats, or making these available for agriculture or for establishing recreational and public facilities such as resorts township, lakes, and sites for ecotourism. Reputed mining companies recognize this and take their environmental obligations seriously as a corporate social responsibility (CSR) activity by adopting good environmental practices. Voluntary efforts are less common in case of small mines whose environmental behavior is often not up to normal expectations. In any case, appropriate legal framework and regulatory mechanisms are essential for ensuring that both large and small mining enterprises take effective measures for the mitigation of environmental impacts of mining operations.

Social issues, probably the most contentious of the three, comprise the third dimension of sustainable development in mining. Mining in many places has affected the traditional livelihood and serene lifestyle of the indigenous people, brought about large-scale displacement and changed the social complexion of a region, making society distrustful and even averse to mining projects. Traditionally, government and mining enterprises were considered only two core stakeholders who negotiated the terms of mining operations. Over time and conflict management, the local community has been recognized the world over as the other important stakeholder in the mining sector. Sustainable development would require mining enterprises to align the interests of local communities with their own in areas where they want to operate and develop mines with those communities on the basis of mature and respectful partnership. This calls for improved engagement between all stakeholders during all stages of mining activity. These sustainability issues have been given the required emphasis in the major mining countries, such as Canada, Australia, South Africa, and Papua New Guinea (PNG) where local stakeholder engagement, socioeconomic development in the mining areas and transparency in communication with all stakeholders have become the commonly accepted norms. In all these countries, the regulatory mining and environmental laws are strictly implemented. The Social and Labour Plan in South Africa, Mineral Development Fund in PNG and Impact Benefit Agreement in Canada are examples of legally binding instruments that make it obligatory for mining companies to carry out local development work in the mining project areas.

9.2.1 Essential Steps in Ensuring Sustainable Development of Mineral Deposits

Mining, as mentioned earlier, has had an ambiguous relationship with sustainable development. The question currently asked is not whether mining should be undertaken, it is rather how it should be carried out: when? by whom? for whose benefit? And at what and whose cost? These are generally uncomfortable questions for the mining industry, triggered by the concept of sustainable development and have been partially answered in the above discussion. For most people, sustainable development is associated with renewable resources. For them it is difficult to conceive of mining, which depletes a nonrenewable resource, to play a role in sustainable development. Any mining activity, right from exploration to extraction to processing, will surely have some negative impact on the physical and social environment. The best the miner or mining industry can do to limit the negative consequences of mining is to display maximum "environmental responsibility" during mining and processing:

- by extracting, processing, and distributing the raw material in the least environmentally damaging way.
- by returning the mining area as nearly as possible to its pre-mining form.
- by providing some other use to the desolation created by mining and
- by developing alternative livelihood for the miners when the mine ceases to work.

To achieve sustainable development of mineral deposits, as closely as possible, the following steps may also be followed:

1. Determining in as much details as possible the size, shape, and disposition of the deposit, including chemistry, mineralogy, and petrography of the ores, processibility of the ores and the estimated ore reserves (of different categories, according to UNFC classification, vide Chap. 1). This will help in projecting the mine life and resource value.
2. Drawing up exploration and mining plan, together with environmental management/restoration plan for exploration, mining and post-mining stages. These should have necessary flexibilities.
3. Involving the local community: their approval (free prior informed consent or FPIC) must be openly sought before any mining project is undertaken (in line with World Bank's extractive industries review, 2003). Once FPIC is obtained, the local community must be involved in planning and execution of the project. At no stage they should feel that they have been deprived of their due in terms of recognition and material benefits. Concept of beneficiaries and benefactors should be discarded. The management and the community, both are stakeholders, in the new perception. Belief in this dictum is likely to inculcate a sense of social equity amongst the people concerned.
4. The project should visibly contribute to the development of human, social, and economic resources of the area. Suitable local people should be trained in mining and metallurgy-related technology (and provided with secondary education if they did not have it already). Qualified and healthy workforce is an asset for any industry. It is a commonsense knowledge, yet economist Nobel Laureate Amartya Sen has been repeating it in the Indian context over the years.
5. Equitable distribution of the socioeconomic gains derived from the project amongst the management, workers, and the community that has been involved.
6. Time to time revision of the estimation of the mineral commodities and modifying the action plan accordingly.
7. Remembering the 3 Rs (*reduce, replace, and recycle*) as a valuable guiding principle in any consideration for sustainable development of mineral deposits.

Metals are now replaced by plastics, ceramics and other materials, that made automobiles lighter. Auto oil/diesel engines are replaced by electric engines, the latter is somewhat costlier though. It must be admitted that the per person throughputs in western world is wasteful. This is accepted even by reasonable people from that part of the world itself. We indeed share the belief of Meadows et al. (2004) that "a good life could be supported by much less destruction of the planet."

Another interesting fact is that in the less developed world people reclaimed and reused materials because of scarce resources. Now the rich communities are relearning how to recycle because of scarcity of sinks! A rule of thumb says that for every ton of garbage at the consumer end of the flow, there was a requirement of 5 tons of waste at the manufacturing stage and 20 tons of waste at the site of the initial resource extraction, i.e., mining, pumping, logging, farming, etc. Besides, the high value of some wastes is being appreciated more and more. In 2014, gold from e-waste formed 11% of world gold production amounting to 10.4 billion Euros. If a recent news item is to be believed, one ton of a popular brand of high-end smartphone would account for 300 times more gold and 6.5 times more silver than an equivalent weight of the respective ores (TOI, 18th Oct., 2016). And yet, barely 10% of such smartphones get recycled. Awareness campaigns are necessary in this regard.

9.3 Some Important Facts/Issues in the Above Context

9.3.1 Evolution of Schemes for Sustainable Development of Mineral Resources

MacDonald and Gibson (2006) have identified four broad eras of public opinion and government policy toward mining over the last several decades that have moulded the present stakeholders' relations (cf. Table 9.2). The first until 1981 was principally an extension of the status quo, focusing on the environmental aspects that had become a concern by that time. The second ranged from 1988 to 1996 when the modernization of the environmental regulations took place in the developed world, vis-à-vis the rapid and relatively unregulated opening of developing countries' mining to foreign direct investment. The 1989 convention of the International Labour Organization (ILO) gave recognition to rights of indigenous people, enshrining the obligation to consult them on the use of their traditional lands. Situation got changed in developing countries also. Many exploration programs in these regions had to interact with local communities, although possibly they had little or no knowledge of their language, culture, or the economy. Mining-focused social activists or NGOs made their presence felt during this period. Their activities centered not only on the opening of mines, but also on their safe closing. The number of abandoned mines in the beginning of this century in the

Table 9.2 Evolution of the scheme of sustainable development (SD) of mineral deposits

Pre-1987 or Pre-SD era	1988–1996	1997 and later
Mining industry was still traditional. Any expense other than for actual mining was thought of as an unnecessary expenditure. Rising clamor for environmental management was treated for long with indifference by the company	The concept of SD was slowly being crystallized. Environmental conservation was willy-nilly accepted by the industry, but not always honored in practice	The rise of civil society fighting for the SD of mineral deposits is a characteristic of this period. It made the corporate sector or any mining company to assume responsibility for improving the life and opportunities of the local people
In the socialist countries and in some developing countries the mining industry was totally/largely controlled by the government. In these countries there were some plans for exploitation, but hardly anything for environmental management	This was a period of economic liberalization or *globalization*, a model of economic growth not readily supporting SD. It conceived the whole world as the field for exploration and exploitation by those who could do it	The mining companies are now expected not only to mine the deposits but also to preserve and restore the environment and share the benefits with local communities
		The governments are obliged to not being overtly sympathetic to the companies, for otherwise there will be social turmoil, as it happened in so many countries of the world, not excluding India

USA ranged from 10,000 to more than 500,000. In the third phase, spread between 1997 and 2002, the NGO activities became transnational, thanks to participatory democracy in the developing world and the development of easier communication system.

These networks demanded from the concerned industry at least transparency, accountability for environmental conservation/restoration, and consultation with the affected communities. This helped some local issues become internationalized and consequently create pressure on the corporate sector and in cases, their government, if they were apparently indifferent.

In the final stage that covers even today, the concept of prior and informed consent has become virtually nonoptional. Both the civil society and concerned people in the World Bank have concluded that "*free prior and informed consent*" (FPIC) is the standard the mining industry must follow, if development is to be truly sustainable. FPIC is a process by which indigenous people, local communities, government, and corporate sector may come closer and agree in a forum that gives affected communities enough leverage to negotiate conditions under which they may proceed and an outcome leaving the community clearly better off.

After what has been discussed above, one cannot possibly ignore an obvious question: even after sustained development, the total reserve of mineral deposits will continue to deplete and it is not renewable in actual terms. So, how can it be truly sustainable? MacDonald and Gibson (2006) argue in this regard that there is the recognition now that natural capital depletion is an inevitable fact, and other forms of capital can replace these losses. Industrial economists and socioeconomic analysts posit that sustainable mining generates more than the mere minerals and metals. It produces at the same time human and physical capital. This point is not absolutely above controversy though. A reasonable compromise could be: follow sustainable development of mineral resources as much and as long as you can. In the meantime, reduce, reuse, and recycle the minerals and metals to the extent it is possible.

9.3.2 Depletion of Reserves and Long-Term Supply

We already know that mineral deposits are nonrenewable and once exhausted cannot be reproduced as such. Not to think of total exhaustion, depletion of their reserves below certain levels will cause great pressure on the present industrial civilization, unless they (metals, minerals) are replaced, recycled, and reduced in use. Recall the anxiety expressed by the Club of Rome (Meadows et al. 1972) in the now famous book "The limits to growth" (LTG). They forewarned the human society of a possible industrial crisis affecting the quality of human life in the context of apparently dwindling mineral resources. Some earth scientists were impressed by the contents of the book, which they accepted as a timely warning against a possible crisis. Others thought differently. They derided the "Club of Rome" people as the "Merchants of doom." Until about this time, people did not bother about the mineral resource positions even in not-too-distant a future. The mineral industry (mining and metallurgy) was happy to make a profit through their business and the governments collecting their revenues. The latter at best dictated some mine safety rules and some countries adopted export–import policies, mostly after the World War II. Today one can easily realize that the authors of LTG were not advocates of doom really. They only wanted the society to be discrete and have the necessary foresight, as in the foreseeable future they may have to deal with a world population of about 10 billion people. Some amongst them realized that proactive measures could be adopted in time to restrain increase in "ecological footprint of humanity beyond the carrying capacity of the planet earth" (Mathis Wackernagel, etc., in Meadows et al. 2004). *Ecological footprint* is defined as the land area that will be required to provide resources (grains, feed, wood, fish, urban land, etc.) and absorb the emissions (CO_2) and wastes of the global society, when compared to the available land. It was also estimated that the resource use by human society was 20% above the global carrying capacity even in 2000 and humanity was last at sustainable levels in the 1980s. Moreover, humans being conscientious cannot exploit all the natural resources for a quality of life leaving the future generations to the uncertainties of the future. Transgenerational interest is an important concern in the concept of Sustainable Development. Add to these our awakened consciousness about the menace of environmental degradation caused by industries, mineral-related industry in particular. Moreover, it is a period of agreed human rights, a doctrine very basic to the human society, but nearly ignored until recently. The principle accepts "*the right to live with dignity*" which means free from hunger, sociopolitical oppression, and retaining age-old culture. Another important development in recent times has been the recognition of the legitimate stake of the local community in the activities, including financial gains of mining companies. This includes their voice in decisions that ultimately affect their interest. The civil society is very vocal about it throughout the world, including India. Many of the mining companies initially did not take kindly to these impositions in the fields which so long they knew to be their own turfs. The problem became compounded when the civil society activists demanded that the additional expenses thus incurred should be internalized. This also, the mineral industry in the developed world has willy-nilly come to accept.

Fast depletion vis-à-vis long-run availability of mineral commodity is a frequent debate that few socioeconomically alert people can afford to neglect. Regarding energy-producing commodities, it is currently estimated that time to depletion is 40–60 years for crude oil and 80–100 years for natural gas. While some people may believe that the advancing technology will be capable of offsetting the adverse effects of mineral depletion and the consequent deterioration of living standards, they may not ignore the social cost that may be prohibitive by that time. Subscribers to the "fixed stock paradigm" believe that as we exploit the commodity, the stock depletes, making some even exhaust before long. Needless to say that profligate use of mineral commodity in developed countries and the rise of population particularly in developing countries are likely to quicken their run out. The proponents of the fixed stock model made some calculations of world reserves of some such commodities, based on finite grades and consumption rates, present and projected into the future, as shown in Table 9.3. Though the data are more than a decade old, it is obvious that the discovery of new deposits, advancement of technology making the low-grade mineral commodities useful, have increased the reserves, and hence, their life expectancies. Recycling and replacement, directly or indirectly, will have similar effect.

Social costs may limit the future use of mineral commodities. The social costs include those of environmental management, sharing the benefits with the local community without disturbing the culture of the latter. In India, as in many other countries, the last one is very important if the project is located in a tribal belt. In a tribal country-side a mineable deposit may be located on a peak or hillock which is very sacred (e.g., *Marang Buru*) to the indigenous community. Interference with such places in connection with mineral exploitation is virtually non-negotiable. Shifting the tribal communities from their normal habitat is also not easy. This issue raised protests in many countries including Australia, Canada, and the United States. The Indian situation will be discussed in some details in a later chapter.

As we have already mentioned, even some in the developed countries (Meadows et al. 2004) believe that per person materials throughputs of the rich nations of today, like their food, water, wood and energy, are wasteful. Today 20% of population living in rich countries consumes 80% of the world's resources. A pertinent

Table 9.3 Life expectancies (world reserves) of some mineral commodities (after Tilton 2006)

Mineral commodity	Year-ended 2003 reserves	2001–2003 average annual primary production	Life expectancies (years) at 3 growth rates for primary production			Avg. annual growth in production 1975–2003 (%)
			0%	2%	5%	
Coal	9.8×10^{11}	4.8×10^{9}	203	81	48	1.4
Crude oil	1.1×10^{12}	2.8×10^{10}	40	30	22	0.9
Natural gas	6.3×10^{15}	9.0×10^{13}	70	44	30	2.9
Aluminum	2.3×10^{10}	1.4×10^{8}	161	72	44	2.4
Copper	4.7×10^{8}	1.4×10^{7}	34	25	20	2.5
Iron	7.0×10^{10}	1.1×10^{9}	64	40	28	0.9
Lead	6.7×10^{7}	3.0×10^{6}	22	18	14	−0.5
Nickel	6.2×10^{7}	1.4×10^{6}	46	32	23	2.0
Silver	2.7×10^{5}	1.9×10^{4}	14	11	10	2.5
Tin	6.1×10^{6}	2.3×10^{5}	27	21	17	−0.2
Zinc	2.2×10^{8}	8.7×10^{6}	25	20	16	2.1

Note Reserves of metals are in terms of metal content, excepting aluminum, which is in terms of bauxitic ores. Reserves and primary production are measured in metric tons, except for crude oil (barrels) and natural gas (cft). Life expectancy is proximate at the given reserves and growth rates of primary production. This may be modified by any one or more of (1) new discoveries (+); (2) rise of Human population (−); (3) profligate consumption (−); (4) finding suitable replacements

question therefore: Why can't a reasonably good life be lived with much less? Here returns the oriental concept of "good life and its simple living"! Well, that was also not only the professed but practical philosophy of the American Philosopher extraordinary, H.D. Thoreau. We end this discussion with the most important slogan of this age, "We the humans have a common future."

References

Ackerman R (1998) Is mining compatible with sustainable development? A world bank perspective. Int Counc Met Environ ICME Newslett 6(1):1–3

Brundtland GH (1987) Our common future: report of UN world commission on environment and development. Oxford Univ Press, p 383

Dadd-Redalia D (1994) Sustaining the earth. Hearst Books, New York, p 352

Hjorth P, Bagheri A (2006) Navigating towards sustainable development: a system dynamic approach. Futures 38:74–92

MacDonald A, Gibson G (2006) The rise of sustainability: changing public concerns and governance approaches towards exploration. Soc Econ Geol Spec Publ 12:127–148

Meadows DH, Meadows DL, Randers J, Behrens WW III (1972) The limits to growth. Universe Books, New York

Meadows DH, Randers J, Meadows DL (2004) Limits to growth—the 30 year update. Chelsea Green Publ Co, Vermont, p 338
Mohanty N, Goyal A (2012) Sustainable development: emerging issues in India's mineral sector. Institute for studies in Ind Development (ISID), Planning Commission Report, p 274
Parris TM, Kates RW (2003) Characterizing and measuring sustainable development. Ann Rev Env Res 559–586
Richards JP (2005a) Challenges for sustainable mineral resources development in the 21st century. Res Geol 55(2):131–137
Richards JP (2005b) The role of minerals in sustainable human development. In: Marker BR, Peterson MG, McEvoy F, Stephenson MH (eds) Sustainable minerals operations in the developing world, vol 250. Geol Soc Lond Spec Publ, pp 25–34
Richards JP (2002) Sustainable development and the minerals industry. Soc Econ Geol Newslett
Tilton JE (2006) Depletion and long-term availability of mineral commodities. Soc Econ Geol Spec Publ 12:61–70

Chapter 10
Mines and Minerals Sector in India and Its Regulatory Regime

Abstract India is a country endowed with rich mineral resources, producing as many as 87 mineral commodities (including energy-producing minerals). The country is self-sufficient in bauxite, chromite, iron and manganese ores, ilmenite, rutile, coal (except coking coal) and lignite, and almost all industrial minerals. Along with barite and limestone, these are among the ten largest mineral reserves of the world. The structure of the Indian M & M sector has evolved considerably in the last two and half decades, since the liberalization of the economy and opening up of the sector to private domestic and foreign investment. The mineral commodities are thus extracted through large and small mines, which belong to either public or private sector enterprises. There is also a large informal sector of artisanal and small-scale mining (ASM), whose status remains rather ambiguous in the present regulatory regime. ASM operations need to be regularized by the government, acknowledging their livelihood and poverty alleviation potential. Mining, thus, is an important sector in Indian economy and contributed about 2.6% to the GDP in 2011. Since then the sector is, however, in a negative growth path with a substantial reduction in the number of operating mines. Various reasons have been ascribed to this decline. They include: regulatory issues, environmental activism and court cases, social unrest due to land acquisition problems and displacement, and Naxalite violence along the "Red corridor" in east-central India. The M & M sector is currently struggling to recover its earlier growth rate by streamlining regulatory and administrative procedures, developing the required infrastructure and taking care of the issues of sustainability. The framework of various regulatory provisions and policies governing this sector need to be understood to appreciate the issues in this regard. The 7th schedule of the Constitution includes Article 246 which categorizes the Union, State and Concurrent lists of responsibilities under which regulations of mines and mineral development fall. Thus, the proprietary control of onshore mineral resources is with the state government while the regulatory powers are with the central government. All mining activities in the country are controlled by the National Mineral Policy while the basic laws governing the mining sector come under the purview of the Mines & Minerals Development and Regulations (MMDR) Act. The other Acts concerned with mining are: Environment Protection Act, Environmental Impact Assessment Act, Forest Conservation Act, Land

M. Deb and S.C. Sarkar, *Minerals and Allied Natural Resources and their Sustainable Development*, Springer Geology,
DOI 10.1007/978-981-10-4564-6_10

Acquisition, and related Acts and Panchayats Act (PESA). The regulatory regime is thus quite wide and complex, with duality of control exercised by the state and central governments. This consequently leads to divided accountability and poor implementation of the laws. The marine mineral resources are exploited keeping the "UN convention on the Law of the Sea" (UNCLOS) in view. It divides the sea into three parts: the territorial sea, the exclusive economic zone (EEZ), and the international area of the seabed. The coastal nations have exclusive rights to exploit marine resources up to their respective EEZs.

Keywords Indian mineral reserves and production · Structure of the Indian mineral industry · Artisanal and small-scale mining in India · National Mineral Policy · MMDR Act · Mining-related Acts in India · Duality of regulatory regimes · Law of the Sea · EEZ

10.1 Mineral Industry in India – An Overview

India has long been recognized as a country endowed with rich mineral potential. The country produces as many as 87 minerals including 4 fuel minerals, 10 metallic minerals, 47 nonmetallic minerals, 3 atomic minerals and 23 minor minerals (including building and other materials). India is ranked 4th among the mineral-producing countries, behind China, USA, and Russia on the basis of volume of production (Table 10.1), and 8th on the value of mineral production (Table 10.2), according to a report of the World Mining Congress.

Minerals produced in the country can be divided into four categories (FICCI 2013). The first concerns *Bulk Minerals* which have high volumes of extraction, transportation, and consumption. Limestone, iron ores, manganese ores, chromite, bauxite, granite, marble, etc., fall in this category. Mining of all these are at shallow depth with considerable amount of overburden removal and waste generation. Production of some of these bulk minerals requires huge input of other resources. For example, 1000 kg of primary aluminum can be produced from 5000 kg of bauxite ore, 13,000 L of fresh water and 15,711 kWh of electricity. Impact of such large-scale mining obviously is not limited to the ore body alone and often goes beyond the carrying capacity of the area. Not surprisingly, therefore, such mining activity faces resistance from environmental, forest, and other clearance issues, to be discussed in a later section. The next category includes the *Base Metal Ores and Associated Strategic Minerals/Metals* such as, copper, lead and zinc along with tin, cobalt, nickel, lithium, germanium, gallium, indium, niobium, beryllium, tantalum, tungsten, bismuth, and selenium. These commodities have limited reserves and fall under the "deficit category". These are extracted both from underground and open cast mining operations. The third category concerns *High Value but Low Volume Commodities* which are precious metals/minerals. This category is represented by gold, silver, platinum, diamond, and precious stones. Though the geological terrain in India is favorable for such minerals, we are highly deficient in their production.

Table 10.1 Contribution and Rank of India in world production of some principal minerals and metals, 2011 (IBM YB 2012, Pt-1)

Commodity	Unit	Production world	Production India[a]	Contribution (%)	India's rank
Mineral fuels					
Coal and lignite	Million tons	7739	582	7.5	3rd
Petroleum (crude)	Million tons	3980	38	1.0	24th
Metallic minerals					
Iron ore	Million tons	3012	167	5.5	4th
Manganese ore	'000 tons	47,300	2349	5.0	6th
Chromite ore	'000 tons	26,300	3764	14.3	3rd
Bauxite ores	'000 tons	248,000	12,877	5.2	6th
Industrial minerals					
Barite	'000 tons	9000	1723	19.1	2nd
Aluminous refractories (kyanite, andalusite, sillimanite)	'000 tons	460	62	13.5	4th
Talc/steatite/pyrophyllite	'000 tons	7800	1198	15.4	2nd
Magnesite	'000 tons	23,100	217	1.0	11th
Apatite and rock phosphate	'000 tons	203,000	2330	1.1	14th
Mica (crude)	Tons	307,000	1807	0.6	15th
Metals					
Aluminum	'000 tons	45,200	1654	3.7	8th
Copper (refined)	'000 tons	19,500	504	2.6	10th
Steel (crude)	Million tons	1516	73.8[b]	4.8	4th
Lead (refined)	'000 tons	10,400	92	0.9	18th
Zinc (slab)	'000 tons	13,000	783	6.0	3rd

Sources World mineral production data compiled from World Mineral Production, 2007–2011, British Geological Survey
[a]Figures relate to 2011–2012
[b]Ministry of steel, Annual Report, 2011–2012

The last group is characterized by *Low Value but High Volume* of such 24 non-metallic minerals. These can be further subdivided into (i) fertilizer minerals; (ii) flux and construction minerals; (iii) glass and ceramic minerals; (iv) refractory minerals; (v) abrasives, pigments, fillers, etc. These are largely produced in small private mines. The country is deficient in some of them, e.g., fertilizer minerals.

In meeting the demand for domestic mineral-based industries, India is self-sufficient in bauxite, chromite, iron and manganese ores, ilmenite and rutile among metallic minerals, coal (with the exception of coking coal required in steel

Table 10.2 Value of mineral production (2008–2009 to 2010–2011) (in Rupees Crores)

Mineral	2008–2009	2009–2010	2010–2011
All minerals (excluding atomic minerals) (100)	174240.31	179384.01	200609.38
Fuel minerals (67.42)	115371.31	124088.33	135243.81
Metallic minerals (20.55)	36067.76	32274.29	41828.44
Nonmetallic minerals (2.40)	4066.79	4286.94	4802.69
Minor minerals (9.33)	18734.45	18734.45	18734.45

Note Figures in brackets are percentages
Source Annual Report 2010–2011, Ministry of Mines, Government of India, 2011

plants) and lignite among fuel minerals, and almost all industrial minerals with very few exceptions. The estimated reserves of major mineral commodities along with their life expectancies at the present rate of consumption are presented in Table 10.3. The reserves of barite, bauxite, chromite, coal, iron ore, limestone, and manganese ore in India are among the ten largest mineral reserves in the world. However, in the future, we will need to focus attention on raw materials for our production of steel, aluminum, titanium, rare earths, base metals, and gold.

At present, nearly half of India's total mineral production (excluding oil and gas), in terms of value is contributed by seven key mining states: Odisha (9.6%), Andhra Pradesh (9%), Rajasthan (7.9%), Chhattisgarh (7.8%), Jharkhand (6.5%), Madhya Pradesh (4.8%), and Karnataka (3.6%). The offshore areas of production had a share of 26% in this mineral production statistics of 2010–2011. The share of India's key mineral reserves in these seven states is presented in Table 10.4. These

Table 10.3 India's position in reserves of major minerals (2010) (after FICCI 2013)

Mineral	Current reserves (million tons)	India's rank in reserves	Reserve life (years) at current rate of consumption
Coal	1,13,000	4th	187
Iron ore	7000	7th	47
Manganese ore	138	5th	47
Chromite	66	3rd	24
Bauxite	900	6th	66
Barite	34	2nd	30
Limestone	12,715	–	55
Zinc metal	11	7th	8
Lead metal	3	7th	26
Copper	4	–	9
Aluminum	2.3	5th	–

Table 10.4 Share of India's key mineral reserves in seven mineral-producing states (after FICCI 2013)

State	Coal	Iron ore	Manganese ore	Chromite	Lead and Zinc	Bauxite
Jharkhand	29%	14%				
Odisha	24%	17%	35%	98%	–	51%
Chhattisgarh	16%	10%	–	–	–	–
Madhya Pradesh	18%	–	10%	–	–	–
Andhra Pradesh	7%	7%	–	–	1%	21%
Rajasthan	–	–	–	–	90%	–
Karnataka	–	41%	29%	1%	–	–
Total	84%	89%	74%	99%	91%	72%

seven states account for a third of India's population but are relatively backward, with the Human Development Index (HDI) in five of these states being the lowest in the country. Sustainable growth of mining in these states could play an important role in economic and social development of these states.

The mining sector is one of the core sectors that drive the economy of a country. It is generally envisaged that a one percent growth in mining results in 1.2–1.4% increment in industrial production and a corresponding increment of 0.3% in India's GDP (FICCI 2013). The data presented above establishes that mining is an important sector in Indian economy and had contributed about 2.6% to our GDP in 2011 (vide Fig. 1.7). However, the contribution of mining and quarrying to India's GDP is on the decline in recent years: it was 3.4% in 1992–93, 3% in 1999–2000 and went down to 2.5–2.6% between 2008–2009 and 2010–2011. After achieving a growth rate of 4.8% over 5 years between 2006–2007 and 2010–2011, the mining sector in India has witnessed a negative growth rate of 0.6% for two consecutive years, i.e., in 2011–12 and 2012–13. This is reflected in the reduction of the number of operating mines (excluding atomic minerals, petroleum and some minor minerals) from 2999 in 2009–2010 to 2628 in 2010–2011. While the number of coal mines (574) has remained the same, mines for both metallic and nonmetallic ores have reduced from 700 to 608 and 1725 to 1446, respectively (Ministry of mines, GOI, Annual Report 2011).

Various reasons can be ascribed to this decline. First, policy paralysis on wide ranging issues at the level of both center and the state has hit the mining sector badly. Mining projects across the country have stalled due to court cases, regulatory issues, environmental activism, social unrest due to land acquisition and displacement, and Naxalite violence along the "Red Corridor" (vide Chap. 11). Add to this, the problem of rampant and large-scale illegal mining of iron ore, coal, and construction material in different parts of India. The National Mineral Policy of 2008 envisaged that liberalization will bring about Foreign Direct Investment

(FDI) and the private sector will take the lead in mineral production. But such a turnaround has not happened due to the interplay of reasons mentioned above, and public sector still plays the dominant role, accounting for 68% of mineral production in 2011–12. It may also indicate that policies and incentives are still not conducive to the private players. The government was looking for more FDI and technology transfer in the mining sector. But people in the business know that the "License Raj" is not over yet in India. Politics, corruption, and intervention of the dense bureaucracy are strong discouragement for most investors. Besides, there are many other serious problems on the ground.

In spite of a favorable geology like that of Australia, South Africa, and Brazil, mineral potential in India remains untapped and the reserves of many important metals are fast depleting while some others like nickel and tin are almost nonexistent. An area of about 0.60 million km^2 of the Indian landmass has been identified by the Geological Survey of India as having obvious geological mineral potential. But our investment in exploration still remains extremely low. Compared to \$124/km^2 in Australia and \$118/km^2 in Canada, it is only \$15/km^2 in India. The possibility of growth in the mineral sector is immense. This is supported by the very low per capita mineral consumption in India. Countries typically go through a mineral consumption curve, where per capita consumption accelerates during industrialization (development) period, gradually stabilizes or declines later (developed phase). In comparison with other BRIC countries, India is way behind in terms of mineral consumption. For example, in 2010, total consumption of iron ores was 148 Mt in India while it was 975 Mt for China; coal consumption was 278 Mtoe (oil equivalent) and it was 1714 Mtoe for China. Projection shows that as India's per capita GDP increases, its mineral consumption will also grow at a rapid pace. Growth potential of key consumer industries like steel, cement, etc., in the 12th plan period (2012–2017) will trigger considerable growth in sectors like construction and power. Consequently, the demand for iron ore, coal, limestone, copper, etc., will drive the growth in the Mines & Minerals Sector. Indian mining industry is also likely to see accelerated growth in export demands for minerals (FICCI 2013).

The supply-side potential of India's mineral reserves has already been pointed out (vide Table 10.3). It is to be noted that at the current rate of consumption, coal reserves will last for almost 200 years while iron ore and limestone, close to 50 years. Such life expectancies of these commodities are bound to increase as more resources are explored and low grade ores become amenable to processing through technological innovations. With appropriate investment in infrastructure and technology, there is a significant potential for the untapped mineral wealth of India.

But in spite of such demand- and supply-side potential of India's mineral wealth, the mining sector is struggling at present to recover its earlier growth rate. This struggle may be ascribed inter alia to three major factors: regulatory and administrative procedures, inadequate infrastructure facilities and finally, issues of sustainability. We shall discuss the first at length in this chapter and various aspects of the others in the succeeding chapter.

10.2 Structure and Evolution of Indian Mining Industry

Large and small mines, belonging to both public and private sector enterprises, as well as the informal sector of artisanal and small-scale mining (ASM), constitute the mining sector in India. In the first category, the innumerable small mines outnumber the few large mines in the country and are responsible for the severe environmental degradation in the major mineral-producing states. Some of the large mines are the captive mines of PSUs like SAIL or large private companies like Tata Steel or Vedanta. The public sector continues to play a dominant role in the production of vital mineral commodities, such as, iron and steel, copper, bauxite and aluminum, coal and lignite, and a major part of petroleum and natural gas. The major Public Sector Undertakings (PSUs) with their percentage share in value of production are: National Mineral Development Corporation Ltd. (NMDC) (diamond and iron ore: 18.54%), Orissa Mining Corporation (chromite, iron ore, Mn ore: 6.09%), Steel Authority of India (SAIL) (iron ore, dolomite, limestone: 2.81%), Manganese Ore India Ltd (Mn ore: 1.6%), Rajasthan State Mines and Minerals Ltd (RSMML) (Fluorite, gypsum, limestone, phosphorite: 1.4%), and Hindusthan Copper Ltd (copper: 1.1%). The large private enterprises include: Sesa Goa (iron ore: 6.99%), Hindusthan Zinc-Vedanta (lead and zinc, phosphorite: 3.97%), Rungta Mines (P) Ltd (iron ore: 3.4%), Tata Iron and Steel (iron ore, Mn ore, chromite, dolomite: 3.28%) (FICCI 2013). Almost all small mines are in the private sector, belonging to public-limited companies or individual proprietors. This large variety of mining enterprises makes the task of sustainable development of mineral industry in India a challenging one.

The structure of the Indian mining industry has evolved considerably in the last two decades in particular. Before 1993, Fe ores, Mn ores, chromite, coal and lignite, zinc-lead, copper and gold were all mined by PSUs through large mines. At the same time, minerals like limestone, baryte, refractory minerals, talc-steatite, mica, etc., were mined by private enterprises through smaller mines. In 1991, liberalization of the economy took place in the country and a National Mineral Policy was enacted in 1993 when mining sector started attracting private domestic and foreign investments. Thirteen commodities became available for private exploitation: Fe, Mn, Cr, Au, Cu, Pb, Zn, Mo, W, Ni, PGE, diamond, and sulfur. Coal and lignite were added to this list in 1999. But no grand change, as expected, was observed: in a full decade (2001–2011), FDI in the mining sector was USD 905.36 million, i.e., 0.63% of the total FDI of USD 1,43,157 million. Only two multinational companies, Vedanta and Rio Tinto got license to mine. In September, 2005, a high-level committee (Hoda Committee) was appointed by the Planning Commission to review the existing mineral policies. Hoda committee submitted its report in

December 2006 and a new National Mineral Policy was formulated in 2008. A draft MMDR bill based on the Hoda committee report was introduced in the Lok Sabha in 2011 and finally passed in 2015.

10.2.1 Artisanal and Small-Scale Mining (ASM)

10.2.1.1 Introduction

Artisanal mining is important because this activity employs a very large number of people. ASM is largely a poverty-driven activity, often associated with conflict, corruption and civil strife, practiced in some of the world's poorest regions in Latin America, Africa, and Asia. If organized properly, however, ASM has the potential to contribute significantly to social and economic development and offers an opportunity to alleviate poverty in rural landscapes. In such mining activities, mechanization is minimal and most stages of mining and processing involve simple tools and human labor.

India has a long historical tradition of artisanal mining and consequently, mining at community scale was one of the main economic activities since ancient times (vide *Kautilya's* advise on the-then mining industry in his *Arthasastra*). With the development of modern mining during British rule in India, artisanal mining activities suffered a setback and slowly became obscure due to the lack of any Government regulations regarding such activities. In recent years, however, there has been a spurt of ASM activities, particularly for low value bulky products, such as, gravels, sand and building stones, besides ASM of high value, low production commodities such as, gemstones, tin ore, and gold. Consequently, some of the traditional practices have continued through the involvement of indigenous communities working in many instances in remote forest areas. ASM activities are considered by and large as unauthorized and illegal, though one could argue that it may be at the most labeled as "nonlegal" since there is no specific law against it. Small-scale mining (SSM) activities are generally carried out by acquiring mining rights over small areas for nonmetallic and industrial minerals. Some part of such activities may be illegal, without a license or after its expiry, or when carried out on old, abandoned mines. It is noteworthy that SSM's contribution to India's mineral production is quite significant.

10.2.1.2 Status of ASM in India

The ASM sector in India is characterized by varied mining practices, size, legality and production, and this has produced much confusion over its definition. The National Mineral Policy of India mentions "Small Deposits" (7.12) only once in passing: It says that efforts would be made to promote small-scale mining of small deposits in a scientific and efficient manner while safeguarding vital environmental and ecological imperatives. In grant of mineral concessions for small deposits in Scheduled Areas, preference shall be given to the Scheduled Tribes. Artisanal mining thus does not figure directly in any of the recognized areas of "mineral development", either in the Act or in the Policy.

Besides the Indian government's indifference to ASM, the classification of minerals into major and minor, also adds to the prevailing confusion in this regard. The minor minerals, in terms of MMRD Act, include "building stones, gravel, ordinary clay, and ordinary sand other than sand used for prescribed purposes". The concerned state governments are vested with the responsibility of developing minor minerals, according to the Act. The Indian Bureau of Mines identified (Rule 42 of Mineral Conservation and Development Rules, 1988) yet another two categories: Category A and Category B, based on employment of labor and the standard of mechanical equipment used. However, SSM is not specifically covered by either of these categories, although small mines falling under category B constitute $\sim 88\%$ of the reported mines which produce $\sim 10\%$ of the total value of mineral production in India. The National Institute of Small Mines (NISM) has categorized mines in India according to their production (Chakravorty 2001).

ASM has been described by some researchers as "those whose production, or excavation quantity is limited in tonnage, mostly manually operated and sometimes employing machines of small capacity". Such mining activities are usually confined to small and shallow deposits and are often seasonal, connected to other means of livelihood, such as agricultural labor. Some of the earnings out of ASM are often usurped by middlemen due to the poor organization of this sector. Such mining activities which do not follow any rules or regulations often produce serious environmental problems like mercury pollution in surface waters and sediments from gold mining (Deb 2016), surface degradation and habitat loss for threatened and endangered species, e.g., extensive sand mining in Chambal river banks has affected the nesting grounds of *Ghariyal* and turtles. However, despite these problems, this type of mining activity provides sustenance to economically backward, often indigenous communities with no social securities. Deposits often considered uneconomic by modern mining are commonly mined by artisanal communities over a long period of time helping alleviation of their poverty to some extent and thus contribute to sustainability.

10.2.1.3 Overview of Some ASM Activities in India

About forty different minerals are exploited by small-scale and artisanal mining in India. Most important of these, in order of their value, are discussed below (cf. Deb et al. 2008).

Gold: Extensive artisanal mining of primary lode and alluvial gold is carried out in the Wynad gold field of Dharwar craton in the contiguous parts of Tamil Nadu and northern Kerala. The abandoned mines and adits located in Pannur and Pandalur-Devala areas of Tamil Nadu are illegally worked by local communities using primitive methods, such as fox holes and vertical pits on the quartz reefs of the area. Artisanal gold mining is carried out extensively in the Nilambur region, south of the Wynad gold field, in the alluvial tracts of the Cheliyar and Punna Puzha Rivers by a local tribe, the Paniyars (Sawarkar 1980). Such mining activity is also prevalent in the nearby Attapady valley in the Palakkad district of Kerala.

In the Hosur area of northern Karnataka, in the northern part of the Dharwar craton, illegal artisanal gold mining is carried out in the abandoned shafts and inclines of the Bharat Gold Mines Ltd. on the auriferous quartz reefs of Gadag greenstone belt. About 200 people (including 70–80 women) take part in such activities. Families are involved in the enterprise in which the earnings in 2006 was up to Rs. 2000 (USD 40) per month per family. Men enter the abandoned mines to collect the quartzose ore, while children help them in the crushing activities and the women are mostly involved in recovering the gold through panning and mercury amalgamation.

In Maru River (a tributary of Waynganga) bed near Bhiwapur in Nagpur district of Maharashtra, intensive artisanal mining of alluvial gold is practiced by local villagers. The primary gold-copper vein occurrences of these alluvial concentrations can be traced to the Pular-Parsori prospects in the region, belonging to the Proterozoic rocks of the Sakoli basin.

In Jharkhand, in the wide alluvial tract of the Subarnarekha River, artisanal mining of gold, especially after the monsoon season, is traditionally practiced by local communities. The activity has intensified after the construction of Subarnarekha dam and consequent displacement of villagers from the reservoir area. Most of the miners are indigenous tribal people and a large part of the work force is women. In a small stretch of the river near Chandil, 33 mining sites were observed.

In the Himalayan foothills near Rampur in Himachal Pradesh, alluvial mining of gold is common in the bed of the meandering Sutlej River between Nugli and Baila, after the monsoon season.

Tin: Primary pegmatitic and alluvial/colluvial types of tin deposits are exploited by artisanal mining in the Bastar region of Chhattisgarh state by local indigenous tribal communities. The miners, large section of who are women, resort to manual digging and hand panning of the cassiterite. But they do not recover the by-products like muscovite, lepidolite, beryl, feldspar, quartz, etc., along with rare metal minerals which are in fact costlier than tin metal (Babu 1994). The tin ore so produced was purchased by two government agencies – Madhya Pradesh State Mining Corporation (MPSMC), Bhopal and Orissa Mining Corporation (OMC), Koraput – and shown in their production figures.

Gemstones: Alluvial and colluvial deposits in gravel beds and paleoplacer deposits in conglomerates have been subjected traditionally to artisanal mining for gemstones in different parts of the country since ancient times. Alluvial diamond is collected in this fashion from many of the river beds in central and south India. Precious and semiprecious gemstones, such as, ruby, sapphire, emerald, corundum, aquamarine, heliodor, apatite, garnet, zircon occur in a number of localities in

Kalahandi, Sambalpur, Bolangir, and Koraput districts of Odisha and in the adjacent Eastern Ghats belt of Andhra Pradesh (Kar and Singh 2001). Most gemstones mined artisanally, occur as discrete grains in bedrocks, or occur in erratic patches and pockets in soil, colluvium, and gravel beds. They are also found as thin zones in the secondary profiles.

In the Eastern Ghats fold belt in parts of Vizianagaram, Vishakhapatanam and east Godavari districts of Andhra Pradesh, several blocks are known to be prospects of gemstones with some being mined illegally (Sridhar and Sumanth 2001). In Chintapalle block located in the reserve forest area, good quality yellowish green and peacock green chrysoberyl cat's eye and honey brown sillimanite cat's eye occur in several places, viz., Meduru, Pinapadu, Poturajuganulu, Gurralagondi, Rintada, and Satyavaram. Here, gemstones are picked up from colluvium and soil or by illegal quarrying of weathered pegmatite up to depths of 10–15 m by making cross adits or inclined shafts. In Addatigala block, 15 people died on one occasion during illegal mining activity in a quarry, mining Alexandrite (variety of chrysoberyl cat's eye) in the Tapasikonda reserve forest. Extensive illegal mining by locals for the yellowish green chrysoberyl cat's eye have been reported from Araku Valley area in Visakhapatnum district.

Gemstones (ruby and sapphire) are collected by the local people in Periyar and Karur districts of Tamil Nadu through rat hole mining after every rainfall. Private parties extract gem quality feldspars in unauthorized manner by digging wells or pits in pegmatites. Local entrepreneurs employ villagers at low wages to engage in rampant illegal mining of known gem-bearing tracts in the state.

Coal: The lowest income states of eastern and northeastern India face problems related to cross-border migration from Bangladesh, Nepal, and Bhutan along with periodic insurgency. Rampant poverty of such migrants and local population pushes them to illegal artisanal mining of coal. The activities help to meet the requirements of fuel need of the families and their livelihood by sale of the coal to small domestic and other consumers (such as sponge iron factories, brick kilns, tea stalls etc.), including part of the production being smuggled across the international border. The total production from ASM of coal appears impressive from the conservative estimate of 2.5 Mt transported on bicycles (Lahiri-Dutt and Williams 2005). In the northeastern states which enjoy different privileges under the Indian Constitution, especially in Meghalaya, the indigenous communities have traditionally enjoyed the ownership of natural resources under their privately owned land. Thus, a large number of diggings for coal and lignite are seen in this part of the country, though without any license. Where the operations are relatively big, the shafts, mostly inclined and shallow, are sunk and the laborers employed are primarily migrant Nepalese. Some amount of environmental degradation is inflicted by such mining to the fragile mountain ecosystem of the Shillong plateau, but it is a major source of employment and income for the local population as well as the migrants.

10.2.1.4 Future Developments and Prospects in the ASM Sector in India

Based on the discussion above, future developments in this sector can be foreseen in the following directions (Deb et al. 2008).

The strongest requirement is to review the existing legal framework of mining in the country as a whole and amend the existing mining laws for artisanal mining in particular, and make them more realistic, unambiguous, and inclusive. Regularization of illegal ASM operations and acknowledging their livelihood potential, instead of marginalizing them, will lead to improvement in lives of miners and their families and the concept of sustainability can begin to be introduced (Marker et al. 2005). Short-term licenses can be granted, with the ASM miner becoming a partner in the scheme. Lack of available official or unofficial data related to ASM practice in India is responsible for the existing confusion about it. There is thus the need to create a broader information base, including a database for the artisanal mining sites. The scientific as well as social data collated together from this database would eventually help policy and decision-makers to organize this sector through clear rules and regulations. The third approach would be to locate the responsible body within the administration to deal with artisanal mining. A possible suggestion in this regard is local governance; in most places in India, the *panchayats* are the local representative bodies of the state. These bodies could be apprised about ASM and the issues thereof, their personnel trained with the legal frameworks, and equipped with necessary powers to deal with the artisanal mining activities under their jurisdiction. Development of small-scale experimental projects under the control of local panchayats for diamond mining in the alluvial tracts of Krishna, Godavari, Mahanadi, and Pennar Rivers in southern India has been suggested (Radhakrishna 2006). Lastly, in terms of the Millennium Development Goals of poverty eradication, suitable education and training programmes could be taken up for the miners and the administrators, instead of turning a blind eye to this sector. This would improve the environmental performance of the artisanal miners and reduce exploitation and poverty within the sector. A concerted effort in this direction has the potentiality to help a large number of currently jobless rural youth in the country to earn their livelihood from ASM activities.

10.3 Regulatory Regimes in the Mines and Minerals Sector

Though mining industry plays an important role in Indian economy, there are numerous controversies associated with mining. This is because mining, forests, tribal population, environmental and displacement issues are closely interrelated in one way or another and have affected a large part of India. Besides, growth and development from mining revenue have been glaringly noninclusive of the communities who live and work in the mining areas during the seven decades since Independence. The problem is exacerbated by the lack of just and efficient administration and self-regulating mining concerns which are socially inclusive, technically efficient, and have environmental integrity. To understand the various issues affecting the mines and minerals sector in India, we need to first look into the framework of various regulatory provisions and policies governing this sector.

Seventh schedule[1] of Indian Constitution includes Article 246 which categorizes Union, State and Concurrent lists of responsibilities and assigns functions and legislative powers to both the centre and the states. The *entry 54 in the Union list* empowers the central government to make laws relating to the "regulation of mines and mineral development to the extent to which such regulation and development under the control of the Union is declared by Parliament as law to be expedient in the public interest". *Entry 23 of the State List* further empowers the state governments also to make laws relating to "regulation of mines and mineral development subject to the provisions of List 1 (Union list) with respect to regulation and development under the control of the Union". In effect, the proprietary control of the onshore mineral resources is with the state government, but the regulatory powers are with the central government. Though the central government's powers in the regulation of mines and mineral development in the Constitution are predominant, the duality of control by the centre and the state often comes in the way of speedy disposal of mining contracts.

The 42nd Amendment to the Constitution, introduced in the year 1976, makes it the responsibility of the state government (under Directive Principles of State Policy, Article 48-A) to protect and improve the environment and to safeguard the forests and wildlife of the country. Besides, Article 51-A (g), makes it the fundamental duty of every citizen to protect and improve the natural environment including forests, lakes, rivers, and wildlife and to have compassion for living creatures.

All mining activities in the country are controlled by the National Mineral Policy, 1993 and the Mines and Minerals (Development and Regulation) (MMDR) Act, 1957, amended in 1999 and 2015. Besides, all mining activities have to comply with the environmental and forest conservation legislations in India. The major laws in this respect are the Environment (Protection) Act and Rules 1986, the Environment Impact Assessment Notification 1994 and the Forest (Conservation) Act 1990. Other relevant Acts for the mining sector are: Land Acquisition Act 1894, amended in 1984, 2015; PESA Act 1996. All mining-related legislations are briefly described in the following pages.

---

[1]Schedules are lists in the Constitution that categorize and tabulate bureaucratic activity and policy of the govt.

10.3.1 National Mineral Policy, 1993 (For Nonfuel and Nonatomic Minerals)

The various aspects of this important national policy, such as, the preamble, objectives, regulations, and strategy of mineral development are presented below. The policy envisages the flow of private investment in this sector as well as the introduction of state-of-the-art technology in exploration and mining of minerals in India.

Preamble

1. Minerals are valuable natural resources being finite and nonrenewable. They constitute the vital raw materials for many basic industries and are a major resource for development. Management of mineral resources has, therefore, to be closely integrated with the overall strategy of development; and exploitation of minerals is to be guided by long-term national goals and perspectives. In this context, the need has been felt to spell out in a statement the different elements of the policy, which has evolved over the years, relating to development of our mineral resources and in regard to areas of concern which have emerged in recent years.
2. The country is not endowed with all the requisite mineral resources. It is, therefore, imperative to achieve the best use of available mineral resources through scientific methods of mining, beneficiation and economic utilization. Simultaneously, it is essential to keep in view the present and future needs of defence and development of the country and strive to ensure indigenous availability of basic and strategic minerals to avoid disruption of core industrial production in times of international strife.
3. The National Mineral Policy, also emphasizes certain new aspects and elements like mineral exploration in the seabed, development of proper inventory, proper linkage between exploitation of minerals and development of mineral industry, preference to members of the Scheduled Tribes for the development of small deposits in scheduled areas, protection of forest, environment, and ecology from the adverse effects of mining, enforcement of mining plan for adoption of proper mining methods and optimum utilization of minerals, export of minerals in value-added form and recycling of metallic scrap and mineral waste.

The basic objectives of the mineral policy are:

(a) To explore for identification of mineral wealth in the land and in offshore areas;
(b) To develop mineral resources taking into account the national and strategic considerations and to ensure their adequate supply and best use keeping in view the present needs and future requirements;
(c) To promote necessary linkages for smooth and uninterrupted development of the mineral industry to meet the needs of the country;
(d) Promote research and development in minerals;

(e) To ensure establishment of appropriate educational and training facilities for human resource development to meet the manpower requirements of the mineral industry;
(f) To minimize adverse effects of mineral development on the forest, environment, and ecology through appropriate protective measures; and
(g) To ensure conduct of mining operations with due regard to safety and health of all concerned.

Regulation of Minerals

1. Management of mineral resources is a function of both central and state government in terms of the 7th schedule of the Constitution which enlists mineral resources in Union List (Entry 54), as well as State List (Entry 23).
2. The Mines and Minerals (Regulation and Development Act, 1957) lay down legal framework for the regulation of the mines and development of all minerals other than petroleum and natural gas.
3. The Central Government has framed the Mineral Concession Rules 1960 for regulating grant of prospecting licenses and mining leases in respect to all minerals other than atomic minerals and minor minerals. It also framed the Mineral Conservation and Development Rules, 1988, for conservation and systematic development of minerals (except for coal, atomic minerals and minor minerals).
4. State government has simultaneously framed regulations for minor minerals.
5. The central government in consultation with the state governments shall continue to formulate the legal measures for the regulation of mines and the development of mineral resources to ensure basic uniformity in mineral administration and to ensure that the development of mineral resources keeps pace, and is in consonance with the national policy goals.

Role of State in Mineral Development

The core functions of the state in mining sector will be to facilitate and regulate exploration and mining activities of investors and entrepreneurs, as well as to ensure provisions of infrastructure and tax collection (Mines and Minerals Regulation and Development Act 1957). In all such activities, there shall be arm's length distance between state agencies (Public Sector Undertakings) that mine and those that regulate. There shall be transparency and fair play in the reservation of ore bodies to state agencies in such areas, where private players are not holding or have not applied for exploration or mining, unless security considerations or specific public interests are involved.

Strategy of Mineral Development

1. Conservation of minerals:

   (a) The strategy for development of any mineral should naturally keep in view its ultimate end uses. The guiding principle in the strategy of development of any mineral deposit at any location shall ordinarily be the economic cost.

The state may, however, undertake the development of any mineral deposit in public interest to ensure unhindered availability of mineral raw material for the realization of national goals.

(b) As minerals are exhaustible and nonrenewable resources, their exploitation has to be done keeping in view not only the present but the long-term needs. The strategy for exploitation and development of each mineral shall be formulated and reviewed periodically on the basis of available resources.

(c) The best use of available mineral resources shall be ensured by adopting, during mining operation, effective measures for conservation and beneficiation, recovery of associated minerals and later by efficient processing of minerals. There shall be adequate and effective legal and institutional framework and commitment to prevent suboptimal and unscientific mining.

(d) Conservation of minerals shall be construed not in the restrictive sense of abstinence from consumption or preservation for future use, but as a positive concept leading to augmentation of resource base through improvement in mining methods, beneficiation, and utilization of low grade ore and rejects, recovery of associated minerals, reduction in the requirements of minerals per unit of material output, etc.

2. Scientific methods of exploitation: Mine development and mineral conservation as governed by the rules and regulations will be on sound scientific basis, with the regulatory agencies closely interacting with R&D organizations, scientific and professional bodies. Conditions of mining leases regarding tenure, size, shape, disposition with reference to geological boundaries, and other mining conditions shall be such as to favorably predispose the leased areas to systematic and complete extraction of minerals.
3. Productivity norms: Studies for fixation of productivity norms and goals will be taken up to promote productivity of men and machines and to improve the consumption norms of fuel and materials. Regular maintenance, replacement, and upgradation of mining machinery and equipment shall be undertaken.
4. Mineral processing and beneficiation: Research organizations, including the National Mineral Processing Laboratories of the Indian Bureau of Mines, should be strengthened for development of processes for beneficiation and mineral and elemental analysis of ores and ore dressing products. There shall be cooperation between and coordination among all organizations in public and private sector engaged in this task.
5. Mining equipment and machinery: Indigenous industry for manufacture of mining equipment and machinery shall be strengthened. Wherever necessary, imports of machinery and equipment may be permitted to improve the efficiency, productivity and economics of mining operations and safety and health of persons in the mines and the surrounding areas.
6. Manpower development: Existing facilities for basic and specialized training shall be constantly reviewed and upgraded from time to time to ensure that adequately trained manpower at all levels is available for the development of mines and minerals industries.

7. Mineral development and protection of the environment: Extraction and development of minerals are closely interlinked with other natural resources like land, water, air, and forest. The areas in which minerals occur often have other resources presenting a choice of utilization of the resources. Both aspects have to be properly coordinated to facilitate and ensure a sustainable development of mineral resources in harmony with environment.
8. Mines safety: Mining operations are hazardous in nature. Accidents happen and often result in the loss of life or limb of persons engaged in it. Efforts must be directed toward the development and adoption of mining methods which would increase the safety of workers and reduce the accidents. Toward this end, participation and cooperation of mine workers shall be secured. Steps will also be taken to minimize the adverse impact of mining on the health of workers and the surrounding population.
9. Planning mine closures: Mineral deposits being exhaustible, once the process of economical extraction of a mine is complete, there is need for its closure, leading not only to job loss but also disruptions of community lives. Thus, whenever mine closure is necessary it should be orderly and systematic and so planned as to help the workers and the dependent community rehabilitate themselves without undue hardship.
10. Rehabilitation of displaced persons: Mining operations often involve acquisition of land held by individuals including those belonging to the weaker sections. While compensation is generally paid to the owner for the acquisition of his land, efforts shall be made to ensure suitable rehabilitation of affected persons especially those belonging to the weaker sections that are likely to be deprived of their means of livelihood as a result of such acquisition.

    Some abandoned mines may be sources of surface and subsurface water, pisiculture, while some may supply pollutants such as, As, Tl, Cd, etc., to the natural water flowage system in the area. These may be taken care of during closures of mines, depending on the situation (Ministry of Mines, GOI).

10.3.2 Mines and Minerals Development and Regulation (MMDR) Act, 1957, 2015

The Mines and Minerals (Development and Regulation) Act, 1957 and the Mines Act, 1952, together with the rules and regulations framed under them constitute the basic laws governing the mining sector in India. The provisions of the MMDR Act are implemented through the promulgation of a number of rules. In this regard, the Mineral Concession Rules (MCR) 1960, the Mineral Conservation and Development Rules (MCDR) 1988, the Granite Conservation and Development Rules 1999, the Marble Development and Conservation Rules 2002 are the important rules in force. The provisions of MCR and MCDR, however, do not apply to coal, atomic minerals, petroleum, and gas. The MCDR 1988 lays down guidelines for ensuring mining on a scientific basis, while at the same time conserving the environment.

The MMDR Act has been amended several times, the first amendment coming in 1958, followed by amendments in 1960, 1972, 1978, 1986, 1994, 1999 and the last amendment in 2015. Some of the important provisions in the Act are discussed below.

The Indian mining industry is subject to various taxes, duties, etc., in force from time to time. There are specific levies for certain mineral rights. Under the MMRD Act, there are provisions of reconnaissance permit fees, prospecting fee, royalties, and dead rent payable to the state governments for minerals (other than minor minerals) for three different forms of concessions, e.g., reconnaissance, prospecting, and mining lease. "Royalty", under the Act is neither a tax nor a fee but is considered in case of a mining lease as a payment made by the lessee to the lessor toward consumption of the subsoil property (mineral) by operating a mine with his labor and investment. The rates of royalty for major minerals are fixed by the central government and specified in Schedule II of the MMRD Act. For major minerals, the revenue thus collected is retained by the state government. For minor minerals, the state governments fix the royalty rates and collect royalty along with the dead rent. The dead rent is payable by the holder of a mining lease for minerals based on the area of the lease and the value of minerals in the form of a minimum guarantee of income to the lessor, irrespective of the fact whether mining operation was carried out or not by the lessee. When any mineral is extracted, the holder of the lease is liable to pay royalty or dead rent whichever is higher.

The latest MMDR Act passed by Parliament in 2015 is focused on a regulatory regime that is conducive to private investment including foreign direct investment (FDI) in the M & M sector of the Indian economy. The new Act also contains various levies on the mineral industry besides royalty and dead rent. These include access on major minerals by the central government, a cess on major and minor minerals by the state government, and an annual amount equal to royalty in case of major minerals (except coal and lignite) and an amount equal to 26% of profit (called profit sharing percentage) in case of coal and lignite, to be paid into the proposed District Mineral Fund. The Act also stipulates application of the concepts of sustainable mineral development at the national and state levels which would involve scientific and sustainable mining throughout the mining life cycle, minimizing and mitigating the adverse impact of mining operations on environment, and also on biodiversity, air, water, ambient noise, land and quality of life of local communities, creating opportunities for socioeconomic development for sustainable livelihood, conservation of mineral resources, minimization of waste generation, promoting restoration and reclamation of mined out land, consultation with stakeholders and transparency in communication through public disclosure of mining operations (Mohanty and Goyal 2012). There are also provisions for corporate social responsibility (CSR) directed toward socioeconomic development in the mining area through a scheme of annual expenditure by the mining lessee. Finally, the Act also envisages benefit-sharing (of mineral revenues) with local population through the District Mineral Fund (DMF) (vide Alaska Fund in Chap. 9) to be constituted by the state government in each mining district.

10.3.3 Environment Protection Act, 1986

As the state of the environment deteriorated and the impact grew, environmental protection became a national priority in 1970s. The Environment Protection Act (EPA) was introduced in 1986 as an umbrella legislation that provides a holistic framework for protection and improvement of the environment. It was enacted under Article 253 of the Constitution of India in the wake of the Bhopal Gas Tragedy and included the provisions of Stockholm Declaration, a United Nations conference on Human Environment. This Act empowers the central government to take all necessary measures protecting and improving the quality of the environment and preventing, controlling and abating environmental pollution. In this Act, the central government is expected to lay down specific standards of quality of the environment and also standards of emission or discharge of environmental pollutants. In the event of contravention of its provisions and requirements, the Act provides for prosecution as well as for strict penalties.

10.3.4 Environmental Impact Assessment, 1994, 2006

The most important environmental requirement for mining (and many other industries) is a comprehensive statutory impact assessment programme (EIA). Such a programme was started in 1994 and was subsequently replaced by a fresh notification on September 14, 2006 (further amended in 2009), which now governs the EIA procedures for mining and other specified industries.

The 2006 notification separates projects into three main categories: the ones having significant effects on the environment (category A projects) will require EIA in all cases and those which either will not require an EIA (category B2 projects) or will require in some circumstances (B1 projects). A category "B" project located within 10 km from the boundary of a protected area and interstate or international boundaries will be treated as category "A" projects. In "mining for minerals", more than 50 ha of mining lease area in respect of non-coal mine lease and 150 ha of lease area in respect of coal mine lease are considered in category "A", along with offshore and onshore, oil and gas exploration and development. Category "B" projects cover mining lease area between 5 and 150 ha in respect of coal and 5–50 ha in respect of other minerals. Mineral prospecting, however, is excluded from prior environment clearance.

While Category "A" projects are assessed by a Central Environment Assessment Committee (EAC) whose recommendations are submitted to the central government (Ministry for Environment, Forests and Climate Change) for final decision, in the case of Category "B" projects (both B1 and B2) the required assessments are made by a State-level Expert Appraisal Committee (SEAC) and final decisions are taken by a State-level Environment Impact Assessment Authority (SEIAA).

The following stages of the EIA process have been laid down in the 2006 notification:

Screening: is carried out to determine whether or not a proposal should be subject to EIA or not (only category B projects are subjected to screening by SEAC which identifies category B1 and B2 projects).

Scoping: comes next to identifying the issues and impacts that are likely to be significant in a project and to establish the terms of reference (TORs) for the preparation of an Environment Impact Assessment (EIA) report in respect of category A and category B1 projects.

Public consultation: is also required to ascertain views of the local, affected/potentially affected population of the project area through public consultation in meetings and/or through web-based responses.

Appraisal: is the final step to examine in detail the EIA report, and the outcome of public consultations. It is determined at this stage whether the report meets the terms of reference, especially with respect to objective description of the impacts of the proposal, the proposed environment management plan outlining measures for mitigation and the significance of the concerns expressed by the local communities affected by the project. Finally, on the basis of the appraisal, supported by all information and reasoning, clear and categorical recommendations are made by the SEAC to the decision-making (regulatory) authority to approve (i.e., give prior environmental clearance) with stipulated terms and conditions or reject the proposal.

Decision-making: at the end, the regulatory authority approves or rejects the proposal and in case of an approved proposal, lays down the terms and conditions of its implementation.

Air (Prevention and Control of Pollution) Act

Passed in 1981, this Act was aimed at improving the air quality by controlling the level of various kinds of atmospheric pollutants generally introduced by different industries, vehicles, power plants, cement plants, etc. The main pollutants in this regard are particulate matter, lead, carbon monoxide, sulfur dioxide (SO_x), nitrogen oxides (NO_x), and volatile organic compounds (VO_x). While source of SO_x is mostly industrial and household fuel, NO_x, ozone, CO, HC are produced mainly by photochemical means through motor vehicle fuel combustion. The chemical effects of the former are reducing, with production of lung and throat irritation, whereas the latter is more oxidizing producing eye and respiratory problems. Air problem is acute in heavily industrialized and urbanized parts of the country. Both central- and state-level pollution control boards have been established to monitor the levels of air pollution. The capital city of Delhi is now considered as one of the most polluted cities in the world in terms of air quality.

The Water (Prevention and Control of Pollution) Act

Formulated in 1974, this Act is meant to control pollution of our water resources through industrial, agricultural, and household sources. Generally, wastewater from any of these sources can pollute vast areas of wetlands, rivers, lakes, and even the sea. These could be point sources such as farm runoff or leachates from waste dump, etc., which are relatively easier to control. However, nonpoint sources like urban runoff or atmospheric deposition are much more difficult for abatement. Conventional water pollutants are biochemical oxygen demand (BOD), nutrients, pH, suspended solids, oil and grease, and pathogenic microorganisms. Toxic pollutants in water include metals (As, Tl, Cd, Pb, Hg, Ni, Cu, Zn) and industrial chemicals (benzene, toluene, pesticides). Nonconventional and nontoxic pollutants are ammonia, chlorides, phenols, Fe, Al, color, and thermal pollution. Fluoride and arsenic pollution of ground water has turned out to be a major concern of public health in parts of India and also in Bangladesh.

10.3.5 Forest Conservation Act

Forests worldwide have been and are being threatened by uncontrolled degradation and conversion to other types of land uses, influenced by increasing human needs for agricultural expansion and environmentally harmful mining and industrial development. The impacts of loss and degradation of forests manifest in the form of soil erosion, loss of biodiversity, damage to wildlife habitats and degradation of watershed areas, deterioration of the environment and quality of life of the forest dwellers, often accompanied by displacement and loss of traditional livelihood of forest dwellers. On the other hand, it is possible to increase the value of forests through non-damaging uses such as ecotourism and the managed supply of genetic materials. Concerted action is required to improve people's perception and awareness about the value and benefits of forests.

The British looked upon the forests as a renewable asset that required management in order to meet the timber needs of the empire. The process of using the forests for commercial activities and the process of alienating the forest dwellers from the forests were legitimized by the *Indian Forest Act, 1865*, which was enacted for the purpose of limiting the rights of the people and transferring ownership of forests to the states. It was subsequently replaced by the *Indian Forest Act of 1927*. With rampant degradation of forests in recent years, the government became more specific regarding use of forests for non-forest purpose and also for granting rights and leases to people for use of forest products. As a result came into force the *Forest (Conservation) Act of 1980*, followed by its *Amendment in 1988*. Further, as an addition to the guidelines and rules mentioned in the Act, came the *Forest Conservation Rules of 2003*, further *amended in 2004*. The *Forest Rights Act of 2006* specifies more clearly the rights that are to be granted to tribals in relation to use of forest products and the conditions thereof.

After India's Independence, forests were placed on the state list of the Constitution. Forest departments of individual states continued to regulate forests in accordance with the Indian Forest Act of 1927, as implemented by state regulations. The Indian Forest Act gives the states jurisdiction over both public and private forests and facilitates the extraction of timber for profit. Public forests are those in which state governments have a proprietary interest and are divided into three categories: *Reserve Forests*, *Protected Forests*, and *Village Forests*.

The debate over how to balance the various demands of the nation on the forests intensified in 1970s. The 42nd Amendment Act of 1976 transferred forests from the State list to the Concurrent list of the Constitution. This transfer empowered the central government to act directly in managing India's forests by forming a new Ministry. The Ministry of Environment and Forests (MoEF) now has administrative jurisdiction over the National Forest Policy, Forestry Department and Indian Forest Service. A revised *National Forest Policy* came into being in 1988. The Framework provided by the Indian Forest Act of 1927, however, continues to guide the forest bureaucracy even today. It is an Act to consolidate the law relating to forests, the transit of forest produce and the duty leviable on timber and other forest produce. It has stipulated "restriction on the de-reservation of forests or use of forest land for non-forest purpose". Mining, including underground mining, is considered as a non-forestry activity. Therefore, prior approval of the central government is essential before a mining lease in respect of any forest area is granted. The Act would apply not only to the surface area which is used in the mining but also to the entire underground mining area beneath the forest. A renewal of an existing mining lease in a forest area also requires the prior approval of the central government.

10.3.6 Land Acquisition and Related Acts

Land is another important issue with regard to mineral development. In general, the right to excavate the minerals from the land of a private owner is based on an agreement between the lessee and the land owner. When mining is to be carried out in areas with considerable tribal population, or if the area is covered with forests, there are additional issues which need to be considered, as discussed in a later section.

With the increasing pressures on land due to urbanization, rapid economic development, increasing infrastructure requirements, dams, SEZs, mining, etc., especially in a fast-growing economy like India, the acquisition of land by the government has become the most controversial issue. Under the "Doctrine of Eminent Domain" and development as a "public purpose", the Government of India since Independence has acquired large tracts of land, displacing millions of people from their homes. This doctrine holds that the state has ultimate sovereignty over all natural resources. Thus the government can take over any land under individual or community ownership in the name of "public purpose". There is no mechanism however to determine if the public purpose has ever been turned into public interest.

The acquisition itself cannot be challenged in a court of law, but only the compensation amount. Statistics reveal that since 1980, 9.8 lakh hectare of forest land have been diverted to 11,282 projects. In Odisha alone, 54–56% of tribal land has been lost to non-tribals over 25–30 years. Moreover, since 1990, 85.39 lakh tribals have been displaced for megaprojects accounting for about 56% of India's total displaced population (Perspectives 2008).

The difficulties that come in the process of land acquisition in India are immense, given the population density and the types of land use in the country. This is evident from the fact that the fundamental issue in a number of headline incidents in the past few years has been the resistance to the process of land acquisition, be it "Narmada Bachao Andolan" or the more recent Nandigram and POSCO issues in West Bengal and Odisha, respectively. With number of state governments demarcating lands as Special Economic Zones (SEZ), the problem is getting only worse, particularly because in most places no land use planning maps are available.

Given the fact that land acquisition falls under the concurrent list both the state government and the central government have amended the law, evolving it with time and according to local needs. The Act was drastically amended in 1984 by the central government. The amended Act has resulted in the setting down of a time limit for the completion of all formalities between the issues of the preliminary notices and the declaration of acquisition. The provision of an opportunity to those dissatisfied with the Collector's award to apply to him for a redetermination of the compensation payable to them is also included. These changes introduced by the amendment are beneficial for they seek to minimize the undue delays that characterize acquisition proceedings and provide for payment of compensation on a realistic scale. Another important aspect of the amended Act relates to the provision dealing with the acquisition of land for domestic and multinational companies: it has simplified the procedure for acquisition of land for a corporation owned or controlled by the state, provided that acquisition for such a body could be regarded as having been done in public interest.

The Land Acquisition Act has been criticized by groups that view the act as weak and ineffective, and also by groups that view the act as draconian. The former argue that the procedure followed is cumbersome and costly, often resulting in inordinate delay in land acquisition. This group argues that the determination of public purpose should be a matter of executive discretion and should not be contestable in a court of law. It has also been argued that the property valuation techniques are flawed and that the land owners get to peg the value higher than the real value, based on "potential value" and "opportunity value" of their property, resulting in, what is claimed as a heavy strain on public finances for development projects. People who argue that the act is draconian claim on the other hand that a number of projects with no public purpose attached, as in the case of SEZs, usurped land from their owners, with the help of the Land Acquisition Act, at what is claimed to be well below the market value of these properties. It is argued that, even in the case of projects that are genuinely for public purposes, there is a considerable difference between the market value of the property and the value that the land acquisition officer pays the land owners. Compensation is defined by the average of registered price in the area for three years. It is however no secret that most land owners register the land at ~40% of the market price to pay less stamp duty. Thus the state saves substantially in compensation amount. It is also argued that the relocation and rehabilitation of land owners displaced by the actions of the Act, is not followed up adequately, and that this is not covered comprehensively in the framework of the Act.

Looking into various problems, government has proposed further amendments in the Act to strictly define the purposes for which land could be acquired. Public purpose would be classified as strategic purposes, public infrastructure purposes and projects useful for the general public. The Land Acquisition, Rehabilitation and Resettlement (LARR) Act of 2013 has recently been passed by the Lok Sabha and has generated a lot of political debate in the country. It centers on land owners' consent and compensation that they and those employed on their land ought to get. The percentage of land owners' consent was set at 70% (80% for private companies). In scheduled areas, the issue was further compounded by other overlapping laws which essentially gave a veto power to each gram panchayat. Since the main opposition is about farmers being snatched of their right to their land, several counter arguments are being put forward. Though much can be done to improve farming, it appears unsustainable for the sector to continue to provide livelihood to 60% of our population when it is supposed to contribute only 14% to the GDP. Further, with industrialization the farmers' children can have alternative career opportunities; they may otherwise be destined to be trapped in ever more fragmented, marginal farming.

10.3.7 Tribal Issues and Legislations

In terms of Article 342 of the Indian Constitution, at present, there exist 697 tribes as notified by the central government. Indian tribal or indigenous people account for 8.14% of the total population of the country, numbering 84.51 million, according to the 2001 census. Article 244 and 244A read with 5th schedule of constitution gave tribal communities right to self-determination and control over land and forests in which they dwelt. Mineral-rich geographical areas are considered under Schedule V of the Constitution. Scheduled areas were identified and demarcated in order to provide special safeguards for areas in which tribal population was dominant. These Indian tribal groups of people have been notified to reside in nine states. More than half of the Indian tribal population is concentrated in the States of Madhya Pradesh, Chhattisgarh, Maharashtra, Odisha, Jharkhand, Himachal Pradesh, Rajasthan, and Gujarat. A large number of tribal populations are concentrated in northeastern part of India as well, but are governed by a different set of provisions under 6th Schedule and Article 371.

During the British rule, the traditional land rights of the tribals were challenged, even usurped. Their habitat – the forests were regarded as property of the government, instead of the people who lived there for generations without any opposition or hindrance. This British policy gave rise to natural resistance from tribals and resulted in confrontation and struggle. The Santhal rebellion (June 1855–January 1856) was spearheaded by four Murmu brothers but was ruthlessly suppressed and thousands of tribals were massacred. Historically, the economy of most tribes was subsistence agriculture or hunting and gathering. Tribal members traded with outsiders for the few necessities they grossly lacked, such as salt and iron. In the early twentieth century, however, large areas fell into the hands of non-tribals, on account of improved transportation and communications. Around 1900, many regions were opened by the government to settlement through a scheme by which inward migrants received ownership of land free in return for cultivating it. For tribal people, however, land was often viewed as a common resource, freely available to whoever needed it. The colonial and post-Independence regimes belatedly realized the necessity of protecting tribals from the predations of outsiders and prohibited the sale of tribal lands. Besides intrusion by non-tribals and British into their habitat, tribals have also faced difficulties due to Government's efforts to reserve the forests. The continuing economic alienation and exploitation of many Adivasis was highlighted as a "Systemic Failure" by the Indian Prime Minister Dr. Manmohan Singh in a 2009 conference of chief ministers of all 29 Indian states, where he also cited this as a major cause of Naxalite unrest that has affected the areas under the Red Corridor.

In 1994, a committee was formed by the government under the chairmanship of Dileep Singh Bhuria to analyze the long standing grievances of the tribal communities. Based on the recommendations of the Bhuria committee, some important legislations were enacted, as discussed below.

10.3.7.1 Panchyats (Extension to Scheduled Areas) Act, 1996 (PESA)

The 73rd Amendment Act of 1992, which inserted part IX into the constitution, recognized the institution of 3-tier (village, block and district) Panchayats in Indian villages in terms of local self-governance, which would implement schemes of economic development and social justice. PESA 1996 extended the part IX of constitution to scheduled areas with modifications and gave special powers to *Gram Sabha*, which was defined as the body of people registered in the electoral roll in a village or villages under the jurisdiction of a *Panchayat*.

The Act stipulates the following powers to the *Gram Sabha* or *Panchayats*, among others:

1. A state legislation shall ensure that the *Panchayats* and *Gram Sabhas* are endowed with the ownership of Minor Forest Produce (MFP) and have the power to prevent alienation of land in scheduled areas and take action to restore any unlawfully alienated land of a scheduled tribe.
2. Every *Gram Sabha* shall be competent to safeguard and preserve the traditions and customs of the people, their cultural identity, community resources, and the customary mode of dispute resolution.

3. Every *Gram Sabha* shall give approval to the plans, programmes, and projects for social and economic development before such plans, programmes, and projects are taken up for implementation by the *Panchayat* at the village level. Every *Panchayat* at the village level shall be required to obtain from the *Gram Sabha* a certification of utilization of funds for the plans, programmes, and projects referred to in the clause.
4. It is mandatory for women, SC and ST members to participate in the *Gram Sabha,* with 50% of seats in every *Panchayat* to be reserved for schedule tribe candidates.
5. The *Gram Sabha* or the *Panchayats* at the appropriate level shall be consulted before making the acquisition of land in the scheduled areas for development projects and before resetting or rehabilitating persons affected by such projects in the Scheduled Areas.
6. The recommendation of the *Gram Sabha* or *Panchayat* at the appropriate level shall be made mandatory prior to grant of Prospecting License (PL) or Mining Lease (ML) for minor minerals.
7. Planning and management of minor water bodies in the scheduled areas shall be entrusted to *Panchayats* at the appropriate level.

10.4 Concluding Statement

The discussion above draws our attention toward the complex structure of the Indian mineral industry. Its legal architecture is characterized by dual control, exercised by both the central and the state government, and consequently, by divided accountability. The proprietary rights of onshore mineral deposits rest with the concerned state government while the regulatory powers are with the central government. In effect, the administration of the mining laws is primarily monitored by the government agencies in the mineral-rich states, such as, the Directorate of Mines and Geology. However, the Indian Bureau of Mines (IBM), a central government agency, has a significant role to play in the governance system for the mineral sector, especially for approval of mining plans and mine closure plans and monitoring their implementation as well as for overseeing if rules are being followed properly for protecting the environment in and around the mining project. It also monitors the progress of reconnaissance and prospecting permits. In addition, regulatory measures and institutional arrangements have also been created under the MOEF in two central legislations, namely the Forest Conservation Act 1980 and the Environment (Protection) Act 1986 for the protection of environment and forests in the mining areas. The Directorate of Mines Safety, another central government agency, also regulates the safety aspects of mines under the provisions of the Mines Act 1952. It is thus clear that multiple state and central government agencies are involved in monitoring different aspects of implementation of the regulatory system in the M & M sector which naturally creates problems of coordination and accountability among them. In fact, due to political reasons, delegation of more powers to the state governments may jeopardize the interests of mineral development.

There are also other regulatory challenges that restrict the growth of mining in India (FICCI 2013). The mineral policies of the state governments are in cases at variance with the National Mineral Policy and the procedures for the grant of mineral concessions also vary from state to state. While exploration activities which are temporary in nature do not require land acquisition, their granting procedures are similar to mining lease. This needs to be changed through legislations and new simpler procedures adopted to attract investment in this field. Prospecting activity needs to be defined clearly in Forest Conservation Act 1980. While oil drilling, transmission towers and 16 bore holes per 10 km^2 for prospecting are permitted in a "forest", trenching/pitting are prohibited. State governments often have a tendency to reserve potential areas for state PSUs which get blocked for long periods without any exploration or development. Current regulatory provisions make it difficult to transfer mining leases. Prospecting licenses are also not transferable. There is no guarantee of obtaining mining lease even if a successful exploration programme is carried out by a company. It seems there are limited incentives for private sector to invest in improvement of technology and equipment in mining projects as the mining industry is one of the most heavily taxed industry in the country (effective tax regime is 45%, compared to Chile 40%, Australia 39%, Canada, Russia 35%, China 32%).

Further, although mineral deposits belong to the state where these are located and the state receives the revenues from their exploitation, they have no control over the determination of the rates of royalty and dead rent, except in respect of minor minerals. This often prompts the mineral-rich states to demand an increase in these rates to meet their rising expenditure as well as deficit. The land acquisition problem is another crucial aspect of mineral development, particularly when there is a considerable proportion of tribal population in the mineral-rich areas. In many areas, the concept of "social license to operate" is invoked and many a viable mining project is dumped or stalled.

Clearly there is no dearth of instruments in the regulatory regime of the mining industry in India, but their poor implementation fails to deliver the expected results. Widespread inefficiency, poor infrastructure, corrupt governance, and absence of coordination between state and central government agencies lead to procedural delays in the granting of mining lease. Along with rising mineral prices and high demand of some essential raw materials, the delay fosters the proliferation of illegal mining activity across India, which as discussed in the next chapter, has serious implications toward the decline of this important sector in Indian economy.

10.5 International Laws for the Exploitation of Marine Mineral Resources

Marine mineral resources occur on the sea beach as well as on the offshore. Exploitation of the beach placers by the host country cannot be objected to by any other country. Take for example, the beach placer deposits that occur along stretches of the west, south, and east coast of India, already discussed. Similar instances may be drawn from many other countries. But the case is not so simple in case of mineral deposits, including oil and natural gas, occurring offshore. When the possibility of exploitation of mineral deposits from the ocean floor came up during the middle of the last century, some of the powerful countries asserted that their exploitation should be transnational. This meant any country that first staked its claim to a mineral deposit on the sea floor will have the right to exploit it, knowing full well that most of the countries in the world did not have the necessary capacity to do it. The rest of the countries naturally could not accept this logic. They initiated debates in the world fora where they asserted that these resources should instead be "supranational", meaning, ethically these should belong to all people in the world. The countries debated for full one decade and in 1982, the UN convention on the Law of the Sea (UNCLOS) was launched. They pressed for equality of mining rights of all resources in the third United Nations Conference on the Law of the Sea (UNCLOS-III). It was accepted and under the 9th session of UNCLOS-III in Geneva in 1980 an "International Sea Bed Authority" (ISBA) was established. 130 nations, including India, signed the document, but the USA, UK, and Germany did not.

The "Law of the Sea" divides the sea into three important parts (Fig. 10.1a): Territorial sea (12 nautical miles from coastline); Exclusive Economic Zone [EEZ: 200 nautical miles (=320 km) from coastline] and the International area of seabed (beyond the EEZ). Coastal nations have exclusive rights to explore and exploit all the resources within their respective EEZ's. Resources in the international area are a common heritage of mankind.

The Law of the Sea is a legal order for seas and oceans, aimed to:

1. Facilitate international communication
2. Promote the peaceful uses of the seas and oceans
3. Equitable and efficient utilization of their resources
4. Conservation of their living resources
5. Protection and preservation of the marine environment

The EEZ of India is shown in Fig. 10.1b. Significant offshore oil and natural gas (including gas hydrate) deposits are located within this exclusive economic zone of India, besides placer and phosphorite deposits.

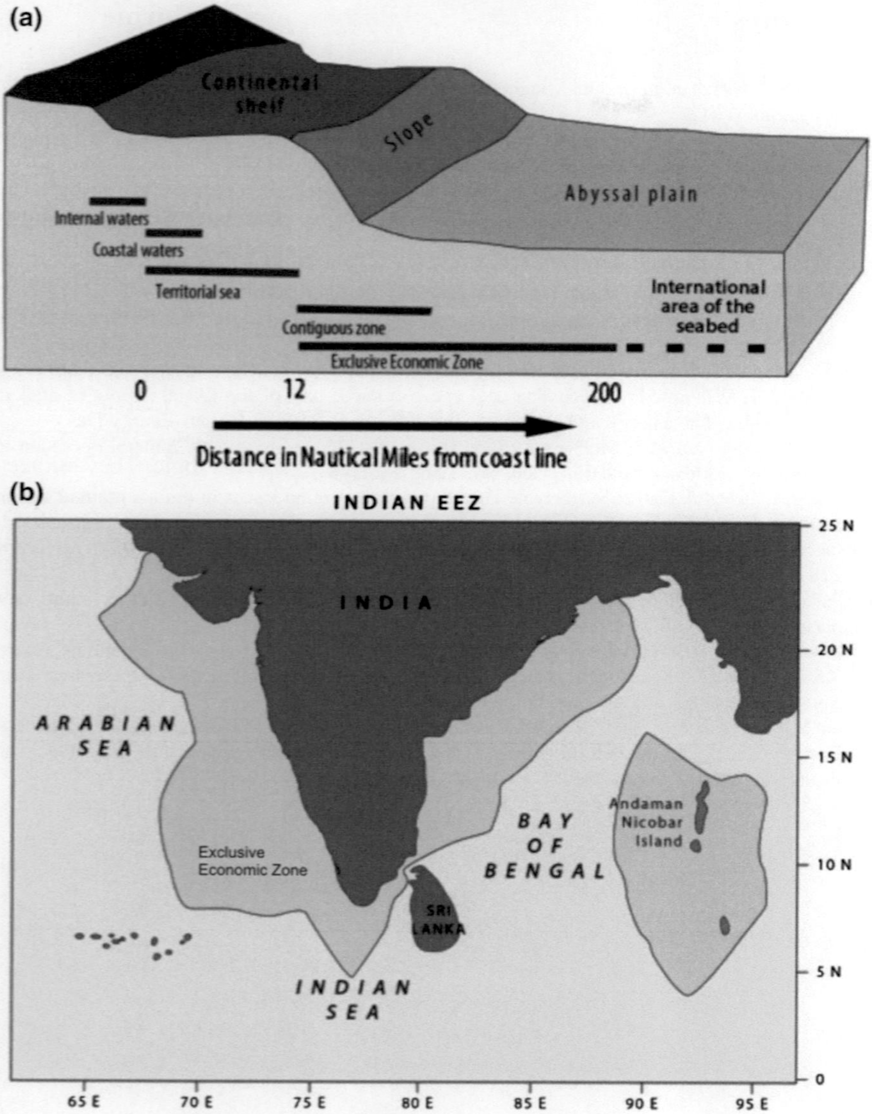

Fig. 10.1 **a** Coastal underwater topography showing the 12 nautical miles extent of territorial waters followed by the extent of exclusive economic zone (EEZ) up to 200 nautical miles. **b** Map showing the exclusive economic zone (EEZ) of India

References

Babu TM (1994) Tin in India. Mineral Resources of India 7. Geological Society of India, Bangalore, p 217

Chakravorty SL (2001) Artisanal and small scale mining in India. Mines, Minerals and Sustainable Development No 68. International Institute for Environment and Development, London

Deb M (2016) Artisanal and small-scale gold mining and related mercury usage in India. In: Environmental geosciences symposium on 35th international geological congress, Cape Town, Paper 805, T 14.4

Deb M, Tiwari G, Lahiri-Dutt K (2008) Artisanal and small scale mining in India: selected studies and an overview of the issues. Int J Min Reclam Environ 22(3):194–209

FICCI (2013) Development of Indian mining industry: the way forward. FICCI Mines & Metals Div, 119 pp

Kar SK, Singh JK (2001) Gemstones in India—a conspectus. Geol Surv India 58:651–657

Lahiri-Dutt K, Williams D (2005) The coal cycle: a small part of the illegal supply of coal in eastern India. The Energy and Resources Institute, New Delhi (J Resour Energy Dev)

Marker BR, Petterson MG, McEvoy F, Stephenson MH (2005) Sustainable mineral operation in developing world: introduction. Geol Soc Lond 250:1–4

Mohanty N, Goyal A (2012) Sustainable Development: emerging issues in India's mineral sector. Institute for Studies in India Development, Planning Commission Report, 274 pp

Perspectives (2008) Abandoned: development and displacement, 2nd edn. The Perspectives Team, 218 pp

Radhakrishna BP (2006) Small scale mining and rural employment generation: some new perspectives. J Geol Soc Ind 67(4):419–422

Sawarkar AR (1980) Geological and geomorphological features of parts of Nilambur valley, Kozikode district, Kerala, with special reference to the alluvial gold deposits in the area. Geol Surv Ind (Geology and geomorphology of Kerala) 5:29–38

Sridar M, Sumanth NJ (2001) Gemstone occurrences of Racheruvalasa-Addatigala gem tract of Eastern Ghats mobile belt in parts of Viziangaram, Vishakhapatanam & East Godavari districts, Andhra Pradesh. Geol Surv Ind 58:659–663

Chapter 11
Issues of Sustainable Development in the Mines and Minerals Sector in India

Abstract India, the third largest economy in the world, has shown phenomenal growth in the last decade or so, but is often criticized for its skewed development. This fact is glaring in the mineral-rich states of east-central India where the tribal population in the mining belts of Jharkhand, Odisha, Chhattisgarh, Madhya Pradesh, and Andhra Pradesh suffer from grinding poverty and malnutrition under a low Human Development Index. It raises a vital question: in rich land with poor people, is sustainable mining possible? We look into this complex socioeconomic and industrial challenge in this concluding chapter in terms of the vital things at stake: natural resources, people, culture and livelihoods, forests and wildlife, water resources, environmental and ecological integrity, and finally the issue of land acquisition for mining. The various issues have snow-balled and affected many large mining projects with very large FDI components. The problem is exacerbated by rampant and large-scale illegal mining of bulk minerals, including coal, which has forced courts to step in and ban iron ore mining in parts of Goa, Karnataka, and Odisha. Environmental degradation and inter-generational equity issues have also been raised by the judiciary to impose the ban. Natural resource utilization is an essential prerequisite for industrial development, and therefore, the three stakeholders, consumers, industry, and government, must join hands to pursue mining under a sustainable development framework which involves scientific mining, environmental protection and mitigation, community stakeholders' engagement, local socioeconomic development along with a high degree of transparency, and accountability. A proper mine closure plan should also be an integral part of any mining proposal. Ecosystem management in a mining area should also be an equally important remedial measure when mining ceases. The other important aspects of sustainability in mining also need to be highlighted. Geogenic toxic elements which are released during mining into the ground and surface waters, like Cr^{VI} in the Sukinda valley in Odisha, have severe adverse effect on human health. Besides, acid mine drainage may be generated from waste rocks, tailing dumps, open pits, and abandoned underground mines of sulfidic ores and flow into rivers, streams, or lakes. Their negative effects need to be controlled and neutralized.

M. Deb and S.C. Sarkar, *Minerals and Allied Natural Resources and their Sustainable Development*, Springer Geology,
DOI 10.1007/978-981-10-4564-6_11

Biological techniques are useful in this regard. Heap leaching of low-grade ores, of Cu for example, has the potentiality for acid water generation too. Cement plants are known to be the worst polluters, both in terms of cement dust and limestone mining. However, situation seems to be improving in modern plants. In the energy front, coal is a necessary evil generating a large part of thermal power but contributing to the global carbon emission and is thus responsible for global warming through greenhouse gas enhancement in the atmosphere. Oil and natural gas, absolutely essential commodities, are potent sources of environmental pollution right from exploration through production, refining, transportation, and use. Uranium ores also affect the environment when the radioactive metal is released into mine water, or is present in effluent of the ore processing plant and tailing dams and is finally passed on to the hydrologic system. Subsequently, it may enter the food chain and serve as a potential carcinogen. Nuclear energy production itself still has challenges of sustainable development as indicated by the recent disaster in Japan. Equally challenging is the issue of nuclear waste management.

Keywords Skewed development in India · Human development index · Issues of the mining sector · Sustainable development framework of mining · Illegal mining in India · Mine closure · Ecosystem management in mining · Toxic elements and human health hazard · Acid mine drainage · Coal and greenhouse gas emission · Paris climate agreement · Environmental pollution of oil and natural gas · Pollution from radionuclide · Hazard of nuclear energy production · Nuclear waste management

11.1 Introduction

Growth and *Development* have been the key words of the Indian economy in recent years with the GDP having touched a phenomenal peak of 9.4% in 2006–2007 and presently (2015–16) remaining around 7.6%. India is poised to become one of the leaders of the world economy, representing one of the BRIC (Brazil, Russia, India, China) countries, an acronym coined by Goldman Sachs to represent the fastest growing and largest market economies in the world. But it is apparent that the development process has been skewed and although India has an ever-expanding and richer urban middle class with improved life style and an ever-increasing number of the richest people in the world, unfortunately the benefits of development have by-passed a lot many. One-third to one-fourth of Indian population remains miserably poor. National Commission for Enterprises in the Unorganized Sector (NSEUS, constituted by GOI) had estimated in 2004–5 that 78% of India's population lived on <Rs. 20 a day. The situation has improved since, though India still ranks high in the Global Hunger Index of 118 countries (2016 score of 28.5

compared to Pakistan 33.4, Bangladesh 27.1 and China 7.7). According to World Bank's revised methodology of 2014, the world has 872.3 million people below the new poverty line, out of which 179.6 million live in India. It is thus clear that the people who steer this country's economy to grow at a frantic pace seem to disregard the fact that the level of "development" and "growth" alone does not provide the answer to the desirable life and living today; how the benefits of growth are distributed, or percolates downward is far more important. In this context, Gandhiji's dictum seems most appropriate: "*The success of development must be judged from the viewpoint of the least advantaged in society*".

The marginalization of a good part of the Indian population is also manifested in the low *Human Development Index* (HDI) of the states in east-central India, rich in natural resources. Some of the mineral-rich regions, such as Koraput, Kalahandi, Malkangiri, Rayagada, near the junction of three states, Andhra Pradesh, Chhattisgarh, and Odisha, are home to the poorest people in the country, 60% of whom happen to be tribals (vide next section for more details). The tribal population in mining areas of Jharkhand, Odisha, Chhattisgarh, Madhya Pradesh, and Andhra Pradesh remain marginalized, in grinding poverty, malnutrition, and deprivation even when they live in the richest (natural resource-wise) part of the country, where mining has been going on since the pre-Independence days. Obviously, the benefits of resource development have by-passed these people along with some others. The relevant debate thus continues: whether the mining industry will bring in development, employment, and growth in these backward areas and improve the lives and livelihood of the indigenous communities, or is it just an illusion provided to the people affected by mining by the Government and the large Indian and multi-national corporate houses. In a rich land with poor people, is sustainable mining possible? Not an impertinent question posed by the Centre for Science and Environment (2008).

It must be clear by now from all the foregoing discussions in this third part of the book that mining in India today is no more a "dig and sell" proposition but is a highly complex socioeconomic and industrial challenge (CSE 2008). In short, the complexity rests on the fact that several key things are at stake. These include the following:

- Natural resources
- People, culture and livelihoods
- Forests and wildlife
- Water resources
- Environmental and ecological integrity
- Land acquisition.

We endeavor in this chapter to discuss issues related to these aspects and then suggest a sustainable development framework for the Mines and Minerals (M & M) sector in India.

11.2 Key Issues

The uncomfortable questions of sustainable development concerning mining were raised earlier (Chap. 9), that is, when should a resource be mined, by whom, for whose benefit and at what and whose cost? These questions are most relevant in the Indian context today when the richest bulk mineral-producing (iron, manganese, chromite, bauxite, and coal) region in the country in east-central India, spreading through southern Bihar, Jharkhand, Chhattisgarh, Odisha, eastern Maharashtra, and northern Andhra Pradesh (Fig. 11.1), is faced with serious sociopolitical, economic, and environmental issues affecting the mining activities.

All the five states have forest cover considerably above the national average of 23.8% (data.worldbank.org): with Chhattisgarh 43%, Jharkhand 30%, Odisha 27%, Madhya Pradesh 26%, and Andhra Pradesh ~22%. The connection between the type of mineral resources mentioned above and the forest cover is a fundamental one. The nutrients in the soil, derived particularly from Fe–Mn deposits, help in the growth of lush vegetation. Such dense forests are also the habitat for a variety of wildlife and indigenous forest dwellers.

India has a large tribal population of ~84 million. They have accommodated themselves in the following habitats:

(a) Flat and normally forested and/or cultivated lands;
(b) Deeply forested lands with limited cultivation (often referred to as *Junglemahal* and people called *Banabasis*). Forest products are their principal source of sustenance; and
(c) Hilly tracts, not necessarily plateaus (for seclusion, if not for anything else). People living here are called *Girijans*.

People living at (a) are socioeconomically relatively advanced, although their identification with the main life stream needs to be improved. People at (b) and (c) are much more isolated. Apparently, they prefer to remain so. Their spatial and social isolation remains a matter of controversy. According to some it is the sense of isolation characteristic of the indigenous people (*adivasis*) the entire world over that kept them apart. According to some others, the adivasis were defeated in the war with the new settlers and took shelter in forests and hills and their evolution was constrained. They have been exploited historically by landlords, forest and mineral exploiters, as the case may be.

A large proportion of these tribals in (b) and (c) now live in these mineral-rich and forested tracts of east-central India. Their livelihood, culture, and economy are closely intertwined with the forest in which they live. Therefore, degradation of the forests due to mining and other developmental projects affects the forest ecosystem and in turn makes the tribal population both socially and economically challenged. As the forest cover is removed the tribals lose most of their traditional livelihood while the destruction of the watershed and hills, revered by the tribals from time

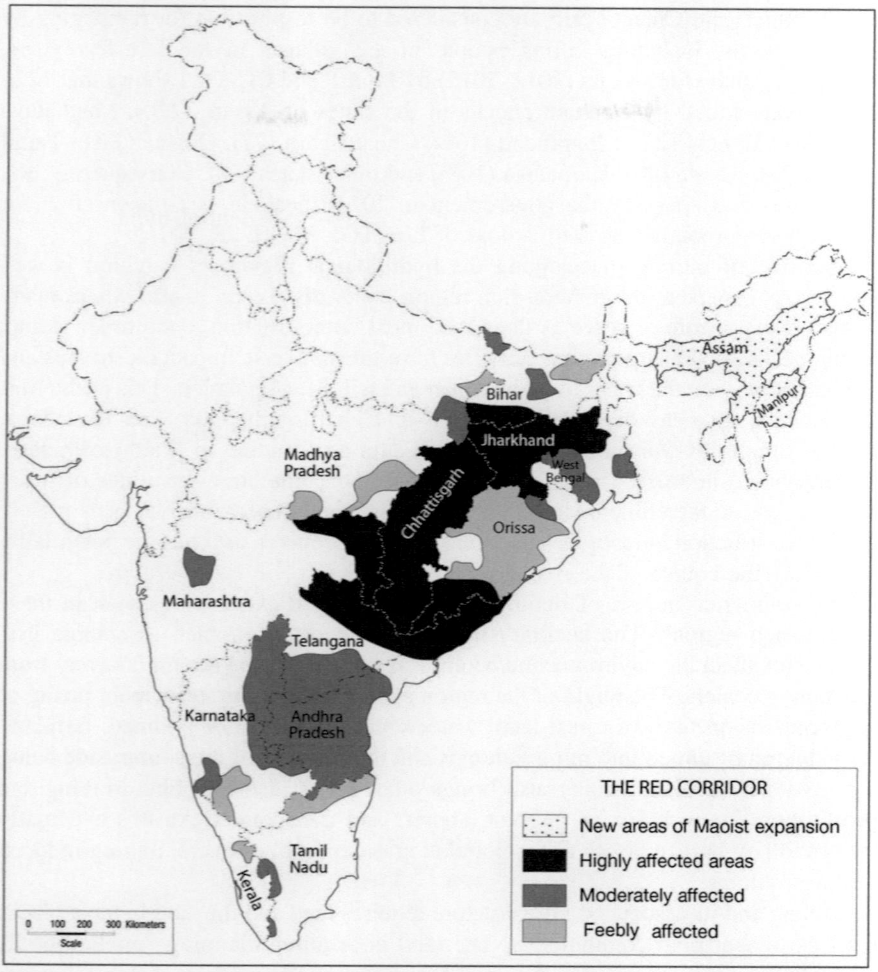

Fig. 11.1 The rough spread of "Red Corridor" in east-central India which is the richest repository of bulk minerals, thickest forest cover in the country and the habitat of a large section of marginalized and poverty-stricken tribal people

immemorial, impacts the tribal culture. According to one Government estimate, in last 25 years, between 1980 and 2005, about 95,000 ha of forest land was diverted for mining in this region (CSE 2008). The tribals have often protested but failed to reverse the situation that affected them. But now they have got a voice through the support of political groups.

As forest land and wildlife habitat is cleared and encroached upon for mining and related development, human–wildlife conflict increases and the wildlife population is also affected. Coal and iron ore mining operations in Jharkhand, Odisha,

West Bengal, and Chhattisgarh are considered to be responsible for rampaging by elephant herds, including killing people in the villages in the last few years. A recent estimate for 3 years (2012–2015) by MOEF and CC, GOI shows that 1226 people were killed by elephant attacks in the states of Assam (220), Meghalaya (14), West Bengal (227), Jharkhand (169), Chhattisgarh (77), Odisha (220), Tamil Nadu (133), Kerala (40), Karnataka (108), and other states (24). Consequently, Rs. 39.8 crores was spent by the government in 2014–15 alone as compensation for loss of crops/properties as well as loss of life (TOI, March 28, 2016).

The role of mining in changing the hydrological regime of a region is well known. As expected, the mineral-rich region under discussion is also where many rivers emanate from or serve as their catchment. Deforestation, overburden dump, tailing heaps, etc. due to mining activities have great adverse impact on streams and rivers that provide the water to both human and wildlife population. This problem is conspicuous in areas of iron ore mining in Chhattisgarh, Goa, and Karnataka, bauxite mining in Odisha and Chhattisgarh, and coal mining in Madhya Pradesh. Mineral-based industries in the region not only consume large quantities of water but also release the effluents into them. Rampant sand mining in river beds mainly for the construction industry is also a cause of great concern as it has the potentiality of altering the course of the river.

Environmental impact of mining can be severe and extensive as seen in these mineral-rich regions. The ancillary industries that crop up, such as sponge iron plants, can affect the environmental/ecological integrity of the region far away from the mining centers. The plight of the region and its inhabitants stems from razing of thousands of hectares of forest land, water sources throttled or polluted, farmland made barren or turned into mining dumps and the fresh air of the countryside being laden with mine dust. Mining also brings other social disorders like drinking and prostitution due to influx of migrant laborers and transporters. All this eventually takes a toll on human health as the population is exposed to several mining-induced health problems.

Mining and its associated infrastructure requires land and the land in most places belongs to the tribal communities. The land is acquired claiming "public good", leading to displacement of the tribal population from their ancestral land. Rehabilitation and compensation are minimal or mostly remain on paper. The uprooted communities lose access to their traditional livelihood and lifestyle and find their economic impoverishment becoming even worse. Thus the communities are left totally marginalized and traumatized. The fall-out is not unexpected: the growing popular anger has been cashed-in by the political extremists, i.e., Naxalites (Maoists) and the like, and today between 160 and 180 of the 274 districts where mining takes place in the country are affected by the movement, which at many a place started by opposing the lopsided development that modern large-scale mining is bringing in.

Economic liberalization in the 90s brought about couple of conspicuous changes in the M & M sector. First, some of the profit-making PSUs were sold off to private parties, e.g., Bharat Aluminium Company and Hindustan Zinc Ltd were both sold to Vedanta Group. Second, India's mineral reserves were opened up to foreign investors and export of minerals saw a phenomenal rise (32.68%) in just 3 years, 2001–2004. The three mineral-rich states, Odisha, Chhattisgarh, and Jharkhand, were in the forefront in this regard. They not only offered captive mines to the multi-national companies but also encouraged integrated mega-projects which involved captive mines, steel factories, townships, rail corridors, and even captive ports. Such projects required large tracts of land, affecting the environment, habitat and biodiversity, and entailed displacement and loss of livelihood to a substantial number of people. Not surprisingly, this led to clashes of the mining concerns with the local population, mostly tribals, environmental activists, and NGOs. State governments used force, while ultra-left Maoists instigated people to take up arms, leading to violence, court cases, and a chaotic situation overall.

A few examples of some of the major projects impacted by some such conflicts in recent years include the following:

(i) Pohang steel company's (POSCO's) US$ 12 billion (Rs. 51,000 crores) investment plan (2005) for integrated steel production in Jagatsinghpur and Keonjhar districts of north Odisha—strong opposition from local people against land acquisition and displacement of 22,000 people has not allowed the project to take off even after a decade.
(ii) Vedanta's proposed US $1.7 billion alumina production (Lanjigarh refinery) and bauxite mining (Niyamgiri hill with bauxite resource of 70 Mt) projects in Odisha faced stiff opposition by local 8000-strong Dongaria Kondh tribe, 12 *Gram Sabhas* across Kalahandi and Raygada districts and NGOs. It led to the eventual withdrawal of forest clearance and turning down of the mining proposal.
(iii) Utkal alumina project, which was a US $ 1 billion joint venture between M/s Hindalco (India) and Alcan (Canada) to mine and refine bauxite was delayed by more than a decade due to opposition to land acquisition.
(iv) Uranium Corporation of India's (UCIL's) two mining projects worth US $200 million and US $ 225 million in Meghalaya and Andhra Pradesh, respectively, could not progress due to opposition from local communities and NGOs who apprehended that the possible radiation would affect their health.
(v) Tata's Tentoposi steel project in western Singhbhum, Jharkhand was stopped by the opposition from local communities.
(vi) ArcelorMittal scrapped its 10 Mt steel plant project in Keonjhar, Odisha.

In the global context, failure of engagement of large mining companies with the local communities has earlier threatened well-known mining operations in Peru and Argentina.

11.3 Illegal Mining

Like some social vices, illegal mining persists in certain measures in some developing countries, including India, where it adds another dimension to issues of sustainable mineral resource development. We briefly discuss the situation here, outlining some case histories from Goa, Karnataka, and Odisha, where such illegal mining was rampant.

Illegal mining/quarrying on small mineral deposits, particularly of nonmetallic minerals has been going on at places in the country for a long time. The state and central governments did not take them seriously. But what has of late become disturbing is illegal mining of some major deposit types, such as iron, manganese, and coal deposits and its common occurrence (Table 11.1).

Illegal mining can be of different types (Mohanty and Goyal 2012):

- Mining may take place in an area for which no mining license has been obtained.
- Mining may take place in leaseholds for which surface right has not been granted.
- Mining operations are being carried out without conforming to the approved mining plan/environmental management plan.
- Production of minerals and their dispatch are in quantities in excess of those approved in the mining plan and/or environmental clearance.
- Ores have been dispatched from leasehold areas in an unauthorized manner (for example, mining lease is for iron ore but production and dispatch are for manganese ore and vice versa are taking place).
- Mining takes place in forest land for which no forest clearance has been obtained.

Areas commonly vulnerable to illegal mining are as follows:

- Virgin areas often reserved for public sector enterprises and/or private companies who intend to set up related industries;
- Areas for which intention to grant lease has been communicated but lease not granted for a long time;
- Areas for which lease has been granted but execution are pending for long;

Table 11.1 Major states with illegal mining in 2009–10 (IBM 2011)

State	No. of illegal mines
Maharashtra	18,638
Andhra Pradesh	15,534
Madhya Pradesh	5275
Rajasthan	5193
Karnataka	3683
Chhattisgarh	2003
Odisha	758
Goa	9

- Areas held under prospecting license but the party's application for mining lease is pending for long;
- Earlier mining areas where no activity has taken place for a long time;
- Surrendered areas lying vacant;
- Areas which have been thrown open for grant of mining lease but applications are pending for long; and
- Abnormal delay in the disposal of applications for mining lease, renewal of lease, forest, and environmental clearance.

The rather long list above clearly points to the main reasons for such illegal mining activity: procedural delays and administrative inefficiency along with rising mineral prices. It also points to the need of an efficient governance framework to curb this menace. A first step should be empowering the State Mines and Geology Directorates with effective laws and manpower.

In recent years there have been several landmark court judgments and reports regarding illegal mining. Honorable Supreme Court, sometimes back, imposed an interim ban on all iron and manganese ore mining in Goa, following a Public Interest Litigation (PIL), citing Justice M.B. Shah Commission Report that stated that illegal mining had cost the state exchequer a huge sum of money, amounting to about Rs. 35,000 crores. The bench also ordered a complete ban on transportation of mine-head material (mined ores) out of the state. The Commission indicted the Digambar Kamat Government for this impropriety. The Court issued notices to the Central Government, the State Government of Goa, the State Pollution Board, and the Ministry of Environment and Forest, after it came to light that illegal mining was going on in majority of the mines. It requested the Court to terminate all mining leases where mining went on illegally and asked the Government to recover the entire money collected by the lease holders through illegal activities. Further they were to revegetate and restore the ecological balance. The Report was placed in the Parliament in September 2012, and the Parliament censured the ex-Chief Minister for his failure to do the needful.

Justice M.B. Shah submitted another report in 2013 on illegal mining in Odisha. The Commission has severely indicted both the Central and Odisha governments on illegal mining of iron and manganese ores. The Commission has also asked the State government to recover about Rs. 60,000 crores from miners and utilize the recovered sum of money in developing the two districts, Keonjhar and Sundergarh, that have suffered most due to the illegal excess mining under discussion. The commission recommended granting all future mining leases via auction route only (fortunately it has been followed in redistribution of about 200 coal mine blocks recently). The commission also found that out of 192 mining leases of iron and manganese ores in the state, 130 lessees were found to be doing production, without lawful authority. Moreover, 94 mines were found to be operating without environmental clearance (EC). Overall, 109 mining leases were working under "deemed extension" and doing production in violation of EPA 1994, 2006. According to the Commission's estimate, iron ore worth over Rs. 45,453 crore and manganese ore worth over Rs. 3089 crore have been exploited illegally (TOI, Sept, 2012 and Jan 3, 2014).

Due to rampant illegal mining of iron ores and attendant environmental damage, ban of iron ore mining has been imposed by the Apex court in parts of Karnataka. Based on the CEC (Central Empowered Committee) reports, the mines involved have been categorized as “A”, “B”, and “C” according to increasing levels of illegalities carried out by them. Subsequently, by its order of September 3, 2012, the court has permitted the resumption of mining in category “A” mining leases. The court has also approved (order dated 28.09.2012) the recommendations of CEC for reopening of category “B” mines, subject to the same conditions as category “A”. The CEC recommendations entail a ceiling of 25 MMT for total production of iron ore from all the mining leases in Bellary district and a ceiling of 5 MMT for production of iron ores from all the mining leases in the districts of Chitradurga and Tumkur in Karnataka. According to the CEC recommendations, the permissible annual production for a mining lease in the districts mentioned above would be based on the following:

(a) Mineral reserves in the lease area.
(b) Area available for overburden/waste dump, subgrade iron ore and other land uses.
(c) Existing transport facilities in relation to the traffic load of the mining lease.
(d) Overall ceiling on the annual production from all mining leases in the district.

Coal is another natural commodity highly essential for the country’s development and well-being, but widely mismanaged. Presently, coal (and lignite) in India is mined mainly by Government agencies, with the exception of captive mines owned by some metallurgical organizations, such as steel plants, belonging to both the public and the private sectors. For further development of the Indian industry, the Government felt coal production in the country should be increased at the earliest. It first requested Coal India, the nodal agency for coal exploitation in the country. The Public sector enterprise excused itself showing heavy work load. The Government then decided to distribute 200 plots containing 600 Mt of coal to private players. There were many contestants, but lucky among them obtained the lease. Before long it was discovered that fair means were not followed in distributing and obtaining the coal blocks. It violated the expected norms in many respects. The Honorable Supreme Court intervened and canceled the earlier distribution of coal blocks and asked the Government to do it again, following a public auction mode. This has been honored, checking mismanagement of this valued natural resource and contributing a handsome fund to the Government exchequer. Another illegal activity with respect to coal is small-scale illegal mining from abandoned or operation-suspended mine. This has become very common particularly in eastern Indian coalfields. Since these operations go on behind people’s eyes, following crude methods, accidental deaths are too many, compared to per capita production. It is a common belief that the administration should be more vigilant in order to control this malpractice.

Rampant quarrying of dolomite in the Bhutan hills, eastern lower Himalaya has adversely affected tea plantation—the principal industry in this part of sub-Himalaya. Over one-third of the local population depends on tea industry for their living. Every time the hills are blasted for the mineral, the ground is covered with a thick layer of dolomitic rock dust, reducing land fertility. Illegal mining has also adversely affected the biodiversity of the Dooars. Quarrying causes landslides, soil erosion, floods, etc. High pH of groundwater due to carbonate dissolution reduces its potability and is also not favorable for tea plantation.

11.3.1 Government's New Policy to Check Illegal Mining

Recently, the Government of India has adopted some new measures in which the public has been involved, to check illegal mining of mineral resources. The recently adopted MSS (Mining Surveillance System) will trigger an alarm whenever there is an instance of illegal mining outside permitted areas. The system developed by the Indian Bureau of Mines in collaboration with Gandhinagar-based Bhaskaracharya Institute for Space Applications and Geo-informatics and the IT industry uses satellite-based remote sensing technology in conjunction with digitized maps of licensed mining areas and global positioning to track operations. The system keeps an eye on areas 500 m beyond the mining lease boundary and can trigger an alarm in case of any unusual activity.

The data setting of the alarm would be studied at a remote sensing control center and transmitted to district-level mining officials for verification on the ground. The program will have other details pertaining to the mine operations, such as the number of miners or workers under employment. Such information is expected to help ensuring the necessary safety and security. Most working mines have been digitized. Soon digitization of nonworking mines would be completed.

According to the information obtained from the Government, 48,467 cases of illegal mining of major and minor minerals were recorded in the April–September period of 2015–16 fiscal. This is against 97,149 cases in 2014–15, 88,689 cases in 2013–14 and 98,597 in 2012–13 (TOI, Kolkata, October 16, 2016).

11.4 Inter-generational Equity Issue

The Court proceedings based on a PIL which led to ban (lifted recently) on iron ore mining in Goa had some interesting issues. It is not only illegal mining and environmental degradation but also the question of inter-generational equity in the small state of Goa that was raised in the PIL.

In the state of Goa, the iron ore resource (as of April, 2010) stood at 927 Mt (hematite) and 214 Mt (magnetite) out of which 51% of hematite ore was of “reserve” category and 49% “resource”, as against 23% “reserve” of magnetite ore and 77% “resource”. It is interesting to note that Goa has a +80 years iron ore mining history (since 1929) and the hematite ore resource estimate in 1971 was only 396 Mt and in spite of an extraction of 500 Mt in the 40-year period (till 2010), and the resource base in 2010 has increased more than the production during this period. This is despite the fact that exploration has remained at a shallow level of 50–60 m and the cut-off grade for iron ore resource is estimated at 60% Fe as against the present threshold of 45% Fe, suggested by IBM. If our steel industry undergoes adequate modification and upgradation of technology to consume this lower grade ore, of course keeping in view the ore characterization ($Al_2O_3 + SiO_2 \leq 5\%$), a substantial increase of the Fe ore resource base will be forthcoming. Besides, if pelletization is incentivized and augmented, the huge quantity of fines, which is an integral part of iron ore mining, will raise the resource base even further. The huge pile of banded iron formation (BIF) now lies as reject, because higher grade ores are available in plenty. These could also add to the resource base of iron ore in the distant future when advanced technology of beneficiation will allow it to be consumed.

It may also be reiterated that India possesses the seventh largest reserve of iron ore which at the current consumption rate will last for about 50 years. But according to the internationally recognized “proven” and “probable” reserves, India has significant quantities of mineral “resources” which are still under various stages of exploration. With more intensive and deeper exploration, many of the now subeconomic resources and marginal reserves may turn out to be commercial “reserves” as the degree of geological assurance increases. Unproven “resources” presently are more than twice the proven “reserves”. Out of an estimated resource of 28 billion tons, 28% is in the “reserve” category, while 72% is in the “resource” category. Obviously, as discussed earlier, the availability of iron ores in India will be several orders of magnitude more than for 50 years.

Considering all the aspects mentioned above, inter-generational equity issues should not be considered as adequate reason to ban iron ore mining in Goa. Also keeping in mind the projected growth of the main consumer industries and the National steel policy (2012), wherein the crude steel production is projected to 192.5 Mt at 6% GDP and 275–300 Mt at 7% GDP in 2025–26, the consequent demand for raw material (Fe ore) at 7% GDP would be 392 Mt in 2025–26. In this background, any cap on mining would seem utterly unreasonable. Instead, all environmental regulations should be enforced rigidly, required infrastructure such as water, power, and transportation facilities developed properly, inclusive growth of the local communities ensured, so that the M & M sector can come out of the present impasse and return to its earlier growth rate.

11.5 Sustainable Development Framework

Allocation of natural resources has been plagued consistently by controversies in India. Nevertheless, its importance in the Indian economy makes proper natural resource utilization an essential requisite for any industrial development in the country. Efficient implementation of existing mining and environmental laws can help to institute a clean and transparent system that will benefit all the three stakeholders: consumers, industry, and Government. The concern is not whether mining should be undertaken. Rather, it is about how it should be undertaken and who all would benefit from it! The goal should be "*common good*", not revenue maximization in terms of *Article 39B* of the Constitution.

In this regard, a sustainable development framework (SDF) along with a set of sustainability indicators is required to enhance public accountability of mining enterprises in India. Its key elements would comprise the following:

(i) *Scientific mining*: Which can help to increase the yield from a mine while keeping the wastes to a minimum. It can reduce the cost of mining and along with development in the field of processing, allow lower grade ores to be utilized, thereby increasing the resource base and prolonging the life of the mine. Undoubtedly, mine safety will also be enhanced.

(ii) *Environmental protection and mitigation*: The negative impacts of mining have to be kept at a minimum, keeping the "carrying capacity" of the area in view. Air and water pollution controls must be in place and impact of mining on human health must be monitored. The problem of acid mine drainage needs to be controlled with regard to the geology of the mining site. Mine closure plan has to be implemented in totality and ecological restoration of the mining site carried out in the best possible way (see more discussion on these aspects below).

(iii) *Community stake-holder engagement*: The local community must be made to feel their stake in the mining activity and consequent development. The *Gram Sabha*[1] should decide and provide the social license to begin and operate the mine.

(iv) *Local socioeconomic development*: A new law on Corporate Business Obligation (CBO) for socioeconomic development related to mining projects should replace CSR (Corporate Social Responsibility) activity, which is voluntary and optional in nature. The benefits may be shared in the form of infrastructure development, social amenities, or direct payment to landowners. Establishment of a district development fund based on mineral revenues is a good option.

(v) *Transparency and accountability*: Should be given the highest priority to establish communities' faith in the mining project.

[1]Gram Sabha is the smallest unit of local self government in present day India.

In conclusion, we may draw the bottom line for the mining sector as follows: Development and Growth alone need not be the goal—how the benefits of growth are distributed is far more important. Here we may also recall the famous comment of the Father of the Nation (M.K. Gandhi): "*The success of development must be judged from the viewpoint of the least advantaged in society*". This should apply for India, as much as for any country that strives to achieve this goal.

11.6 Why and How Should We Close a Mine?

Closure is an inevitable stage in the history of a mine. It is an important step in environ-friendly exploitation of a mineral deposit. But until some years ago it was, so to say, an ignored step. Mine owners were interested only in exploitation of the economic commodity by whatever means possible and hardly bothered about any potential hazard they left behind in the way.

The major hazards that may visit the area post-mining are

(1) Chemical,
(2) Thermal,
(3) Mechanical, and
(4) Landscape degradational.

Chemical degradation may take place by several ways. One such is production of acid waters by the oxidation of iron sulfides (pyrite, marcasite, pyrrhotite) inside the mine or in the waste rock heaps at the mine head (Fig. 11.2a). Abandoned underground coal mines may be the sources of CO_2, CH_4 in the neighborhood, whereas sulfidic coal mines could be sources of acid waters.

Today, thousands of coal mine fires are burning (involving left out coals) in abandoned coal mines throughout the world (Stracher 2004). The hazard they cause is unimaginable to the lay people. They destroy a nonrenewable natural resource; threaten the health and safety of the local people; and degrade the ecology in general. Ash produced is considerably lower than the coal burnt and as a result roof collapse is not uncommon. The underground fires in coal mines may start by self-ignition (by FeS_2 oxidation) or human action. Because the mine is abandoned, fire at the early stage usually goes unnoticed, and when advanced, it may be noticed but by then became unmanageable or difficultly manageable. In the Raniganj–Jharia coalfields of West Bengal–Jharkhand, India, about 2000 km^2 of land and about 1 million people have been affected by coal fires, some of which have been there for a century (since 1916) (Micholski and Grey 2001). Many early mine owners did not leave behind (or did not have?) detailed mine plans, which by default caused some disasters. Some of these are addressed in the next paragraph. Before that it may be mentioned that if the pits of open cast mines are not redressed before closure, then they are bound to degrade the landscape.

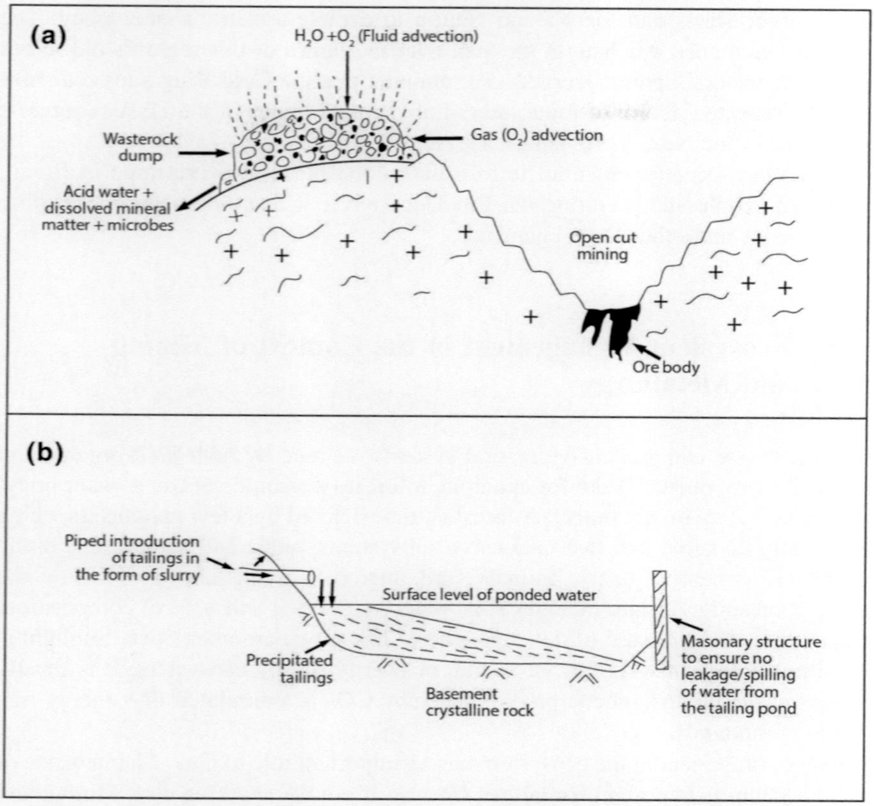

Fig. 11.2 **a** Sketch showing acid water generation from waste rock dump (after Sarkar 2002). **b** A sectional view of a hypothetical tailing pond with impoundment

Severe environmental crises caused by collapse of abandoned mines are not a few. A couple of them are discussed below by way of elucidation.

A gold mine called Gold King was opened in 1890 and within a decade it became the most important mine in Colorado San Juan country and hence the name. But unfortunately the high hopes that were associated with this deposit were found to be not very well-founded before long and the mine had to be abandoned during the days of recession in early 1920s. However, the open tunnels filled with snow melts and rain water and ultimately turned acidic leaving behind a toxic legacy that this region was trying to clean up for decades. On August 5, 2015, the Gold King split open by accident when a team appointed by the Environment Protection Agency (EPA) was investigating the source of the leak. From 2011 Gold King was spitting metal-bearing wastes at an average rate of 176 gallons/min. As a result, the creek below the mine reached Cd and Co contents 10 times the maximum Federal standard and Zn >40 times this standard. The accident let out a yellow plume south into the Animus River and turned western waterways into a “mustard ribbon”, forcing three states and the Navajo Nation to declare a state of emergency. The accident heightened a debate in the area over the future of this region's old mines and, in the general opinion, served as a reminder that the Gold King's toxic demise could be repeated at any of thousands of abandoned mines in the USA (Source: a news item in the New York Times, 17 August, 2015).

In another recent event mud from a dam burst at an iron ore mine in Brazil, killing 12 people and polluting the Rio Doce River with toxic matter, containing heavy metals and other toxic chemicals.

11.7 Ecosystem Management in the Context of Mining and Metallurgy

Ecosystems are comprehensive natural systems, defined by both their organic and inorganic components. Take for example, a forest (we could choose a water body, forest land, a bay or any other). A forest is to be defined by a few parameters which are broadly divisible into two interactive subsystems, biotic and abiotic. The biotic comprises numerous plants, animals, and microbes, living above or below the surface (subsoil). Abiotic includes soil, water (surface or subsurface) composition, temperature, and pressure of the atmosphere. These two are interactive. Sunlight is the ultimate and only significant source of energy of any ecosystem. It is mostly utilized in the photosynthetic process whereby CO_2 is assimilated into energy-rich carbon compounds.

Every component in the ecosystem has an important role to play. Maintenance of the ecosystem is important for human interest, if not for anything else. Mining and metallurgy may adversely affect the ecosystem in an area. If the mining camp is located in a forest land, deforestation over a large area will be inevitable. It will affect the local climate involving humidity and rainfall. Depending on the type of mining the air in the locality will be dust-laden. The sound of mine blasting will frighten away the forest-dwelling animals and the ground vibration will disturb some beneficial microorganisms in the soil.

Biodiversity is a very important indicator of a healthy ecosystem in a locality/region. Biodiversity is indicated by the number and variety of bio-populations, presence of food chains, life-saving medicinal raw materials, as well as nutrient cycles in the ecosystem. A minimum level of biodiversity is needed for proper functioning of the ecosystem. Mining (and accompanying metallurgical plant, if any) is likely to adversely affect the ecosystem and by default, the biodiversity.

Suggested remedial measures include the following steps:

(i) Landscape must be restored as much as possible after mining ceases.
(ii) The open stopes of coal mines must be filled up (back-filled) by rock wastes. Fire in the abandoned coal mines in Jharia, going on for many years now, has severely affected the local ecosystems. In some large underground metal mines the roof is allowed to cave in. It saves expense but is not "full-proof".

(iii) Execute reforestation in the area; it will increase the rainfall and ultimately restore the biodiversity to some extent. Full restoration is not possible though. Grass cover to the land surface should be restored to the possible extent. It will reduce soil erosion.

(iv) In India 600 odd national parks and sanctuaries are inviolable, i.e., free of human interference, except what is allowed by the recent Forest Rights Act.

11.8 Some Other Important Aspects of Sustainability

11.8.1 Human Health Hazards Caused by Intake of Some Toxic Elements

Some toxic elements contributed by both the geogenic and industrial processes are commonly ingested by human beings. Their primary sources are geogenic (Fig. 11.3). Industrial effluents may also contribute these elements to the environment when they become part of the geogenic system. Their intake by human beings and domestic animals is through food (*food chain*) and drinking water.

It may be noted that the oxidation state may be the determinants of toxicity of some elements. For example, Mo(VI) and Cr(VI) are much more toxic than Mo(II) and Cr(III). The toxicity of arsenic varies with the state of oxidation in the order: $As^0 < As^{5+} < As^{3+}$.

Both Cd and Hg are often hosted by zincblende ores. They are liberated from the host through natural process(es) or during industrial processing.

Expert like Ross (1999) believe that any mineral dust when inhaled in large enough doses may cause nonmalignant respiratory disease and in cases, even cancer. Therefore, there is nothing to be surprised to note that workers heavily exposed to dust from serpentinite, taconite, slate, coal, bentonite, Fuller's earth, diatomaceous earth and the very common rock granite may show a greater than otherwise expected morbidity and mortality.

We may mention the case of environmental pollution by Cr(III) in the Sukinda valley, Odisha, where a couple of large deposits of chromite ores occur. Cr in natural waters can occur as both Cr(III) and Cr(VI). Cr(III) is insoluble, immobile, and nontoxic. Cr(VI) is just opposite; it is soluble, mobile, and toxic. The latter can be reduced to Cr(III) by Fe(II), sulfides, and organic material. Mn oxides oxidize Cr (III) to Cr(VI). Recent studies show that Cr in local surface waters isotopically heavy relative to the soils indicating that during oxidative weathering isotopically heavy Cr(VI) is leached from soils to the runoff (Paulukat et al. 2015). These authors show that microbes have the capacity to reduce and immobilize Cr(VI) which could be a method of controlling the hazardous impact of Cr(VI). Mining activities in the Sukinda valley reportedly adversely affected the health of the local people. Discharge of mine water and also leaching of oxidized Cr and other heavy metals from dumps polluted the surface and groundwater with Cr(VI) causing health hazard to many local people (Dhakate and Singh 2008). The human tolerance limits of some metals and compounds are presented in Table 11.2.

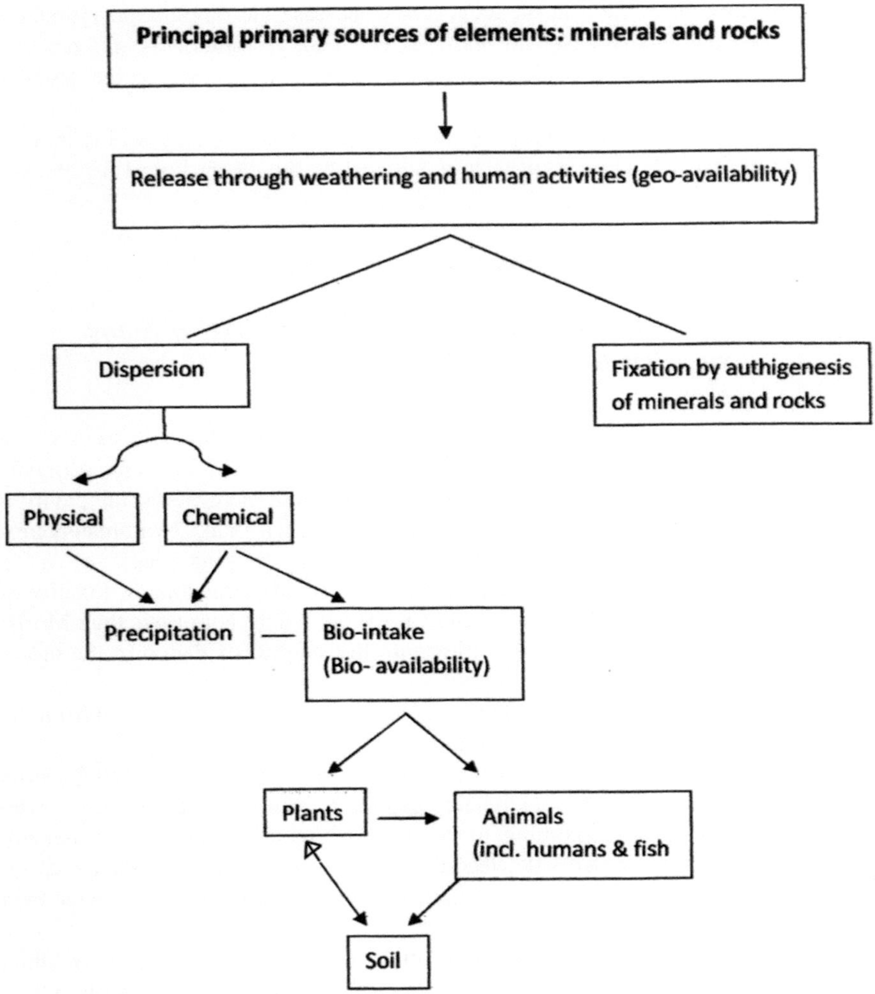

Fig. 11.3 Exogenic cycle of chemical elements in the ecosystem

11.8.2 Acid Production in Mines and Its Neutralization

Natural processes of acid production in mineral belts, particularly those of base metal-rich sulfidic ores and coal formations, are fairly well known to the people living in and around those places. The iron sulfide pyrite (FeS_2) plays a very important role in the process. To be a little more elaborate, the natural process of pyrite oxidation is crucial to the supergene alteration of ore deposits, the formation of acid sulfate soils, lowering of pH and the mobilization of metals in natural waters. The process is called "acid rock drainage", if the acidic liquid forms from

Table 11.2 Human tolerance limits for some metals/compounds

Metal/compound	Ill Effect	Tolerance limit (mg/l)
Arsenic	Affects dermal and nervous systems; even carcinogenic	0.01[a]
Barium	Affects circulatory system	1.00
Cadmium	Affects kidney	0.01
Chromium	Affects liver and kidney	0.05
Fluoride	Skeletal damage	1.80 (at 20 °C)
Lead	Damages central and peripheral nervous systems and kidney. Highly toxic to infants and expectant mothers	0.05
Mercury	Affects central nervous system	0.002
Nitrite and Nitrate	Blue baby syndrome	10.00
Selenium	Causes gastro-intestinal effects	0.01
Silver	Skin discolouration (argyria)	0.05

[a]Surprisingly this limit in India is 0.05

chemical disintegration of pyrite in rocks, or "acid mine water", when it forms in a mine. These waters, varying in pH, being at places as low as 2–4, are generated in waste rock (Figs. 11.2a and 11.4b), tailing dumps, open pits, and abandoned parts of underground mines and flow into rivers, streams, or lakes. They may be added to the groundwater as well. These acid waters besides killing biota (fish, etc.) in these waters may harm livestocks and spoil crops. The presence of flues (high in As, Zn and Cd), slag piles, soils, and rocks contaminated by smelter fumes, when spread by erosion, may also be the cause of harm to flora and fauna, including human beings (Nordstrom and Alpers 1999). We in India have hardly any quantitative estimate of environmental impairment by acid waters. Kleinman (1989) estimated that 19,300 km of rivers and streams and more than 180,000 acres of lakes and reservoirs in continental USA have been seriously damaged by acid mine drainage. Mine areas are worse polluting sources, compared to the unmined mineralized areas, because the former have greater accessibility of air (+H_2O), through mine workings, waste dumps, and tailings. The natural acid production needs pyrite, oxygen, and water as the principal components. Their mutual relationship may be presented by the following simple equation:

$$4FeS_{2(S)} + 15O_{2(g)} + 14H_2O_{(L)} \rightarrow 4Fe(OH)_{3(S)} + 8H_2SO_{4(aq)},$$

which in fact is rather complex and is controlled by several rate-controlling steps. When pyrite oxidizes, two species get oxidized, Fe(II) and sulfidic sulfur. It may be noted that geobiochemical activities play an important role in the process. It has been suggested that bacteria increases the ferrous iron oxidation rate by 10^5 over the abiotic rate (Nordstrom and Alpers 1999 and references therein). O_2 and Fe(III) do not equally well work with all the sulfides as oxidants, as is shown in Table 11.3.

Fig. 11.4 **a** Tailing pond of Zawar mine, Udaipur district, Rajasthan (Photo: S.C. Sarkar). **b** Waste dump at the edge of Malanjkhand mine – apparently a source of acid water (Photo: S.C. Sarkar), **c** Bauxite mine, Bagru Hills, Lohardaga, Jharkhand in the process of landscape restoration (Photo: S.C. Sarkar)

Table 11.3 Sulfide oxidation characteristics (from Nordstrom and Alpers 1999)

Mineral	Formula	Oxidant	pH
Pyrite	FeS_2	O_2, Fe^{3+}, H_2O_2	0–10
Marcasite	FeS_2	O_2, Fe^{3+}	2–3
Pyrrhotite	$Fe_{1-x}S$	O_2	2–6
Sphalerite	(Zn, Fe)S	O_2, Fe^{3+}	2–7
Galena	PbS	O_2, Fe^{3+}	2.0
Chalcopyrite	$CuFeS_2$	O_2, Fe^{3+}	1.2–2.5
Arsenopyrite	FeAsS	O_2, Fe^{3+}	2.0
Covellite	CuS	O_2, Fe^{3+}	2.0
Chalcocite	Cu_2S	O_2	2–4.8
Greenockite	CdS	O_2	2.3
Millerite	NiS	O_2	2.3
Cobalt sulfide	CoS	O_2	2.3
Cinnabar	HgS	Fe^{3+}	2.0
Energite	Cu_3AsS_4	O_2	3.0
Orpiment	As2S3	O_2	
Bornite	Cu_5FeS_4	O_2	
Molybdenite	MoS_2	O_2	2.5
Stibnite	Sb_2S_3	O_2	
Pentlandite	$(Fe, Ni)_9S_8$	O_2	

There are acid-consuming minerals also. The acid-generating minerals often occur in association with, or in close proximity to a variety of minerals, dominated by carbonates and alumino-silicates, that can react and consume acid generated during sulfide oxidation. Calcite ($CaCO_3$), dolomite [(Ca, Mg)CO_3], and magnesite ($MgCO_3$) typically react with acids according to the following reaction:

$$MCO_{3(S)} + H^+ = M^{2+} + HCO_3^-.$$

However, if hydrous iron or manganese oxides form as a result of dissolution of their carbonates, then a net generation of acid results. Carbonates of zinc (smithsonite) and copper (malachite, azurite) occur in oxidized zones of sulfide deposits in dry climates. These are also effective acid consumers (Plumlee 1999).

In the above context we briefly discuss the acid generation potential and its neutralization in major ore types, such as the VMS deposits, SEDEX deposits, MVT deposits, Porphyry Cu deposits, etc. In VMS deposits, Fe sulfides are generally present in considerable amounts. In wet countries they will try to produce acids which in turn are likely to be buffered by calcic feldspars present in the host rock(s). SEDEX deposits will have little or no advantage in that respect and therefore may contain a large number of toxic elements in the acidic fluid. Carbonate-hosted sulfide ores, commonly known as Mississippi valley type (MVT) and its subtypes, Irish, Penine and Alpine, are unlikely to be affected by mine water, as it will be acid-neutralized. Water from porphyry copper deposits may not be considerably neutralized as the associated feldspar is not as calcic as in the VMS deposits. Mine waters in iron and manganese deposits are almost neutral and are therefore expected to be free of toxic elements.

There are some current practices to isolate sulfide-bearing (and uranium-bearing) wastes from the surface oxidants (O_2, Fe(III)) and water (Sarkar 2002). Capping of such waste dumps with impermeable barriers made of such material as clay helps in containing the flow of oxygenated water through the dump. Tailing ponds (Figs. 11.2b and 11.4a) are first dried and then filled with soil and covered with grass and finally redeemed by planting trees (Fig. 11.4c).

Blending of acid potential mine wastes with carbonate rocks or lime is practiced in some mines. Wetlands, both natural and man-made, are Nature's factories, where microbial activities are available to break sulfate, increase the pH, and precipitate metals.

Simple dilution of acid mine waters by surface waters may liberate some toxic elements by breaking the complexes in which they were so long stable. These ions will then try to get adsorbed to suitable particles (Fe- and Mn-hydroxides, clay, etc.). It will be a good riddance.

Mine air composition with respect to mineral dusts and harmful gases must be kept within tolerance limits by proper ventilation, and if necessary by wetting. The latter is likely to add to the acid mine water in cases and cause inconvenient increase of humidity inside the mine. The decision should be situation specific.

Summing up, the acid generation potential of an ore deposit is dependent on the mineral composition, and textures and structures of ores and immediate host rocks and of course the climate. In certain cases, the geology of the watershed around an ore deposit may also be important. Carefully monitored, the negative effects of mineral-related acid generation can be kept under control to a large extent.

Mill tailings and the associated water

Mill tailings are the left over material after the commodity of interest is extracted physically and chemically. This is discharged in the form of slurry in a tailings impoundment (Fig. 11.2b), not in the immediate neighborhood of the plant. It is generally transported to the impoundment through pipes. Compositionally the tailings are not much different from the waste rocks, except that it is more ore-free. The solid particles settle down and the overlying watery part is processed for reuse, disposal, and even allowed to evaporate. The composition of the liquid in the impoundment usually contains objectionable quantities of such chemicals as detergent-like surfactants, cyanide, and other reagents used to increase mineral recovery. With time these fluids become diluted by mixing with rainwater, or degraded by bacterial activity photolytic decomposition or other processes. If the liquid is acidic, getting suitable rocks in the impoundment walls may help chemical reactions and increase the pH, lowering metal contents (Plumlee and Logsdon 1999). It is particularly worth noting that both waters and solids from uranium mill tailings, as we have mentioned earlier, may have higher concentrations of radioactive daughter products such as radon and radium. Since the metals, Co, Ni,

Mo, and Se, are not uncommonly associated with uranium ores, they, therefore, are components, minor though, of the total dissolved solids in those systems.

The agencies responsible for designing and maintaining the tailing "pond" must see that the impoundment does not fail. Geotechnical characteristics of the impoundment site and its surroundings should be fully investigated in advance with particular respect to possible seepage, probable earthquakes, landslides and flash floods, etc. A common method of tailings utilization from environment point of view is the backfill of tailings and waste mixture in underground workings. This practice primarily provides ground support, but may also help in neutralizing acid, if it is already there in the mine water, as cement acts as a base chemically.

Organic chemicals and other reagents in tailings waters may be treated with as variety of chemicals, with or without bacterial support. Acid and metals in the tailing pond are usually treated in the same way as the acid mine water is treated (Plumlee and Logsdon 1999).

In case the tailings are not utilized in the backfill of underground mines, then the impoundment after being full with tailing solids is allowed to dry up, covered with soil or even forested, as already mentioned.

Heap leaching in low-grade ores pay

Low-grade Cu ores and Au ores are these days being treated by heap leaching at places. Sulfuric acid is used for heap leaching Cu ores and cyanide for the heap leaching of Au ores. Copper ores from Malanjkhand deposit, India, particularly those from the oxidized zone were treated by this method, about 15 years ago. The target metal is dissolved from the ores by reacting with H_2SO_4 and then recovered by the solvent extraction–electro-winning method. The solution is repeatedly recycled through the heap and the metal is extracted. At the end of the process the acid generation potential of the heap may still remain, at least locally. In that case, these have to be treated as mine wastes. Heap leaching is unsuitable during the wet season.

Bacterial extraction of metals from low-grade ores

Mining has not changed much through time. In the meantime the demand of metals increased, quality of ores generally decreased, and the inputs for their extraction such as energy and chemical agents such as charcoal have become costlier. So the industry (Mining and Metallurgy) is increasingly turning to bacteria so that the metals can be extracted at low cost from the low-grade ores. Using bacteria a mining company can extract up to 85% of a metal from ores, having mean metal concentration of $<1\%$ by simply seeding waste heaps with microbes and irrigating it with dil. acid. Inside the heap *Acidithiobacillus* or *Leptospirillum* bacteria oxidize iron and sulfur for energy. In the process they generate reactive ferric iron and sulfuric acid, which degrade the ores being processed, and free the targeted metal (Fecht 2011).

Biological techniques are also being used to clean up acidic runoff from old mines, extracting at the same time the last precious bits of metals that still remained.

Bio-mining has an unprecedented growth in recent years. About 20% of world's copper came from bio-mining in 2011. "As the world moves on to a less carbon-dependant society, we have to look for doing things that are less energy-demanding and more natural" (Johnson in Fecht 2011).

11.8.3 Cement Industry Is no More that Polluting?

India presently is the second largest producer of cement in the World, having a total installed capacity of 300 Mt per year. There are as many as 112 plants with a million or more tons of capacity. During 2011–12 about 230 million tons of cement was produced in the country and by the end of the 12th 5-year plan the estimated production has been ~400 Mt. Three cement plants having a total capacity of 990,000 tons/year are producing white cement. Most of these plants are modern and are based on the energy-efficient dry process technology.

Cement industry was previously branded as one of the worst polluters of the environment, cement dust spreading out all around. Now it meets pollution standards and contributes to environmental cleanliness by consuming fly ash from thermal power plants and slag produced by steel manufacturing units (IBM YB 2012(1)). Dolomitic limestone in the neighboring deposits remains the principal feed, however. In underdeveloped/developing countries carbonate rocks are burnt at small scales and contribute CO_2 to the atmosphere. However, limestone mines in many parts of the country, both working and abandoned, are known to cause huge environmental impacts due to lack of proper reclamation procedures being followed. Cement plants in many states have unfortunately eaten into productive farmlands (Chandra Bhusan et al. 2006). That is, however, not an intrinsic fault of the industry, as much as it is the responsibility of the management, even administration, in those cases.

11.9 Environmental Issues of Energy Resources

11.9.1 Coal as a Necessary Evil

Coal is the major source of secondary energy (the primary source being the Sun itself), as we raise steam by burning coal and produce electricity using the steam to turn turbines. At the same time it is held responsible for polluting the environment (even threatening life) on so many counts. These are enumerated below.

1. Discharging greenhouse gases, such as CO_2 and CH_4 into the atmosphere, thereby contributing to the global warming. We do not know as yet for certain what we can reasonably do with CO_2, being produced all around the world. Practical suggestions are in short supply in this context. Still worse has been the

reluctance of most of the developed countries to be seriously involved in the solution of the problem.

2. Sulfur in coal, whether bound in the organic macerals, or occurring as iron sulfides, pyrite, and marcasite (FeS_2), goes out into the system where exposed to wet/moist atmosphere, the Fe-sulfide ultimately converting to H_2SO_4. During burning of the coal, sulfur goes into the atmosphere as SO_2 and comes down as acid rain.
3. Surface mining affects the landscape, making some areas inhabitable and uncultivable.
4. Roofs of subsurface mines may cave in, if left unsupported, endangering life and property.
5. In earlier days much coal was left out in the pillars and the mined-out space was not filled up by sand and rock waste. As a result, fire was produced due to spontaneous combustion, or by accident. Extinguishing this fire is a big problem. In the Ranigunj and Jharia coalfields of West Bengal and Jharkhand, more than 65 major underground mine fires are going on for many decades now, spoiling more than a billion tons of good quality coal, causing subsidence and belching out smoke, hot air, and other unhygienic gases. The area is gradually becoming difficult to negotiate for common use.
6. In mining history, much more people died working in coal mines than all other mines, added together. Respiratory diseases are common in people working in coal mines. Even today, news of coal mine accidents are common from China, India, Russia, South Africa, and even USA.
7. Blasting for mining make the atmosphere dust-laden over a large area adversely affecting human health. Stunning sound and severe ground shaking during blasting affect the ecology.
8. Limited studies on trace elements in Indian coals show that these are enriched in As, Cd, Sb, Mo, Hg, and Bi (Mukherjee et al. 1982; Banerjee et al. 2000; Parui 2005). Some of these have the potential to cause health hazard. Amongst these Hg is most toxic for humans and aquatic fauna.
9. *Fly ash* is a part of coal combustion products (CCPs). The amount of coal fly ash generation depends on gross coal-fired electricity generation, thermal efficiency of the power plant and the calorific value and ash content of the coal used. The total coal fly ash production in India was projected to be 154 Mt in 2011–2012. Though it is an important resource, a major problem is the storage and utilization of fly ash and the environmental hazard of various trace metals contained in them (Table 11.4), particularly Hg and U. India is yet to catch up with developed countries where >70% of fly ash is used for the manufacture of bricks, concrete, road construction, land fill application, agriculture, and many other sectors. Utilization in India remains restricted to three sectors, namely, cement production, construction of road, and brick manufacturing.

Mercury speciation studies show that fly ash can capture both oxidized and elemental Hg. The mean Hg in Indian coal fly ash is estimated at 0.528 mg kg^{-1} which translates into 81.3 t of Hg in coal fly ash in India in 2012. It is known that

Table 11.4 Trace metal concentrations (mg kg^{-1}) in Indian coal fly ash (from Mukherjee and Zevenhoven 2006)

Cu	Ni	Co	Pb	Mn	Zn	Cd	Cr	Se	As	Hg
42	58	45	87	281	60	7.2	46.9	2	0.003	2.1[a] 0.244–0.821[b]

[a] Santaldih thermal power plant, West Bengal
[b]Singrauli thermal power plant, U.P

Hg^0 is not soluble in water but Hg^{2+} in ash is water soluble. Unless stored properly, a good part of this large amount of Hg can interact with soil, enter groundwater and can have serious negative effects on human health, agriculture, and the natural ecosystem in general. Due to the alkaline nature of fly ash (pH 9.29) and pond ash (pH 8.86) in India, however, the risk of leaching from ash pond may be low (Mukherjee and Zevenhoven 2006).

Some Indian coal have significant amount of radioactive elements (cf. Chap. 5). Bricks made of fly ash from such coal, when used for house construction, can give hazardous exposure to the occupants over long periods of time.

11.9.2 Greenhouse Gas (GHG) Containment

Containing the GHG has been a highly controversial problem for sometime past. 20% of the world's population is responsible for the production of 80% of the GHG in the world. Total global carbon emission during 2013 has been as follows: China 28%, USA 14%, European Union 10%, India 7%. Per capita US emission was 4.5 tons while Indian emission was 0.5 ton. For two decades countries hovered around a reasonable solution of the problem but could reach none.

World community felt relieved when an agreement was reached in the Paris Climate Conference in December, 2015. 196 countries participated in the Conference and agreed to transform the fossil fuel-rich economics within decades and constrain the global warming. Governments of 55 countries causing 55% of global emission ratified it and presumably finally inked it at the United Nations Headquarters in New York. The Agreement is expected to be operative from 2020. The essential tenets of the Agreement are as follows:

1. The Global warming will be held well below 2 °C, desirably at 1.5 °C. The developed countries will take "enhanced action" on mitigation, adaptation, climate finance, technology transfer, capacity building, and transparency.
2. The developed countries will contribute USD 1000 billion/year for enhanced action. There is no legal binding on them on this, however.
3. The developed countries will review actions first in 2023 and then at intervals of 5 years. The developing countries' action on this will be voluntary.

4. The countries committed to limit the green house gases (GHG), emitted by human activities to the same levels as the trees, soils, and oceans can absorb naturally at some point of time between 2050 and 2100.

The agreement, as such, is an improvement on earlier ones. But it may not make the developing countries anxiety-free, until it is effectively executed (Sarkar 2015). India submitted its instrument of ratification of the Paris Agreement on October 2, 2016 making the following declaration: "The Government of India declares its understanding that, as per its national laws, keeping in view its development agenda, particularly the eradication of poverty and provision of basic needs for all its citizens, coupled with its commitment to following the low carbon path to progress, and on the assumption of unencumbered availability of cleaner sources of energy, technology and financial resources from around the world, and based on a fair and ambitious assessment of global commitment to combating climate change, it is ratifying the Paris Agreement." Eleven other countries including China brought in some relevant conditions while ratifying the Agreement.

USA being one of the major emitters of GHG and at the same time a responsible country needs to apply its mind and soul to the agreement for its own benefit and benefit of the world. The new Government at Washington does not appear to appreciate its urgency. This has been a cause of great concern for the world community, which believes good senses will ultimately prevail and the USA will revise its views and agree with the world opinion before long.

11.9.3 Environmental Hazards in Exploration and Exploitation of Oil and Natural Gas

Petroleum and natural gas, however, important to the modern society are potent sources of environmental pollution right from exploration through production, refining, transportation, and use. In seismic studies during exploration the rocks are vibrated affecting the ecosystem. Drilling, following successful geological and geophysical investigations, is a necessary step that involves use of heavy machineries, construction of platforms, movement of drilling rigs, barges, etc., particularly in the case of off-shore projects. Environmental hazards such activities cause are as follows:

1. Noise pollution.
2. Emission of toxic gases, such as, CO_2, NO_2, SO_2, H_2SO_4, etc.
3. Dumping of the cement slurry in the neighborhood.
4. Passing of saline water onto the adjacent agricultural land.
5. Degradation of surface and subsurface waters by dumping of toxic wastes.

During production through wells, hazards are almost the same. Refining of oil has its share of hazards that may involve gas flares, leakage of oil and gas, and dumping of toxic solids.

Major hazards involving natural oil and gas industry relate to transportation. There may be flares and fires while transporting by road/railroad. But environmental disasters take place when large capacity sea going oil tankers by accident or negligence spill hundreds of thousands of liters of crude or refined oil on to the sea. Because of their physical property, they spread very fast and adversely affect the marine ecosystem and the environment of the beaches, including life of the beach dwellers. One estimate says that about 6 million tons of crude oil are dumped into the oceans every year either through accidents, cleaning of large tankers, or transfer of oil at ports.

Use of petroleum as a fuel in about 700 million automobiles is a major cause of atmospheric pollution. Their number is rising everyday and there is no prospect of its abating until we are able to produce a safe and affordable alternative energy source for the vehicles and/or make more use of public transport systems.

11.9.4 Geochemical Contamination of Environment by Radionuclides

Uranium ore prospecting, mining, and processing are done carefully. Even then a part, however, small, goes out into the environment along with mine water, effluent of the ore processing plant, and tailings dam.

Presence of iron sulfides in the ore zone renders the mine water variably acidic. Most of the uranium in the water is U(VI) or some complex thereof. Uranium in the mine waters forms complex ions depending on the ligands in the system and its pH. These complex ions may be sulfate-, carbonate-, phosphate-, and fluorine-bearing besides being uranyl (Fig. 11.5). Dissolved uranium passing on to the hydrologic system may ultimately enter the food chain. For human beings, chronic ingestion of uranium may lead to kidney dysfunction. Uranium may enter the bone structure and behave as a potential carcinogen.

The most important processes that decrease uranium concentration in mine runoff waters are adsorption, co-precipitation, and dilution. The most effective absorbents are iron oxy-hydroxides, clay minerals, micas, and organic material. The complexes $UO_2(OH)^+$ and $(UO_2)_3(OH)_5^+$ commonly absorb onto negatively charged sites on Fe (III)-hydroxides.

Although about 90–95% of uranium in the ore is extracted in the milling processes, bulk of the uranium daughter products, such as radium, remains with the tailings. Radium undergoes natural decay to radon $[Rn]^{222}$ gas. Because of its high radio toxicity, radon is supposed to be most hazardous in environmental management related to uranium mill tailing. Radium from aqueous solutions may be absorbed by some minerals. Their sorption ability (at pH = 7.5–8.0) are as follows:

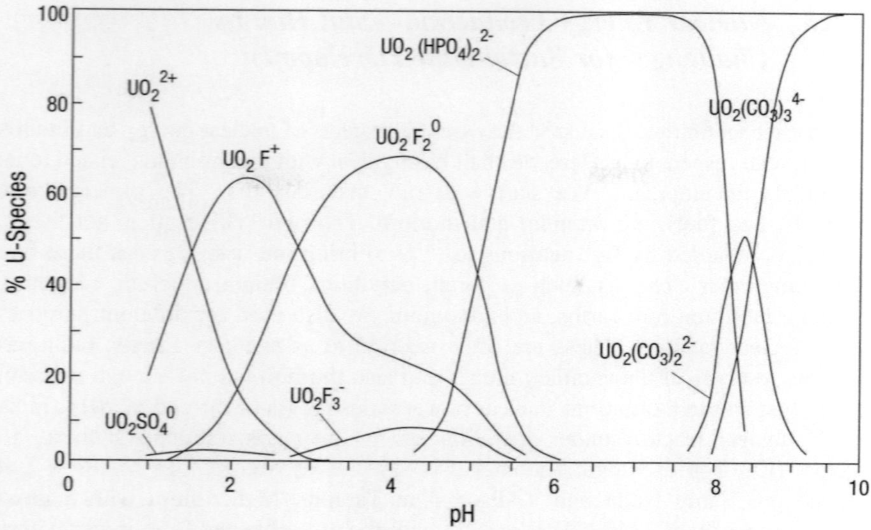

Fig. 11.5 Stability of different U-species in terms of mole % dissolved U (VI) species and pH of the environment (after Nash et al. 1981)

$$\text{Muscovite} \gg \text{albite} \gg \text{montmorillonite} \gg \text{kaolinite} \gg \text{quartz}.$$

Tailing dam seepage may pollute the environment as it did at Mary Kathleen uranium mine, Australia (Lottermoser and Ashley 2005). These seepages, besides radionuclides, may contain metals and metalloids as dissolved species and insoluble and potentially soluble solids.

Recently, a Washington-based news organization, the Centre for Public Integrity, has released a report on radiation hazard at the Jaduguda uranium mine, in Jharkhand's East Singhbhum district. This mine operating since 1967 has seen extraction of 1000 tons of uranium ore per day which accounts for about 20% of the raw material for India's nuclear power generation. The damning report is based on accounts of locals, activists, and researchers since 1990s and claims filed in the court to suggest that mining has exposed workers and villagers to radiation, heavy metals, and other carcinogens including arsenic. Toxins leaching into subsurface aquifers and Subarnarekha River through the Garrah Nallah could be contaminating the food chain from fish to vegetables. It highlights that epidemiological and medical surveys warn of high incidence of infertility and congenital defects in women living near the industrial facility. Further, some studies show that water samples of Subarnarekha River and nearby wells have levels of radioactive alpha particles 160% higher than WHO's safe limits. However, the Uranium Corporation of India Ltd. and Atomic Energy Commission have refuted these claims.

11.9.5 Nuclear Energy Production—Still Has Its Challenges for Sustainable Development

In an earlier section we discussed the essential aspects of nuclear energy and India's position with respect to it. Here we shall briefly deal with the problems related to its sustainable development. We start with raw materials first. The principal raw materials, i.e., fuels, are uranium and thorium. Thorium (Th) itself is not fissile. ^{232}Th is bombarded by fast neutrons to ^{233}U to bring into use. Besides these fuel feed, some other elements, such as boron, beryllium, lithium, niobium, cadmium, tantalum, hafnium, rare earths, and zirconium, are also used for different purposes in the nuclear industry. These are often referred to as ancillary metals. India has adequate reserves of the ancillary metals and also thorium but not enough uranium ores to sustain their ambitious nuclear power industry. As of the end of 2014, India has 21 nuclear reactors under operation and 6 under construction. Of these, 10 reactors (KG 1–4 at Kaiga, Karnataka; NAPs 1–2 at Narora, UP; MAPs 1–2 at Kalpakkam, Tamil Nadu and TAPs 3–4 at Tarapur, Maharashtra with a gross installed capacity of 2840 MWe) are run with the ores obtained from mines within the country. They are beyond the supervision of International Atomic Energy Agency (IAEA). For the rest we are depending on the uranium ore supplied by Russia, Kazakhstan, and France. We expect ore supply from Australia and Canada before long. Australia has large reserves of uranium ores, but no reactor. So at least for sometime supply of nuclear raw materials is assured, if not for the entire period of 30–40 years, which is the assumed lifetime of these reactors. The problem will ease out if India in the meantime discovers one or more large deposits of uranium ores.

How does uranium act upon the human system? Chronic ingestion of uranium in the human system may affect the normal functioning of kidney. Uranium may enter the bone structure and as is well known, act as a potential carcinogen. It may pass on to the food chain being dissolved (U^{VI}) in surface/groundwater. Major source of this U-pollution is nearby mine working. This, however, may be contained using absorbents such as Fe oxy-hydroxides, clay minerals, micas, and organic materials. Although about 90–95% uranium in ores is extracted in the milling processes, bulk of its daughter products (^{226}Ra principally) remains with the tailings. ^{226}Ra is supposed to be most hazardous in environmental management related to uranium mill tailings. About 98% of ^{226}Ra in the original uranium ores pass on to the tailings. Radium from aqueous solutions may be absorbed in slightly alkaline condition (pH = 7.5–8.0).

From the above discussion, an inescapable conclusion is that human settlements should desirably be located not close to uranium mines, tailing dumps, and concentration plants. A comprehensive study of the tectonic situation of the proposed reactor site should be conducted in advance. The proposed Jaitapur nuclear plant in western India was embroiled in such a controversy for quite some time.

Appendix C

Schneiderhohn's classification of ore deposits (1962)

I	First magmatic principal group: magmatic deposits
	1. Chromite deposits
	2. Diamond deposits
	3. Platinoid deposits
I/II	Magmatic-pneumatolytic transition group deposits
	1. Magnetite-apatite deposits
	2. Apatite-nepheline deposits
II	Second magmatic principal group: pegmatite-pneumatolytic deposits
	1. Pegmatite deposits
	2. Sn–W–Molybdenite deposits
	3. Au–Cu–Pb–Ag–Bi deposits
	4. Contact metasomatic-pneumatolytic Sn–W–Molybdenite deposits
	5. Contact metasomatic-pneumatolytic Fe and Mn deposits
	6. Contact metasomatic-pneumatolytic Au–Cu–Pb–Ag–Co-Zn deposits
	7. Other Contact metasomatic-pneumatolytic deposits
II/III	Pneumatolytic-hydrothermal transition group deposits
III	Third major magmatogenous group: hydrothermal deposits
	1. Au and Au–Ag deposits
	2. Pyritic Cu deposits
	3. Pb–Ag–Zn deposits
	4. Ag–Co–Ni–Bi–U deposits
	5. Sn–Ag–W–Bi deposits
	6. Sb–Hg (As–Se) deposits
	7. Oxide deposits of Fe–Mn–Mg deposits
	8. Fluorite–barite–witherite–quartz deposits

(continued)

M. Deb and S.C. Sarkar, *Minerals and Allied Natural Resources and their Sustainable Development*, Springer Geology,
DOI 10.1007/978-981-10-4564-6

(continued)

IV	Fourth major magmatogenous group: exhalation deposits
	Exhalation and solfateric deposits
V	Submarine exhalation-sedimentary deposits
	1. Mainly magmatogenous oxide deposits of Fe and Mn
	2. Mainly sedimentary deposits of pyrite and Cu–Ni sulfide deposits
VI	Sedimentary deposits
	1. Weathered zone deposits
	2. Placers
	3. Residual deposits
	4. Organically or inorganically deposited metallic deposits
	5. Salts
	6. Fuels
	7. Ground water precipitated deposits
VII	Metamorphism-related deposits
	1. Metamorphosed ore deposits
	2. Metamorphic deposits

Appendix D

Jensen and Bateman's (1981) classification of ore deposits

	Process		Deposits
I	Magmatic concentration	1	Early magmatic
	High *T* and *P*		A. Disseminated crystallization
			B. Seggregation
			C. Injection
		2	Late magmatic
			A. Residual liquid seggregation
			B. Residual liquid injection
			C. Immiscible liquid seggregation
			D. Immiscible liquid injection
II	Sublimation (low *T* & *P*)		Sublimates (unimportant)
III	Contact metamorphism Int. low high *T* & *P*		Contact metasomatic Fe, Cu, Au etc.
IV	Hydrothermal processes (*T* & *P* conditions low to high)		Different kinds of cavity filling and replacement deposits
	1. Telethermal		
	2. Epithermal		
	3. Leptothermal		
	4. Mesothermal		
	5. Hypothermal		
V	Sedimentation (exclusive of evaporation)		Sedimentary products of Fe, Mn, Phosphate etc. Sulfides (sedimentary exhalative)
VI	Bacteriogenic		Bacterial products of reduction
VII	Submarine exhalative volcanism, low to high *T* & *P*		Submarine volcanic (sulfides of Cu, Zn, Pb)

(continued)

M. Deb and S.C. Sarkar, *Minerals and Allied Natural Resources and their Sustainable Development*, Springer Geology,
DOI 10.1007/978-981-10-4564-6

(continued)

	Process		Deposits
VIII	Evaporation, low *T* & *P*		Sulfate (barite, gypsum), chloride (rock salt)
IX	Residual and mechanical concentration, low *T* & *P*		
	1. Residual concentration		Residual deposits of Fe, Mn, Al etc.
	2. Mechanical concentration		Placer deposits of Au, PGE, ilmenite, monazite
X	Surficial oxidation and supergene enrichment, low *T* & *P*		Oxidised supergene sulfides
XI	Metamorphism, Int. to high *T* & *P*		1. Metamorphosed deposits 2. Metamorphic deposits